SIMPLIFIED ENGINEERING FOR ARCHITECTS AND BUILDERS

Tenth Edition

JAMES AMBROSE
University of Southern California

with

PATRICK TRIPENY
University of Utah

WILEY

JOHN WILEY & SONS, INC.

This book is printed on acid-free paper. ♾

Published by John Wiley & Sons, Inc., Hoboken, New Jersey

Published simultaneously in Canada

Library of Congress Cataloging-in-Publication Data:

Ambrose, James E.
Simplified engineering for architects and builders / James Ambrose, with Patrick Tripeny.—10th ed.
p. cm.
Includes bibliographical references and index.
ISBN-13: 978-0-471-67607-2 (cloth)
ISBN-10: 0-471-67607-1 (cloth)
1. Structural engineering. I. Tripeny, Patrick. II. Title.
TA633.A43 2005
624—dc22 2005026251

Printed in the United States of America

10 9 8 7 6 5 4

CONTENTS

PREFACE TO THE TENTH EDITION

This book treats the topic of design of structures for buildings. As with previous editions, the book materials have been prepared for persons lacking formal training in engineering. Mathematical work is limited mostly to simple algebra. It is thus well suited for programs in architecture and building construction.

However, because most programs in civil engineering offer little opportunity for study of the general fields of building planning and construction, this book may well be useful as a supplement to engineering texts. The emphasis here is the development of practical design, which typically involves a relatively small effort in structural investigation and a lot of consideration for circumstantial situations relating to the existence of the building structure.

Changes that occur in reference sources and in design and construction practices make it necessary to revise the materials in this book periodically. This edition has indeed received such an updating, although the reader is advised that these changes are continuous, so that it is inevitable that some materials present here will be outdated in a short time. However, the concentration in this work is on fundamental concepts and processes of investigation and design; thus, the use of specific data is of less concern to the learning of the fundamental materials. For use in any actual design work, data should be obtained from current references.

In addition to updating, each new edition affords an opportunity to reconsider the organization, presentation, and scope of the materials contained in the book. This new edition therefore offers some minor alterations of the basic content of previous editions, although just about everything contained in the previous edition is here somewhere. Some trimming has occurred, largely in order to add some new materials without significantly increasing the size of the book.

A major change in this edition is the shift to the load and resistance factor design method (LRFD) for steel and concrete structures. This change reflects the quite well established practices in structural engineering. However, the allowable stress design method (ASD) has been retained for work in wood structures, where it is still highly favored by designers. This affords an opportunity for the reader to learn the basic procedures of both methods

In recent editions, it has been the practice to provide answers for all of the computational exercise problems. However, this book receives considerable use as a course text, and several teachers have requested that some problems be reserved for use without given answers. To accommodate this request in this edition, additional exercise problems have been provided, with answers given only to alternate problems. There remains, however, at least one problem—relating to each text demonstration problem—for which an answer is provided; this is to accommodate readers using this book for a self-study program.

For text demonstrations, as well as for the exercise problems, it is desirable to have some data sources contained in this book. I am grateful to various industry organizations for their permission to use excerpts from these data sources; acknowledgment for which is provided where data is provided.

Major contributions have been made by Professor Patrick Tripeny of the University of Utah to the work for this edition. His work has been critical to the task of redevelopment of the book parts on design of steel and concrete structures-in particular to the use of the LRFD method.

Both personally, as the author of this edition, and as a representative of the academic and professional communities, I must express my gratitude to John Wiley & Sons for its continued publication of this highly utilized reference source. For this edition in particular, I am grateful for the sympathetic and highly competent support provided by the Wiley editors and production staff.

I must also express, once again, my gratitude for the encouragement, support, and considerable assistance provided by my wife, Peggy, throughout the preparation of this edition. Like many other creative efforts, book preparation is about 5% inspiration and 95% drudgery, and my wife helps enormously to get me through both kinds of effort.

James Ambrose
September 2005

PREFACE TO THE FIRST EDITION

(The following is an excerpt from Professor Parker's preface to the first edition.)

To the average young architectural draftsman or builder, the problem of selecting the proper structural member for given conditions appears to be a difficult task. Most of the numerous books on engineering which are available assume that the reader has previously acquired a knowledge of fundamental principles and, thus, are almost useless to the beginner. It is true that some engineering problems are exceedingly difficult, but it is also true that many of the problems that occur so frequently are surprisingly simple in their solution. With this in mind, and with a consciousness of the seeming difficulties in solving structural problems, this book has been written.

In order to understand the discussions of engineering problems, it is essential that the student have a thorough knowledge of the various terms which are employed. In addition, basic principles of forces in equilibrium must be understood. The first section of this book, "Principles of Mechanics," is presented for those who wish a brief review of the subject. Following this section are structural problems involving the most commonly used building materials, wood, steel, reinforced concrete, and roof trusses. A major portion of the book is devoted to numerous problems and their solution, the purpose of which is to explain practical procedure in the design of structural members. Similar examples are given to be solved by the student. Although handbooks published by the manufacturers are necessities to the more advanced student, a great number of appropriate tables are presented herewith so that sufficient data are directly at hand to those using this book.

Care has been taken to avoid the use of advanced mathematics, a knowledge of arithmetic and high school algebra being all that is required to follow the discussions presented. The usual formulas employed in the solution of structural problems are given with explanations of the terms involved and their application, but only the most elementary of these formulas are derived. These derivations are given to show how simple they are and how the underlying principle involved is used in building up a formula that has practical application.

No attempt has been made to introduce new methods of calculation, nor have all the various methods been included. It has been the desire of the author to present to those having little or no knowledge of the subject simple solutions of everyday problems. Whereas thorough technical training is to be desired, it is hoped that this presentation of fundamentals will provide valuable working knowledge and, perhaps, open the doors to more advanced study.

HARRY PARKER

INTRODUCTION

The principal purpose of this book is to develop the topic of *structural design*. However, to do the necessary work for design, various methods of *structural investigation* must be used. The work of investigation consists of the considering the tasks required of a structure and the evaluating the responses of the structure in performing these tasks. Investigation may be performed in various ways; the principal methods are using mathematical models and constructing physical models. For the designer, a major first step in any investigation is the visualization of the structure and the force actions to which it must respond. This book uses extensive illustrations to encourage the reader in the development of the habit of first clearly *seeing* what is happening before proceeding with the essentially abstract procedures of mathematical investigation.

STRUCTURAL MECHANICS

The branch of physics called *mechanics* concerns the actions of forces on physical bodies. Most of engineering design and investigation is based on applications of the science of mechanics. *Statics* is the branch of mechanics that deals with bodies held in a state of unchanging motion by the balanced nature (called *static equilibrium*) of the forces acting on them. *Dynamics* is the branch of mechanics that concerns bodies in motion or in a process of change of shape due to actions of forces. A static condition is essentially unchanging with regard to time; a dynamic condition implies a time-dependent action and response.

When external forces act on a body, two things happen. First, internal forces that resist the actions of the external forces are set up in the body. These internal forces produce *stresses* in the material of the body. Second, the external forces produce *deformations*, or changes in shape, of the body. *Strength of materials*, or mechanics of materials, is the study of the properties of material bodies that enable them to resist the actions of external forces, of the stresses within the bodies, and of the deformations of bodies that result from external forces.

Taken together, the topics of applied mechanics and strength of materials are often given the overall designation of *structural mechanics* or *structural analysis*. This is the fundamental basis for structural investigation, which is essentially an analytical process. On the other hand, *design* is a progressive refining process in which a structure is first generally visualized; then it is investigated for required force responses and its performance is evaluated. Finally, possibly after several cycles of investigation and modification, an acceptable form is derived for the structure.

UNITS OF MEASUREMENT

Early editions of this book have used U.S. units (feet, inches, pounds, etc.) for the basic presentation. In this edition, the basic work is developed with U.S. units with equivalent metric unit values in brackets [thus]. Although the building industry in the United States is now in process of changing to metric units, our decision for the presentation here is a pragmatic one. Most of the references used for this book are still developed primarily in U.S. units and most readers educated in the United States will have acquired use of U.S units as their "first language," even if they now also use metric units.

Table 1 lists the standard units of measurement in the U.S. system with the abbreviations used in this work and a description of common usage in structural design work. In similar form, Table 2 gives the corresponding units in the metric system. Conversion factors to be used for shifting from one unit system to the other are given in Table 3. Direct use of

TABLE 1 Units of Measurement: U.S. System

Name of Unit	Abbreviation	Use in Building Design
Length		
Foot	ft	Large dimensions, building plans, beam spans
Inch	in.	Small dimensions, size of member cross sections
Area		
Square feet	ft^2	Large areas
Square inches	$in.^2$	Small areas, properties of cross sections
Volume		
Cubic yards	yd^3	Large volumes, of soil or concrete (commonly called simply "yards")
Cubic feet	ft^3	Quantities of materials
Cubic inches	$in.^3$	Small volumes
Force, Mass		
Pound	lb	Specific weight, force, load
Kip	kip, k	1000 pounds
Ton	ton	2000 pounds
Pounds per foot	lb/ft, plf	Linear load (as on a beam)
Kips per foot	kips/ft, klf	Linear load (as on a beam)
Pounds per square foot	lb/ft^2, psf	Distributed load on a surface, pressure
Kips per square foot	k/ft^2, ksf	Distributed load on a surface, pressure
Pounds per cubic foot	lb/ft^3	Relative density, unit weight
Moment		
Foot-pounds	ft-lb	Rotational or bending moment
Inch-pounds	in.-lb	Rotational or bending moment
Kip-feet	kip-ft	Rotational or bending moment
Kip-inches	kip-in.	Rotational or bending moment
Stress		
Pounds per square foot	lb/ft^2, psf	Soil pressure
Pounds per square inch	$lb/in.^2$, psi	Stresses in structures
Kips per square foot	$kips/ft^2$, ksf	Soil pressure
Kips per square inch	$kips/in.^2$, ksi	Stresses in structures
Temperature		
Degree Fahrenheit	°F	Temperature

TABLE 2 Units of Measurement: SI System

Name of Unit	Abbreviation	Use in Building Design
Length		
Meter	m	Large dimensions, building plans, beam spans
Millimeter	mm	Small dimensions, size of member cross sections
Area		
Square meters	m^2	Large areas
Square millimeters	mm^2	Small areas, properties of member cross sections
Volume		
Cubic meters	m^3	Large volumes
Cubic millimeters	mm^3	Small volumes
Mass		
Kilogram	kg	Mass of material (equivalent to weight in U.S. units)
Kilograms per cubic meter	kg/m^3	Density (unit weight)
Force, Load		
Newton	N	Force or load on structure
Kilonewton	kN	1000 Newtons
Stress		
Pascal	Pa	Stress or pressure (1 pascal = 1 N/m^2)
Kilopascal	kPa	1000 pascals
Megapascal	MPa	1,000,000 pascals
Gigapascal	GPa	1,000,000,000 pascals
Temperature		
Degree Celsius	°C	Temperature

the conversion factors will produce what is called a *hard conversion* of a reasonably precise form.

In this book, many of the unit conversions presented are *soft conversions*, meaning the converted value is rounded off to produce an approximate equivalent value of some slightly more relevant numerical significance to the unit system. Thus, a wood 2 × 4 (actually 1.5 × 3.5 inches in the U.S. system) is precisely 38.1 mm × 88.9 mm in the metric system. However, the metric equivalent "2 × 4" is more likely to be

TABLE 3 Factors for Conversion of Units

To Convert from U.S. Units to SI Units, Multiply by:	U.S. Unit	SI Unit	To Convert from SI Units to U.S. Units, Multiply by:
25.4	in.	mm	0.03937
0.3048	ft	m	3.281
645.2	in.^{2}	mm^2	1.550×10^{-3}
16.39×10^3	in.^{3}	mm^3	61.02×10^{-6}
416.2×10^3	in.^{4}	mm^4	2.403×10^{-6}
0.09290	ft^2	m^2	10.76
0.02832	ft^3	m^3	35.31
0.4536	lb (mass)	kg	2.205
4.448	lb (force)	N	0.2248
4.448	kip (force)	kN	0.2248
1.356	ft-lb (moment)	N-m	0.7376
1.356	kip-ft (moment)	kN-m	0.7376
16.0185	lb/ft^3 (density)	kg/m^3	0.06243
14.59	lb/ft (load)	N/m	0.06853
14.59	kip/ft (load)	kN/m	0.06853
6.895	psi (stress)	kPa	0.1450
6.895	ksi (stress)	MPa	0.1450
0.04788	psf (load or pressure)	kPa	20.93
47.88	ksf (load or pressure)	kPa	0.02093
$0.566 \times (°F - 32)$	°F	°C	$(1.8 \times °C) + 32$

made 40 × 90 mm, which is close enough for most purposes in construction work.

For some of the work in this book, the units of measurement are not significant. What is required in such cases is simply to find a numerical answer. The visualization of the problem, the manipulation of the mathematical processes for the solution, and the quantification of the answer are not related to specific units, only to their relative values. In such situations, the use of dual units in the presentation is omitted in order to reduce the potential for confusion for the reader.

ACCURACY OF COMPUTATIONS

Structures for buildings are seldom produced with a high degree of dimensional precision. Exact dimensions are difficult to achieve, even for the most diligent of workers and builders. Add this to considerations for the lack of precision in predicting loads for any structure, and the significance of highly precise structural computations becomes moot. This is not to be used for an argument to justify sloppy mathematical work, overly sloppy construction, or use of vague theories of investigation of behaviors. Nevertheless, it makes a case for not being highly concerned with any numbers beyond about the second digit (103 or 104; who cares?).

Even though most professional design work these days is likely to be done with computer support, most of the work illustrated here is quite simple and was actually performed with a hand calculator (the eight-digit, scientific type is adequate). Rounding off of these primitive computations is done with no apologies.

With the use of the computer, accuracy of computational work is a somewhat different matter. Still, it is the designer (a person) who makes judgments based on the computations and who knows how good the input to the computer was and what the real significance of the degree of accuracy of an answer is.

SYMBOLS

The following shorthand symbols are frequently used.

Symbol	Reading
$>$	greater than
$<$	less than
$\geq$	equal to or greater than
$\leq$	equal to or less than
$6'$	6 feet
$6''$	6 inches
Σ	the sum of
ΔL	change in L

NOMENCLATURE

Notation used in this book complies generally with that used in the building design field. A general attempt has been made to conform to usage in the reference standards commonly used by structural designers. The following list includes all of the notation used in this book that is general and is related to the topic of the book. Specialized notation is used by various groups, especially as related to individual materials: wood, steel, masonry, concrete, and so on. The reader is referred to basic references for notation in special fields. Some of this notation is explained in later parts of this book.

Building codes use special notation that is usually carefully defined by the code, and the reader is referred to the source for interpretation of these definitions. When used in demonstrations of computations, such notation is explained in the text of this book.

A_g = gross area of a section, defined by the outer dimensions

A_n = net area

C = compressive force

E = modulus of elasticity (general)

I = moment of inertia

L = length (usually of a span)

M = bending moment

P = concentrated load

S = section modulus

T = tension force

W = (1) total gravity load; (2) weight, or dead load of an object; (3) total wind load force; (4) total of a uniformly distributed load or pressure due to gravity

a = unit area

e = eccentricity of a nonaxial load, from point of application of the load to the centroid of the section

f = computed stress

h = effective height (usually meaning unbraced height) of a wall or column

l = length, usually of a span

s = spacing, center to center

I

PRINCIPLES OF STRUCTURAL MECHANICS

This part consists of basic concepts and applications from the field of applied mechanics as they have evolved in the process of investigation of the behavior of structures. The purpose of studying this material is twofold. First is the general need for a comprehensive understanding of what structures do and how they do it. Second is the need for some factual, quantified basis for the exercise of judgment in the processes of structural design. If it is accepted that the understanding of a problem is the necessary first step in its solution, this analytical study should be seen as the cornerstone of any successful, informed design process.

Although considerable use is made of mathematics in this work, it is mostly only a matter of procedural efficiency. The concepts, not the mathematical manipulations, are the critical concerns.

1

INVESTIGATION OF FORCES AND FORCE ACTIONS

Loads deriving from the tasks of a structure produce forces. The tasks of the structure involve the transmission of the load forces to the supports for the structure. Applied to the structure, these external load and support forces produce a resistance from the structure in terms of internal forces that resist changes in the shape of the structure. This chapter treats the basic properties and actions of forces.

1.1 PROPERTIES OF FORCES

The idea of force is one of the fundamental concepts of mechanics and does not yield to simple, precise definition. An accepted definition of force is that which produces, or tends to produce, motion or a change in the state of motion of objects. A type of force is the effect of gravity, by which all objects are attracted toward the center of the earth.

What causes the force of gravity on an object is the mass of the object, and in U.S. units, this force is quantified as the weight of the body. Gravity forces are thus measured in pounds (lb), or in some other unit such as tons (T) or kips (one kilopound, or 1000 pounds). In the metric system, force is measured in a more purely scientific manner as directly related to the mass of objects; the mass of an object being a constant, whereas weight is proportional to the precise value of the acceleration of gravity, which varies from place to place. Force in metric units is measured in newtons (N) or in kilonewtons (kN) or in meganewtons (mN), whereas weight is measured in grams or in kilograms.

Vectors

A quantity that involves magnitude, direction (vertical, for example), and sense (up, down, etc.) is a *vector quantity*, whereas a *scalar quantity* involves only magnitude and sense. Force, velocity, and acceleration are vector quantities, whereas energy, time, and temperature are scalar quantities. A vector can be represented by a straight line, leading to the possibility of constructed graphical solutions in some cases; a situation that will be demonstrated later. Mathematically, a scalar quantity can be represented completely as +50 or –50, whereas a vector must somehow have its direction represented as well (50 vertical, horizontal, etc.).

Identifying a Force

In order to completely identify a force, it is necessary to establish the following:

Magnitude, or the amount of the force, which is measured in weight units such as pounds or tons.

Direction of the force, which refers to the orientation of its path, called its *line of action*. Direction is usually described by the angle that the line of action makes with some reference, such as the horizontal.

Sense of the force, which refers to the manner in which it acts along its line of action (e.g., up or down, right or left, etc.). Sense is usually expressed algebraically in terms of the sign of the force, either plus or minus.

Forces can be represented graphically in terms of these three properties by the use of an arrow, as shown in Figure 1.1*a*. Drawn to some scale, the length of the arrow represents the magnitude of the force. The angle of inclination of the arrow represents the direction of the force. The location of the arrowhead represents the sense of the force. This form of representa-

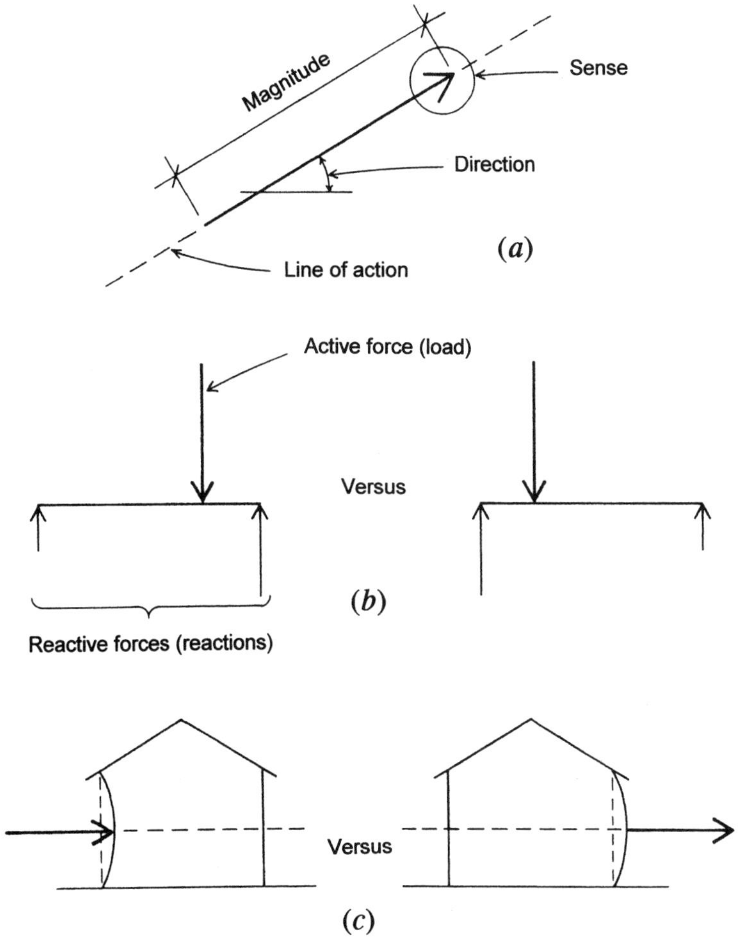

Figure 1.1 Representation of forces and force actions.

tion can be more than merely symbolic, because actual mathematical manipulations may be performed using the vector representation that the force arrows constitute. In the work in this book, arrows are used in a symbolic way for visual reference when performing algebraic computations, and in a truly representative way when performing graphical analyses.

In addition to the basic properties of magnitude, direction, and sense, some other concerns that may be significant for certain investigations are

The *position of the line of action* of the force with respect to the lines of action of other forces or to some object on which the force operates, as shown in Figure 1.1*b*. For the beam, shifting of the location of the load (active force) affects changes in the forces at the supports (reactions).

The *point of application* of the force along its line of action may be of concern in analyzing for the specific effect of the force on an object, as shown in Figure 1.1*c*.

When forces are not resisted, they tend to produce motion. An inherent aspect of static forces is that they exist in a state of *static equilibrium*, that is, with no motion occurring. In order for static equilibrium to exist, it is necessary to have a balanced system of forces. An important consideration in the analysis of static forces is the nature of the geometric arrangement of forces in a given set of forces that constitute a single system. The usual technique for classifying force systems involves consideration of whether the forces in the system are

Coplanar. All acting in a single plane, such as the plane of a vertical wall.

Parallel. All having the same direction.

Concurrent. All having their lines of action intersect at a common point.

Using these three considerations, the possible variations are given in Table 1.1 and illustrated in Figure 1.2. Note that variation 5 in the table

TABLE 1.1 Classification of Force Systems [a]

	Qualifications		
System Variation	Coplanar	Parallel	Concurrent
1	Yes	Yes	Yes
2	Yes	Yes	No
3	Yes	No	Yes
4	Yes	No	No
5	No[b]	Yes	Yes
6	No	Yes	No
7	No	No	Yes
8	No	No	No

[a] See Fig. 1.2.

[b] Not possible—parallel, concurrent forces are essentially coplanar.

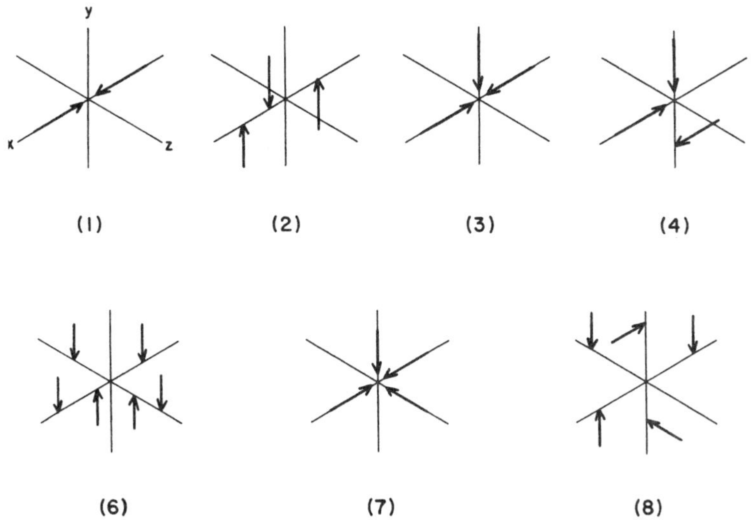

Figure 1.2 Types of force systems.

is really not possible because a set of coacting forces that is parallel and concurrent cannot be noncoplanar; in fact, the forces all fall on a single line of action and are called *collinear*.

It is necessary to qualify a set of forces in the manner just illustrated before proceeding with any analysis, whether it is to be performed algebraically or graphically.

1.2 STATIC EQUILIBRIUM

As stated previously, an object is in *equilibrium* when it is either at rest or has uniform motion. When a system of forces acting on an object produces no motion, the system of forces is said to be in *static equilibrium*.

A simple example of equilibrium is illustrated in Figure 1.3*a*. Two equal, opposite, and parallel forces, having the same line of action, P_1 and P_2, act on a body. If the two forces balance each other, the body does not move and the system of forces is in equilibrium. These two forces are *concurrent*. Put another way, if the lines of action of a system of forces have a point in common, the forces are concurrent.

Another example of forces in equilibrium is illustrated in Figure 1.3*b*. A vertical downward force of 300 lb acts at the midpoint in the length of

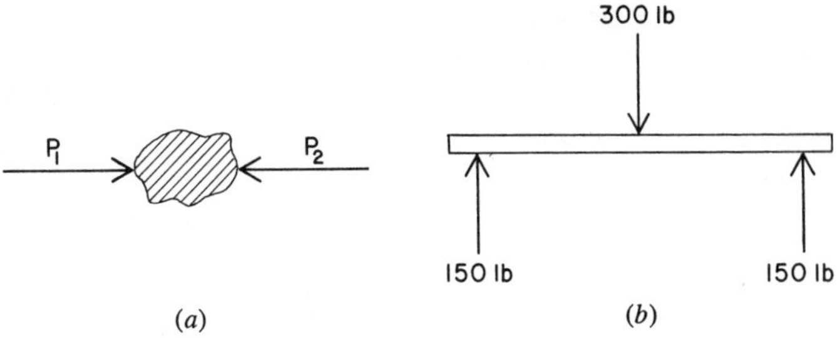

Figure 1.3 Equilibrium of forces.

a beam. The two upward vertical forces of 150 lb each (the reactions) act at the ends of the beam. The system of three forces is in equilibrium. The forces are parallel and, not having a point in common, are *nonconcurrent.*

1.3 FORCE COMPONENTS AND COMBINATIONS

Individual forces may interact and be combined with other forces in various situations. Conversely, a single force may have more than one effect on an object, such as a vertical action and a horizontal action simultaneously. This section considers both of these issues: adding up of forces (combination) and breaking down of single forces into components (resolution).

Resultant of Forces

The *resultant* of a system of forces is the simplest system (usually a single force) that has the same effect as the various forces in the system acting simultaneously. The lines of action of any system of two coplanar nonparallel forces must have a point in common, and the resultant of the two forces will pass through this common point. The resultant of two coplanar, nonparallel forces may be found graphically by constructing a *parallelogram of forces.*

In constructing a parallelogram of two forces, the forces are drawn at any scale (of so many pounds to the inch) with both forces pointing toward, or both forces pointing away, from the point of intersection of their lines of action. A parallelogram is then produced with the two forces as adjacent sides. The diagonal of the parallelogram passing through the common point is the resultant in magnitude, direction, and line of action, the direction of the resultant being similar to that of the given forces, toward or away from the point in common. In Figure 1.4*a*, P_1 and P_2 represent two nonparallel forces whose lines of action intersect at point *O*. The parallelogram is drawn, and the diagonal *R* is the resultant of the given system. In this illustration note that the two forces point *away* from the point in common; hence, the resultant also has its direction away from point *O*. It is a force upward to the right. Notice that the resultant of forces P_1 and P_2 shown in Figure 1.4*b* is *R*; its direction is toward the point in common.

Forces may be considered to act at any points on their lines of action. In Figure 1.4*c* the lines of action of the two forces P_1 and P_2 are extended until they meet at point *O*. At this point the parallelogram of forces is constructed, and *R*, the diagonal, is the resultant of forces P_1 and P_2. In determining the magnitude of the resultant, the scale used is, of course, the same scale used in drawing the given system of forces.

Example 1. A vertical force of 50 lb and a horizontal force of 100 lb, as shown in Figure 1.4*d*, have an angle of 90° between their lines of action. Determine the resultant.

Solution: The two forces are laid off from their point of intersection at a scale of 1 in. = 80 lb. The parallelogram is drawn, and the diagonal is the resultant. Its magnitude scales approximately 112 lb, its direction is

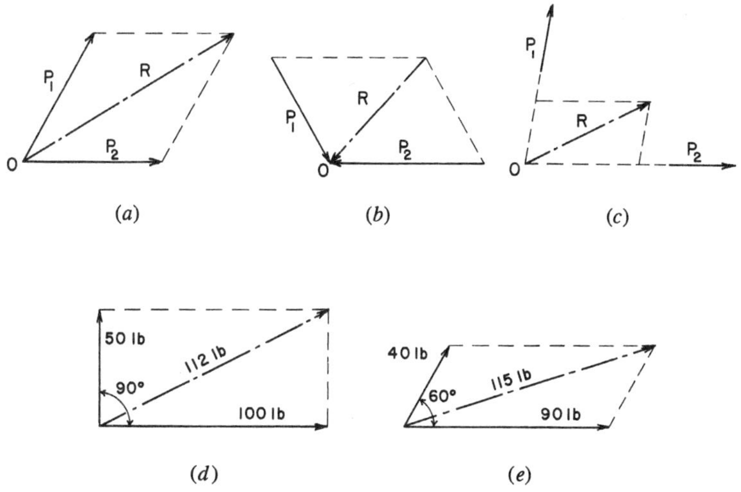

Figure 1.4 Consideration of the resultant of a set of forces.

upward to the right, and its line of action passes through the point of intersection of the lines of action of the two given forces. By use of a protractor, it is found that the angle between the resultant and the force of 100 lb is approximately 26.5°.

Example 2. The angle between two forces of 40 and 90 lb, as shown in Figure 1.4*e*, is 60°. Determine the resultant.

Solution: The forces are laid off from their point of intersection at a scale of 1 in. = 80 lb. The parallelogram of forces is constructed, and the resultant is found to be a force of approximately 115 lb, its direction is upward to the right, and its line of action passes through the common point of the two given forces. The angle between the resultant and the force of 90 lb is approximately 17.5°.

Attention is called to the fact that these two problems have been solved graphically by the construction of diagrams. Mathematics might have been employed. For many practical problems, graphical solutions give sufficiently accurate answers and frequently require far less time. Do not make diagrams too small, as greater accuracy is obtained by using larger parallelograms of forces.

Problems 1.3.A–F. By constructing the parallelogram of forces, determine the resultants for the pairs of forces shown in Figures 1.5*a* to *f*.

Components of a Force

In addition to combining forces to obtain their resultant, it is often necessary to replace a single force by its *components*. The components of a force are the two or more forces that, acting together, have the same effect as the given force. In Figure 1.4*d*, if we are *given* the force of 112 lb, its vertical component is 50 lb and its horizontal component is 100 lb. That is, the 112-lb force has been *resolved* into its vertical and horizontal components. Any force may be considered as the resultant of its components.

Combined Resultants

The resultant of more than two nonparallel forces may be obtained by finding the resultants of pairs of forces and finally the resultant of the resultants.

Example 3. Let it be required to find the resultant of the concurrent forces P_1, P_2, P_3, and P_4 shown in Figure 1.6.

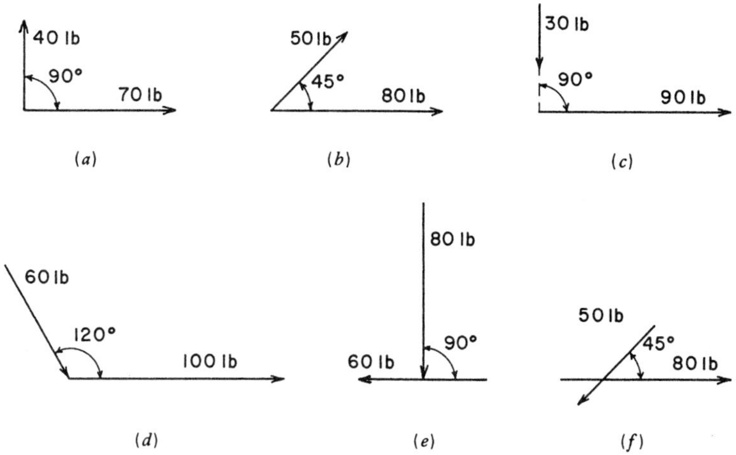

Figure 1.5 Reference for Problem 1.3, part 1.

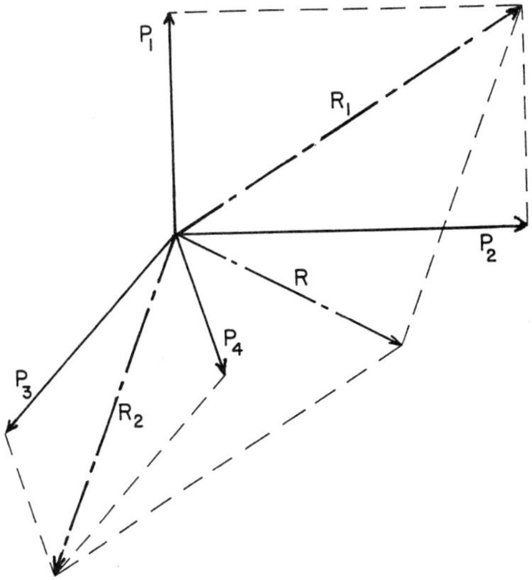

Figure 1.6 Finding a resultant by successive pairs.

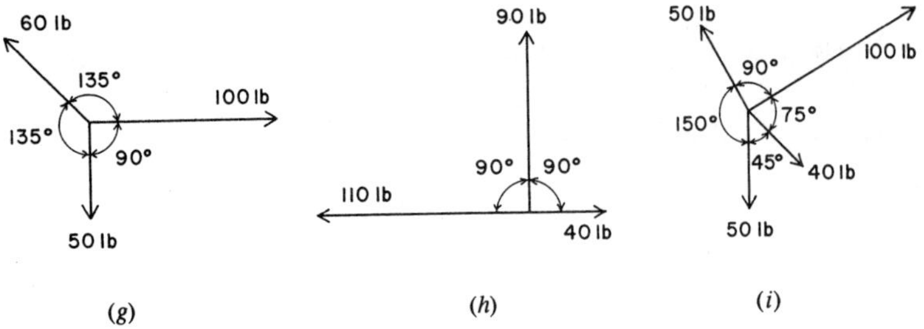

Figure 1.7 Reference for Problem 1.3, part 2.

Solution: By constructing a parallelogram of forces, the resultant of P_1 and P_2 is found to be R_1.

Simarily, the resultant of P_3 and P_4 is R_2. Finally, the resultant of R_1 and R_2 is R, the resultant of the four given forces.

Problems 1.3.G–I. Using graphical methods, find the resultant of the systems of concurrent forces shown in Figure 1.7.

Equilibrant

The force required to maintain a system of forces in equilibrium is called the *equilibrant* of the system. Suppose that we are required to investigate the system of two forces, P_1 and P_2, as shown in Figure 1.8 The parallelogram of forces is constructed, and the resultant is found to be R. The system is not in equilibrium. The force required to maintain equilibrium is force E, shown by the dotted line. E, the equilibrant, is the same as the resultant in magnitude and direction, but is opposite in sense. The three forces, P_1 and P_2 and E, constitute a system in equilibrium.

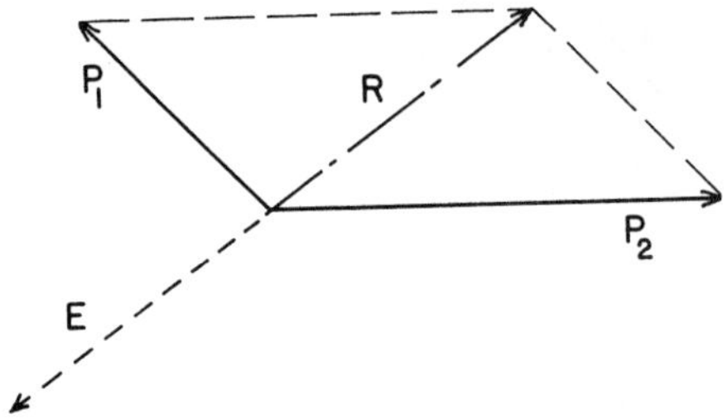

Figure 1.8 Finding an equilibrant.

If two forces are in equilibrium, they must be equal in magnitude, opposite in sense, and have the same direction and line of action. Either of the two forces may be said to be the equilibrant of the other. The resultant of a system of forces in equilibrium is zero.

1.4 GRAPHICAL ANALYSIS OF FORCES

Force Polygon

The resultant of a system of concurrent forces may be found by constructing a *force polygon.* To draw the force polygon, begin with a point and lay off, at a convenient scale, a line parallel to one of the forces, with its length equal to the force in magnitude and having the same sense. From the termination of this line, draw similarly another line corresponding to one of the remaining forces, and continue in the same manner until all the forces in the given system are accounted for. If the polygon does not close, the system of forces is not in equilibrium, and the line required to close the polygon *drawn from the starting point* is the resultant in magnitude and direction. If the forces in the given system are concurrent, the line of action of the resultant passes through the point they have in common.

If the force polygon for a system of concurrent forces closes, the system is in equilibrium and the resultant is zero.

Example 4. Let it be required to find the resultant of the four concurrent forces P_1, P_2, P_3, and P_4 shown in Figure 1.9*a*. This diagram is called

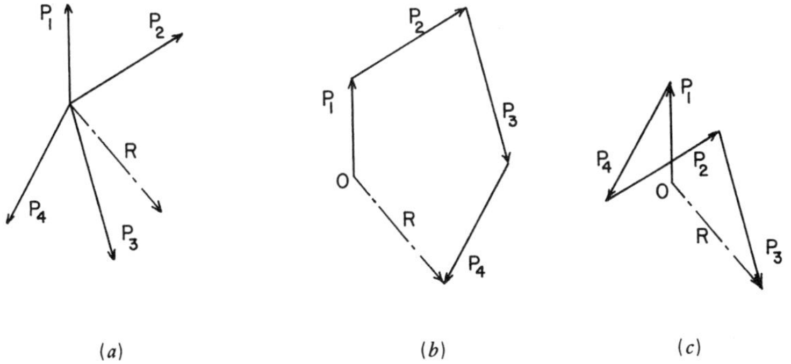

Figure 1.9 Finding a resultant by continuous vector addition of forces.

the *space diagram;* it shows the relative positions of the forces in a given system.

Solution: Beginning with some point such as O, shown in Figure 1.9*b*, draw the upward force P_1. At the upper extremity of the line representing P_1, draw P_2, continuing in a like manner with P_3 and P_4. The polygon does not close; therefore, the system is not in equilibrium. The resultant R, shown by the dot-and-dash line, is the resultant of the given system. Note that its directions is *from* the starting point O, downward to the right. The line of action of the resultant of the given system shown in Figure 1.9*a* has its line of action passing through the point they have in common, its magnitude and direction having been found in the force polygon.

In the drawing of the force polygon, the forces may be taken in any sequence. In Figure 1.9*c* a different sequence is taken, but the resultant R is found to have the same magnitude and direction as previously found in Figure 1.9*b*.

Bow's Notation

Thus far, forces have been identified by the symbols P_1, P_2, and so on. A system of identifying forces, known as Bow's notation, affords many advantages. In this system letters are placed in the space diagram on each side of a force, and a force is identified by two letters. The sequence in which the letters are read is important. Figure 1.10*a* shows the space diagram of five concurrent forces. Reading about the point in common *in a clockwise manner* the forces are *AB, BC, CD, DE,* and *EA*. When a force in the force polygon is represented by a line, a letter is placed at each end of the line. As an example, the vertical upward force in Figure 1.10*a* is read *AB* (note that this is read clockwise about the common point); in the force polygon (Fig. 1.10*b*) the letter *a* is placed at the bottom of the line representing the force *AB* and the letter *b* is at the top. Use capital letters to identify the forces in the space diagrams and lowercase letters in the force polygon. From point *b* in the force polygon, draw force *bc*, then *cd*, and continue with *de* and *ea*. Because the force polygon closes, the five concurrent forces are in equilibrium.

In reading forces, a clockwise manner is used in all the following discussions. It is important that this method of identifying forces be thoroughly understood. To make this clear, suppose that a force polygon is drawn for the five forces shown in Figure 1.10*a*, reading the forces in sequence in a counterclockwise manner. This will produce the force poly-

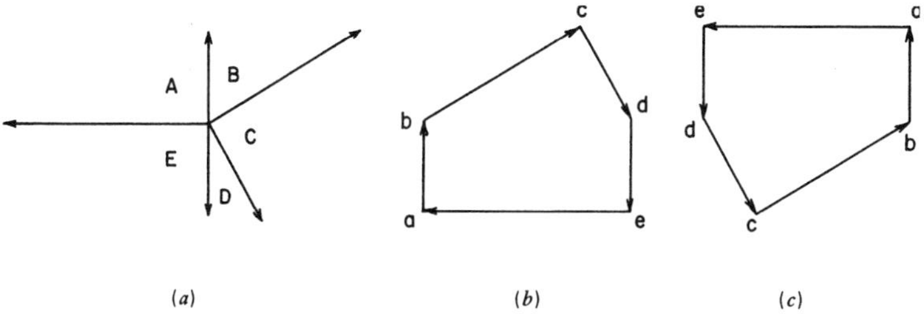

Figure 1.10 Construction of a force plygon.

gon shown in Figure 1.10*c*. Either method may be used, but for consistency, the method of reading clockwise is used here.

Use of the Force Polygon

Two ropes are attached to a ceiling and their lower ends are connected to a ring, making the arrangement shown in Figure 1.11*a*. A weight of 100 lb is suspended from the ring. Obviously, the force in the rope *AB* is 100 lb, but the magnitudes of the forces in ropes *BC* and *CA* are unknown.

The forces in the ropes *AB, BC,* and *CA* constitute a concurrent force system in equilibrium (Figure 1.11*b*). The magnitude of only one of the forces is known—it is 100 lb in rope *AB*. Because the three concurrent forces are in equilibrium, their force polygon must close, and this fact makes it possible to find their magnitudes. Now, at a convenient scale,

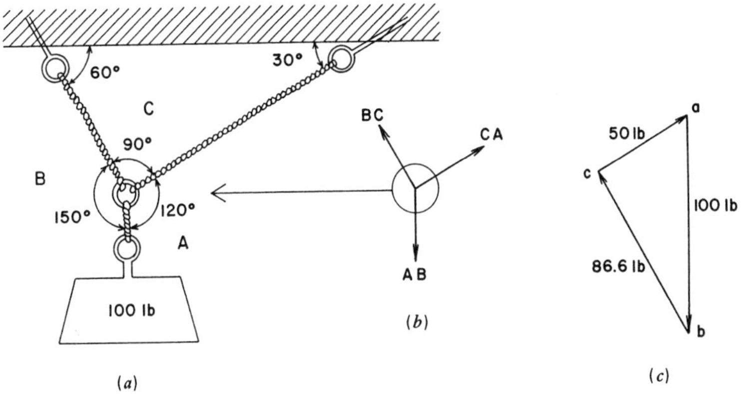

Figure 1.11 Solution of a concentric force system.

draw the line *ab* (Figure 1.11*c*) representing the downward force *AB*, 100 lb. The line *ab* is one side of the force polygon. From point *b* draw a line parallel to rope *BC*; point *c* will be at some location on this line. Next, draw a line through point *a* parallel to rope *CA*; point *c* will be at some position on this line. Because point *c* is also on the line through *b* parallel to *BC*, the intersection of the two lines determines point *c*. The force polygon for the three forces is now completed; it is *abc*, and the lengths of the sides of the polygon represent the magnitudes of the forces in ropes *BC* and *CA*, 86.6 lb and 50 lb, respectively.

Particular attention is called to the fact that the lengths of the ropes in Figure 1.11*a* are not an indication of magnitude of the forces within the ropes; the magnitudes are determined by the lengths of the corresponding sides of the force polygon (Figure 1.11*c*). Figure 1.11*a* merely determines the geometric layout for the structure.

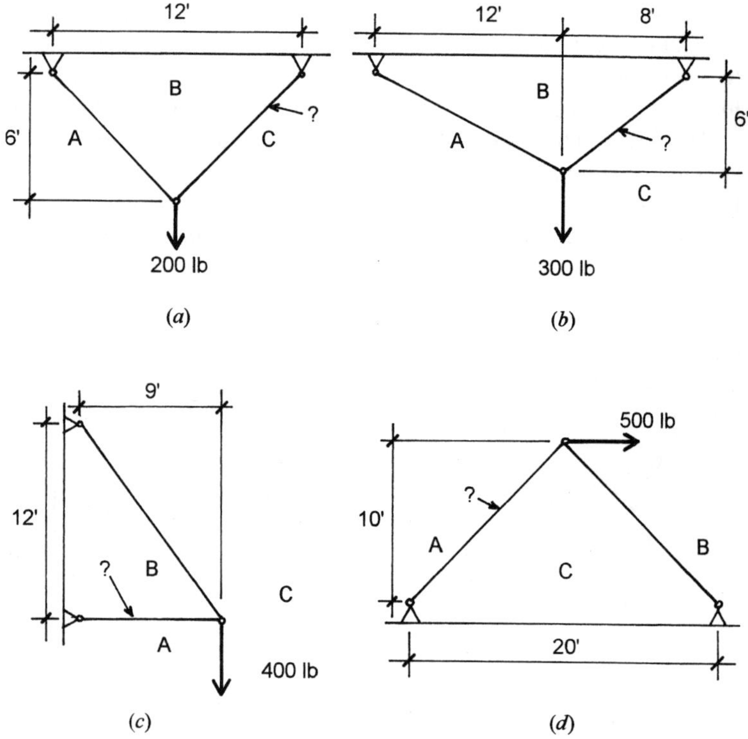

Figure 1.12 Reference for Problem 1.4.

Problems 1.4.A–D. Find the sense (tension or compression) and magnitude of the internal force in the member indicated in Figure 1.12 using graphical methods.

1.5 GRAPHICAL ANALYSIS OF PLANAR TRUSSES

Planar trusses, composed of linear elements assembled in triangulated frameworks, have been used for spanning structures in buildings for many centuries. Investigation for internal forces in trusses is typically performed by the basic methods illustrated in the preceding sections. In this section these procedures are demonstrated using both algebraic and graphical methods of solution.

When we use the so-called *method of joints*, finding the internal forces in the members of a planar truss consists of solving a series of concurrent force systems. Figure 1.13 shows a truss with the truss form, the loads, and the reactions displayed in a space diagram. Below the space diagram is a figure consisting of the free body diagrams of the individual joints of

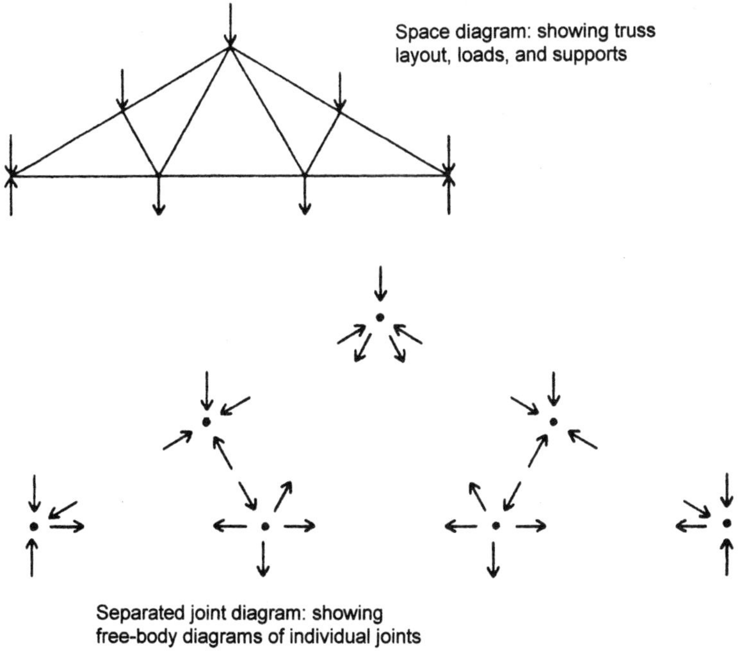

Figure 1.13 Examples of diagrams used to represent trusses and their actions.

the truss. These are arranged in the same manner as they are in the truss in order to show their interrelationships. However, each joint constitutes a complete concurrent planar force system that must have its independent equilibrium. Solving the problem consists of determining the equilibrium conditions for all of the joints. The procedures used for this solution will now be illustrated.

Figure 1.14 shows a single-span, planar truss subjected to vertical gravity loads. This example will be used to illustrate the procedures for determining the internal forces in the truss, that is, the tension and com-

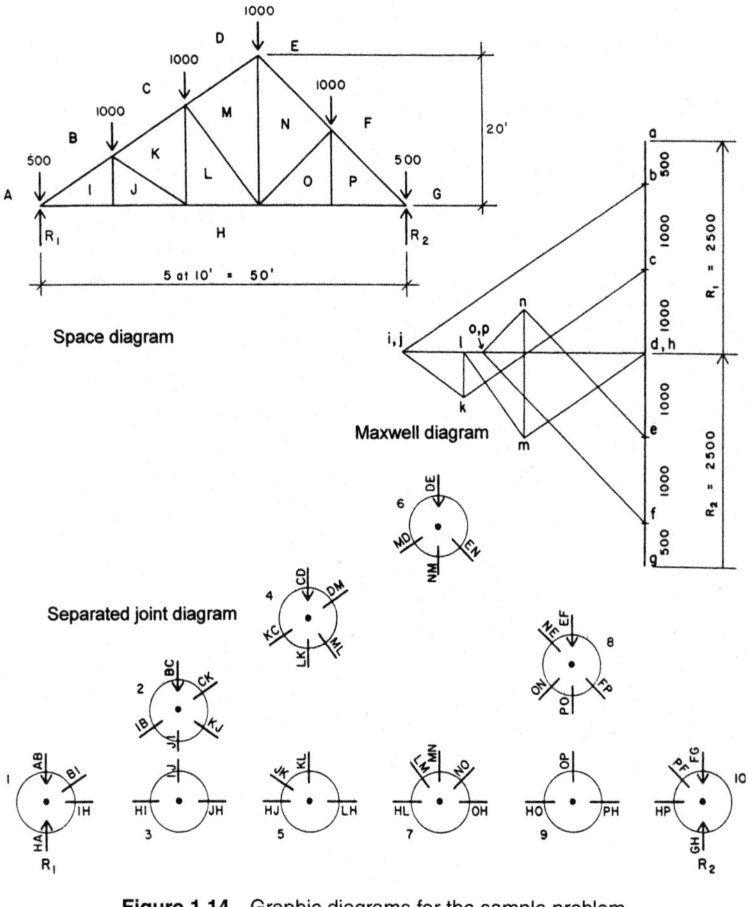

Figure 1.14 Graphic diagrams for the sample problem.

pression forces in the individual members of the truss. The space diagram in the figure shows the truss form and dimensions, the support conditions, and the loads. The letters on the space diagram identify individual forces at the truss joints, as discussed in Section 1.4. The sequence of placement of the letters is arbitrary, the only necessary consideration being to place a letter in each space between the loads and the individual truss members so that each force at a joint can be identified by a two-letter symbol.

The separated joint diagram in the figure provides a useful means for visualization of the complete force system at each joint as well as the interrelation of the joints through the truss members. The individual forces at each joint are designated by two-letter symbols that are obtained by simply reading around the joint in the space diagram in a clockwise direction. Note that the two-letter symbols are reversed at the opposite ends of each of the truss members. Thus, the top chord member at the left end of the truss is designated as *BI* when shown in the joint at the left support (joint 1) and is designated as *IB* when shown in the first interior upper chord joint (joint 2). The purpose of this procedure will be demonstrated in the following explanation of the graphical analysis.

The third diagram in Figure 1.14 is a composite force polygon for the external and internal forces in the truss. It is called a Maxwell diagram after one of its early promoters, James Maxwell, a British engineer. The construction of this diagram constitutes a complete solution for the magnitudes and senses of the internal forces in the truss. The procedure for this construction is as follows.

1. *Construct the force polygon for the external forces.* Before this can be done, the values for the reactions must be found. There are graphic techniques for finding the reactions, but it is usually much simpler and faster to find them with an algebraic solution. In this example, although the truss is not symmetrical, the loading is, and it may simply be observed that the reactions are each equal to one half of the total load on the truss, or $5000 \div 2 = 2500$ lb. Because the external forces in this case are all in a single direction, the force polygon for the external forces is actually a straight line. Using the two-letter symbols for the forces and starting with the letter *A* at the left end, we read the force sequence by moving in a clockwise direction around the outside of the truss. The loads are thus read as *AB, BC, CD, DE, EF,* and *FG,* and the two reactions are read as *GH* and *HA*. Beginning at *A* on the Maxwell diagram, the force vector

sequence for the external forces is read from *A* to *B*, *B* to *C*, *C* to *D*, and so on, ending back at *A*, which shows that the force polygon closes and the external forces are in the necessary state of static equilibrium. Note that we have pulled the vectors for the reactions off to the side in the diagram to indicate them more clearly. Note also that we have used lowercase letters for the vector ends in the Maxwell diagram, whereas uppercase letters are used on the space diagram. The alphabetic correlation is thus retained (*A* to *a*), preventing any possible confusion between the two diagrams. The letters on the space diagram designate open spaces, and the letters on the Maxwell diagram designate points of intersection of lines.

2. *Construct the force polygons for the individual joints.* The graphic procedure for this consists of locating the points on the Maxwell diagram that correspond to the remaining letters, *I* through *P*, on the space diagram. When all the lettered points on the diagram are located, the complete force polygon for each joint may be read on the diagram. In order to locate these points, we use two relationships. The first is that the truss members can resist only forces that are parallel to the members' positioned directions. Thus, we know the directions of all the internal forces. The second relationship is a simple one from plane geometry: A point may be located at the intersection of two lines. Consider the forces at joint 1, as shown in the separated joint diagram in Figure 1.14. Note that there are four forces and that two of them are known (the load and the reaction) and two are unknown (the internal forces in the truss members). The force polygon for this joint, as shown on the Maxwell diagram, is read as *ABIHA*. *AB* represents the load; *BI* the force in the upper chord member; *IH* the force in the lower chord member; and *HA* the reaction. Thus, the location of point *i* on the Maxwell diagram is determined by noting that *i* must be in a horizontal direction from *h* (corresponding to the horizontal position of the lower chord) and in a direction from *b* that is parallel to the position of the upper chord.

The remaining points on the Maxwell diagram are found by the same process, using two known points on the diagram to project lines of known direction whose intersection will determine the location of an unknown point. Once all the points are located, the diagram is complete and can be used to find the magnitude and sense of each internal force. The process for construction of the Maxwell diagram typically consists of moving from joint to joint along the truss. Once one of the letters for an internal space is determined on the Maxwell diagram, it may be used as a known point for finding the letter for an adjacent space on the space diagram. The only limitation of the process is that it is not possible to find more than one unknown point on the Maxwell diagram for any single joint. Consider joint 7 on the separated joint diagram in Figure 1.14. To solve this joint first, knowing only the locations of letters *a* through *h* on the Maxwell diagram, it is necessary to locate four unknown points: *l, m, n,* and *o*. This is three more unknowns than can be determined in a single step, so three of the unknowns must be found by using other joints.

Solving for a single unknown point on the Maxwell diagram corresponds to finding two unknown forces at a joint, because each letter on the space diagram is used twice in the force identification for the internal forces. Thus, for joint 1 in the previous example, the letter *I* is part of the identity of forces *BI* and *IH*, as shown on the separated joint diagram. The graphic determination of single points on the Maxwell diagram, therefore, is analogous to finding two unknown quantities in an algebraic solution. As discussed previously, two unknowns are the maximum that can be solved for the equilibrium of a coplanar, concurrent force system, which is the condition of the individual joints in the truss.

When the Maxwell diagram is completed, the internal forces can be read from the diagram as follows:

1. The magnitude is determined by measuring the length of the line in the diagram, using the scale that was used to plot the vectors for the external forces.
2. The sense of individual forces is determined by reading the forces in clockwise sequence around a single joint in the space diagram and tracing the same letter sequences on the Maxwell diagram.

Figure 1.15*a* shows the force system at joint 1 and the force polygon for these forces as taken from the Maxwell diagram. The forces known initially are shown as solid lines on the force polygon, and the unknown forces are shown as dashed lines. Starting with letter *A* on the force system, we read the forces in a clockwise sequence as *AB, BI, IH,* and *HA*. Note that on the Maxwell diagram, moving from *a* to *b* is moving in the order of the sense of the force—that is, from tail to end of the force vector that represents the external load on the joint. With this sequence on the Maxwell diagram, this force sense flow will be a continuous one. Thus, reading from *b* to *i* on the Maxwell diagram is reading from tail to

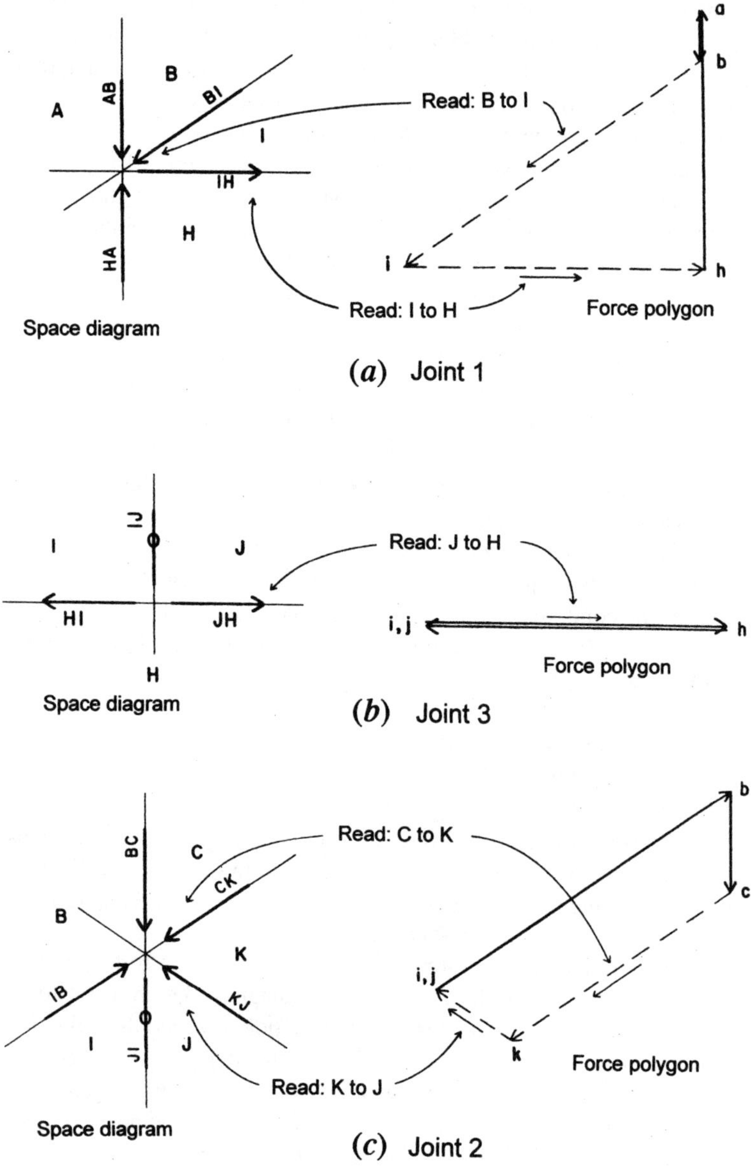

Figure 1.15 Graphic solutions for joints 1, 2, and 3.

head of the force vector, which indicates that force *BI* has its head at the left end. Transferring this sense indication from the Maxwell diagram to the joint diagram indicates that force *BI* is in compression; that is, it is pushing, rather than pulling, on the joint. Reading from *i* to *h* on the Maxwell diagram shows that the arrowhead for this vector is on the right, which translates to a tension effect on the joint diagram.

Having solved for the forces at joint 1 as described, the fact that the forces in truss members *BI* and *IH* are known can be used to consider the adjacent joints, 2 and 3. However, it should be noted that the sense reverses at the opposite ends of the members in the joint diagrams. Referring to the separated joint diagram in Figure 1.14, note that if the upper chord member shown as force *BI* in joint 1 is in compression, its arrowhead is at the lower left end in the diagram for joint 1, as shown in Figure 1.15*a*. However, when the same force is shown as *IB* at joint 2, its pushing effect on the joint will be indicated by having the arrowhead at the upper right end in the diagram for joint 2. Similarly, the tension effect of the lower chord is shown in joint 1 by placing the arrowhead on the right end of the force *IH*, but the same tension force will be indicated in joint 3 by placing the arrowhead on the left end of the vector for force *HI*.

If the solution sequence of solving joint 1 and then joint 2 is chosen, it is now possible to transfer the known force in the upper chord to joint 2. Thus, the solution for the five forces at joint 2 is reduced to finding three unknowns because the load *BC* and the chord force *IB* are now known. However, it is still not possible to solve joint 2 because there are two unknown points on the Maxwell diagram (*k* and *j*) corresponding to the three unknown forces. An option, therefore, is to proceed from joint 1 to joint 3, at which there are presently only two unknown forces. On the Maxwell diagram the single unknown point *j* can be found by projecting vector *IJ* vertically from *i* and projecting vector *JH* horizontally from point *h*. Because point *i* is also located horizontally from point *h*, this shows that the vector *IJ* has zero magnitude, because both *i* and *j* must be on a horizontal line from *h* in the Maxwell diagram. This indicates that there is actually no stress in this truss member for this loading condition and that points *i* and *j* are coincident on the Maxwell diagram. The joint force diagram and the force polygon for joint 3 are as shown in Figure 1.15*b*. In the joint force diagram, place a zero, rather than an arrowhead, on the vector line for *IJ* to indicate the zero stress condition. In the force polygon in Figure 1.15*b*, the two force vectors are slightly separated for clarity, although they are actually coincident on the same line.

Having solved for the forces at joint 3, proceed to joint 2, because there remain only two unknown forces at this joint. The forces at the joint and the force polygon for joint 2 are shown in Figure 1.15*c*. As for joint 1, read the force polygon in a sequence determined by reading clockwise around the joint: *BCKJIB*. Following the continuous direction of the force arrows on the force polygon in this sequence, we can establish the sense for the two forces *CK* and *KJ*.

It is possible to proceed from one end and to work continuously across the truss from joint to joint to construct the Maxwell diagram in this example. The sequence in terms of locating points on the Maxwell diagram would be *i-j-k-l-m-n-o-p*, which would be accomplished by solving the joints in the following sequence: 1,3,2,5,4,6,7,9,8. However, it is advisable to minimize the error in graphic construction by working from both ends of the truss. Thus, a better procedure would be to find points *i-j-k-l-m*,

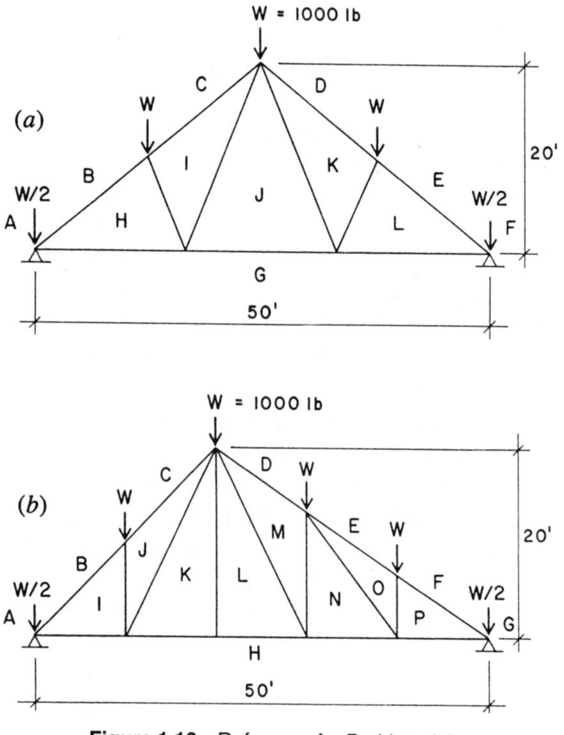

Figure 1.16 Reference for Problem 1.5.

working from the left end of the truss, and then to find points *p-o-n-m*, working from the right end. This would result in finding two locations for *m*, whose separation constitutes the error in drafting accuracy.

Problems 1.5.A, B. Using a Maxwell diagram, find the internal forces in the truss in Figure 1.16.

1.6 ALGEBRAIC ANALYSIS OF PLANAR TRUSSES

Graphical solution for the internal forces in a truss using the Maxwell diagram corresponds essentially to an algebraic solution by the *method of joints*. This method consists of solving the concentric force systems at the individual joints using simple force equilibrium equations. The process will be illustrated using the previous example.

As with the graphic solution, first determine the external forces, consisting of the loads and the reactions. Then proceed to consider the equilibrium of the individual joints, following a sequence as in the graphic solution. The limitation of this sequence, corresponding to the limit of finding only one unknown point in the Maxwell diagram, is that only two unknown forces at any single joint can be found in a single step. (Two conditions of equilibrium produce two equations.) Refer to Figure 1.17; the solution for joint 1 is as follows.

The force system for the joint is drawn with the sense and magnitude of the known forces shown, but with the unknown internal forces represented by lines without arrowheads because their senses and magnitudes initially are unknown. For forces that are not vertical or horizontal, replace the forces with their horizontal and vertical components. Then consider the two conditions necessary for the equilibrium of the system: The sum of the vertical forces is zero and the sum of the horizontal forces is zero.

If the algebraic solution is performed carefully, the sense of the forces will be determined automatically. However, it is recommended that whenever possible the sense be predetermined by simple observations of the joint conditions, as will be illustrated in the solutions.

The problem to be solved at joint 1 is as shown in Figure 1.17*a*. In Figure 1.17*b* the system is shown with all forces expressed as vertical and horizontal components. Note that although this now increases the number of unknowns to three (*IH*, BI_v, and BI_h), there is a numeric relationship between the two components of *BI*. When this condition is added to the two algebraic conditions for equilibrium, the number of usable relationships totals three, so that the necessary conditions to solve for the three unknowns are present.

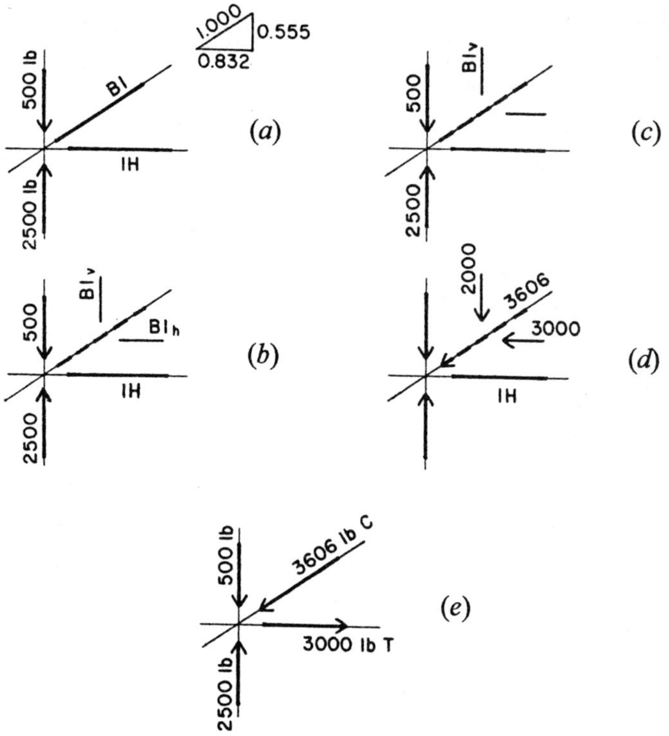

Figure 1.17 Algebraic solution for joint 1.

The condition for vertical equilibrium is shown at (*c*) in Figure 1.17. Because the horizontal forces do not affect the vertical equilibrium, the balance is between the load, the reaction, and the vertical component of the force in the upper chord. Simple observation of the forces and the known magnitudes makes it obvious that force BI_v must act downward, indicating that BI is a compression force. Thus, the sense of BI is established by simple visual inspection of the joint, and the algebraic equation for vertical equilibrium (with upward force considered positive) is

$$\Sigma F_v = 0 = +2500 - 500 - BI_v$$

From this equation, BI_v is determined to have a magnitude of 2000 lb. Using the known relationships between BI, BI_v, and BI_h, the values of these three quantities can be determined if any one of them is known. Thus:

$$\frac{BI}{1.000} = \frac{BI_v}{0.555} = \frac{BI_h}{0.832}$$

from which

$$BI_h = \left(\frac{0.832}{0.555}\right)(2000) = 3000 \text{ lb}$$

and

$$BI = \left(\frac{1.000}{0.555}\right)(2000) = 3606 \text{ lb}$$

The results of the analysis to this point are shown at (*d*) in Figure 1.17, from which it may be observed that the conditions for equilibrium of the horizontal forces can be expressed. Stated algebraically (with force sense toward the right considered positive), the condition is

$$\Sigma F_h = 0 = IH - 3000$$

from which it is established that the force in *IH* is 3000 lb.

The final solution for the joint is then as shown at (*e*) in the figure. On this diagram the internal forces are identified as to sense by using *C* to indicate compression and *T* to indicate tension.

As with the graphic solution, proceed to consider the forces at joint 3. The initial condition at this joint is as shown at (*a*) in Figure 1.18, with the single known force in member *HI* and the two unknown forces in *IJ* and *JH*. Because the forces at this joint are all vertical and horizontal, there is no need to use components. Consideration of vertical equilibrium

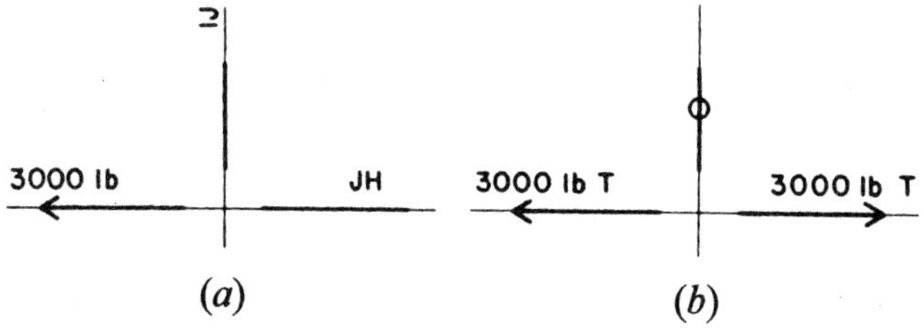

Figure 1.18 Algebraic solution for joint 3.

makes it obvious that it is not possible to have a force in member *IJ*. Stated algebraically, the condition for vertical equilibrium is

$$\Sigma F_v = 0 = IJ \text{ (because } IJ \text{ is the only force)}$$

It is equally obvious that the force in *JH* must be equal and opposite to that in *HI* because they are the only two horizontal forces. That is, stated algebraically

$$\Sigma F_v = 0 = JH - 3000$$

The final answer for the forces at joint 3 is as shown at (*b*) in Figure 1.18. Note the convention for indicating a truss member with no internal force.

Now proceed to consider joint 2; the initial condition is as shown at (*a*) in Figure 1.19. Of the five forces at the joint only two remain unknown. Following the procedure for joint 1, first resolve the forces into their vertical and horizontal components, as shown at (*b*) in Figure 1.19.

Because the sense of forces *CK* and *KJ* is unknown, use the procedure of considering them to be positive until proven otherwise. That is, if they are entered into the algebraic equations with an assumed sense, and the solution produces a negative answer, then the assumption was wrong. However, be careful to be consistent with the sense of the force vectors, as the following solution will illustrate.

Arbitrarily assume that force *CK* is in compression and force *KJ* is in tension. If this is so, the forces and their components will be as shown at (*c*) in Figure 1.19. Then consider the conditions for vertical equilibrium; the forces involved will be those shown at (*d*) in Figure 1.19, and the equation for vertical equilibrium will be

$$\Sigma F_v = 0 = -1000 + 2000 - CK_v - KJ_v$$

or

$$0 = +1000 - 0.555CK - 0.555KJ \qquad (1.6.1)$$

Now consider the conditions for horizontal equilibrium; the forces will be as shown at (*e*) in Figure 1.19, and the equation will be

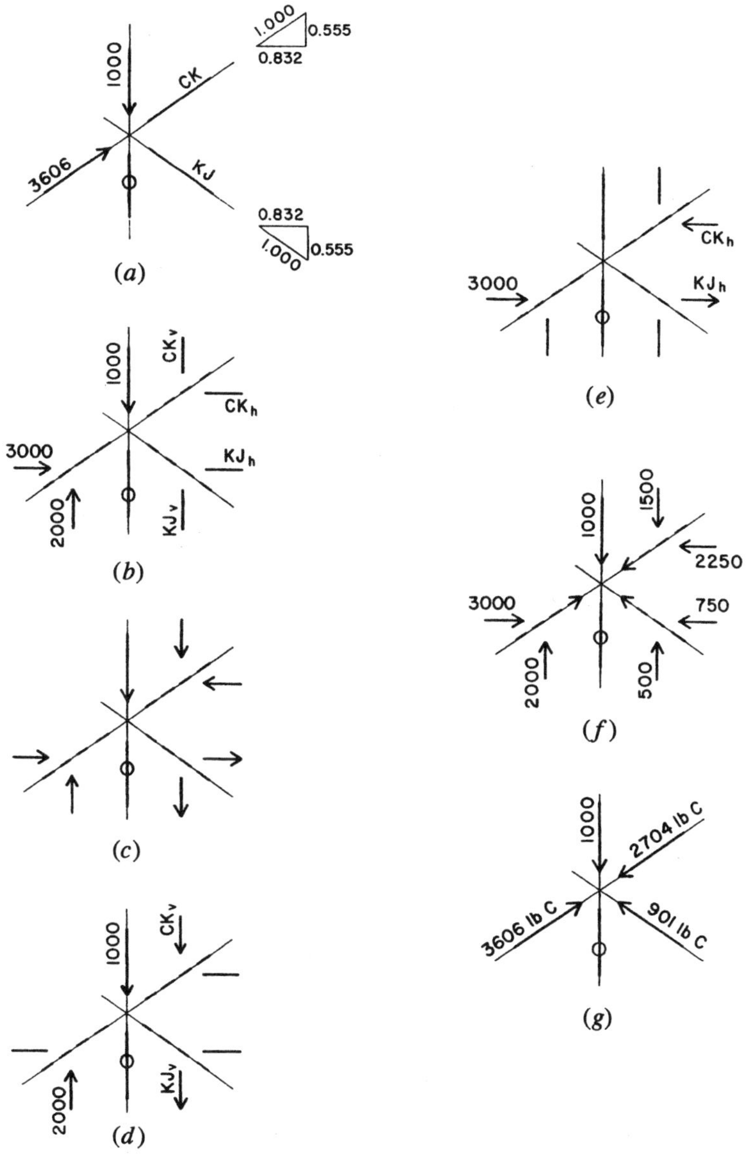

Figure 1.19 Algebraic solution for joint 2.

$$\Sigma F_h = 0 = +\ 3000 - CK_h + KJ_h$$

or

$$0 = +\ 3000 - 0.832CK + 0.832KJ \qquad (1.6.2)$$

Note the consistency of the algebraic signs and the sense of the force vectors, with positive forces considered as upward and toward the right. Now solve these two equations simultaneously for the two unknown forces as follows

1. Multiply equation (1.6.1) by 0.832/0.555

$$0 = \left(\frac{0.832}{0.555}\right)(+\ 1000) + \left(\frac{0.832}{0.555}\right)(-\ 0.555CK) + \left(\frac{0.832}{0.555}\right)(-\ 0.555KJ)$$

or

$$0 = +\ 1500 - 0.832CK - 0.832KJ$$

2. Add this equation to equation (1.6.2) and solve for *CK*.

$$0 = +\ 4500 - 1.664CK, \qquad CK = \frac{4500}{1.664} = 2704 \text{ lb}$$

Note that the assumed sense of compression in *CK* is correct because the algebraic solution produces a positive answer. Substituting this value for *CK* in equation (1.6.1),

$$0 = +\ 1000 - 0.555(2704) - 0.555(KJ)$$

and

$$KJ = \frac{-500}{0.555} = -901 \text{ lb}$$

Because the algebraic solution produces a negative quantity for *KJ*, the assumed sense for *KJ* is wrong and the member is actually in compression.

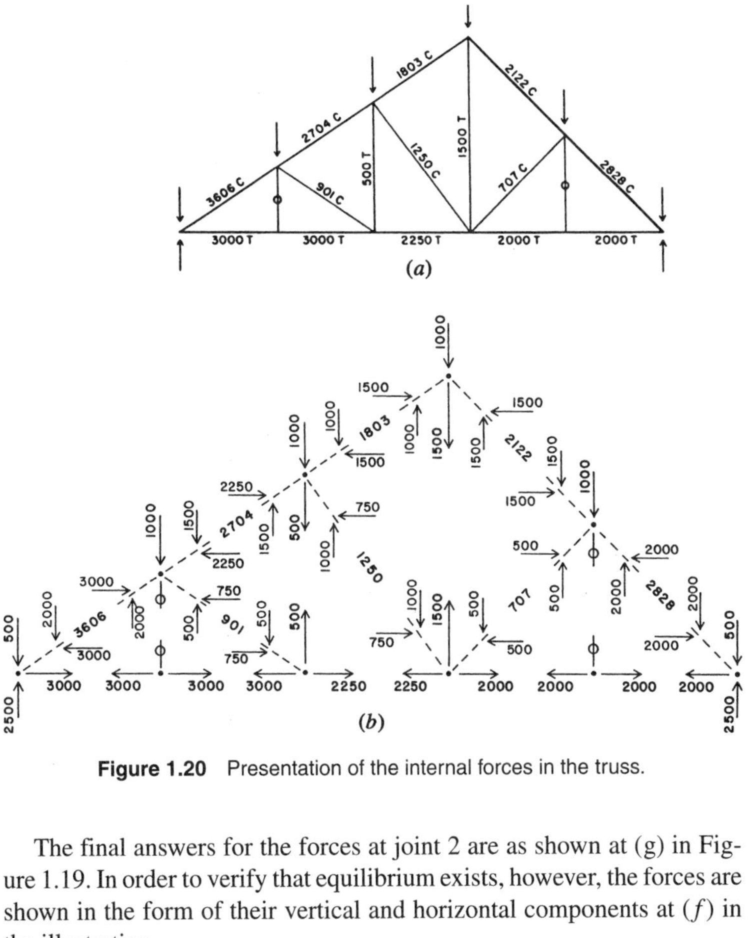

Figure 1.20 Presentation of the internal forces in the truss.

The final answers for the forces at joint 2 are as shown at (g) in Figure 1.19. In order to verify that equilibrium exists, however, the forces are shown in the form of their vertical and horizontal components at (*f*) in the illustration.

When all of the internal forces have been determined for the truss, the results may be recorded or displayed in a number of ways. The most direct way is to display them on a scaled diagram of the truss, as shown in Figure 1.20*a*. The force magnitudes are recorded next to each member with the sense shown as *T* for tension or *C* for compression. Zero stress members are indicated by the conventional symbol consisting of a zero placed directly on the member.

When you are solving by the algebraic method of joints, the results may be recorded on a separated joint diagram as shown in Figure 1.20*b*.

If the values for the vertical and horizontal components of force in sloping members are shown, it is a simple matter to verify the equilibrium of the individual joints.

Problem 1.6.A, B. Using the algebraic method of joints, find the internal forces in the truss in Figure 1.16.

2

FORCE ACTIONS

This chapter presents two considerations for force actions. The first treats the development of resistance within a structure to the external force actions on it, involving the mechanisms of stresses and deformations. The second consideration deals with force actions of a dynamic character—as opposed to a static character—in which the factor of time is a significant issue.

2.1 FORCES AND STRESSES

Direct Stress

Figure 2.1*a* represents a block of metal weighing 6400 lb supported on a wooden block having an 8-×-8-in. cross section. The wooden block is in turn supported on a base of masonry. The gravity force of the metal block exerted on the wood is 6400 lb, or 6.4 kips. Ignoring its own weight, the

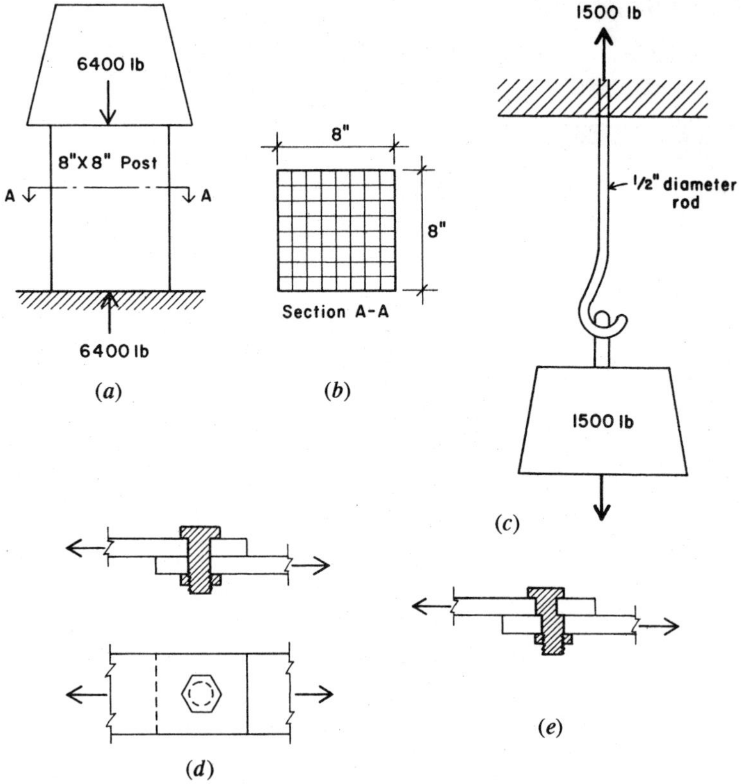

Figure 2.1 Direct force actions and stresses.

wooden block in turn transmits a force of equal magnitude to the masonry base. If there is no motion (a state described as equilibrium), there must be an equal upward force developed by the supporting masonry. Thus, the wooden block is acted on by a set of balanced forces consisting of the applied (or active) downward load of 6400 lb and the resisting (called reactive) upward force of 6400 lb.

To resist being crushed, the wooden block develops an internal force of compression through stress in the material; stress is defined as internal force per unit area of the block cross section. For the situation shown, each square inch of the block cross section must develop a stress equal to $6400/64 = 100$ lb/sq. in. (psi). See Figure 2.1*b*.

Direct Shear Stress

Consider the two steel bars held together by a 0.75-in.-diameter bolt, as shown in Figure 2.1*d*, and subjected to a tension force of 5000 lb. The tension force in the bars becomes a shear force on the bolt, described as a direct shear force. There are many results created by the force in Figure 2.1*d*, including tensile stress in the bars and bearing on the sides of the hole by the bolt. For now we are concerned with the slicing action on the bolt, described as direct shear stress. The bolt cross section has an area of $3.1416(0.375)^2 = 0.4418$ in.^{2}, and the shear stress in the bolt is thus equal to $5000/0.4418 = 11{,}317$ psi. Note that this type of stress is visualized as acting in the plane of the bolt cross section, as a slicing or sliding effect, while both compressive and tensile stresses are visualized as acting perpendicular to a stressed cross section.

Shear in Beams

The foregoing manipulations of the direct-stress formula can, of course, be carried out also with the shearing stress formula $f_v = P/A$. However, it must be borne in mind that the shearing stress acts transversely to the cross section—not at right angles to it. Furthermore, while the shearing stress equation applies directly to the situation illustrated by Figure 2.1*d* and *e*, it requires modification for application to beams, as shown in Figure 2.2*b*. The latter situation will be considered in more detail in Section 3.8.

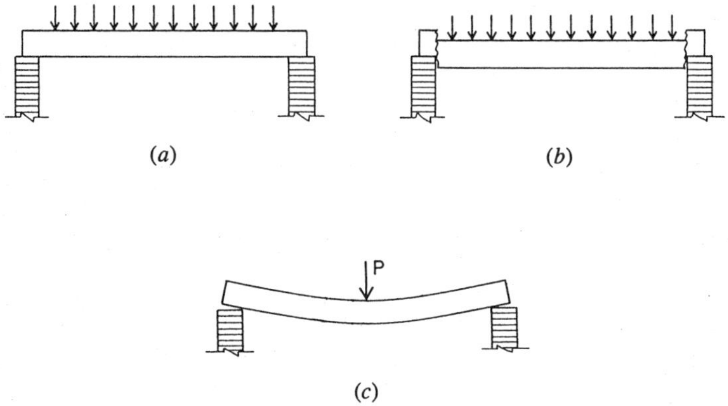

Figure 2.2 Force actions in beams: (*a*) typical loading condition; (*b*) beam shear; (*c*) bending.

Design Use of Direct Stress

The fundamental relationship for simple direct stress may be stated as

$$f = \frac{P}{A} \quad \text{or} \quad P = fA \quad \text{or} \quad A = \frac{P}{f}$$

The first form is used for stress determinations with stress defined as *force per unit area* or simply *unit stress*; the second form for finding the total load (total force) capacity of a member; the third form for determining the required cross-sectional area of a member for a required load with a defined limiting stress condition (called the *allowable stress* or the *working stress*).

In the examples and problems dealing with the direct-stress equation, differentiation is made between the unit stress developed in a member sustaining a given load, $f_v = P/A$, and the *allowable unit stress* used when determining the size of a member required to carry a given load, $A = P/f$. The latter form of the equation is, of course, the one used in design. The procedures for establishing allowable unit stresses in tension, compression, shear, and bending are different for different materials and are prescribed in industry-prepared specifications. A sample of such data is presented in Table 2.1.

TABLE 2.1 Selected Values for Common Structural Materials

	Common Values	
Material and Property	(in psi)	(in kPa)
Structural Steel		
Yield strength	36,000	248,220
Allowable tension	22,000	151,690
Modulus of elasticity, E	29,000,000	200,000,000
Concrete		
f'_c (specified compressive strength)	3,000	20,685
Usable compression in bearing	900	6,206
Modulus of elasticity, E	3,000,000	21,374,500
Structural Lumber (Douglas fir-larch, select structural grade, posts and timbers)		
Compression, parallel to grain	1,150	7,929
Modulus of elasticity, E	1,600,000	11,032,000

In actual design work, the building code governing the construction of buildings in the particular locality must be consulted for specific requirements. Many municipal codes are revised infrequently and, consequently, may not be in agreement with current editions of the industry-recommended allowable stresses.

Except for shear, the stresses discussed so far have been direct or axial stresses. This means they are assumed to be uniformly distributed over the cross section. The examples and problems presented fall under three general types: first, the design of structural members ($A = P/f$); second, the determination of safe loads ($P = fA$); third, the investigation of members for safety ($f_v = P/A$). The following examples will serve to fix in mind each of these types.

Example 1. Design (determine the size of) a short, square post of Douglas fir, select structural grade, to carry a compressive load of 30,000 lb [133,440 N].

Solution: Referring to Table 2.1, the allowable unit compressive stress for this wood parallel to the grain is 1150 psi [7929 kPa]. The required area of the post is

$$A = \frac{P}{F} = \frac{30{,}000}{1150} = 26.09 \text{ in.}^2 \; [16{,}829 \text{ mm}^2]$$

From Table A.8 an area of 30.25 in. ^{2} [19,517 mm^2] is provided by a 6 $\times$ 6 in. post with a dressed size of 5.5 $\times$ 5.5 in. [139.7 mm].

Example 2. Determine the safe axial compressive load for a short, square concrete pier with a side dimension of 2 ft [0.6096 m].

Solution: The area of the pier is 4 ft^2 or 576 in.^{2} [0.3716 m^2]. Table 2.1 gives the allowable unit compressive stress for concrete as 900 psi [6206 m]. Therefore, the safe load on the pier is

$$P = (f)(A) = (900)(576) = 528{,}400 \text{ lb} \; [206 \text{ kN}]$$

Example 3. A running track in a gymnasium is hung from the roof trusses by steel rods, each of which supports a tensile load of 11,2000 lb [49,818 N]. The round rods have a diameter of ⅞ in. [22.23 mm] with the ends *upset*, that is, made larger by forging. This upset allows the full cross-sectional area of the rod (0.601 in.^{2}) [388 mm^2] to be utilized;

otherwise, the cutting of the threads will reduce the cross section of the rod. Investigate this design to determine whether it is safe.

Solution: Because the gross area of the hanger rod is effective, the unit stress developed is

$$f = \frac{P}{A} = \frac{11{,}200}{0.601} = 18{,}636 \text{ psi } [128{,}397 \text{ kPa}]$$

Table 2.1 gives the allowable unit tensile stress for steel as 22,000 psi [151,690 kPa], which is greater than that developed by the loading. Therefore, the design is safe.

Problem 2.1.A. A wrought iron bar sustains a tensile force of 40 kips. If the allowable tensile unit stress is 12 ksi, what is the required cross-sectional area of the bar?

Problem 2.1.B. What axial compression load may be placed on a short timber post whose cross-sectional dimensions are 9.5 in. square, if the allowable unit compressive stress is 1100 psi?

Problem 2.1.C. What should be the diameter of the bolt shown in Figure 2.1*d* if the shearing force is 9000 lb and the allowable unit shearing stress is 15 ksi?

Problem 2.1.D. The allowable compressive bearing capacity of a soil is 8000 psf. What should be the length of the side of a square footing if the total load (including the weight of the footing) is 240 kips?

Problem 2.1.E. If a steel bolt with a diameter of $1\frac{1}{4}$ in. is used for the fastener shown in Figure 2.1*d*, find the total shearing force that can be transmitted across the joint if the allowable unit shearing stress is 15 ksi.

Problem 2.1.F. A short, hollow, cast-iron column is circular in cross section, the outside diameter being 10 in. and the thickness of the shell $\frac{3}{4}$ in. If the allowable unit compressive stress is 9 ksi, what total load will the column support?

Problem 2.1.G. Determine the minimum cross-sectional area of a steel bar required to support a tensile force of 50 kips if the allowable unit tensile stress is 20 ksi.

Problem 2.1.H. A short, square timber post supports a load of 115 kips. If the allowable unit compressive stress is 1000 psi, what nominal size square timber should be used? (See Table A.8.)

Bending

Figure 2.2*c* illustrates a simple beam with a concentrated load P at the center of the span. This is an example of the force action of *bending*, or *flexure*. The materials in the upper part of the beam are in compression, while those in the lower part are in tension. These stresses are not uniformly distributed over the beam cross section, as in the case of direct-stress situations; they cannot be determined by the direct-stress formula. The formula used to compute bending stress is known as the *beam formula* or *flexural formula* and is considered in Section 3.7.

2.2 DEFORMATION

Whenever a force acts on a body, there is an accompanying change in shape or size of the body. In structural mechanics this is called *deformation*. Regardless of the magnitude of the force, some deformation is always present, although often it is so small that it is difficult to measure even with the most sensitive instruments. In the design of structures, it is often necessary to know what the deformation in certain members will be. A floor joist, for instance, may be large enough to support a given load safely but may *deflect* (the term for deformation that occurs with bending) to such an extent that the plaster ceiling below will crack, or the floor may feel excessively springy to persons walking on it. For the usual cases we can readily determine what the deformation will be.

Stress is a major issue, primarily for determination of the strength of structures. However, deformation due to stress is often of concern, and the relation of stress to strain is one that must be quantitatively established. These relations and the issues they raise are discussed in this section.

Hooke's Law

As a result of experiments with clock springs, Robert Hooke, a mathematician and physicist working in the seventeenth century, developed the theory that "deformations are directly proportional to stresses." In other words, if a force produces a certain deformation, twice the force will produce twice the amount of deformation. This law of physics is of utmost importance in structural engineering, although, as we shall find, Hooke's law holds true only up to a certain limit.

Elastic Limit and Yield Point

Suppose that a bar of structural steel with a cross-sectional area of 1 sq in. is placed into a machine for making tension tests. Its length is accurately measured, and then a tensile force of 5000 lb is applied, which, of course, produces a unit tensile stress of 5000 psi in the bar. Measuring the length again, it is found that the bar has lengthened a definite amount, call it x in. On applying 5000 lb more, the amount of lengthening is now $2(x)$, or twice the amount noted after the first 5000 lb. If the test is continued, it will be found that for each 5000 lb increment of additional load, the length of the bar will increase the same amount as noted when the initial 5000 lb was applied; that is, the deformations (length changes) are directly proportional to the stresses. So far, Hooke's law has held true, but when a unit stress of about 36,000 psi is reached, the length increases more than x for each additional 5000 lb of load. This unit stress is called the *elastic limit*, or the *yield stress*. Beyond this stress limit, Hooke's law will no longer apply.

Another phenomenon may be noted in this connection. In the test just described, it will be observed that when any applied load that produces a unit stress *less* than the elastic limit is removed, the bar returns to its original length. If the load producing a unit stress *greater* than the elastic limit is removed, it will be found that the bar has permanently increased its length. This permanent deformation is called the *permanent set.* This fact permits another way of defining the elastic limit: It is that unit stress beyond which the material does not return to its original length when the load is removed.

If this test is continued beyond the elastic limit, a point is reached where the deformation increases without any increase in the load. The unit stress at which this deformation occurs is called the *yield point*; it has a value only slightly higher than the elastic limit. Because the yield point, or yield stress, as it is sometimes called, can be determined more accurately by test than the elastic limit, it is a particularly important unit stress. Nonductile materials such as wood and cast iron have poorly defined elastic limits and no yield point.

Ultimate Strength

After passing the yield point, the steel bar of the test described in the preceding discussion again develops resistance to the increasing load. When the load reaches a sufficient magnitude, rupture occurs. The unit stress in the bar just before it breaks is called the *ultimate strength.* For the grade of steel assumed in the test, the ultimate strength may occur at a stress as high as about 80,000 psi.

Steel members are designed so that stresses under normal service conditions will not exceed the elastic limit, even though there is considerable reserve strength between this value and the ultimate strength. This procedure is followed because deformations produced by stresses above the elastic limit are permanent and hence change the shape of the structure in a permanent manner.

Factor of Safety

The degree of uncertainty that exists, with respect to both actual loading of a structure and uniformity in the quality of materials, requires that some reserve strength be built into the design. This degree of reserve strength is the *factor of safety.* Although there is no general agreement on the definition of this term, the following discussion will serve to fix the concept in mind.

Consider a structural steel that has an ultimate tensile unit stress of 58,000 psi, a yield-point stress of 36,000 psi, and an allowable stress of 22,000 psi. If the factor of safety is defined as the ratio of the ultimate stress to the allowable stress, its value is 58,000 ÷ 22,000, or 2.64. On the other hand, if it is defined as the ratio of the yield-point stress to the allowable stress, its value is 36,000 ÷ 22,000, or 1.64. This is a considerable variation, and because deformatioin failure of a structural member begins when it is stressed beyond the elastic limit, the higher value may be misleading. Consequently, the term *factor of safety* is not employed extensively today. Building codes generally specify the allowable unit stresses that are to be used in design for the grades of structural steel to be employed.

If one should be required to pass judgment on the safety of a structure, the problem resolves itself into considering each structural element, finding its actual unit stress under the existing loading conditions, and comparing this stress with the allowable stress prescribed by the local building regulations. This procedure is called *structural investigation.*

Modulus of Elasticity

Within the elastic limit of a material, deformations are directly proportional to the stresses. The magnitude of these deformations can be computed by use of a number (ratio), called the *modulus of elasticity*, that indicates the degree of *stiffness* of a material.

A material is said to be stiff if its deformation is relatively small when the unit stress is high. As an example, a steel rod 1 in.^{2} in cross-sectional area and 10 ft long will elongate about 0.008 in. under a tensile load of 2000 lb. But a piece of wood of the same dimensions will stretch about

0.24 in. with the same tensile load. The steel is said to be stiffer than the wood because, for the same unit stress, the deformation is not so great.

Modulus of elasticity is defined as the unit stress divided by the unit deformation. *Unit deformation* refers to the percent of deformation and is usually called strain. It is dimensionless because it is expressed as a ratio, as follows:

$$\text{Strain} = s = \frac{e}{L}$$

where s = the strain, or the unit deformation
e = the actual dimensional change
L = the original length of the member

The modulus of elasticity for direct stress is represented by the letter E, expressed in pounds per square inch, and has the same value in compression and tension for most structural materials. Letting f represent the unit stress and s the strain, then, by definition:

$$E = \frac{f}{s}$$

From Section 2.1, $f = P/A$. It is obvious that, if L represents the length of the member and e the total deformation, then s, the deformation per unit of length, must equal the total deformation divided by the length, or $s = e/L$. Now by substituting these values in the equation determined by definition:

$$E = \frac{f}{s} = \frac{\left(\dfrac{P}{A}\right)}{\left(\dfrac{e}{L}\right)} = \frac{PL}{Ae}$$

This can also be written in the form

$$e = \frac{PL}{AE}$$

where e = total deformation in inches
P = force in pounds
L = length in inches
A = cross-sectional area in square inches
E = modulus of elasticity in pounds per square inch

Note that E is expressed in the same units as f (pounds per square inch) because in the equation, $E = f/s$, s is a dimensionless number. For steel, E = 29,000,000 psi [200,000,000 kPa], and for wood, depending on the species and grade, it varies from something less than 1,000,000 psi [6,895,000 kPa] to about 1,900,000 psi [13,100,000 kPa]. For concrete, E ranges from about 2,000,000 psi [13,790,000 kPa] to about 5,000,000 psi [34,475,000 kPa] for common structural grades.

Example 4. A 2-in. [50.8 mm] diameter round steel rod 10 ft [3.05 m] long is subjected to a tensile force of 60 kips [266.88 kN]. How much will it elongate under the load?

Solution: The area of the 2-in. rod is 3.1416 in.^{2} [2027 mm^2]. Checking to determine whether the stress in the bar is within the elastic limit, we find that

$$f = \frac{P}{A} = \frac{60}{3.1416} = 19.1 \text{ ksi } [131{,}663 \text{ kPa}]$$

which is within the elastic limit of ordinary structural steel (36 ksi), so the formula for finding the deformation is applicable. From data, P = 60 kips, L = 120 (length in inches), A = 3.1416, and E = 29,000,000. Substituting these values, we calculate the total lengthening of the rod as

$$e = \frac{PL}{AE} = \frac{(60{,}000)(120)}{(3.1416)(29{,}000{,}000)} = 0.079 \text{ in.}$$

$$[2.0 \text{ mm}]$$

Problem 2.2.A. What force must be applied to a steel bar, 1 in. [25.4 mm] square and 2 ft [610 mm] long, to produce an elongation of 0.016 in. [0.4064 mm]?

Problem 2.2.B. How much will a nominal 8 × 8 in. [actually 190.5 mm] Douglas fir post, 12 ft [3.658m] long, shorten under an axial load of 45 kips [200 kN]?

Problem 2.2.C. A routine quality control test is made on a structural steel bar that is 1 in. [25.4 mm] square and 16 in. [406 mm] long. The data developed during the test show that the bar elongated 0.0111 in [0.282 mm] when subjected to a tensile force of 20.5 kips [91.184 kN]. Compute the modulus of elasticity of the steel.

Problem 2.2.D. A ½ in. [12.7 mm] diameter round steel rod 40 ft [12.19 m] long supports a load of 4 kips [17.79 kN]. How much will it elongate?

2.3 ASPECTS OF DYNAMIC BEHAVIOR

A good lab course in physics should provide a reasonable understanding of the basic ideas and relationships involved in dynamic behavior. A better preparation is a course in engineering dynamics that focuses on the topics in an applied fashion, dealing with their applications in various engineering problems. The material in this section consists of a brief summary of basic concepts in dynamics that will be useful to those with a limited background and that will serve as a refresher for those who have studied the topic before.

The general field of dynamics may be divided into the areas of *kinetics* and *kinematics.* Kinematics deals exclusively with motion, that is, with time-displacement relationships and the geometry of movements. Kinetics adds the consideration of the forces that produce or resist motion.

Kinematics

Motion can be visualized in terms of a moving point, or in terms of the motion of a related set of points that constitute a body. The motion can be qualified geometrically and quantified dimensionally. In Figure 2.3 the point is seen to move along a path (its geometric character) a particular distance. The distance traveled by the point between any two separate locations on its path is called *displacement* (s). The idea of motion is that this displacement occurs over time, and the general mathematical expression for the time-displacement function is

$$s = f(t)$$

Velocity (v) is defined as the rate of change of the displacement with respect to time. As an instantaneous value, the velocity is expressed as the

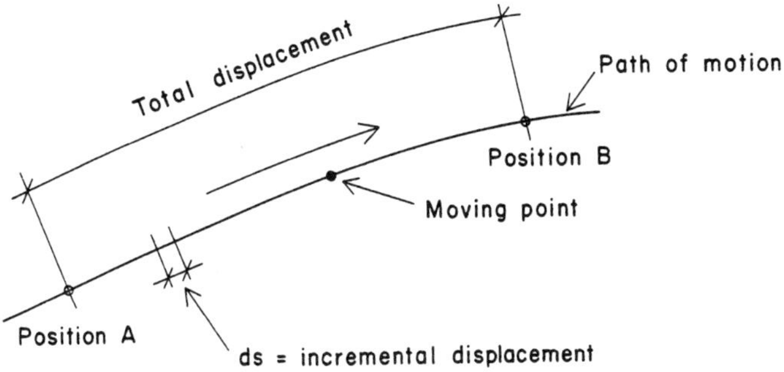

Figure 2.3 Motion of a point.

ratio of an increment of displacement (*ds*) divided by the increment of time (*dt*) elapsed during its displacement. Using the calculus, the velocity is thus defined as

$$v = \frac{ds}{dt}$$

That is, the velocity is the first derivative of the displacement.

If the displacement occurs at a constant rate with respect to time, it is said to have constant velocity. In this case the velocity may be expressed more simply without the calculus as

$$v = \frac{\text{Total displacement}}{\text{Total elapsed time}}$$

When the velocity changes over time, its rate of change is called the *acceleration* (*a*). Thus, as an instantaneous change

$$a = \frac{dv}{dt} = \frac{d^2s}{dt^2}$$

That is, the acceleration is the first derivative of the velocity or the second derivative of the displacement with respect to time.

Motion

A major aspect of consideration in dynamics is the nature of motion. Although building structures are not really supposed to move (as opposed to machine parts), their responses to force actions involve consideration of motions. These motions may actually occur in the form of very small deformations, or may be a failure response that the designer must visualize. The following are some basic forms of motion.

1. *Translation.* This occurs when an object moves in simple linear displacement, with the displacement measured as simple change of distance from some reference point.
2. *Rotation.* This occurs when the motion can be measured in the form of angular displacement; that is, in the form of revolving about a fixed reference point.
3. *Rigid-body motion.* A rigid body is one in which no internal deformation occurs and all particles of the body remain in fixed relation to each other. Three types of motion of such a body are possible. Translation occurs when all the particles of the body move in the same direction at the same time. Rotation occurs when all points in the body describe circular paths about some common fixed line in space, called the *axis of rotation.* Plane motion occurs when all the points in the body move in planes that are parallel. Motion within the planes may be any combination of translation or rotation.
4. *Motion of deformable bodies.* In this case motion occurs for the body as a whole, as well as for the particles of the body with respect to each other. This is generally of more complex form than rigid-body motion, although it may be broken down into simpler component motions in many cases. This is the nature of the motion of fluids and of elastic solids. The deformation of elastic structures under load is of this form, involving both the movement of elements from their original positions and changes in their shapes.

Kinetics

As stated previously, kinetics includes the additional consideration of the forces that cause motion. This means that added to the variables of displacement and time is the consideration of the mass of the moving objects. From Newtonian physics the simple definition of mechanical force is

$$f = ma \text{ (mass times acceleration)}$$

Mass is the measure of the property of inertia, which is what causes an object to resist change in its state of motion. The more common term for dealing with mass is *weight,* which is a force defined as

$$W = mg$$

in which g is the constant acceleration of gravity (32.3 ft/sec^2).

Weight is literally a dynamic force, although it is the standard means of measurement of force in statics, when the velocity is assumed to be zero. Thus, in static analysis force is simply expressed as

$$F = W$$

and in dynamic analysis, when using weight as the measure of mass, force is expressed as

$$F = ma = \frac{W}{g}a$$

Work, Power, Energy, and Momentum

If a force moves an object, work is done. *Work* is defined as the product of the force multiplied by the displacement (distance traveled). If the force is constant during the displacement, work may be simply expressed as

$$w = Fs = \text{force} \times \text{total distance traveled}$$

Energy may be defined as the capacity to do work. Energy exists in various forms: heat, mechanical, chemical, and so on. For structural analysis the concern is with mechanical energy, which occurs in one of two forms. Potential energy is stored energy, such as that in a compressed spring or an elevated weight. Work is done when the spring is released or the weight is dropped. Kinetic energy is possessed by bodies in motion; work is required to change their state of motion, that is, to slow them down or speed them up.

In structural analysis energy is considered to be indestructible, that is, it cannot be destroyed, although it can be transferred or transformed. The potential energy in the compressed spring can be transferred into kinetic energy if the spring is used to propel an object. In a steam engine the chemical energy in the fuel is transformed into heat and then into pressure of the steam and finally into mechanical energy delivered as the engine's output.

An essential idea is that of the conservation of energy, which is a statement of its indestructibility in terms of input and output. This idea can be stated in terms of work by saying that the work done on an object is totally used and that it should therefore be equal to the work accomplished plus any losses due to heat, air friction, and so on. In structural analysis this concept yields a "work equilibrium" relationship similar to the static force equilibrium relationship. Just as all the forces must be in balance for static equilibrium, so the work input must equal the work output (plus losses) for "work equilibrium."

Harmonic Motion

A special problem of major concern in structural analysis for dynamic effects is that of *harmonic motion.* The two elements generally used to illustrate this type of motion are the swinging pendulum and the bouncing spring. Both the pendulum and the spring have a neutral position where they will remain at rest in static equilibrium. If either of them is displaced from this neutral position, by pulling the pendulum sideways or compressing or stretching the spring, they will tend to move back to the neutral position when released. Instead of stopping at the neutral position, however, they will be carried past it by their momentum to a position of displacement in the opposite direction. This sets up a cyclic form of motion (swinging of the pendulum; bouncing of the spring) that has some basic characteristics.

Figure 2.4 illustrates the typical motion of a bouncing spring. Using the calculus and the basic motion and force equations, the displacement-time relationship may be derived as

$$s = A \cos Bt$$

The cosine function produces the basic form of the displacement-time graph, as shown in Figure. 2.4*b*. The maximum displacement from the neutral position is called the *amplitude.* The time elapsed for one full cycle is called the *period.* The number of full cycles in a given unit of time is called the *frequency* (usually expressed in cycles per second) and is equal to the inverse of the period. Every object subject to harmonic motion has a fundamental period (also called natural period), which is determined by its weight, stiffness, size, and so on.

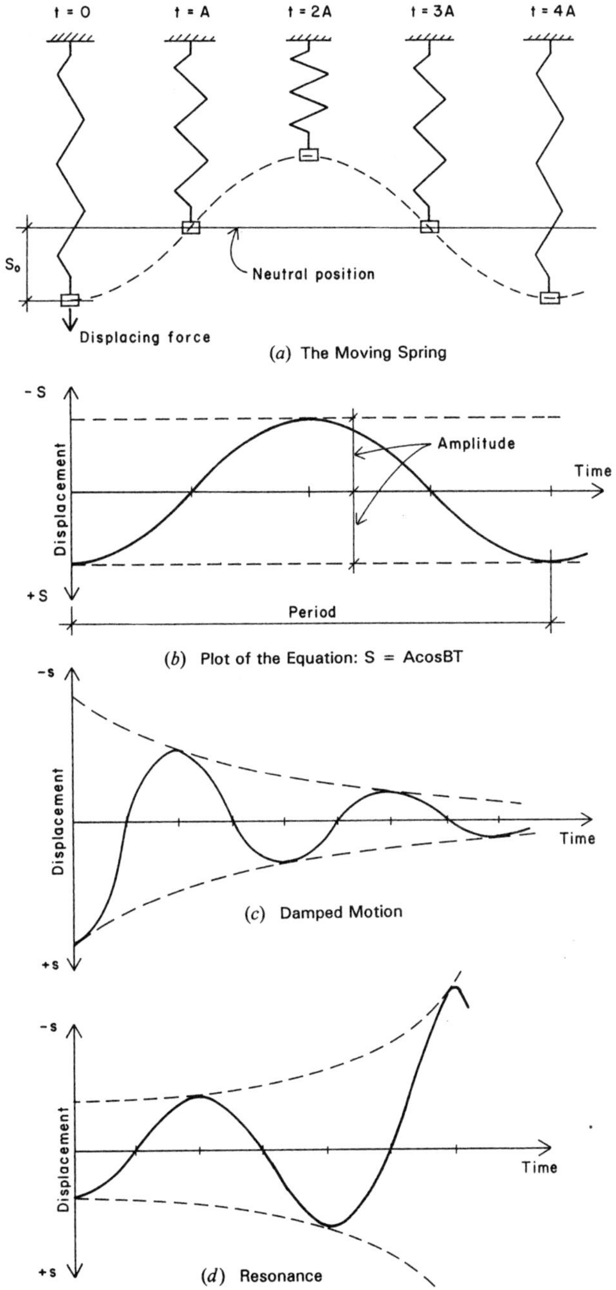

Figure 2.4 Considerations for harmonic motion.

Any influence that tends to reduce the amplitude in successive cycles is called a *damping effect.* Heat loss in friction, air resistance, and so on are natural damping effects. Shock absorbers, counterbalances, cushioning materials, and other devices can also be used to damp the amplitude. Figure 2.4*c* shows the form of a damped harmonic motion, which is the normal form of most such motions, because perpetual motion is not possible without a continuous reapplication of the original displacing force.

Resonance is the effect produced when the displacing effort is itself harmonic with a cyclic nature that corresponds with the period of the impelled object. An example is someone bouncing on a diving board in rhythm with the board's fundamental period, thus causing a reinforcement, or amplification, of the board's free motion. This form of motion is illustrated in Figure 2.4*d.* Unrestrained resonant effects can result in intolerable amplitudes, producing destruction or damage of the moving object or its supports. A balance of damping and resonant effects can sometimes produce a constant motion with a flat profile of the amplitude peaks.

Loaded structures tend to act like springs. Within the elastic stress range of the materials, they can be displaced from a neutral (unloaded) position and, when released, will go into a form of harmonic motion. The fundamental period of the structure as a whole, as well as the periods of its parts, are major properties that affect responses to dynamic loads.

Equivalent Static Effects

Use of equivalent static effects essentially permits simpler analysis and design by eliminating the complex procedures of dynamic analysis. To make this possible, the load effects and the structure's responses must be translated into static terms.

For wind load the primary translation consists of converting the kinetic energy of the wind into an equivalent static pressure, which is then treated in a manner similar to that for a distributed gravity load. Additional considerations are made for various aerodynamic effects, such as ground surface drag, building shape, and suction, but these do not change the basic static nature of the work.

For earthquake effects, the primary translation consists of establishing a hypothetical horizontal static force that is applied to the structure to simulate the effects of sideward motions during ground movements. This force is calculated as some percentage of the dead weight of the building, which is the actual source of the kinetic energy loading once the building is in motion—just as the weight of the pendulum and the spring keeps them moving after the initial displacement and release. The specific percentage used is determined by a number of factors, including some of the dynamic response characteristics of the structure.

An apparently lower safety factor is used when designing for the effects of wind and earthquake because an increase is permitted in allowable stresses. This is actually not a matter of a less-safe design but is merely a way of compensating for the fact that one is actually adding static (gravity) effects and *equivalent* static effects. The total stresses thus calculated are really quite hypothetical because in reality one is adding static strength effects to dynamic strength effects, in which case 2 + 2 does not necessarily make 4.

Regardless of the number of modifying factors and translations, there are some limits to the ability of an equivalent static analysis to account for dynamic behavior. Many effects of damping and resonance cannot be accounted for. The true energy capacity of the structure cannot be accurately measured in terms of the magnitudes of stresses and strains. There are some situations, therefore, in which a true dynamic analysis is desirable, whether it is performed by mathematics or by physical testing. These situations are actually quite rare, however. The vast majority of building designs present situations for which a great deal of experience exists. This experience permits generalizations on most occasions that the potential dynamic effects are really insignificant or that they will be adequately accounted for by design for gravity alone or with use of the equivalent static techniques.

2.4 SERVICE VERSUS ULTIMATE CONDITIONS

Use of allowable stress as a design condition relates to the classic method of structural design known as the *allowable stress method.* Design loads used for this method are generally those described as *service loads*; that is, they are related to the service (use) of the structure. Deformation limits are also related to this load condition.

Even from the earliest times of use of stress methods, it was known that for most materials and structures the true ultimate capacity was not predictable by use of elastic stress methods. Compensating for this with the allowable stress method was mostly accomplished by considerations for the establishing of the limiting design stresses. For more accurate predictions of true failure limits, however, it was necessary to abandon elastic methods and to use true ultimate strength behaviors. This led eventually to the so-called *strength method* for design, presently described as the LRFD method, or *load and resistance factor design* method.

The procedures of the stress method are still applicable in many cases—especially for design for deformation limitations. However, the LRFD methods are now very closely related to more accurate use of test data and risk analysis, and purport to be more realistic with regard to true structural safety. Considerations for use of the LRFD method are discussed further in Chapter 4 and in Parts III and IV.

3

INVESTIGATION OF BEAMS AND FRAMES

This chapter presents considerations that are made in the investigation of the behavior of beams, columns, and simple frames. For basic explanation of relationships, the units used for forces and dimensions are of less significance than their numeric values. For this reason, and for sake of brevity and simplicity, most numerical computations in the text have been done using only U.S. units. For readers who wish to use metric units, however, the exercise problems have been provided with dual units.

3.1 MOMENTS

The term *moment of a force* is commonly used in engineering problems It is fairly easy to visualize a length of 3 ft, an area of 26 sq in., or a force of 100 lb. A moment, however, is less easily understood; it is a force multiplied by a distance. *A moment is the tendency of a force to cause rotation about a given point or axis.* The magnitude of the moment of a force about a given point is the magnitude of the force (pounds, kips, etc.) multiplied by the distance (feet, inches, etc.) from the force to the point of rotation. The point is called the center of moments, and the distance, which is called the *lever arm* or *moment arm*, is measured by a line drawn through the center of moments perpendicular to the line of action of the force. Moments are expressed in compound units such as foot-pounds and inch-pounds or kip-feet and kip-inches. In summary:

$$\text{Moment of force} = \text{Magnitude of force} \times \text{Moment arm}$$

Consider the horizontal force of 100 lb shown in Figure 3.1*a*. If point A is the center of moments, the lever arm of the force is 5 ft. Then the moment of the 100 lb force with respect to point A is $100 \times 5 = 500$ ft-lb. In this illustration the force tends to cause a *clockwise* rotation (shown by the dashed-line arrow) about point A and is called a *positive moment*. If point B is the center of moments, the moment arm of the force is 3 ft. Therefore, the moment of the 100 lb force about point B is $100 \times 3 = 300$ ft-lb. With respect to point B, the force tends to cause *counterclockwise* rotation; it is called a negative moment. It is important to remember that you can never consider the moment of a force without having in mind the particular point or axis about which it tends to cause rotation.

Figure 3.1*b* represents two forces acting on a bar that is supported at point A. The moment of force P_1 about point A is $100 \times 8 = 800$ ft-lb, and it is clockwise or positive. The moment of force P_2 about point A is $200 \times 4 = 800$ ft-lb. The two moment values are the same, but P_2 tends to produce a counterclockwise, or negative, moment about point A. The positive and negative moments are equal in magnitude and are in *equilibrium*; that is, there is no motion. Another way of stating this is to say that the sum of the positive and negative moments about point A is zero, or:

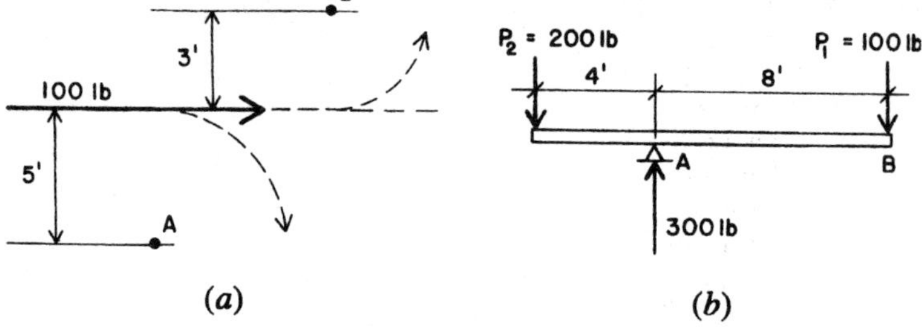

Figure 3.1 Development of moments.

$$\Sigma M_A = 0$$

Stated more generally, *if a system of forces is in equilibrium, the algebraic sum of the moments is zero.* This is one of the laws of equilibrium.

In Figure 3.1*b* point *A* was taken as the center of the moments, but the fundamental law holds for any point that might be selected. For example, if point *B* is taken as the center of moments, the moment of the upward supporting force of 300 lb acting at *A* is clockwise (positive) and that of P_2 is counterclockwise (negative). Then

$$(300 \times 8) - (200 \times 12) = 2400 - 2400 = 0$$

Note that the moment of force P_1 about point *B* is $100 \times 0 = 0$; it is therefore omitted in writing the equation. The reader should be satisfied that the sum of the moments is zero also when the center of moments is taken at the left end of the bar under the point of application of P_2.

Laws of Equilibrium

When an object is acted on by a number of forces, each force tends to move the object. If the forces are of such magnitude and position that their combined effect produces no motion of the object, the forces are said to be in *equilibrium* (Section 1.2). The three fundamental laws of static equilibrium for a general set of coplanar forces are as follows:

1. The algebraic sum of all the vertical forces equals zero.
2. The algebraic sum of all the horizontal forces equals zero.
3. The algebraic sum of the moments of all the forces about any point equals zero.

These laws, sometimes called the conditions for equilibrium, may be expressed as follows (the symbol Σ indicates a summation, i.e., an algebraic addition of all similar terms involved in the problem):

$$\Sigma V = 0 \qquad \Sigma H = 0 \qquad \Sigma M = 0$$

The law of moments, $\Sigma M = 0$, was presented in the preceding discussion.

The expression $\Sigma V = 0$ is another way of saying that *the sum of the downward forces equals the sum of the upward forces.* Thus, the bar of Figure 3.1*b* satisfies $\Sigma V = 0$ because the upward force of 300 lb equals the sum of P_1 and P_2.

Moments of Forces on a Beam

Figure 3.2*a* shows two downward forces of 100 lb and 200 lb acting on a beam. The beam has a length of 8 ft between the supports; the supporting forces, which are called *reactions*, are 175 lb and 125 lb. The four forces are parallel and for equilibrium to exist; therefore, the two laws, $\Sigma V = 0$ and $\Sigma M = 0$, apply.

First, because the forces are in equilibrium, the sum of the downward forces must equal the sum of the upward forces. The sum of the downward forces, the loads, is $100 + 200 = 300$ lb; and the sum of the upward forces, the reactions, is $175 + 125 = 300$ lb. Thus, the force summation is zero.

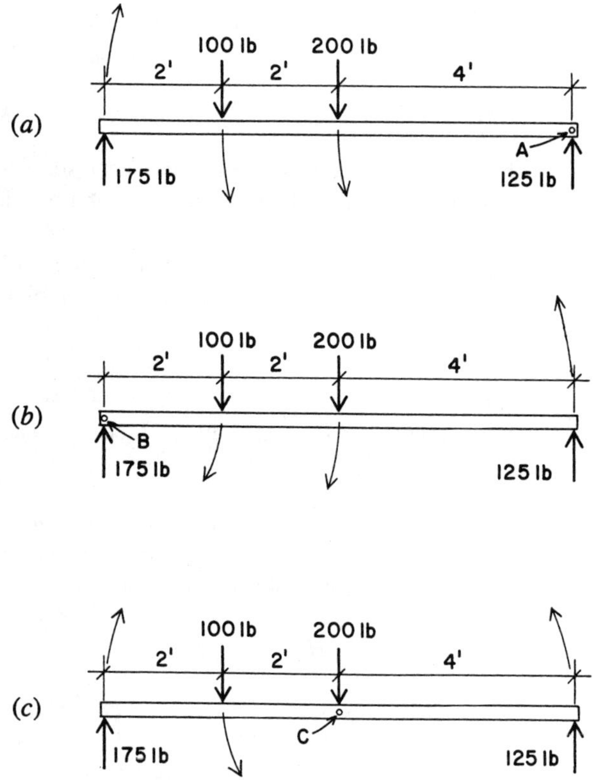

Figure 3.2 Summations of moments about selected points.

Second, because the forces are in equilibrium, the sum of the moments of the forces tending to cause clockwise rotation (positive moments) must equal the sum of the moments of the forces tending to produce counterclockwise rotation (negative moments) about any center of moments. Consider an equation of moments about point A at the right-hand support. The force tending to cause clockwise rotation (shown by the curved arrow) about this point is 175 lb; its moment is $175 \times 8 = 1400$ ft-lb. The forces tending to cause counterclockwise rotation about the *same point* are 100 lb and 200 lb, and their moments are (100×6) and (200×4) ft-lb. Thus, if $\Sigma M_A = 0$, then

$$(175 \times 8) = (100 \times 6) + (200 \times 4)$$

$$1400 = 600 + 800$$

$$1400 \text{ ft-lb} = 1400 \text{ ft-lb}$$

which is true.

The upward force of 125 lb is omitted from the preceding equation because its lever arm about point A is 0 ft, and consequently its moment is zero. A force passing through the center of moments does not cause rotation about that point.

Now select point B at the left support as the center of moments (see Figure 3.2*b*). By the same reasoning, if $\Sigma M_B = 0$, then

$$(100 \times 2) + (200 \text{ x } 4) = (125 \text{ x } 8)$$

$$200 + 800 = 1000$$

$$1000 \text{ ft-lb} = 1000 \text{ ft-lb}$$

Again, the law holds. In this case the force of 175 lb has a lever arm of 0 ft about the center of moments and its moment is zero.

The reader should verify this case by selecting any other point, such as point C in Figure 3.2*c* as the center of moments, and confirming that the sum of the moments is zero for this point.

Problem 3.1.A. Figure 3.3 represents a beam in equilibrium with three loads and two reactions. Select five different centers of moments and write the equation of moments for each, showing that the sum of the clockwise moments equals the sum of the counterclockwise moments.

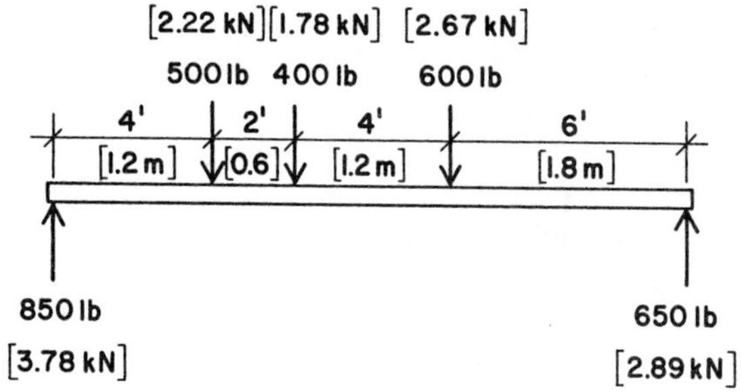

Figure 3.3 Reference for Problem 3.1.

3.2 BEAM LOADS AND REACTION FORCES

A beam is a structural member that resists transverse loads. The supports for beams are usually at or near the ends, and the supporting upward forces are called *reactions*. The loads acting on a beam tend to *bend* it rather than shorten or lengthen it. *Girder* is the name given to a beam that supports smaller beams; all girders are beams insofar as their structural action is concerned. For construction usage beams carry various names, depending on the form of construction; these include *purlin, joist, rafter, lintel, header*, and *girt.*

There are, in general, five types of beams identified by the number, kind, and position of the supports. Figure 3.4 shows diagrammatically the different types and also the shape each beam tends to assume as it bends (deforms) under the loading. In ordinary steel or reinforced concrete beams, these deformations are not usually visible to the eye, but some deformation is always present.

A *simple beam* rests on a support at each end, the ends of the beam being free to rotate (Figure 3.4*a*).

A *cantilever beam* is supported at one end only. A beam embedded in a wall and projecting beyond the face of the wall is a typical example (Figure 3.4*b*).

An *overhanging beam* is a beam whose end or ends project beyond its supports. Figure 3.4*c* indicates a beam overhanging one support only.

A *continuous beam* rests on more than two supports (Figure 3.4*d*). Continuous beams are commonly used in reinforced concrete and welded steel construction.

A *restrained beam* has one or both ends restrained or fixed against rotation (Figure 3.4*e*).

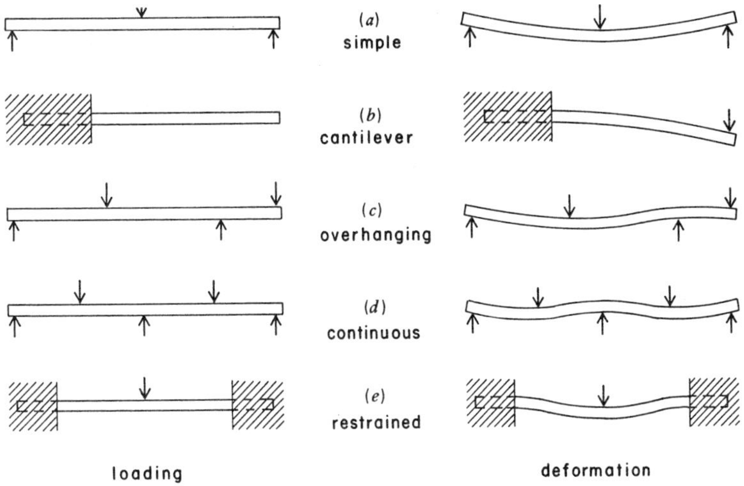

Figure 3.4 Types of beams.

Beams are acted on by external forces that consist of the loads and the reaction forces developed by the beam's supports. The two types of loads that commonly occur on beams are called *concentrated* and *distributed.* A concentrated load is assumed to act at a definite point; such a load is that caused when one beam supports another beam. A distributed load is one that acts over a considerable length of the beam; such a load is one caused by a floor deck supported directly by a beam. If the distributed load exerts a force of equal magnitude for each unit of length of the beam, it is known as a *uniformly distributed load.* The weight of a beam is a uniformly distributed load that extends over the entire length of the beam. However, some uniformly distributed loadings supported by the beam may extend over only a portion of the beam length.

Beam Reactions

Reactions are the upward forces acting at the supports that hold in equilibrium the downward forces or loads. The left and right reactions of a simple beam are usually called R_1 and R_2, respectively. Determination of reactions for simple beams is achieved with the use of equilibrium conditions for parallel force systems.

If a beam 18 ft long has a concentrated load of 9000 lb located 9 ft from the supports, it is readily seen that each upward force at the supports will be equal and will be one half the load in magnitude, or 4500 lb. But

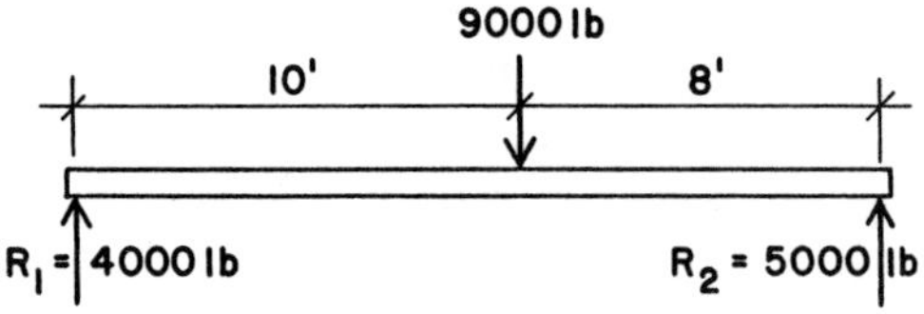

Figure 3.5 Beam reactions for a single load.

consider, for instance, the 9000-lb load placed 10 ft from one end, as shown in Figure 3.5. What will the upward supporting forces be? This is where the principle of moments can be used. Consider a summation of moments about the right-hand support R_2. Thus:

$$\Sigma M = 0 = +(R_1 \times 18) - (9000 \times 8)$$

$$R_1 = \frac{72000}{18} = 4000 \text{ lb}$$

Then, considering the equilibrium of vertical forces:

$$\Sigma V = 0 = +R_1 + R_2 - 9000 \text{ or } R_2 = 9000 - 4000 = 5000 \text{ lb}$$

The accuracy of this solution can be verified by taking moments about the left-hand support. Thus:

$$\Sigma M = 0 = -(R_2 \times 18) + (9000 \times 10), \qquad R_2 = \frac{90{,}000}{18} = 5000 \text{ lb}$$

Example 1. A simple beam 20 ft long has three concentrated loads, as indicated in Figure 3.6. Find the magnitudes of the reactions.

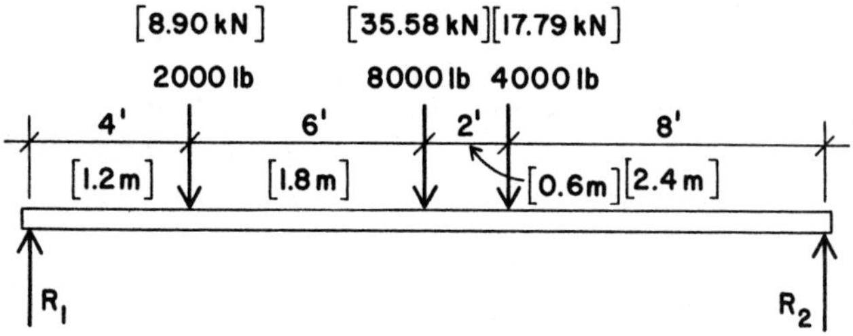

Figure 3.6 Reference for Example 1.

Solution: Using the right-hand support as the center of moments:

$$\Sigma M = +(R_1 \times 20) - (2000 \times 16) - (8000 \times 10) - (4000 \times 8)$$

From which

$$R_1 = \frac{(32{,}000 + 80{,}000 + 32{,}000)}{20} = 7200 \text{ lb}$$

From a summation of the vertical forces:

$$\Sigma V = 0 = +R_2 + 7200 - 2000 - 8000 - 4000, \qquad R_2 = 6800 \text{ lb}$$

With all forces determined, a summation about the left-hand support—or any point except the right-hand support—will verify the accuracy of the work.

The following example demonstrates a solution with uniformly distributed loading on a beam. A convenience in this work is to consider the total uniformly distributed load as a concentrated force placed at the center of the distributed load.

Example 2. A simple beam 16 ft long carries the loading shown in Figure 3.7*a*. Find the reactions.

Solution: The total uniformly distributed load may be considered as a single concentrated load placed at 5 ft from the right-hand support; this loading is shown in Figure 3.7*b*. Considering moments about the right-hand support:

$$\Sigma M = 0 = +(R_1 \times 16) - (8000 \times 12) - (14{,}000 \times 5)$$

from which:

$$R_1 = \frac{166{,}000}{16} = 10{,}375 \text{ lb}$$

And from a summation of vertical forces:

$$R_2 = (8000 + 14{,}000) - 10{,}375 = 11{,}625 \text{ lb}$$

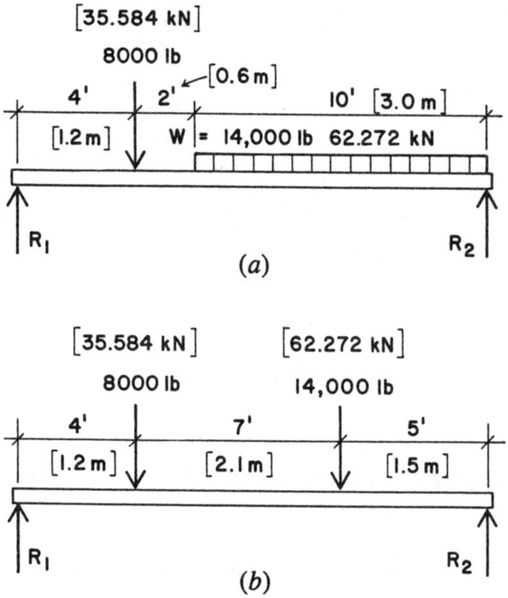

Figure 3.7 Reference for Example 2.

Again, a summation of moments about the left-hand support will verify the accuracy of the work.

In general, any beam with only two supports that each develops only vertical reaction forces will be statically determinate. This includes the simple span beams in the preceding examples as well as beams with overhanging ends.

Problems 3.2.A–F. Find the reactions for the beams shown in Figure 3.8.

3.3 SHEAR IN BEAMS

Figure 3.9*a* represents a simple beam with a uniformly distributed load over its entire length. Examination of an actual beam so loaded probably would not reveal any effects of the loading on the beam. However, there are three distinct major tendencies for the beam to fail. Figures 3.9*b* to *d* illustrate the three phenomena.

First, there is a tendency for the beam to fail by dropping between the supports (Figure 3.9*b*). This is called *vertical shear*. Second, the beam

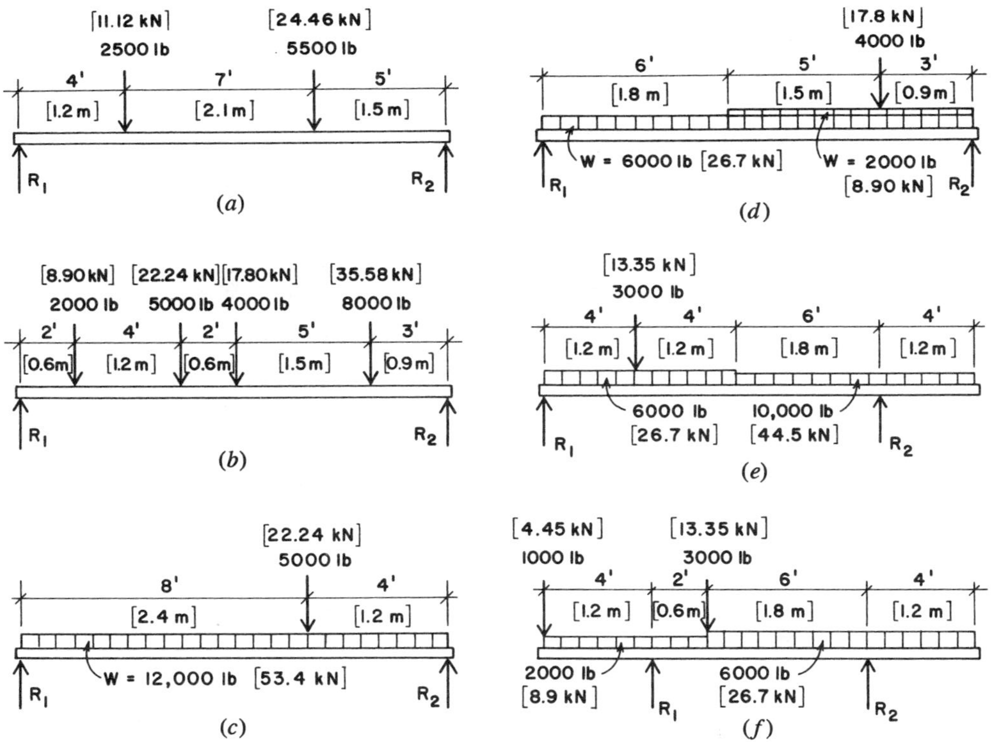

Figure 3.8 Reference for Problem 3.2.

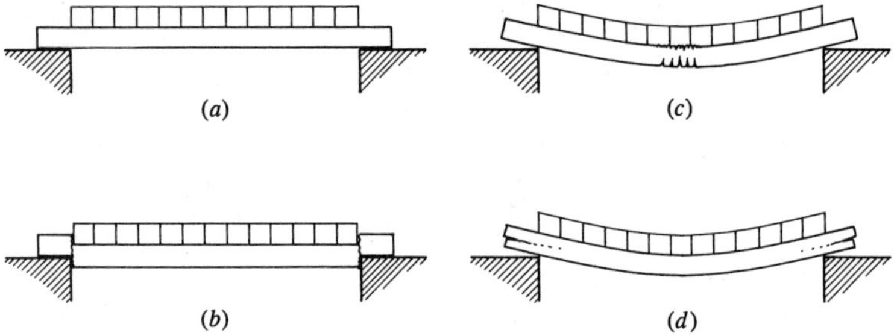

Figure 3.9 Stress failures in beams: (*b*) bending, (*c*) vertical shear, (*d*) horizontal shear.

may fail by bending (Figure 3.9*c*). Third, there is a tendency in wood beams for the fibers of the beam to slide past each other in a horizontal direction (Figure 3.9*d*), an action described as horizontal shear. Naturally, a beam properly designed does not fail in any of the ways just mentioned, but these tendencies to fail are always present and must be considered in structural design.

Vertical Shear

Vertical shear is the tendency for one part of a beam to move vertically with respect to an adjacent part. The magnitude of the shear force at any section in the length of a beam is equal to the algebraic sum of the vertical forces on either side of the section. Vertical shear is usually represented by the letter V. In computing its values in the examples and problems, consider the forces to the left of the section, but keep in mind that the same resulting force magnitude will be obtained with the forces on the right. To find the magnitude of the vertical shear at any section in the length of a beam, simply add up the forces to the right or the left of the section. It follows from this procedure that the maximum value of the shear for simple beams is equal to the greater reaction.

Example 3. Figure 3.10*a* illustrates a simple beam with concentrated loads of 600 lb and 1000 lb. The problem is to find the value of the vertical shear at various points along the length of the beam. Although the weight of the beam constitutes a uniformly distributed load, it is neglected in this example.

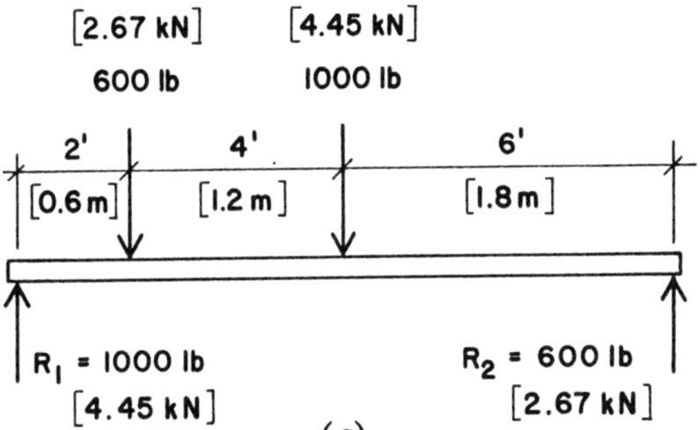

(*a*)

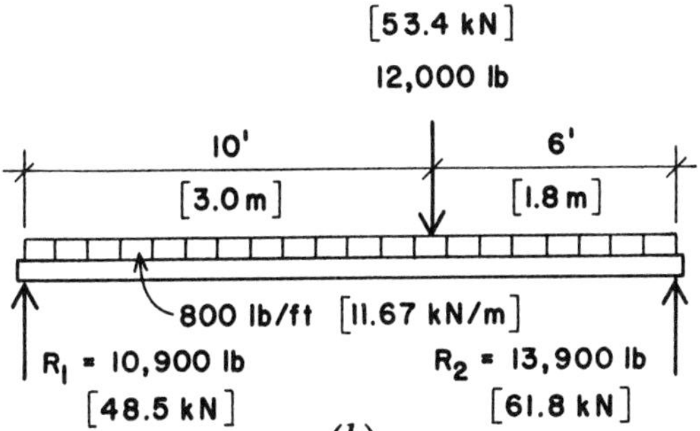

(*b*)

Figure 3.10 Reference for Examples 3 and 4.

Solution: The reactions are computed as previously described, and are found to be $R_1 = 1000$ lb and $R_2 = 600$ lb.

Consider next the value of the vertical shear V at an infinitely short distance to the right of R_1. Applying the rule that the shear is equal to the reaction minus the loads to the left of the section, we write

$$V = R_1 - 0, \quad \text{or} \quad V = 1000 \text{ lb}$$

The zero represents the value of the loads to the left of the section, which of course, is zero. Now take a section 1 ft to the right of R_1; again

$$V_{(x=1)} = R_1 - 0, \quad \text{or} \quad V_{(x=1)} = 1000 \text{ lb}$$

The subscript ($x = 1$) indicates the position of the section at which the shear is taken, the distance of the section from R_1. At this section the shear is still 1000 lb and has the same magnitude up to the 600-lb load.

The next section to consider is a very short distance to the right of the 600 lb load. At this section:

$$V_{(x = 2+)} = 1000 - 600 = 400 \text{ lb}$$

Because there are no loads intervening, the shear continues to be the same magnitude up to the 1000-lb load. At a section a short distance to the right of the 1000-lb load:

$$V_{(x = 6+)} = 1000 - (600 + 1000) = -600 \text{ lb}$$

This magnitude continues up to the right-hand reaction R_2.

Example 4. The beam shown in Figure 3.10*b* supports a concentrated load of 12,000 lb located 6 ft from R_2 and a uniformly distributed load of 800 pounds per linear foot (lb/ft) over its entire length. Compute the value of vertical shear at various sections along the span.

Solution: By use of the equations of equilibrium, the reactions are determined to be $R_1 = 10{,}900$ lb and $R_2 = 13{,}900$ lb. Note that the total distributed load is $800 \times 16 = 12{,}800$ lb. Now consider the vertical shear force at the following sections at a distance measured from the left support.

$$V_{(x=0)} = 10{,}900 - 0 = 10{,}900 \text{ lb}$$

$$V_{(x=1)} = 10{,}900 - (800 \times 1) = 10{,}100 \text{ lb}$$

$$V_{(x=5)} = 10{,}900 - (800 \times 5) = 6900 \text{ lb}$$

$$V_{(x=10-)} = 10{,}900 - (800 \times 10) = 2900 \text{ lb}$$

$$V_{(x=10+)} = 10{,}900 - [(800 \times 10) + 12{,}000)] = -9100 \text{ lb}$$

$$V_{(x=16)} = 10{,}900 - [(800 \times 16) + 12{,}000)] = -13{,}900 \text{ lb}$$

Shear Diagrams

In the two preceding examples the value of the shear at several sections along the length of the beams was computed. To visualize the results, it is common practice to plot these values on a diagram, called the *shear diagram*, which is constructed as explained in the following.

To make such a diagram, first draw the beam to scale and locate the loads. This has been done in Figures 3.11*a* and *b* by repeating the load diagrams of Figures 3.10*a* and *b*, respectively. Beneath the beam, draw a horizontal base line representing zero shear. Above and below this line,

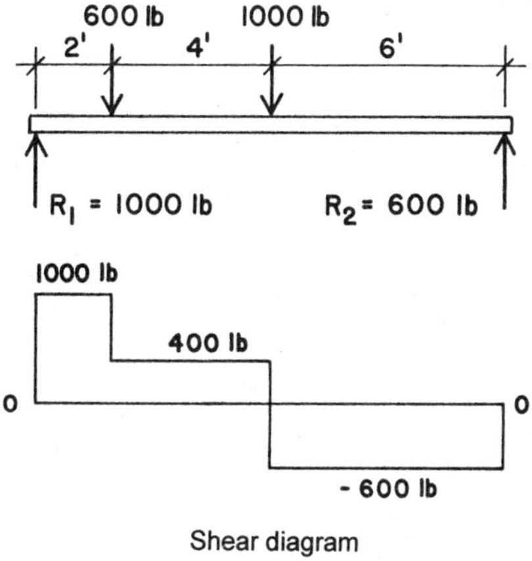

(*a*)

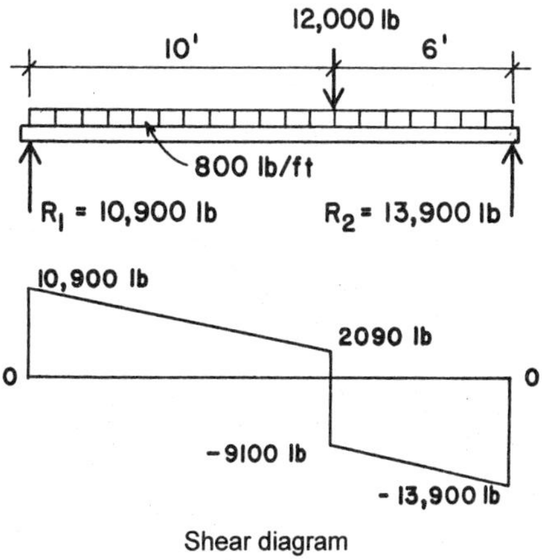

(*b*)

Figure 3.11 Construction of shear diagrams.

plot at any convenient scale the values of the shear at the various sections; the positive, or plus, values are placed above the line and the negative, or minus, values below. In Figure 3.11*a*, for instance, the value of the shear at R_1 is + 1000 lb. The shear continues to have the same value up to the load of 600 lb, at which point it drops to 400 lb. The same value continues up to the next load, 1000 lb, where it drops to –600 lb and continues to the right-hand reaction. Obviously, to draw a shear diagram, it is necessary to compute the values at significant points only. Having made the diagram, we may readily find the value of the shear at any section of the beam by scaling the vertical distance in the diagram. The shear diagram for the beam in Figure 3.11*b* is made in the same manner.

There are two important facts to note concerning the vertical shear. The first is the maximum value. The diagrams in each case confirm the earlier observation that the maximum shear is at the reaction having the greater value, and its magnitude is equal to that of the greater reaction. In Figure 3.11*a* the maximum shear is 1000 lb, and in Figure 3.11*b* it is 13,900 lb. We disregard the positive or negative signs in reading the maximum values of the shear, for the diagrams are merely conventional methods of representing the absolute numerical values.

Another important fact to note is the point at which the shear changes from a plus to a minus quantity. We call this the point at which the shear passes through zero. In Figure 3.11*a* it is under the 1000-lb load, 6 ft from R_1. In Figure 3.11*b* it is under the 12,000-lb load, 10 ft from R_1. A major concern for noting this point is that it indicates the location of the maximum value of bending moment in the beam, as discussed in the next section.

Problems 3.3.A–F. For the beams shown in Figure 3.12, draw the shear diagrams and note all critical values for shear. Note particularly the maximum value for shear and the point at which the shear passes through zero.

3.4 BENDING MOMENTS IN BEAMS

The forces that tend to cause bending in a beam are the reactions and the loads. Consider the section *X-X*, 6 ft from R_1 (Figure 3.13). The force R_1, or 2000 lb, tends to cause a clockwise rotation about this point. Because the force is 2000 lb and the lever arm is 6 ft, the moment of the force is $2000 \times 6 = 12{,}000$ ft-lb. This same value may be found by considering the forces to the right of the section *X-X*: R_2, which is 6000 lb, and the load 8000 lb, with lever arms of 10 and 6 ft, respectively. The moment of

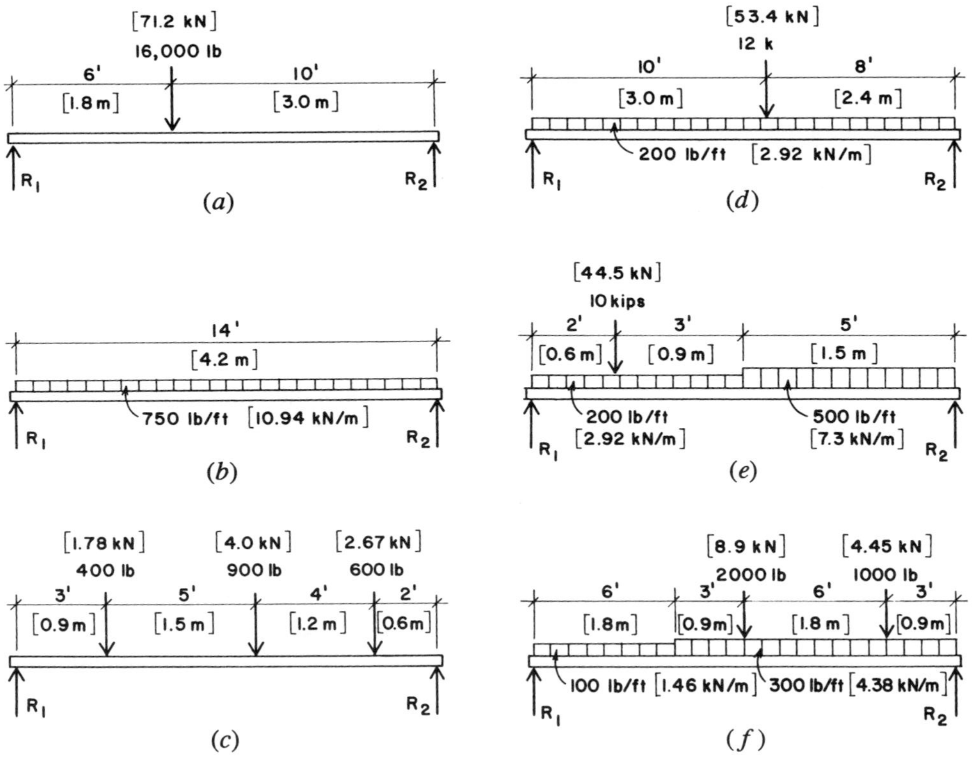

Figure 3.12 Reference for Problem 3.3.

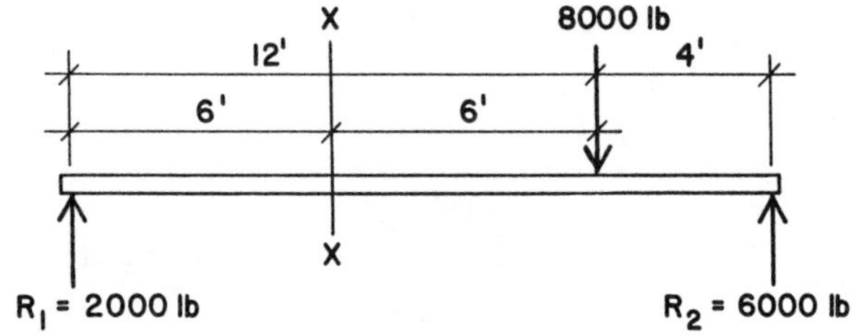

Figure 3.13 Internal bending at a selected beam cross section.

the reaction is 6000 × 10 = 60,000 ft-lb, and its direction is counterclockwise with respect to the section *X-X*. The moment of force 8000 lb is 8000 × 6 = 48,000 ft-lb, and its direction is clockwise. Then 60,000 ft-lb – 48,000 ft-lb = 12,000 ft-lb, the resultant moment tending to cause counterclockwise rotation about the section *X-X*. This is the same magnitude as the moment of the forces on the left, which tend to cause a clockwise rotation.

Thus, it makes no difference whether use is made of the forces to the right of the section or the left; the magnitude of the moment is the same. It is called the *bending moment* (or the *internal bending moment*), because it is the moment of the forces that cause bending stresses in the beam. Its magnitude varies throughout the length of the beam. For instance, at 4 ft from R_1, it is only 2000 × 4, or 8000 ft-lb. The bending moment is the algebraic sum of the moments of the forces on either side of the section. For simplicity, take the forces on the left; then the bending moment at any section of a beam is equal to the moments of the reactions minus the moments of the loads to the left of the section. Because the bending moment is the result of multiplying forces by distances, the denominations are foot-pounds or kip-feet.

Bending Moment Diagrams

The construction of bending moment diagrams follows the procedure used for shear diagrams. The beam span is drawn to scale, showing the locations of the loads. Below this, and usually below the shear diagram, a horizontal base line is drawn representing zero bending moment. Then the bending moments are computed at various sections along the beam span, and the values are plotted vertically to any convenient scale. In simple beams all bending moments are positive and therefore are plotted above the base line. In overhanging or continuous beams there are also negative moments, and these are plotted below the base line.

Example 5. The load diagram in Figure 3.14 shows a simple beam with two concentrated loads. Draw the shear and bending moment diagrams.

Solution: R_1 and R_2 are first computed and are found to be 16,000 lb and 14,000 lb, respectively. These values are recorded on the load diagram.

The shear diagram is drawn as described in Section 3.3. Note that in this instance it is necessary to compute the shear at only one section (between the concentrated loads) because there is no distributed load, and we know that the shear at the supports is equal in magnitude to the reactions.

Because the value of the bending moment at any section of the beam is equal to the moments of the reactions minus the moments of the loads to the left of the section, the moment at R_1 must be zero, for there are no forces to the left. Other values in the length of the beam are computed as follows. The subscripts ($x = 1$, etc.) show the distance from R_1 at which the bending moment is computed.

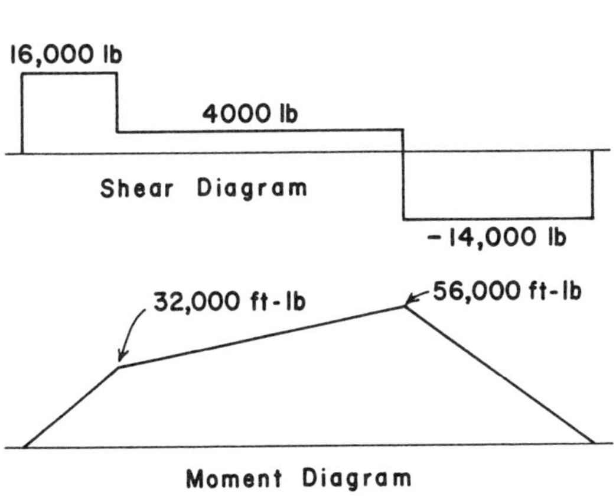

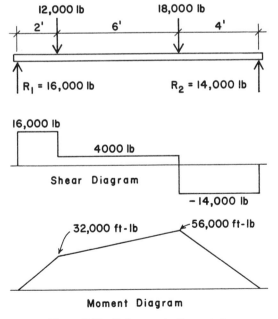

Figure 3.14 Reference for Example 5.

$$M_{(x=1)} = (16{,}000 \times 1) = 16{,}000 \text{ ft-lb}$$

$$M_{(x=2)} = (16{,}000 \times 2) = 32{,}000 \text{ ft-lb}$$

$$M_{(x=5)} = (16{,}000 \times 5) - (12{,}000 \times 3) = 44{,}000 \text{ ft-lb}$$

$$M_{(x=8)} = (16{,}000 \times 8) - (12{,}000 \times 6) = 56{,}000 \text{ ft-lb}$$

$$M_{(x=10)} = (16{,}000 \times 10) - [(12{,}000 \times 8) + (18{,}000 \times 2)] = 28{,}000 \text{ ft-lb}$$

$$M_{(x=12)} = (16{,}000 \times 12) - [(12{,}000 \times 10) + (18{,}000 \times 4)] = 0 \text{ ft-lb}$$

The result of plotting these values is shown in the bending moment diagram of Figure 3.14. More moments were computed than were necessary. We know that the bending moments at the supports of simple beams are zero, and in this instance only the bending moments directly under the loads were needed.

Relations Between Shear and Bending Moment

In simple beams the shear diagram passes through zero at some point between the supports. As stated earlier, an important principle in this respect is that the bending moment has a maximum magnitude wherever the shear passes through zero. In Figure 3.14 the shear passes through zero under the 18,000-lb load, that is, at $x = 8$ ft. Note that the bending moment has its greatest value at this same point, 56,000 ft-lb.

Example 6. Draw the shear and bending moment diagrams for the beam shown in Figure 3.15, which carries a uniformly distributed load of 400 lb per lin ft and a concentrated load of 21,000 lb located 4 ft from R_1.

Solution: Computing the reactions, we find $R_1 = 17{,}800$ lb and $R_2 = 8800$ lb. By use of the process described in Section 3.3, the critical shear values are determined and the shear diagram is drawn as shown in the figure.

Although the only value of bending moment that must be computed is that where the shear passes through zero, some additional values are determined in order to plot the true form of the moment diagram. Thus:

$$M_{(x=2)} = (17{,}800 \times 2) - (400 \times 2 \times 1) = 34{,}800 \text{ ft-lb}$$

$$M_{(x=4)} = (17{,}800 \times 4) - (400 \times 4 \times 2) = 68{,}000 \text{ ft-lb}$$

$$M_{(x=8)} = (17{,}800 \times 8) - [(400 \times 8 \times 4) + (21{,}000 \times 4)] = 45{,}600 \text{ ft-lb}$$

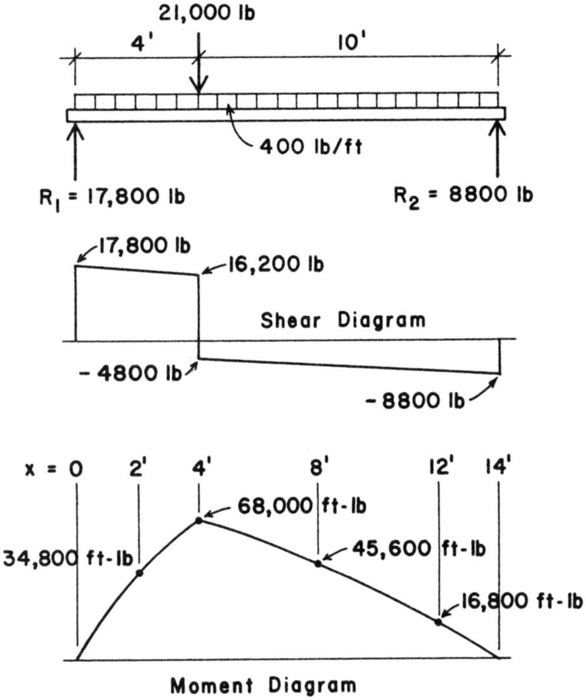

Figure 3.15 Reference for Example 6.

$$M_{(x=12)} = (17{,}800 \times 12) - [(400 \times 12 \times 6) + (21{,}000 \times 8)]$$
$$= 16{,}800 \text{ ft-lb}$$

From the two preceding examples (Figures 3.14 and 3.15), it will be observed that the shear diagram for the parts of the beam on which no loads occur is represented by horizontal lines. For the parts of the beam on which a uniformly distributed load occurs, the shear diagram consists of straight inclined lines. The bending moment diagram is represented by straight inclined lines when only concentrated loads occur and by a curved line if the load is distributed.

Occasionally, when a beam has both concentrated and uniformly distributed loads, the shear does not pass through zero under one of the concentrated loads. This frequently occurs when the distributed load is relatively large compared with the concentrated loads. Because it is necessary in designing beams to find the maximum bending moment, we must know the point at which it occurs. This, of course, is the point where the shear passes through zero, and its location is readily determined by the procedure illustrated in the following example.

Example 7. The load diagram in Figure 3.16 shows a beam with a concentrated load of 7000 lb, applied 4 ft from the left reaction, and a uniformly distributed load of 800 lb per lin ft extending over the full span. Compute the maximum bending moment on the beam.

Solution: The values of the reactions are found to be $R_1 = 10{,}600$ lb and $R_2 = 7600$ lb and are recorded on the load diagram.

The shear diagram is constructed and it is observed that the shear passes through zero at some point between the concentrated load of 7000 lb and the right reaction. Call this distance x ft from R_2. The value of the shear at this section is zero; therefore, an expression for the shear for this point, using the reaction and loads, is equal to zero. This equation contains the distance x:

$$V_{(\text{at } x)} = -7600 + 800x = 0, \qquad X = \frac{7600}{800} = 9.5 \text{ ft}$$

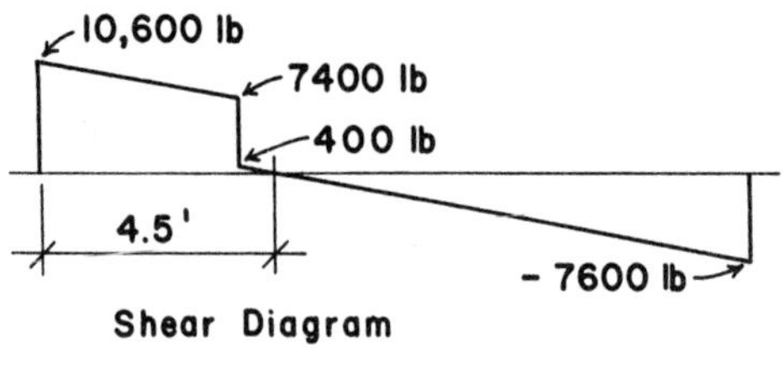

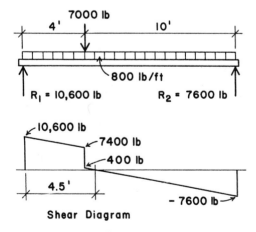

Figure 3.16 Reference for Example 7.

The zero shear point is thus at 9.5 ft from the right support and (as shown in the diagram) at 4.5 ft from the left support. This location can also be determined by writing an equation for the summation of shear from the left of the point, which should produce the answer of 4.5 ft.

Following the convention of summing up the moments from the left of the section, the maximum moment is determined as

$$M_{(x=4.5)} = +(10{,}600 \times 4.5) - (7000 \times 0.5) - \left[800 \times 4.5 \times \left(\frac{4.5}{2}\right)\right]$$

$$= 36{,}100 \text{ ft-lb}$$

Problems 3.4.A–F. Draw the shear and bending moment diagrams for the beams in Figure 3.12, indicating all critical values for shear and moment and all significant dimensions. (Note: These are the beams for Problem 3.3, for which the shear diagrams were constructed.)

3.5 SENSE OF BENDING IN BEAMS

When a simple beam bends, it has a tendency to assume the shape shown in Figure 3.17*a*. In this case the fibers in the upper part of the beam are in compression. For this condition the bending moment is considered as positive. Another way to describe a positive bending moment is to say that it is positive when the curve assumed by the bent beam is concave upward. When a beam projects beyond a support (Figure 3.17*b*), this portion of the beam has tensile stresses in the upper part. The bending moment for this condition is called negative; the beam is bent concave downward. When constructing moment diagrams, following the method previously described, the positive and negative moments are shown graphically.

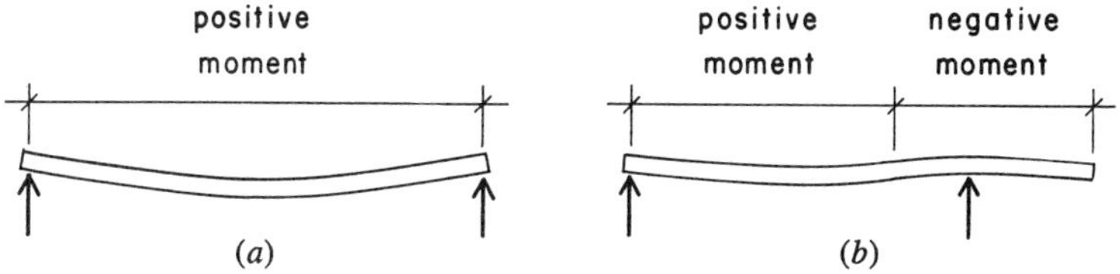

Figure 3.17 Sense of bending moment in beams.

Example 8. Draw the shear and bending moment diagrams for the overhanging beam shown in Figure 3.18.

Solution: Computing the reactions

$$\text{From } \Sigma M \text{ about } R_1\text{: } R_2 \times 12 = 600 \times 16 \times 8,\ R_2 = 6400 \text{ lb}$$

$$\text{From } \Sigma M \text{ about } R_2\text{: } R_1 \times 12 = 600 \times 16 \times 4,\ R_1 = 3200 \text{ lb}$$

With the reactions determined, the construction of the shear diagram is quite evident. For the location of the point of zero shear, considering its distance from the left support as x:

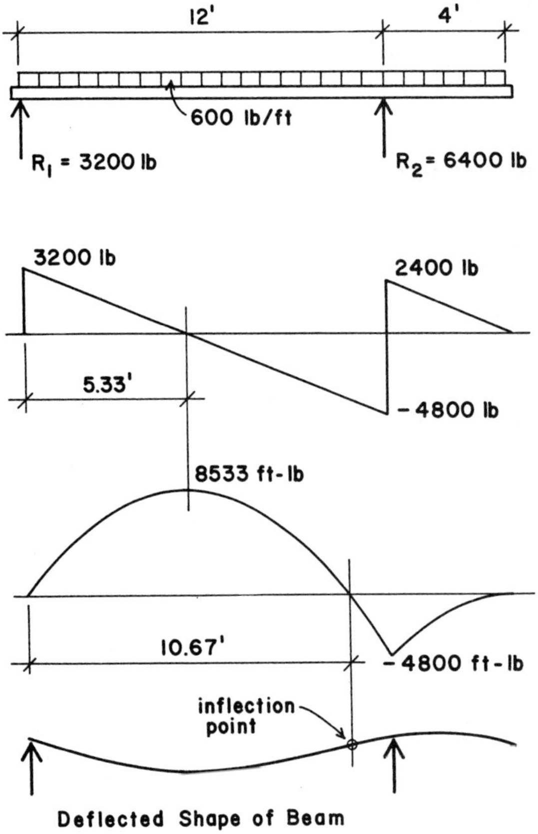

Figure 3.18 Reference for Example 8.

$$3200 - 600x = 0, \qquad x = 5.33 \text{ ft}$$

For the critical values needed to plot the moment diagram:

$$M_{(x=5.33)} = +(3200 \times 5.33) - \left[600 \times 5.33 \times \left(\frac{5.33}{2}\right)\right] = 8533 \text{ ft-lb}$$

$$M_{(x=12)} = (3200 \times 12) - (600 \times 12 \times 6) = -4800 \text{ ft-lb}$$

The form of the moment diagram for the distributed loading is a curve (parabolic), which may be verified by plotting some additional points on the graph.

For this case the shear diagram passes through zero twice, both of which points indicate peaks of the moment diagram—one positive and one negative. As the peak in the positive portion of the moment diagram is actually the apex of the parabola, the location of the zero moment value is simply twice the value previously determined as x. This point corresponds to the change in the form of curvature on the elastic curve (deflected shape) of the beam; this point is described as the *inflection point* for the deflected shape. The location of the point of zero moment can also be determined by writing an equation for the sum of moments at the unknown location. In this case, calling the new unknown point x:

$$M = 0 = +(3200 \times x) - \left[600 \times x \times \left(\frac{x}{2}\right)\right]$$

Solution of this quadratic equation should produce the value of $x = 10.67$ ft.

Example 9. Compute the maximum bending moment for the overhanging beam shown in Figure 3.19.

Solution: Computing the reactions, $R_1 = 3200$ lb and $R_2 = 2800$ lb. As usual, the shear diagram can now be plotted as the graph of the loads and reactions, proceeding from left to right. Note that the shear passes through zero at the location of the 4000 lb load and at both supports. As usual, these are clues to the form of the moment diagram.

With the usual moment summations, values for the moment diagram can now be found at the locations of the supports and all of the concen-

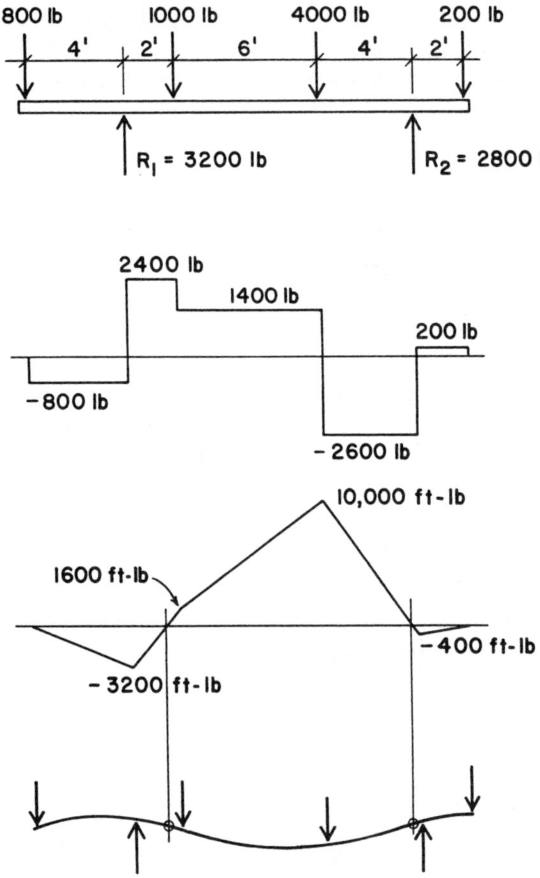

Figure 3.19 Reference for Example 9.

trated loads. From this plot it will be noted that there are two inflection points (locations of zero moment). As the moment diagram is composed of straight-line segments in this case, the locations of these points may be found by writing simple linear equations for their locations. However, use can also be made of some relationships between the shear and moment graphs. One of these has already been used, relating to the correlation of zero shear and maximum moment. Another relationship is that the change of the value of moment between any two points along the beam is equal to the total area of the shear diagram between the points. If the value of moment is known at some point, it is thus a simple matter to find values at other points. For example, starting from the left end, the value of moment is known to be zero at the left end of the beam; then the value of the moment at the support is the area of the rectangle on the shear diagram with base of 4 ft and height of 800 lb—the area being 4 × 800 = 3200 ft-lb.

Now, proceeding along the beam to the point of zero moment (call it x distance from the support), the change is again 3200, which relates to an area of the shear diagram that is $x \times 2400$. Thus

$$2400x = 3200, \qquad x = \frac{3200}{2400} = 1.33 \text{ ft}$$

And, now calling the distance from the right support to the point of zero moment x

$$2600x = 400, \qquad x = \frac{400}{2600} = 0.154 \text{ ft}$$

Problems 3.5.A–D. Draw the shear and bending moment diagrams for the beams in Figure 3.20, indicating all critical values for shear and moment and all significant dimensions.

Cantilever Beams

In order to keep the signs for shear and moment consistent with those for other beams, it is convenient to draw a cantilever beam with its fixed end to the right, as shown in Figure 3.21. We then plot the values for the shear and moment on the diagrams as before, proceeding from the left end.

Example 10. The cantilever beam shown in Figure 3.21*a* projects 12 ft from the face of the wall and has a concentrated load of 800 lb at the unsupported end. Draw the shear and moment diagrams. What are the values of the maximum shear and maximum bending moment?

Solution: The value of the shear is – 800 lb throughout the entire length of the beam. The bending moment is maximum at the wall; its value is 800 × 12 = – 9600 ft-lb. The shear and moment diagrams are as shown in Figure 3.21*a*. Note that the moment is negative for the entire length of the cantilever beam, corresponding to its concave downward shape throughout its length.

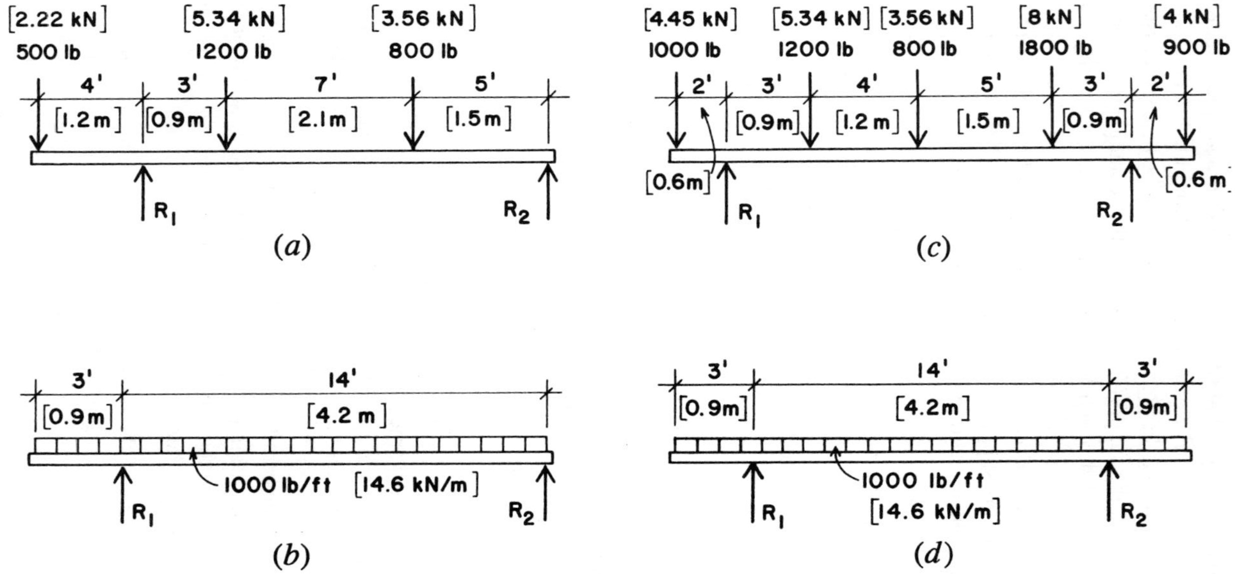

Figure 3.20 Reference for Problem 3.5, part 1.

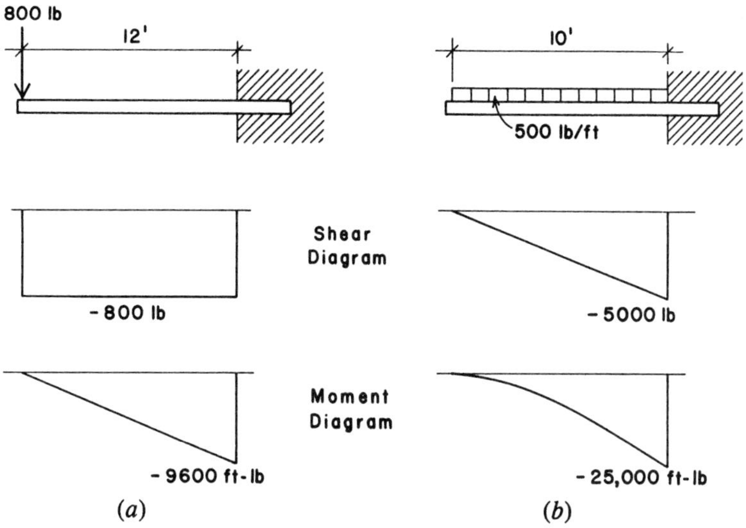

Figure 3.21 Reference for Examples 10 and 11.

Although they are not shown in the figure, the reactions in this case are a combination of an upward force of 800 lb and a clockwise resisting moment of 9600 ft-lb.

Example11. Draw shear and bending moment diagrams for the beam in Figure 3.21*b,* which carries a uniformly distributed load of 500 lb/ft over its full length.

Solution: The total load is 500 × 10 = 5000 lb. The reactions are an upward force of 5000 lb and a moment determined as

$$M = -500 \times 10 \times \left(\frac{10}{2}\right) = -25{,}000 \text{ ft-lb}$$

which, it may be noted, is also the total area of the shear diagram between the outer end and the support.

Example 12. The cantilever beam indicated in Figure 3.22 has a concentrated load of 2000 lb and a uniformly distributed load of 600 lb per lin ft at the positions shown. Draw the shear and bending moment dia-

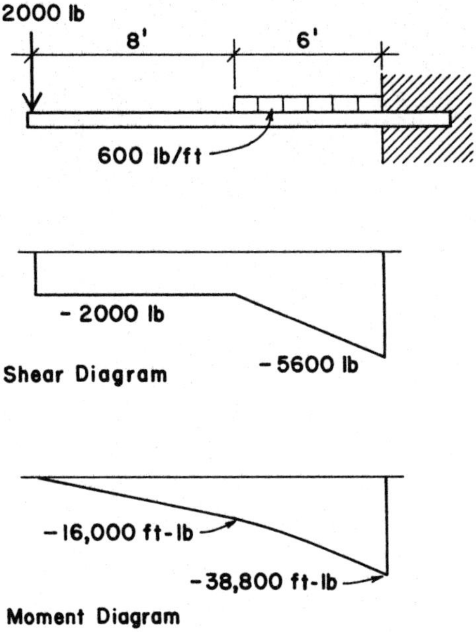

Figure 3.22 Reference for Example 12.

grams. What are the magnitudes of the maximum shear and maximum bending moment?

Solution: The reactions are actually *equal* to the maximum shear and bending moment. Determined directly from the forces, they are

$$V = 2000 + (600 \times 6) = 5600 \text{ lb}$$

$$M = -(2000 \times 14) - \left[600 \times 6 \times \left(\frac{6}{2}\right)\right] = -38{,}800 \text{ ft-lb}$$

The diagrams are quite easily determined. The other moment value needed for the moment diagram can be obtained from the moment of the concentrated load or from the simple rectangle of the shear diagram: $2000 \times 8 = 16{,}000$ ft-lb.

Note that the moment diagram has a straight-line shape from the outer end to the beginning of the distributed load, and becomes a curve from this point to the support.

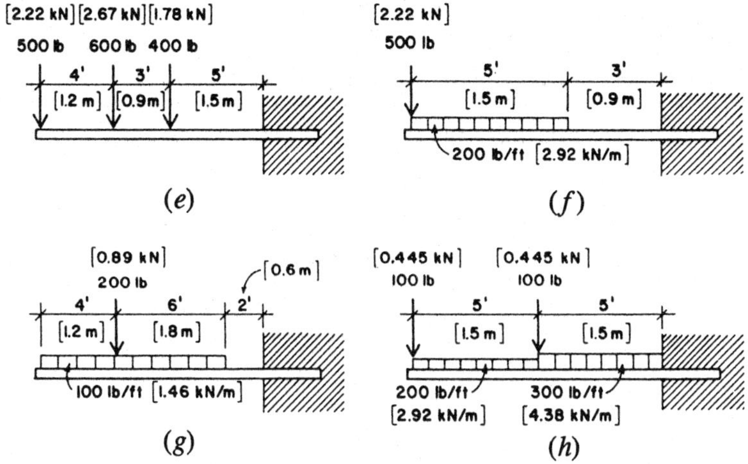

Figure 3.23 Reference for Problem 3.5, part 2.

It is suggested that Example 10 be reworked with Figure 3.22 reversed, left for right. All numerical results will be the same, but the shear diagram will be positive over its full length.

Problems 3.5.E–H. Draw the shear and bending moment diagrams for the beams in Figure 3.23, indicating all critical values for shear and moment and all significant dimensions.

3.6 TABULATED VALUES FOR BEAM BEHAVIOR

Bending Moment Formulas

The methods of computing beam reactions, shears, and bending moments presented thus far in this chapter make it possible to find critical values for design under a wide variety of loading conditions. However, certain conditions occur so frequently that it is convenient to use formulas that give the maximum values directly. Structural design handbooks contain many such formulas; two of the most commonly used formulas are derived in the following examples.

Simple Beam, Concentrated Load at Center of Span

A simple beam with a concentrated load at the center of the span occurs very frequently in practice. Call the load P and the span length between supports L, as indicated in the load diagram of Figure 3.24*a*. For this symmetrical loading each reaction is $P/2$, and it is readily apparent that the shear will pass through zero at distance $x = L/2$ from R_1. Therefore, the maximum bending moment occurs at the center of the span, under the load. Computing the value of the bending moment at this section:

$$M = \frac{P}{2} \times \frac{L}{2} = \frac{PL}{4}$$

Example 13. A simple beam 20 ft in length has a concentrated load of 8000 lb at the center of the span. Compute the maximum bending moment.

Solution: As just derived, the formula giving the value of the maximum bending moment for this condition is $M = PL/4$. Therefore:

$$M = \frac{PL}{4} = \frac{8000 \times 20}{4} = 40{,}000 \text{ ft-lb}$$

Simple Beam, Uniformly Distributed Load

This is probably the most common beam loading; it occurs time and again. For any beam, its own dead weight as a load to be carried is usually of

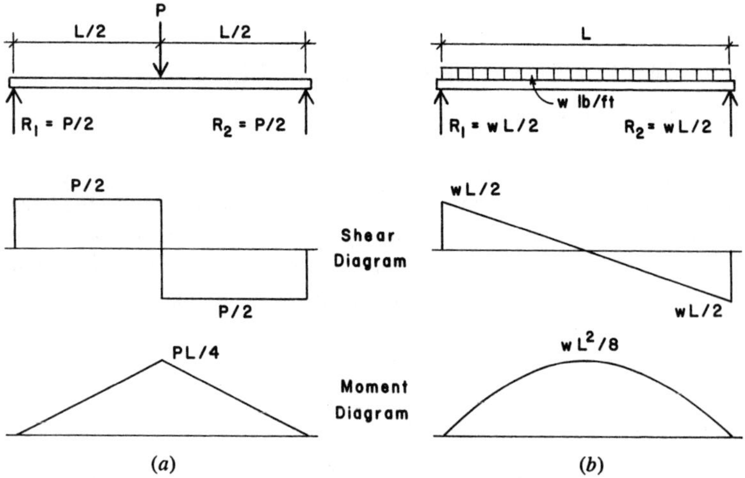

Figure 3.24 Values for simple beam loadings.

this form. Call the span L and the unit load w as indicated in Figure 3.24*b*. The total load on the beam is $W = wL$; hence, each reaction is $W/2$ or $wL/2$. The maximum bending moment occurs at the center of the span at distance $L/2$ from R_1. Writing the value of M for this section:

$$M = +\left[\frac{wL}{2} \times \frac{L}{2}\right] - \left[w \times \left(\frac{L}{2}\right) \times \left(\frac{L}{4}\right)\right] = \frac{wL^2}{8}, \quad \text{or} \quad \frac{WL}{8}$$

Note the alternative use of the unit load w or the total load W in this formula. Both forms will be seen in various references. It is important to carefully identify the use of one or the other.

Example 14. A simple beam 14 ft long has a uniformly distributed load of 800 lb/ft. Compute the maximum bending moment.

Solution: As just derived, the formula that gives the maximum bending moment for a simple beam with uniformly distributed load is $M = wL^2/8$. Substituting these values:

$$M = \frac{wL^2}{8} = \frac{800 \times 14^2}{8} = 19{,}600 \text{ ft-lb}$$

or, using the total load of $800 \times 14 = 11{,}200$ lb:

$$M = \frac{WL}{8} = \frac{11{,}200 \times 14}{8} = 19{,}600 \text{ ft-lb}$$

Use of Tabulated Values for Beams

Some of the most common beam loadings are shown in Figure 3.25. In addition to the formulas for the reactions R, for maximum shear V, and for maximum bending moment M, expressions for maximum deflection Δ are given also. Discussion of deflections formulas will be deferred for the time being but will be considered in Section 5.5.

In Figure 3.25 if the loads P and W are in pounds or kips, the vertical shear V will also be in units of pounds or kips. When the loads are given in pounds or kips and the span is given in feet, the bending moment M will be in units of foot-pounds or kip-feet.

Also given in Figure 3.25 are values designated ETL, which stands for *equivalent tabular load.* These may be used to derive a hypothetical uni-

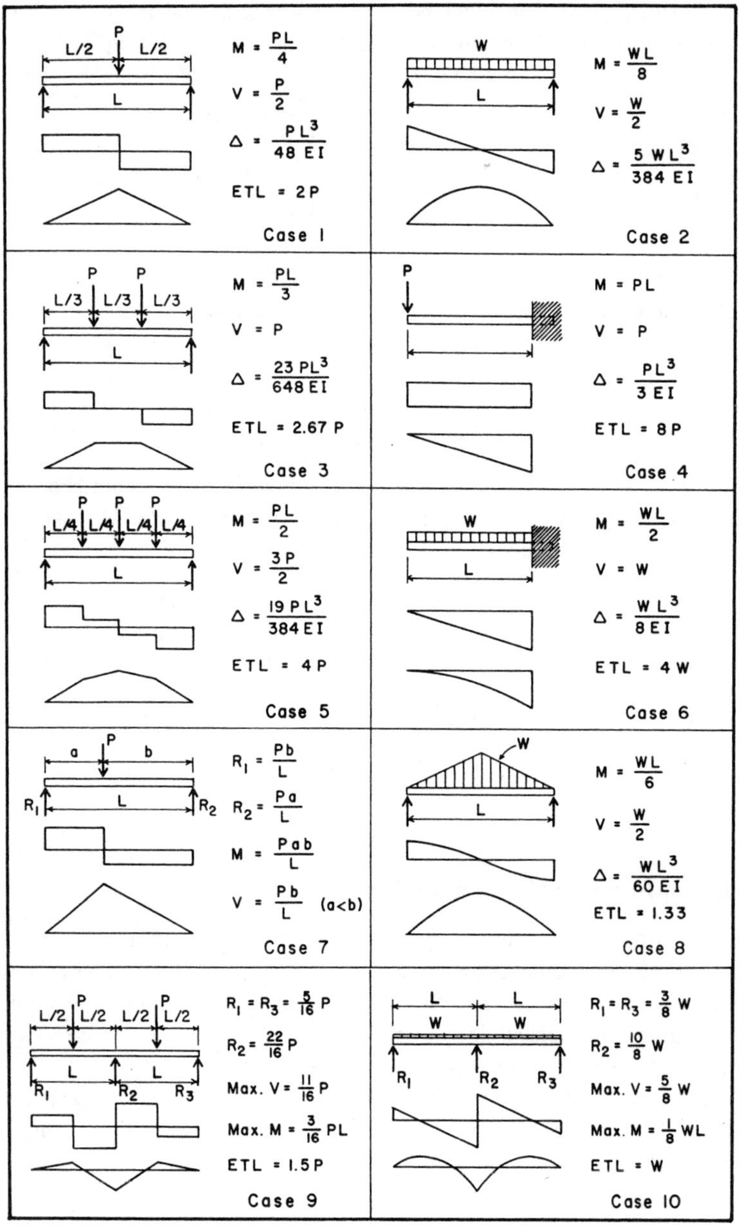

Figure 3.25 Values for typical beam loadings and support conditions.

formly distributed load that when applied to the beam will produce the same magnitude of maximum bending moment as that for the given case of loading. Use of these factors is illustrated in later parts of the book.

Problem 3.6.A. A simple span beam has two concentrated loads of 4 kips [17.8 kN], each placed at the third points of the 24-ft [7.32-m] span. Find the value for the maximum bending moment in the beam.

Problem 3.6.B. A simple span beam has a uniformly distributed load of 2.5 kips/ft [36.5 kN/m] on a span of 18 ft [5.49 m]. Find the value for the maximum bending moment in the beam.

Problem 3.6.C. A simple beam with a span of 32 ft [9.745 m] has a concentrated load of 12 kips [53.4 kN] at 12 ft [3.66 m] from one end. Find the value for the maximum bending moment in the beam.

Problem 3.6.D. A simple beam with a span of 36 ft has a distributed load that varies from a value of 0 at its ends to a maximum of 1000 lb/ft at its center (Case 8 in Figure 3.25). Find the value for the maximum bending moment in the beam.

3.7 DEVELOPMENT OF BENDING RESISTANCE

As developed in the preceding sections, bending moment is a measure of the tendency of the external forces on a beam to deform it by bending. The purpose of this section is to consider the action within the beam that resists bending and is called the *resisting moment.*

Figure 3.26*a* shows a simple beam, rectangular in cross section, supporting a single concentrated load *P*. Figure 3.26*b* is an enlarged sketch of the left-handed portion of the beam between the reaction and section *X-X*. It is observed that the reaction R_1 tends to cause a clockwise rotation about point *A* in the section under consideration; this is defined as the bending moment at the section. In this type of beam the fibers in the upper part are in compression, and those in the lower part are in tension. There is a horizontal plane separating the compressive and tensile stresses; it is called the *neutral surface,* and at this plane there are neither compressive nor tensile stresses with respect to bending. The line in which the neutral surface intersects the beam cross section (Figure 3.26*c*) is called the *neutral axis*, NA.

Call *C* the sum of all the compressive stresses acting on the upper part of the cross section, and call *T* the sum of all the tensile stresses acting on the lower part. It is the sum of the moments of those stresses at the sec-

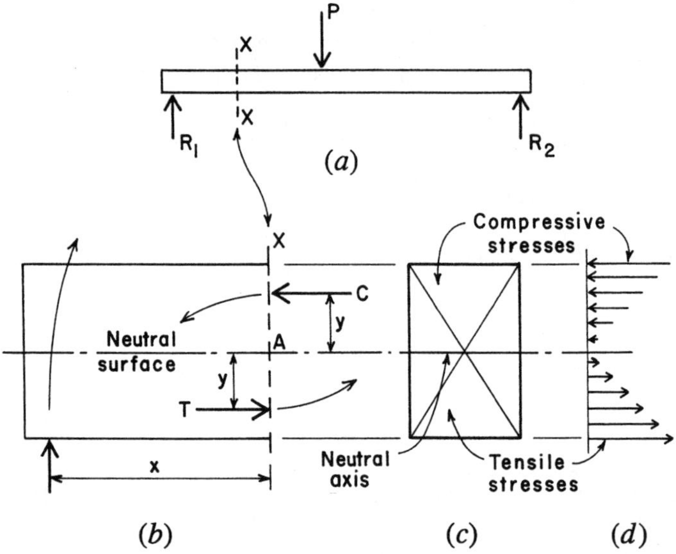

Figure 3.26 Development of bending stress in beams.

tion that holds the beam in equilibrium; this is called the *resisting moment* and is equal to the bending moment in magnitude. The bending moment about A is $R_1 \times x$, and the resisting moment about the same point is $(C \times y) + (T \times y)$. The bending moment tends to cause a clockwise rotation, and the resisting moment tends to cause a counterclockwise rotation. If the beam is in equilibrium, these moments are equal, or

$$R_1 \times x = (C \times y) + (T \times y)$$

that is, the bending moment equals the resisting moment. This is the theory of flexure (bending) in beams. For any type of beam, it is possible to compute the bending moment, and to design a beam to withstand this tendency to bend; this requires the selection of a member with a cross section of such shape, area, and material that it is capable of developing a resisting moment equal to the bending moment.

The Flexure Formula

The flexure formula, $M = fS$, is an expression for resisting moment that involves the size and shape of the beam cross section (represented by S in

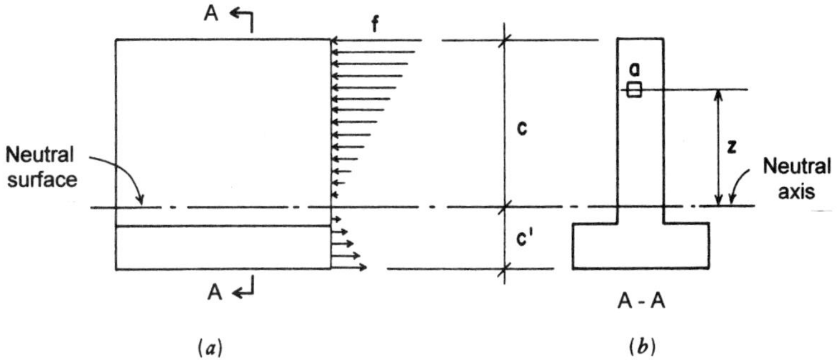

Figure 3.27 Distribution of bending stress on a beam cross section.

the formula) and the material of which the beam is made (represented by f). It is used in the design of all homogeneous beams, that is, beams made of one material only, such as steel, or wood. The following brief derivation is presented to show the principles on which the formula is based.

Figure 3.27 represents a partial side elevation and the cross section of a homogeneous beam subjected to bending stresses. The cross section shown is unsymmetrical about the neutral axis, but this discussion applies to a cross section of any shape. In Figure 3.27*a* let c be the distance of the fiber farthest from the neutral axis, and let f be the unit stress on the fiber at distance c. If f, the extreme fiber stress, does not exceed the elastic limit of the material, the stresses in the other fibers are directly proportional to their distances from the neutral axis. That is to say, if one fiber is twice the distance from the neutral axis than another fiber, the fiber at the greater distance will have twice the stress. The stresses are indicated in the figure by the small lines with arrows, which represent the compressive and tensile stresses acting toward and away from the section, respectively. If c is in inches, the unit stress on a fiber at 1-in. distance is f/c. Now imagine an infinitely small area a at z distance from the neutral axis. The unit stress on this fiber is $(f/c) \times z$, and because this small area contains a square inches, the total stress on fiber a is $(f/c) \times z \times a$. The moment of the stress on fiber a at z distance is

$$\frac{f}{c} \times z \times a \times z \quad \text{or} \quad \frac{f}{c} az^2$$

There is an extremely large number of these minute areas. Using the symbol Σ to represent the sum of this very large number:

$$\sum\left[\frac{f}{c}az^2\right]$$

which means the sum of the moments of all the stresses in the cross section with respect to the neutral axis. This is the *resisting moment*, and it is equal to the bending moment.
Therefore

$$M_R = \frac{f}{c}\sum az^2$$

The quantity Σaz^2 may be read as "the sum of the products of all the elementary areas times the square of their distances from the neutral axis." This is called the *moment of inertia* and is represented by the letter I (see Section A.2). Therefore, substituting in the above:

$$M_R = \frac{f}{c}I \quad \text{or} \quad M_R = \frac{fI}{c}$$

This is known as the *flexure formula* or *beam formula*, and by its use it is possible to design any beam that is composed of a single material. The expression may be simplified further by substituting S for I/c, called the *section modulus*, a term that is described more fully in Section A.4. Making this substitution, the formula becomes

$$M = fS$$

Use of the flexural formula is discussed in Chapter 5 for wood beams, in Chapter 9 for steel beams, and in Chapter 13 for reinforced concrete beams.

Inelastic Stress Conditions

At the limits of bending resistance, the preceding descriptions of stress and deformation do not generally represent true conditions. The limit states are described as the *ultimate resistance* and design based on the limit states is called *ultimate strength design*. These conditions and the procedures for their use in design are described in Part III for steel members and in Part IV for reinforced concrete members.

3.8 SHEAR STRESS IN BEAMS

Shear is developed in beams in direct resistance to the vertical force at a beam cross section. Because of the interaction of shear and bending in the beam, the exact nature of stress resistance within the beam depends on the form and materials of the beam. For an example, in wood beams the wood grain is normally oriented in the direction of the span and the wood material has a very low resistance to horizontal splitting along the grain. An analogy to this is represented in Figure 3.28, which shows a stack of loose boards subjected to a beam loading. With nothing but minor friction between the boards, the individual boards will slide over each other to produce the loaded form indicated in the bottom figure. This is the failure tendency in the wood beam, and the shear phenomenon for wood beams is usually described as one of *horizontal shear.*

Shear stresses in beams are not distributed evenly over the cross section of the beam, as was assumed for the case of simple direct shear (see Section 2.1). From observations of tested beams and derivations considering the equilibrium of beam segments under combined actions of shear and bending, the following expression has been obtained for shear stress in a beam.

$$f_v = \frac{VQ}{Ib}$$

where: $V =$ shear force at the beam section

$Q =$ moment about the neutral axis of the portion of the cross-section area between the edge of the section and the point where stress is being computed

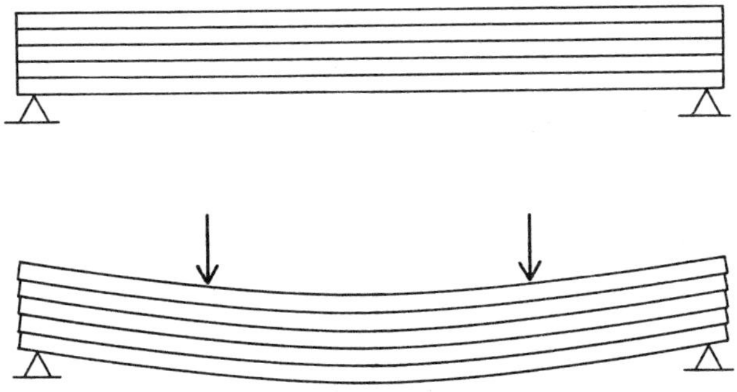

Figure 3.28 Nature of horizontal shear in beams.

I = moment of inertia of the section with respect to the neutral centroidal) axis

b = width of the section at the point where stress is being computed

It may be observed that the highest value for Q, and thus for shear stress, will occur at the neutral axis and that shear stress will be zero at the top and bottom edges of the section. This is essentially opposite to the form of distribution of bending stress on a section. The form of shear distribution for various geometric shapes of beam sections is shown in Figure 3.29.

The following examples illustrate the use of the general shear stress formula.

Example 15. A rectangular beam section with a depth of 8 in. and width of 4 in. Sustains a shear force of 4 kips. Find the maximum shear stress. (See Figure 3.30*a*.)

Solution: For the rectangular section the moment of inertia about the centroidal axis is as follows. (See Figure A.13.)

$$I = \frac{bd^3}{12} = \frac{4 \times 8^3}{12} = 170.7 \text{ in.}^4$$

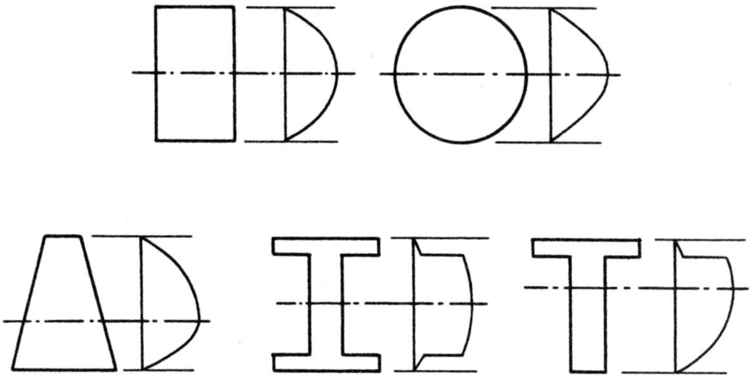

Figure 3.29 Distribution of shear stress in beams with various shapes of cross sections.

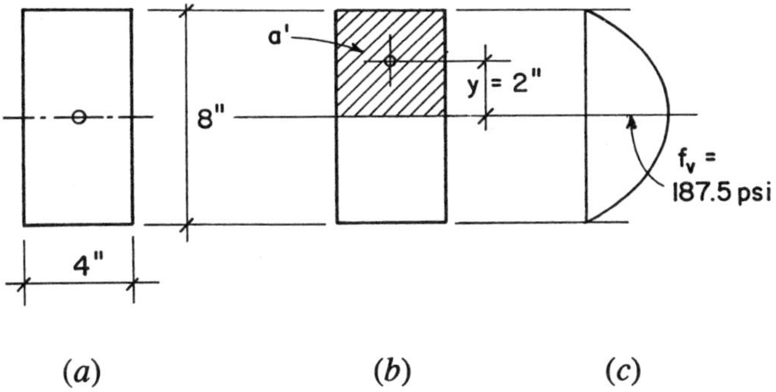

Figure 3.30 Reference for Example 15.

The static moment (Q) is the product of the area a' and its centroidal distance from the neutral axis of the section (y as shown in Figure 3.30*b*). This is the greatest value that can be obtained for Q and will produce the highest shear stress for the section. Thus:

$$Q = a' y = (4 \times 4)(2) = 32 \text{ in.}^3$$

and

$$f_v = \frac{VQ}{Ib} = \frac{4000 \times 32}{170.7 \times 4} = 187.5 \text{ psi}$$

The distribution of shear stress on the beam cross section is as shown in Figure 3.30*c*.

Example 16. A beam with the T-section shown in Figure 3.31*a* is subjected to a shear force of 8 kips. Find the maximum shear stress and the value of shear stress at the location of the juncture of the web and the flange of the T.

Solution: Because this section is not symmetrical with respect to its horizontal centroidal axis, the first steps for this problem consist of locating the neutral axis and determining the moment of inertia for the section with respect to the neutral axis. To save space, this work is not shown here, although it is performed as Examples 1 and 8 in the appendix. From that work it is found that the centroidal neutral axis is located at 6.5 in.

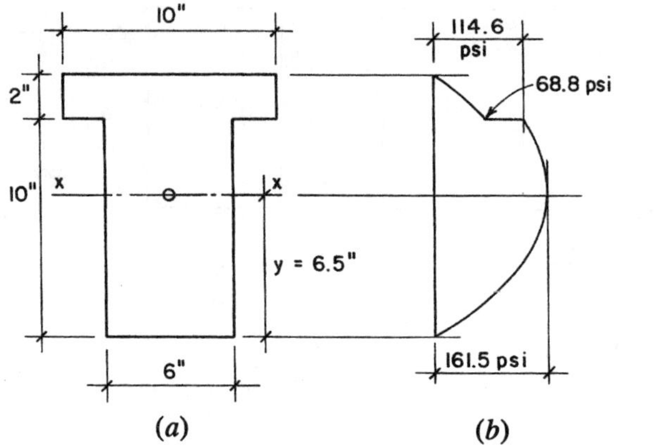

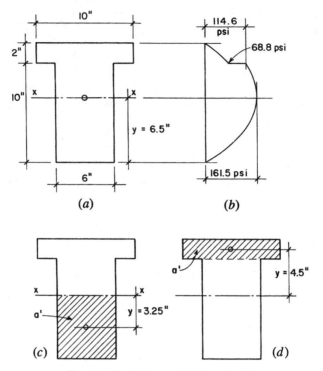

Figure 3.31 Reference for Example 16.

from the bottom of the T and the moment of inertia about the neutral axis is 1046.7 in.^{4}.

For computation of the maximum shear stress at the neutral axis, the value of Q is found by using the portion of the web below the neutral axis, as shown in Figure 3.31*c*. Thus:

$$Q = a'y = (6.5 \times 6) \times \left(\frac{6.5}{2}\right) = 126.75 \text{ in.}^3$$

and the maximum stress at the neutral axis is thus

$$f_v = \frac{VQ}{Ib} = \frac{(8000 \times 126.75)}{(1046.7 \times 6)} = 161.5 \text{ psi}$$

For the stress at the juncture of the web and flange, Q is determined using the area shown in Figure 3.31*d*. Thus

$$Q = (2 \times 10)(4.5) = 90 \text{ in.}^3$$

And the two values for shear stress at this location, as displayed in Figure 3.31*b*, are

$$f_v = \frac{8000 \times 90}{1046.7 \times 6} = 114.6 \text{ psi} \quad \text{(in the web)}$$

$$f_v = \frac{8000 \times 90}{1046.7 \times 10} = 68.8 \text{ psi} \quad \text{(in the flange)}$$

In many situations it is not necessary to use the complex form of the general expression for shear stress in a beam. For wood beams, the sections are mostly simple rectangles, for which the following simplification can be made.

For the simple rectangle, from Figure A.13, $I = bd^3/12$. And

$$Q = \left[b \times \frac{d}{2}\right]\frac{d}{4} = \frac{bd^2}{8}$$

thus

$$f_v = \frac{VQ}{Ib} = \frac{V\left(\frac{bd^2}{8}\right)}{\left(\frac{bd^3}{12}\right)b} = 1.5\frac{V}{bd}$$

This is the formula specified by design codes for investigation of shear in wood beams.

The somewhat more complex investigations for shear stress are discussed in Chapter 9 for steel beams and in Chapter 13 for reinforced concrete beams.

Problem 3.8.A. A beam has an I-shaped cross section with an overall depth of 16 in. [400 mm], a web thickness of 2 in. [50 mm], and flanges that are 8 in. wide [200 mm] and 3 in. [75 mm] thick. Compute the critical shear stresses and plot the distribution of shear stress on the cross section if the beam sustains a shear force of 20 kips [89 kN].

Problem 3.8.B. A T-shaped beam cross section has an overall depth of 18 in. [450 mm], a web thickness of 4 in. [100 mm], a flange width of 8 in. [200 mm], and a flange thickness of 3 in. [75 mm]. Compute the critical shear stresses and plot the distribution of shear stress on the cross section if the beam sustains a shear force of 12 kips [53.4 kN].

3.9 CONTINUOUS AND RESTRAINED BEAMS

Continuous Beams

It is beyond the scope of this book to give a detailed discussion of bending in members continuous over supports, but the material presented in this section will serve as an introduction to the subject. A *continuous beam* is a beam that rests on more than two supports. Continuous beams are characteristic of site-cast concrete construction but occur less often in wood and steel construction.

The concepts underlying continuity and bending under restraint are illustrated in Figure 3.32. Figure 3.32*a* represents a single beam resting on three supports and carrying equal loads at the centers of the two spans. If the beam is cut over the middle support as shown in Figure 3.32*b*, the result will be two simple beams. Each of these simple beams will deflect as shown. However, when the beam is made continuous over the middle support, its deflected form is as indicated by the dashed line in Figure 3.32*a*.

It is evident that there is no bending moment developed over the middle support in Figure 3.32*b*, while there must be a moment over the support in Figure 3.32*a*. In both cases there is positive moment at the midspan; that is, there is tension in the bottom and compression in the top of the beam at these locations. In the continuous beam, however, there is a negative moment over the middle support; that is, there is tension in the top and compression in the bottom of the beam. The effect of the negative moment over the support is to reduce the magnitudes of both maximum bending moment and deflection at midspan, which is a principal advantage of continuity.

Values for reaction forces and bending moments cannot be found for continuous beams by use of the equations for static equilibrium alone. For example, the beam in Figure 3.32*a* has three unknown reaction forces, which constitute a parallel force system with the loads. For this condition there are only two conditions of equilibrium, and thus only two available equations for solving for the three unknowns. This presents a

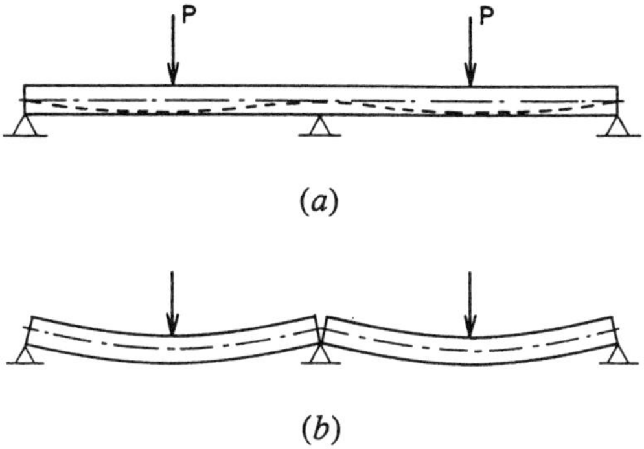

Figure 3.32 Continuous versus simple beams.

situation in algebra that is qualified as *indeterminate*, and the structure so qualified is said to be *statically indeterminate*.

Solutions for investigation of indeterminate structures require additional conditions to supplement those available from simple statics. These additional conditions are derived from the deformation and the stress mechanisms of the structure. Various methods for investigation of indeterminate structures have been developed. Of particular interest now are those that yield to application to computer-aided processes. Just about any structure, with any degree of indeterminacy, can now be investigated with readily available programs.

A procedural problem with highly indeterminate structures is that something about the structure must be determined before an investigation can be performed. Useful for this purpose are shortcut methods that give reasonably approximate answers without an extensive investigation. One of these approximation methods is demonstrated in the investigation of a rigid frame structure in Chapter 20.

Theorem of Three Moments

One method for determining reactions and constructing the shear and bending moment diagrams for continuous beams is based on the *theorem of three moments*. This theorem deals with the relation among the bending moments at any three consecutive supports of a continuous beam. Application of the theorem produces an equation, called the *three-moment equation*. The three-moment equation for a continuous beam of two spans with uniformly distributed loading and constant moment of inertia is

$$M_1 L_1 + 2M_2(L_1 + L_2) + M_3 L_2 = -\frac{w_1 L_1^3}{4} - \frac{w_2 L_2^3}{4}$$

in which the various terms are as shown in Figure 3.33. The following examples demonstrate the use of this equation.

Continuous Beam with Two Equal Spans. This is the simplest case, with the formula reduced by the symmetry plus the elimination of M_1 and M_2 due to the discontinuity of the beam at its outer ends. The equation is reduced to

$$4M_2 = -\frac{wL^2}{2}$$

With the loads and spans as given data, a solution for this case is reduced to solving for M_2, the negative moment at the center support. Transforming the equation produces a form for direct solution of the unknown moment, thus

$$M_2 = -\frac{wL^2}{8}$$

With this moment determined, it is possible to now use the available conditions of statics to solve the rest of the data for the beam. The following example demonstrates the process.

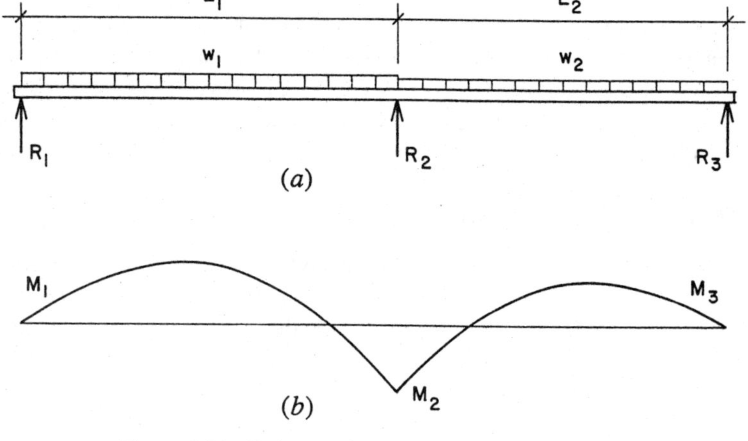

Figure 3.33 Reference for the three-moment equation.

Example 17. Compute the values for the reactions and construct the shear and moment diagrams for the beam shown in Figure 3.34*a*.

Solution: With only two conditions of statics for the parallel force system, it is not possible to solve directly for the three unknown reactions. However, use of the equation for the moment at the middle support yields a condition that can be used as shown in the following work:

$$M_2 = -\frac{wL^2}{8} = -\frac{100 \times (10)^2}{8} = -1250 \text{ ft-lb}$$

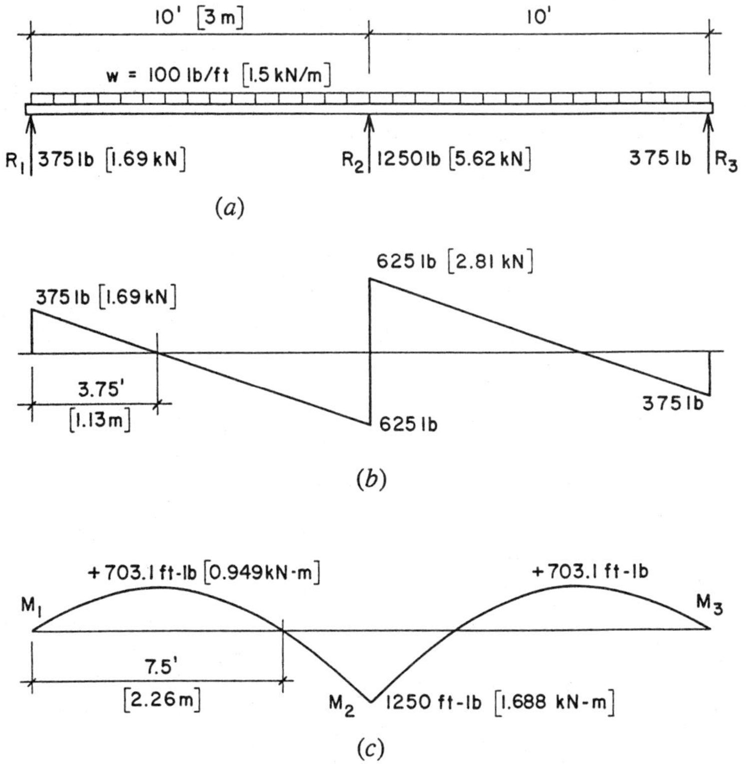

Figure 3.34 Reference for Example 17.

Next, an equation for the bending moment at 10 ft to the right of the left support is written in the usual manner, and is equated to the now known value of 1250 ft-lb.

$$M_{(x=10)} = (R_1 \times 10) - (100 \times 10 \times 5) = -1250 \text{ ft-lb}$$

From which

$$10R_1 = 3750, \qquad R_1 = 375 \text{ lb}$$

By symmetry, this is also the value for R_3. The value for R_2 can then be found by a summation of vertical forces, thus:

$$\Sigma F_V = 0 = (375 + 375 + R_2) - (100 \times 20), \qquad R_2 = 1250 \text{ lb}$$

Sufficient data has now been determined to permit the complete construction of the shear diagram, as shown in Figure 3.34*b*. The location of zero shear is determined by the equation for shear at the unknown distance x from the left support

$$375 - (100 \times x) = 0, \qquad x = 3.75 \text{ ft}$$

The maximum value for positive moment at this location can be determined with a moment summation or by finding the area of the shear diagram between the end and the zero shear location.

$$M = \frac{375 \times 3.75}{2} = 703.125 \text{ ft-lb}$$

Because of symmetry, the location of zero moment is determined as twice the distance of the zero shear point from the left support. Sufficient data is now available to plot the moment diagram as shown in Figure 3.34*c*.

Problems 3.9.A, B. Using the Three Moment Equation, find the bending moments and reactions and draw the complete shear and moment diagrams for the following beams that are continuous over two equal spans and carry uniformly distributed loadings.

	Span Length in ft	Load in lbs/ft
A	16	200
B	24	350

Continuous Beam with Unequal Spans. The following example shows the slightly more complex problem of dealing with unequal spans.

Example 18. Construct the shear and moment diagrams for the beam in Figure 3.35*a*.

Solution: In this case the moments at the outer supports are again zero, which reduces the task to solving for only one unknown. Applying the given values to the equation:

$$2M_2(14 + 10) = -\frac{1000 \times (14)^3}{4} - \frac{1000 \times (10)^3}{4}$$

$$M_2 = -19{,}500 \text{ ft-lb}$$

Writing a moment summation about a point 14 ft to the right of the left end support, using the forces to the left of the point:

$$14R_1 - (1000 \times 14 \times 7) = -19{,}500, \qquad R_1 = 5607 \text{ lb}$$

Then writing a moment summation about a point 10 ft to the *left* of the right end, using the forces to the *right* of the point:

$$10R_3 - (1000 \times 10 \times 5) = -19{,}500, \qquad R_3 = 3050 \text{ lb}$$

A vertical force summation will yield the value of $R_2 = 15{,}343$ lb. With the three reactions determined the shear values for completing the shear diagram are known. Determination of the points of zero shear and zero moment and the values for positive moment in the two spans can be done as demonstrated in Example 17. The completed diagrams are shown in Figure 3.35.

Problems 3.9.C, D. Find the reactions and draw the complete shear and moment diagrams for the following continuous beams with two unequal spans and uniformly distributed loading.

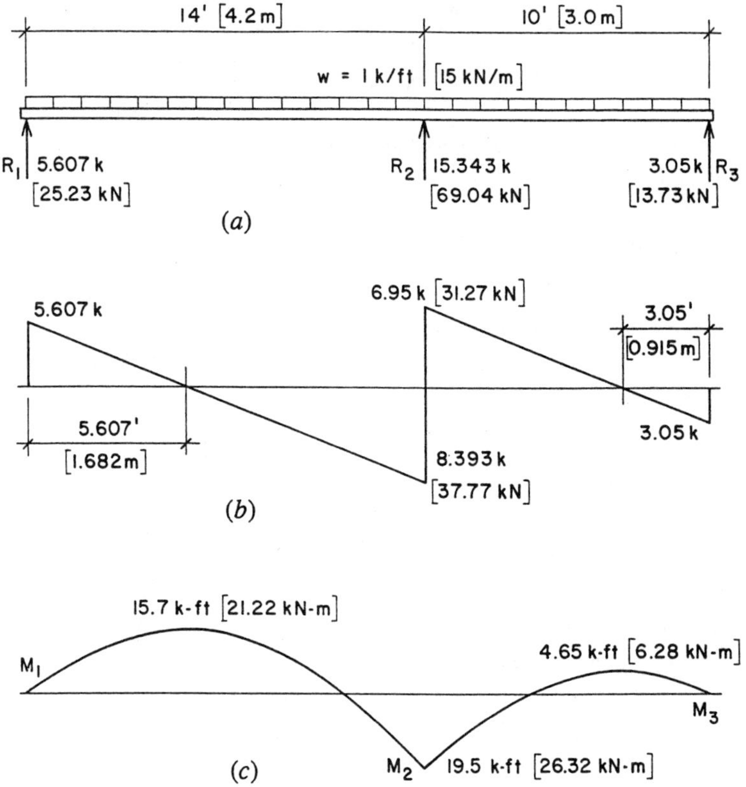

Figure 3.35 Reference for Example 18.

	First Span, ft	Second Span, ft	Load, lbs/ft
C	12	16	2000
D	16	20	1200

Continuous Beam with Concentrated Loads. In the previous examples the loads were uniformly distributed. Figure 3.36*a* shows a two-span beam with a single concentrated load in each span. The shape for the moment diagram for this beam is shown in Figure 3.36*b*. For these conditions, the form of the three-moment equation is

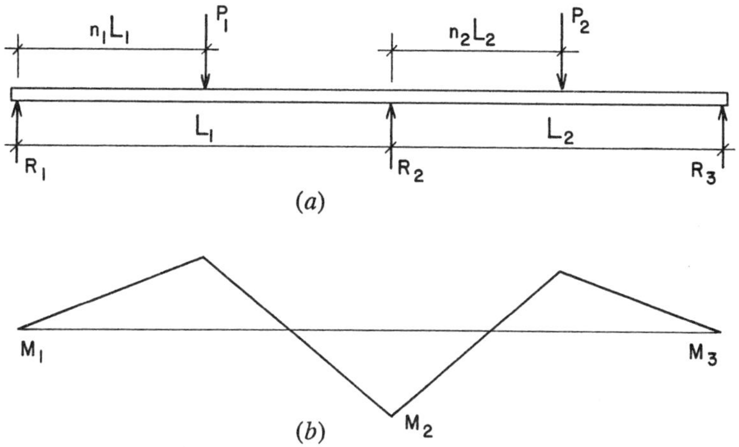

Figure 3.36 Two-span beam with concentrated loads.

$$M_1L_1 + 2M_2(L_1 + L_2) + M_3L_2 =$$
$$-P_1L_1^2[n_1(1 - n_1)(1 + n_1)] - P_2L_2^2[n_2(1 - n_2)(2 - n_2)]$$

in which the various terms are as shown in Figure 3.36.

Example 19. Compute the reactions and construct the shear and moment diagrams for the beam in Figure 3.37*a*.

Solution: For this case note that $L_1 = L_2$, $P_1 = P_2$, $M_1 = M_3 = 0$, and both n_1 and $n_2 = 0.5$. Substituting these conditions and given data into the equation

$$2M_2(20 + 20) = -4000(20^2)(0.5 \times 0.5 \times 1.5)$$
$$-4000(20^2)(0.5 \times 0.5 \times 1.5)$$

from which $M_2 = 15{,}000$ ft-lb.

The value of moment at the middle support can now be used as in Examples 17 and 18 to find the end reaction, from which it is determined that the value is 1250 lb. Then a summation of vertical forces will determine the value of R_2 to be 5500 lb. This is sufficient data for construction of the shear diagram. Note that points of zero shear are evident on the diagram.

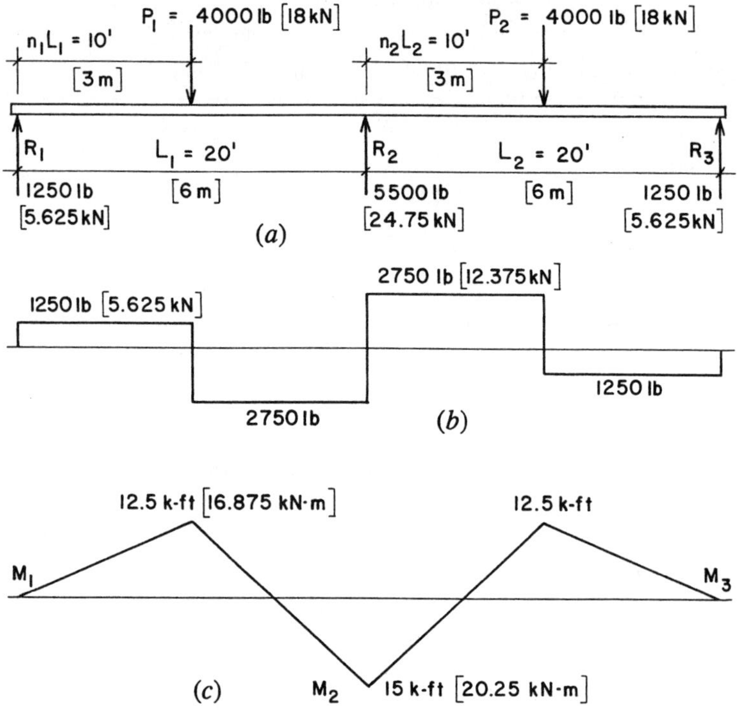

Figure 3.37 Reference for Example 19.

The values for maximum positive moment can be determined from moment summations at the sections or simply from the areas of the rectangles in the shear diagrams. The locations of points of zero moment can be found by simple proportion, because the moment diagram is composed of straight lines.

Problems 3.9.E, F. Find the reactions and draw the complete shear and moment diagrams for the following continuous beams with two equal spans and a single concentrated load at the center of each span.

	Span Length, ft	Load, kips
E	24	3
F	32	2.4

Continuous Beam with Three Spans. The preceding examples demonstrate that the key operation in investigation of continuous beams is the determination of negative moment values at the supports. Use of the three-moment equation has been demonstrated for a two-span beam, but the method may be applied to any two adjacent spans of a beam with multiple spans. For example, when applied to the three-span beam shown in Figure 3.38*a,* it would first be applied to the left span and the middle span, and next to the middle span and right span. This would produce two equations involving the two unknowns: the negative moments at the two interior supports. In this example case, the process would be simplified by the symmetry of the beam, but the application is a general one, applicable to any arrangement of spans and loads.

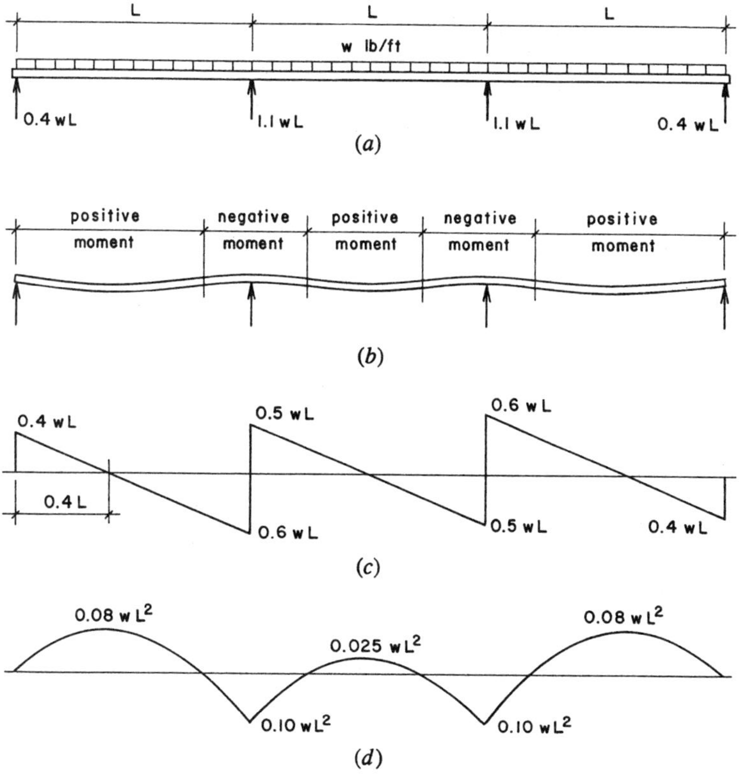

Figure 3.38 Three-span beam with distributed loading.

As with simple beams and cantilevers, common situations of spans and loading may be investigated and formulas for beam behavior values derived for subsequent application in simpler investigation processes. Thus, the values of reactions, shears, and moments displayed for the beam in Figure 3.38 may be used for any such support and loading conditions. Tabulations for many ordinary situations are available from various references.

Example 20. A continuous beam has three equal spans of 20 ft [6 m] each and a uniformly distributed load of 800 lb/ft [12 kN/m] extending over the entire length of the beam. Compute the maximum bending moment and the maximum shear.

Solution: Referring to Figure 3.38*d*, the maximum positive moment ($0.08wL^2$) occurs near the middle of each end span, and the maximum negative moment ($0.10wL^2$) occurs over each of the interior supports. Using the larger value, the maximum bending moment on the beam is

$$M = -0.10wL^2 = -(0.10 \times 800 \times 20 \times 20)$$
$$= -32{,}000 \text{ ft-lb } [43.2 \text{ kN-m}]$$

Figure 3.38*c* shows that the maximum shear occurs at the face of the first interior support and is

$$V = 0.6wL = (0.6 \times 800 \times 20) = 9600 \text{ lb } [43.2 \text{ kN}]$$

Using this process, it is possible to find the values of the reactions and then to construct the complete shear and moment diagrams, if the work at hand warrants it.

Problem 3.9.G, H. For the following continuous beams with three equal spans and uniformly distributed loading, find the reactions and draw the complete shear and moment diagrams.

	Span Length, ft	Load, lbs/ft
G	24	1000
H	32	1600

Restrained Beams

A simple beam was previously defined as a beam that rests on a support at each end, there being no restraint against bending at the supports; the

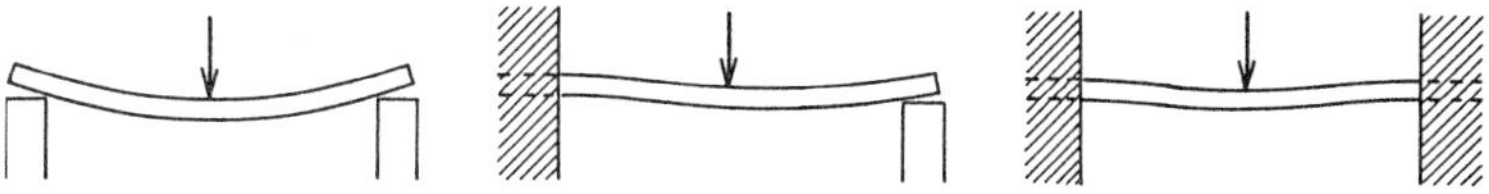

Figure 3.39 Behavior of beams with various forms of rotational restraint at supports: (*a*) no restraint; (*b*) one end restrained; (*c*) both ends restrained.

ends are *simply supported*. The shape a simple beam tends to assume under load is shown in Figure 3.39*a*. Figure 3.39*b* shows a beam whose left end is *restrained* or *fixed*, meaning that free rotation of the beam end is prevented. Figure 3.39*c* shows a beam with both ends restrained. End restraint has an effect similar to that caused by the continuity of a beam at an interior support: A negative bending moment is induced in the beam. The beam in Figure. 3.39*b* has a profile with an inflection point, indicating a change of sign of the moment within the span. This span behaves in a manner similar to one of the spans in the two-span beam.

The beam with both ends restrained has two inflection points, with a switch of sign to negative bending moment near each end. Although values are slightly different for this beam, the general form of the deflected shape is similar to that for the middle span in the three-span beam (see Figure 3.38).

Although they have only one span, the beams in Figures 3.39*b* and *c* are both indeterminate. Investigation of the beam with one restrained end involves finding three unknowns: the two reactions plus the restraining moment at the fixed end. For the beam in Figure 3.39*c*, there are four unknowns. There are, however, only a few ordinary cases that cover most common situations, and tabulations of formulas for these ordinary cases are readily available from references. Figure 3.40 gives values for the beams with one and two fixed ends under both uniformly distributed load and a single concentrated load at center span. Values for other loadings are also available from references.

Example 21. Figure 3.41*a* represents a 20-ft span beam with both ends fixed and a total uniformly distributed load of 8 kips. Find the reactions and construct the complete shear and moment diagrams.

Solution: Despite the fact that this beam is indeterminate to the second degree (four unknowns; only two equations of static equilibrium), its

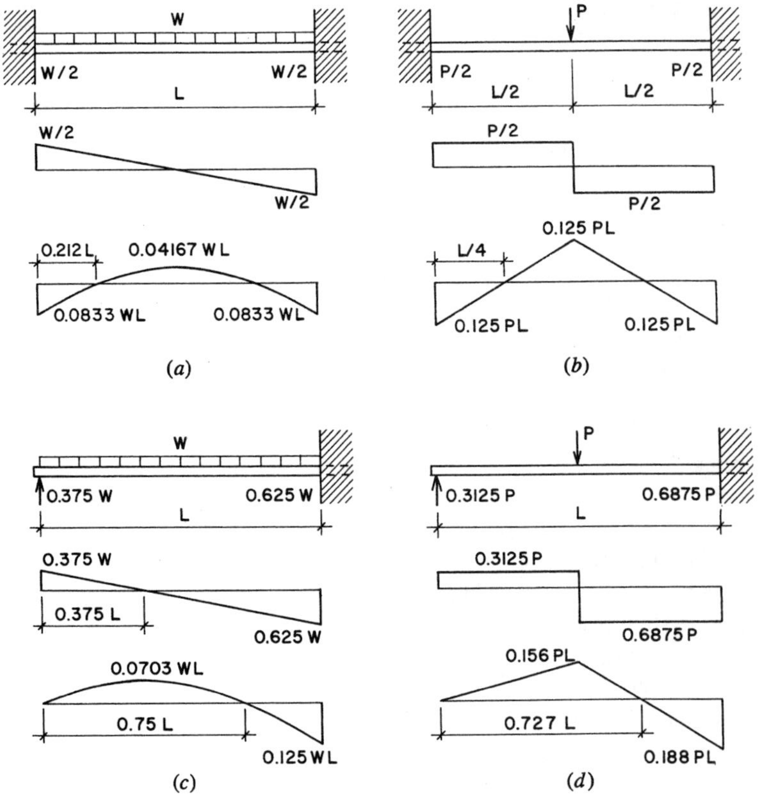

Figure 3.40 Values for restrained beams.

symmetry makes some investigation data self-evident. It can be observed that the two vertical reaction forces, and thus the two end shear values, are each equal to one half of the total load, or 4000 lb. Symmetry also indicates that the location of the point of zero moment and thus the point of maximum positive bending moment is at the center of the span. Also, the end moments, although indeterminate, are equal to each other, leaving only a single value to be determined.

From data in Figure 3.40*a*, the negative end moment is 0.0833 *WL* (actually 1/12 W*L*) = (8000 × 20)/12 = 13,333 ft-lb. The maximum positive moment at midspan is 0.04167 *WL* (actually 1/24 *WL*) = (8000 × 20)/24 = 6667 ft-lb. And the point of zero moment is 0.212L = (0.212)

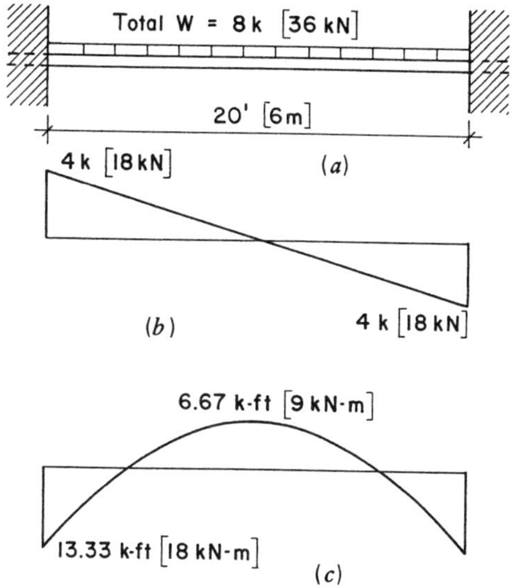

Figure 3.41 Reference for Example 21.

(20) = 4.24 ft from the beam end. The complete shear and moment diagrams are as shown in Figure 3.41.

Example 22 A beam fixed at one end and simply supported at the other end has a span of 20 ft and a total uniformly distributed load of 8000 lb (see Figure 3.42*a*). Find the reactions and construct the shear and moment diagrams.

Solution: This is the same span and loading as in the preceding example. Here, however, one end is fixed and the other simply supported (loading case *c* in Figure 3.40). The beam vertical reactions are equal to the end shears; thus from the data in Figure 3.40*c*:

$$R_1 = V_1 = 0.375(8000) = 3000 \text{ lb}$$

$$R_2 = V_2 = 0.625(8000) = 5000 \text{ lb}$$

and for the maximum moments:

$$+M = 0.0703(8000 \times 20) = 11{,}248 \text{ ft-lb}$$

$$-M = 0.125(8000 \times 20) = 20{,}000 \text{ ft-lb}$$

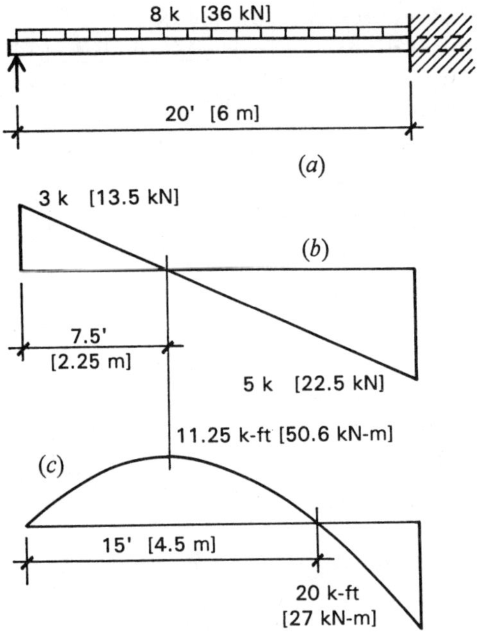

Figure 3.42 Reference for Example 22.

The point of zero shear is at 0.375(20) = 7.5 ft from the left end, and the point of zero moment is at twice this distance, 15 ft, from the left end. The complete shear and moment diagrams are shown in Figure 3.42.

Problem 3.9.I. A 22-ft [6.71 m] span beam is fixed at both ends and carries a single concentrated load of 16 kips [71.2 kN] at midspan. Find the reactions and construct the complete shear and moment diagrams.

Problem 3.9.J. A 16-ft [4.88 m] span beam is fixed at one end and simply supported at the other end. A single concentrated load of 9600 lb [42,7 kN] is placed at the center of the span. Find the vertical reactions and construct the complete shear and moment diagrams.

3.10 STRUCTURES WITH INTERNAL PINS

In many structures conditions exist at supports or within the structure that modify the behavior of the structure, often eliminating some potential components of force actions. Qualification of supports as fixed or pinned (not rotation-restrained) has been a situation in most of the structures presented in this work. We now consider some qualification of conditions *within* the structure that modify its behavior.

Internal Pins

Consider the structure shown in Figure 3.43*a*. It may be observed that there are four potential components of the reaction forces: A_x, A_y, B_x, and B_y. These are all required for the stability of the structure for the loading shown, and there are thus four unknowns in the investigation of the external forces. Because the loads and reactions constitute a general planar force system, there are three conditions of equilibrium (See Section 3.1; for example: $\Sigma F_x = 0$, $\Sigma F_y = 0$, $\Sigma M_p = 0$). As shown in Figure 3.43*a*, therefore, the structure is statically indeterminate, not yielding to complete investigation by use of static equilibrium conditions alone.

If the two members of the structure in Figure 3.43 are connected to each other by a pinned joint, as shown in *b*, the number of reaction components is not reduced, and the structure is still stable. However, the internal pin establishes a fourth condition that may be added to the three equilibrium conditions. There are then four conditions that may be used to find the four reaction components. The method of solution for the reactions of this type of structure is illustrated in the following example.

Example 23. Find the components of the reactions for the structure shown in Figure 3.44*a*.

Solution: It is possible to write four equilibrium equations and to solve them simultaneously for the four unknown forces. However, it is always easier to solve these problems if a few tricks are used to simplify the

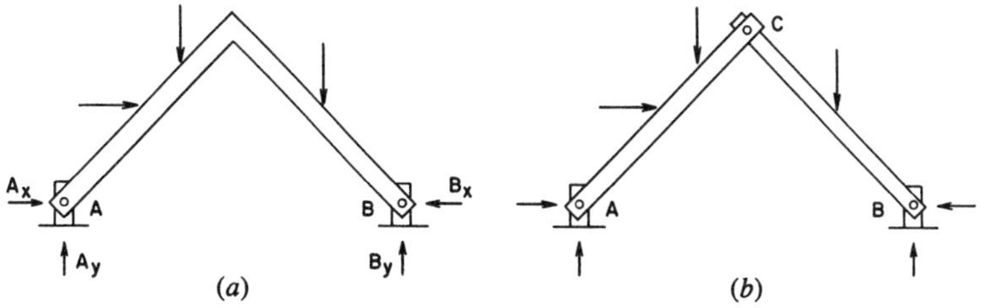

Figure 3.43 Effect of an internal pin.

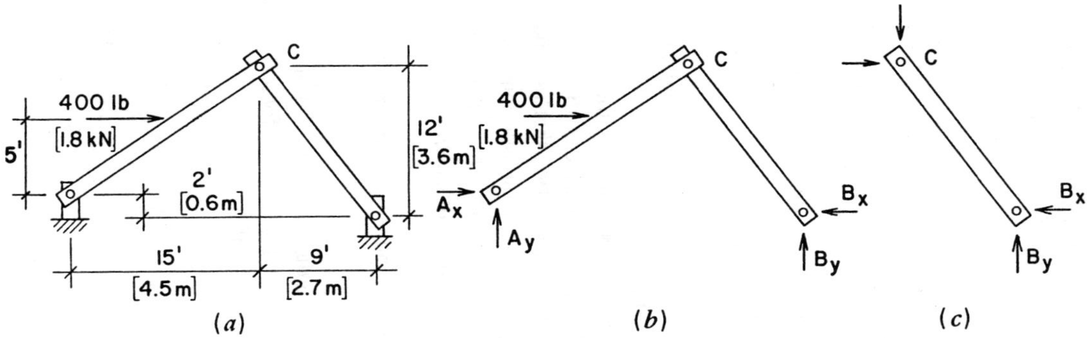

Figure 3.44 Reference for Example 23.

equations. One trick is to write moment equations about points that eliminate some of the unknowns, thus reducing the number of unknowns in a single equation. Consider the free body of the entire structure, as shown in Figure 3.44*b*.

$$\Sigma M_A = 0 = +(400 \times 5) + (B_x \times 2) - (B_y \times 24)$$

$$24B_y - 2B_x = 2000, \text{ or, } 12B_y - B_x = 1000 \qquad (1)$$

Now consider the free-body diagram of the right member, as shown in Figure 3.44*c*.

$$\Sigma M_C = 0 = +(B_x \times 12) - (B_y \times 9)$$

thus

$$B_x = \frac{9}{12} B_y = 0.75B_y \qquad (2)$$

Substituting equation (2) in equation (1)

$$12B_y - 0.75B_y = 1000, \quad B_y = \frac{1000}{11.25} = 88.89 \text{ lb}$$

Then, from equation (2)

$$B_x = 0.75B_y = 0.75 \times 88.89 = 66.67 \text{ lb}$$

Referring to Figure 3.44b:

$$\Sigma F_x = 0 = A_x + B_x - 400 = A_x + 66.67 - 400, \quad A_x = 333.33 \text{ lb}$$

$$\Sigma F_y = 0 = A_y + B_y = A_y + 88.89, \quad A_y = 88.89 \text{ lb}$$

Note that the condition stated in equation (2) is true in this case because the right member behaves as a two-force member. This is not the case if load is directly applied to the member, and the solution of simultaneous equations would be necessary.

Note also that the assumed sense of A_x and A_y as shown in Figure 3.44*b* is incorrect, as determined by the force summations. If B_y acts upward, then A_y must act downward, as they are the only two vertical forces. If the sum of A_x and B_x adds up to resist the load, then A_x acts toward the left.

Continuous Beams with Internal Pins

The actions of continuous beams are discussed in Section 3.9. It is observed that a beam such as that shown in Figure 3.45*a* is statically indeterminate, having a number of reaction components (3) that exceed the conditions of equilibrium by one too many unknowns for the parallel force system (2). The continuity of such a beam results in the deflected

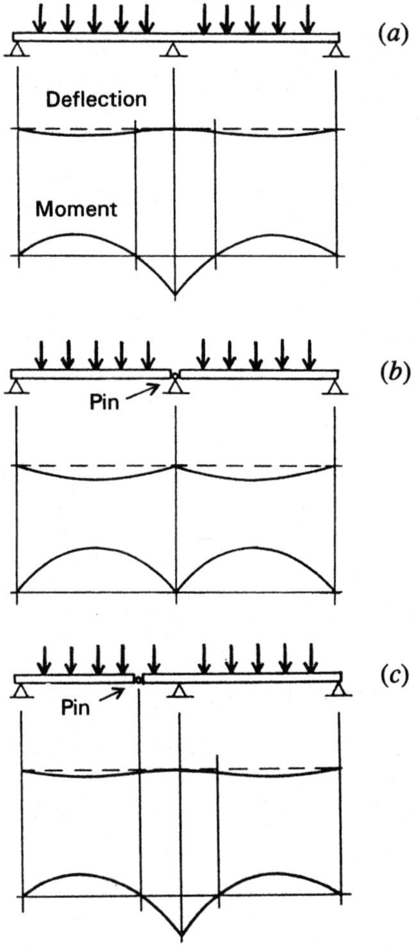

Figure 3.45 Effect of internal pins in multiple-span beams.

shape and variation of moment as shown beneath the beam in Figure 3.45*a*. If the beam is made discontinuous at the middle support, as shown in Figure 3.45*b*, the two spans each behave independently as simple beams, with the deflected shapes and moment as shown.

If a multiple-span beam is made internally discontinuous at some point off of the supports, its behavior may emulate that of a truly continuous beam. For the beam shown in Figure 3.45*c*, the internal pin is located at the point where the continuous beam inflects. Inflection of the deflected shape is an indication of zero moment, and thus the pin does not actually change the continuous nature of the structure. The deflected shape and moment variation for the beam in Figure 3.45*c* is therefore the same as for the beam in Figure 3.45*a*. This is true, of course, only for the single loading pattern that results in the inflection point at the same location as the internal pin.

In the first of the following examples, the internal pin is deliberately placed at the point where the beam would inflect if it was continuous. In the second example, the pins are placed slightly closer to the support, rather than in the location of the natural inflection points. The modification in the second example results in slightly increasing the positive moment in the outer spans, while reducing the negative moments at the supports; thus, the values of maximum moment are made closer. If it is desired to use a single-size beam for the entire length, the modification in Example 25 permits design selection of a slightly smaller size.

Example 24. Investigate the beam shown in Figure 3.46*a*. Find the reactions, draw the shear and moment diagrams, and sketch the deflected shape.

Solution: Because of the internal pin, the first 12 ft of the left-hand span acts as a simple beam. Its two reactions are therefore equal, being one half the total load, and its shear, moment, and deflected shape diagrams are those for a simple beam with a uniformly distributed load. (See Case 2, Figure 3.25). As shown in *b* and *c* in Figure 3.46, the simple beam reaction at the right end of the 12-ft portion of the left span becomes a 6-kip concentrated load at the left end of the remainder of the beam. This beam (Figure 3.46*c*) is then investigated as a beam with one overhanging end, carrying a single concentrated load at the cantilevered end and the total distributed load of 20 kips. (Note that on the diagram the total uniformly distributed load is indicated in the form of a single force, representing its resultant.) The second portion of the beam is statically determinate, and its reactions can now be determined by statics equations.

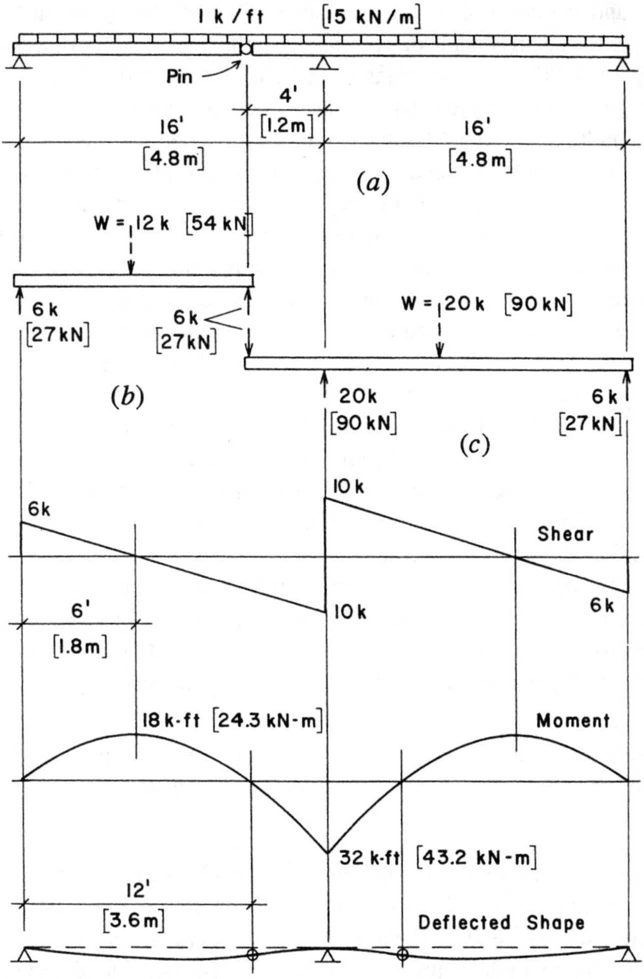

Figure 3.46 Reference for Example 24.

With the reactions known, the shear diagram can be completed. Note the relation between the point of zero shear in the span and the location of maximum positive moment. For this loading the positive moment curve is symmetrical, and thus the location of the zero moment (and beam inflection) is at twice the distance from the end as the point of zero shear. As noted previously, the pin in this example is located exactly at the inflection point of the continuous beam. (For comparison, see Section 3.9, Example 17.)

Example 25. Investigate the beam shown in Figure 3.47.

Solution: The procedure is essentially the same as for Example 23. Note that this beam with four supports requires two internal pins to become statically determinate. As before, the investigation begins with the consideration of the end portion acting as a simple beam. The second step is to consider the center portion as a beam with two overhanging ends.

Problems 3.10.A, B. Find the components of the reactions for the structures shown in Figure 3.48*a* and *b*.

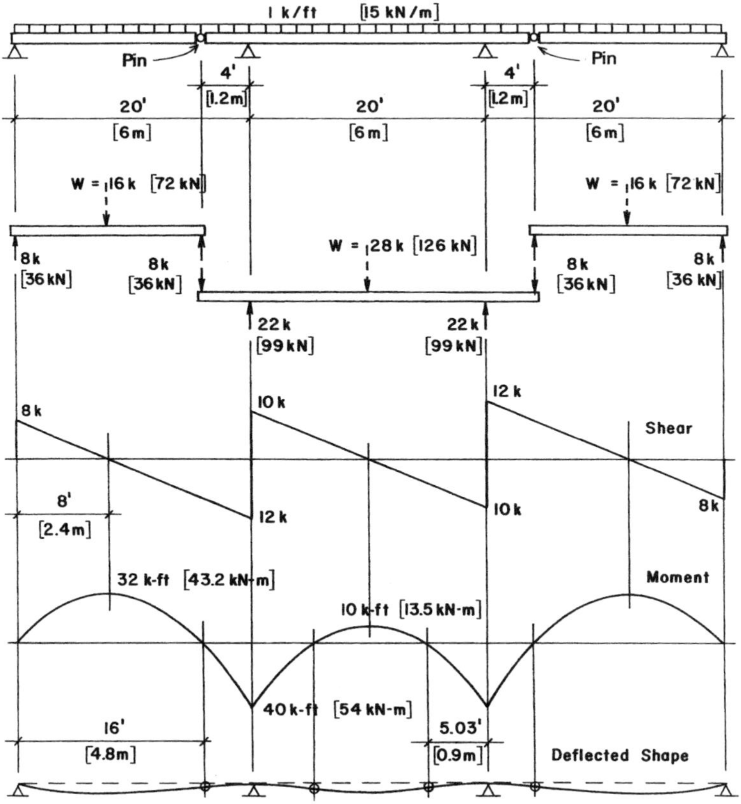

Figure 3.47 Reference for Example 25.

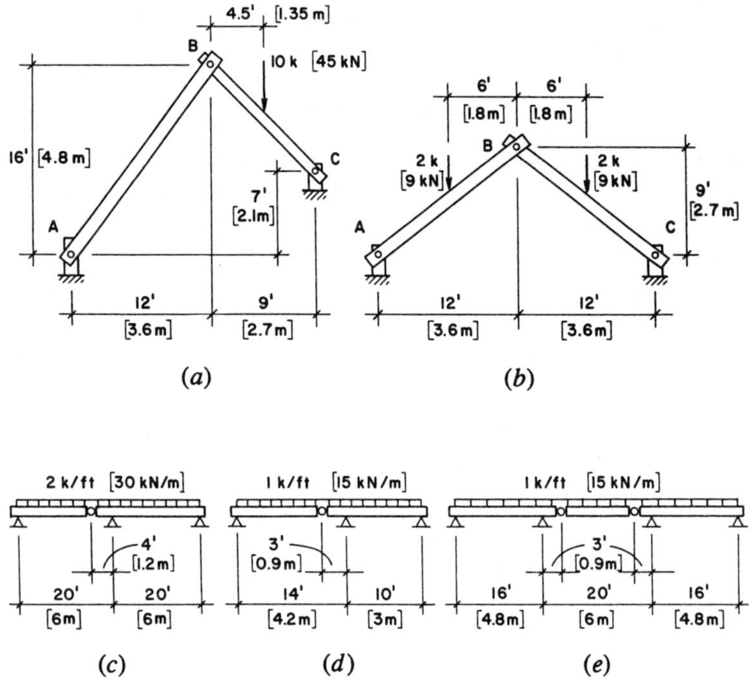

Figure 3.48 Reference for Problem 3.10.

Problems 3.10.C, D, E. Investigate the beams shown in Figure 3.48*c to e*. Find the reactions and draw the shear and moment diagrams, indicating all critical values. Sketch the deflected shape and determine the locations of any inflection points not related to the internal pins. (*Note:* Problem 3.10.D has the same spans and loading as Example 18 in Section 3.9.)

3.11 COMPRESSION MEMBERS

Compression is developed in a number of ways in structures, including the compression component that accompanies the development of internal bending. In this section consideration is given to elements whose primary purpose is resistance of compression. This includes truss members, piers, bearing walls, and bearing footings, although major treatment here is given to columns, which are linear compression members. Building columns may be freestanding architectural elements, with the structural column itself exposed to view. However, for fire or weather protection the structural column must often be incorporated into other construction and may in some cases be fully concealed from view.

Slenderness Effects

Structural columns are often quite slender, and the aspect of slenderness (called *relative slenderness*) must be considered (see Figure 3.49). At the extremes the limiting situations are those of the very stout or short column that fails by crushing, and the very slender or tall column that fails by lateral buckling.

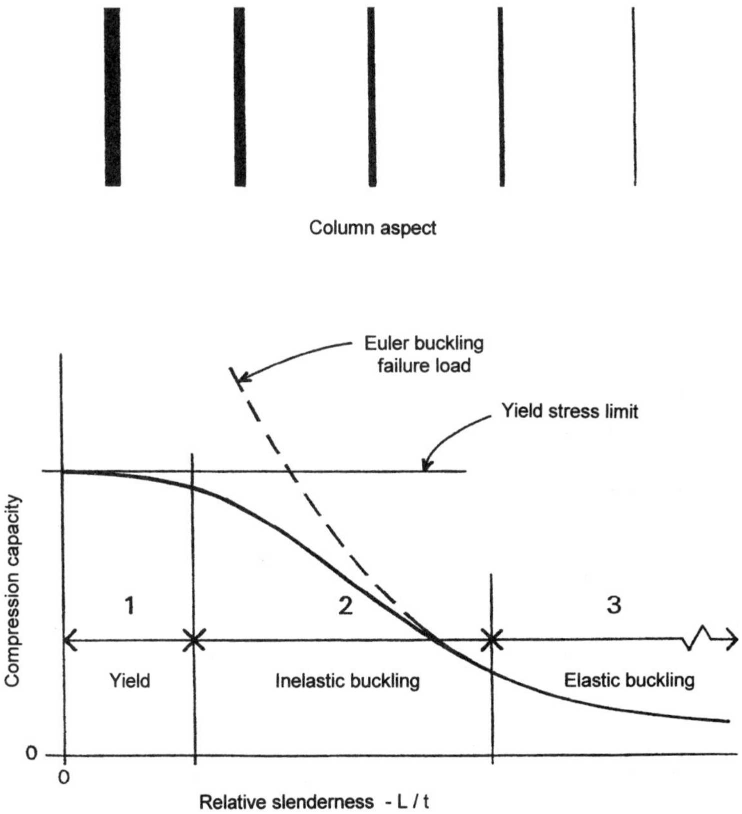

Figure 3.49 Effect of column slenderness on axial compression capacity.

The two basic limiting response mechanisms—crushing and buckling—are entirely different in nature. Crushing is a stress resistance phenomenon, and its limit is represented on the graph in Figure 3.49 as a horizontal line, basically established by the compression resistance of the material and the amount of material (area of the cross section) in the compression member. This behavior is limited to the range labeled zone 1 in Figure 3.49.

Buckling actually consists of lateral deflection in bending, and its extreme limit is affected by the bending stiffness of the member, as related to the stiffness of the material (modulus of elasticity) and to the geometric property of the cross section directly related to deflection—the moment of inertia of the cross-section area. The classic expression for elastic buckling is stated in the form of the equation developed by Euler

$$P = \frac{\pi^2 EI}{L^2}$$

The curve produced by this equation is of the form shown in Figure 3.49. It closely predicts the failure of quite slender compression members in the range labeled zone 3 in Figure 3.49.

Most building columns fall somewhere between very stout and very slender, in other words in the range labeled zone 2 in Figure 3.49. Their behavior, therefore, is one of an intermediate form, somewhere between pure stress response and pure elastic buckling. Predictions of structural response in this range must be established by empirical equations that somehow make the transition from the horizontal line to the Euler curve. Equations currently used are explained in Chapter 6 for wood columns and in Chapter 10 for steel columns.

Buckling may be affected by constraints, such as lateral bracing that prevents sideways movement, or support conditions that restrain the rotation of the member's ends. Figure 3.50*a* shows the case for the member that is the general basis for response as indicated by the Euler formula. This form of response can be altered by lateral constraints, as shown in Figure 3.50*b*, that result in a multimode deflected shape. The member in Figure 3.50*c* has its ends restrained against rotation (described as a fixed end). This also modifies the deflected shape, and thus the value produced from the buckling formula. One method used for adjustment is to modify the column length used in the buckling formula to that occurring between inflection points; thus, the *effective buckling length*

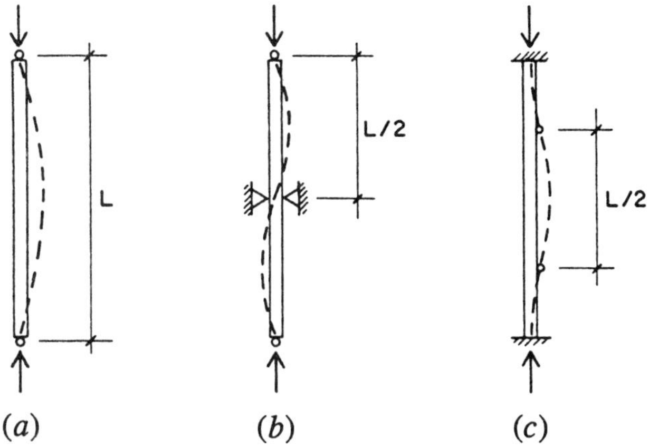

Figure 3.50 Form of buckling of a column as affected by various end conditions and lateral restraint.

for the columns in both Figures 3.50*b* and *c* would be one half that of the true column total length. Inspection of the Euler formula will indicate the impact of this modified length on buckling resistance.

Development of Bending in Columns

Bending moments can be developed in structural members in a number of ways. When a member is subjected to an axial compression force, there are various ways in which the compression effect and any bending present can relate to each other.

Figure 3.51*a* shows a very common situation that occurs in building structures when an exterior wall functions as a bearing wall or contains a column. The combination of vertical gravity load and lateral load due to wind or seismic action can result in the loading shown. If the member is quite flexible, an additional bending is developed as the axis of the member deviates from the action line of the vertical compression load. This added bending is the product of the load and the member deflection; that is, P times Δ, as shown in Figure 3.51*d*. It is thus referred to as the *P-delta effect*.

There are various other situations that can result in the P-delta effect. Figure 3.51*b* shows an end column in a rigid frame structure, where moment is induced at the top of the column by the moment-resistive connection to the beam. Although slightly different in its profile, the column response is similar to that in Figure 3.51*a*.

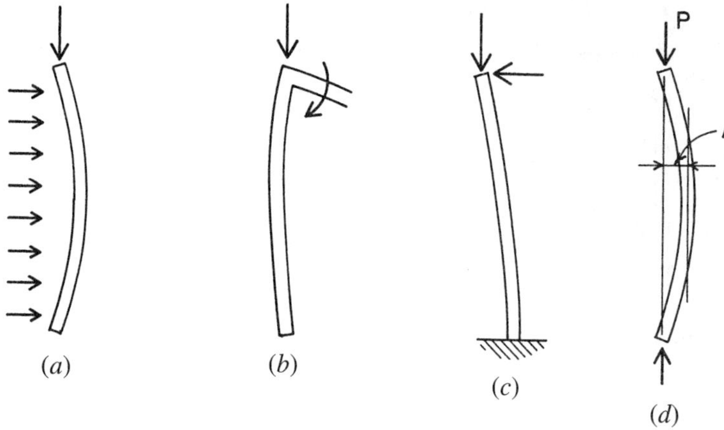

Figure 3.51 Development of bending in columns.

Figure 3.51*c* shows the effect of a combination of gravity and lateral loads on a vertically cantilevered structure that supports a sign or a tank at its top.

In any of these situations, the *P*-delta effect may or may not be critical. The major factor that determines its seriousness is the relative stiffness of the structure as it relates to the magnitude of deflection produced. However, even a significant deflection may not be of concern if the vertical load (*P*) is quite small. In a worst-case scenario, a major *P*-delta effect may produce an accelerating failure, with the added bending producing more deflection, which in turn produces more bending, and so on.

Interaction of Bending and Axial Compression

There are a number of situations in which structural members are subjected to the combined effects that result in development of axial compression and internal bending. Stresses developed by these two actions are both of the direct stress type (tension and compression) and can be combined for consideration of a net stress condition. This is useful for some cases, as is considered following this discussion, but for columns with bending, the situation involves two essentially different actions: column action in compression and beam behavior. For column investigation, therefore, it is the usual practice to consider the combination by what is called *interaction*.

The classic form of interaction is represented by the graph in Figure 3.52*a*. Referring to the notation on the graph:

1. The maximum axial load capacity of the member (with no bending) is P_o.
2. The maximum bending capacity of the member (without compression) is M_o.
3. At some compression load below P_o (indicated as P_n), the member is assumed to have some tolerance for a bending moment (indicated as M_n) in combination with the axial load.
4. Combinations of P_n and M_n are assumed to fall on a line connecting points P_o and M_o. The equation of this line has the form expressed as

$$\frac{P_n}{P_o} + \frac{M_n}{M_o} = 1$$

A graph similar to that in Figure 3.52*a* can be constructed using stresses rather than loads and moments. This is the procedure used for wood and steel members; the graph taking the form expressed as

$$\frac{f_a}{F_a} + \frac{f_b}{F_b} \leq 1$$

where f_a = computed stress due to axial load
F_a = allowable column-action stress in compression
f_b = computed stress due to bending
F_b = allowable beam-action stress in flexure

For various reasons, real structures do not adhere strictly to the classic straight-line form of response shown in Figure 3.52*a*. Figure 3.52*b* shows a form of response characteristic of reinforced concrete columns. In the mid range, there is some approximation of the theoretical straight-line behavior, but at the terminal ends of the response graph, there is considerable variation. This has to do with the nature of the ultimate failure of the reinforced concrete materials in both compression (upper end) and in tension (lower end), as explained in Chapter 15.

Steel and wood members also have various deviations from the straight-line interaction response. Special problems include inelastic behavior,

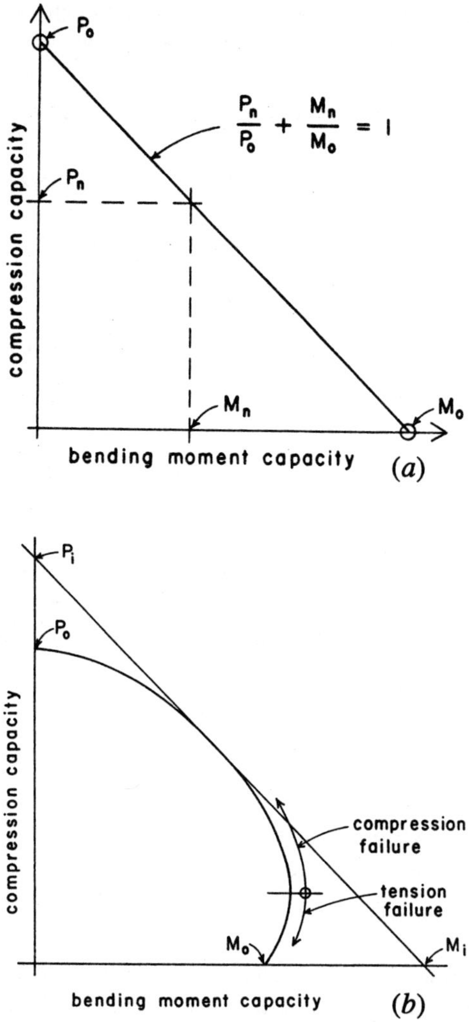

Figure 3.52 Interaction of axial compression and bending in a column: (*a*) classic form of the interaction formula; (*b*) form of interaction response in a reinforced concrete column.

effects of lateral stability, geometry of member cross-sections, and lack of initial straightness of members. Wood columns are discussed in Chapter 6 and steel columns in Chapter 10.

Combined Stress: Compression Plus Bending

Combined actions of compression plus bending produce various effects on structures. In some situations the actual stress combinations may of themselves be critical, one such case being the development of bearing stress on soils. At the contact face of a bearing footing and its supporting soil, the "section" for stress investigation is the contact face—that is, the bottom surface of the footing. The following investigation presents an approach to this investigation.

Figure 3.53 illustrates the situation of combined direct force and bending moment at a cross section. In this case the "cross section" is the contact face of the footing bottom with the soil. Whatever produces the combined load and moment, a transformation is made to an equivalent eccentric force that produces the same effect. The value for the hypothetical eccentricity e is established by dividing the moment by the load, as shown in the figure. The net, or combined, stress distribution at the section is visualized as the sum of separate stresses caused by the load and the bending moment. For the limiting stresses at the edges of the section the general equation for the combined stress is

$$P = \text{Direct stress} \pm \text{Bending stress}$$

or

$$p = \frac{N}{A} \pm \frac{Nec}{I}$$

Four cases for this combined stress are shown in the figure. The first case occurs when e is small, resulting in very little bending stress. The section is thus subjected to all compressive stress that varies from a maximum value at one edge to a minimum value at the opposite edge.

The second case occurs when the two stress components are equal, so that the minimum stress is zero. This is the boundary condition between the first and third cases because an increase in e will produce some reversal stress—in this situation some tension stress.

The second stress case is significant for a footing because tension stress is not possible between the footing and soil. Case 3 is only possible

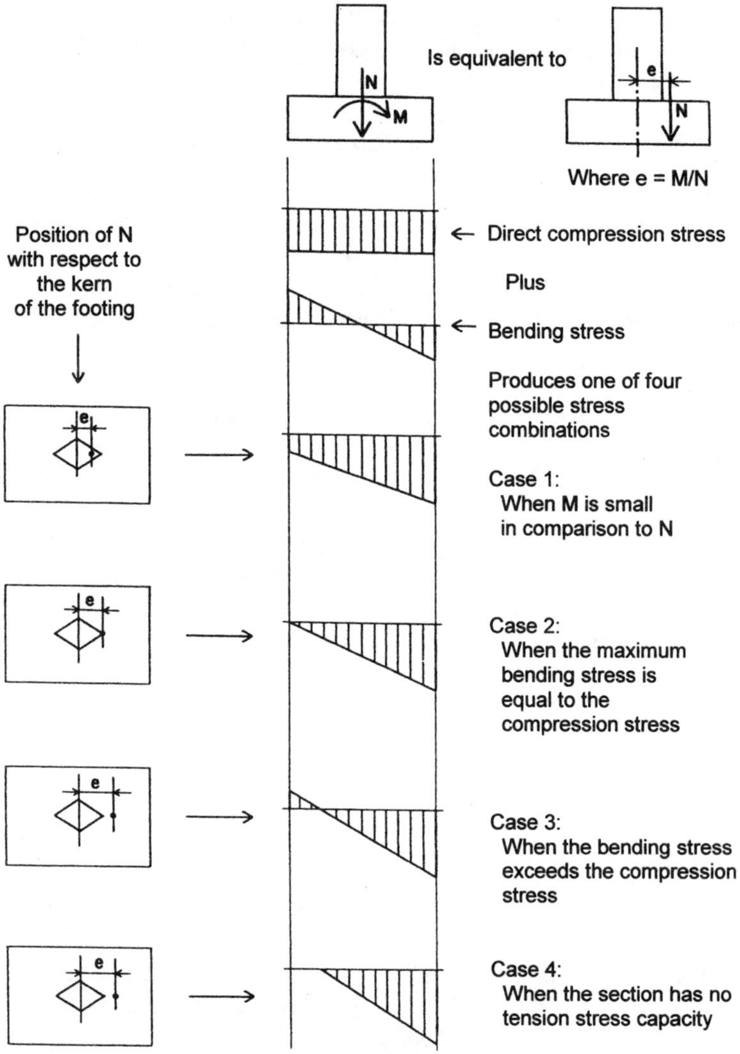

Figure 3.53 Combinations of compression and bending stress considered as generated by an eccentric compression force.

for a beam or column or some other continuously solid element. The value for e that produces Case 2 can be derived by equating the two stress components as follows:

$$\frac{N}{A} = \frac{Nec}{I}, \quad \text{and thus} \quad e = \frac{I}{AC}$$

This value for e establishes what is known as the *kern limit* of the section, which defines a zone around the centroid of the section within which an eccentric force will not cause reversal stress on the section. The form and dimensions for this zone can be established for any geometric shape by use of the derived equation for e. The kern limit zones for three common geometric shapes are shown in Figure 3.54.

When tension stress is not possible, larger eccentricities of the compression normal force will produce a so-called *cracked section* as shown for Case 4 in Figure 3.53. In this situation some portion of the cross section becomes unstressed, or cracked, and the compressive stress on the remainder of the section must develop the entire resistance to the loading effects of the combined force and moment.

Figure 3.55 shows a technique for the analysis of a cracked section, called the *pressure wedge method*. The wedge is a volume that represents the total compressive force as developed by the soil pressure; its volumetric unit is in force units produced by multiplying stress times area. Analysis of the static equilibrium of this wedge produces two relationships that can be used to establish the dimensions of the wedge. These relationships are as follows:

1. The volume of the wedge is equal to the vertical force.
2. The centroid of the wedge is located on a vertical line that coincides with the location of the hypothetical eccentric force.

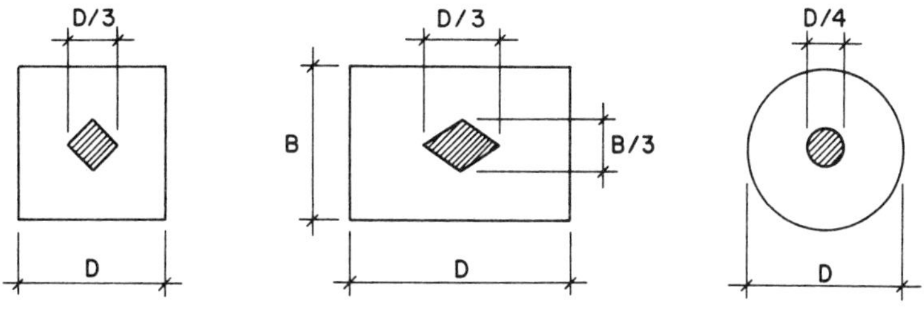

Figure 3.54 Form of the kern for common shapes of cross sections.

Referring to Figure 3.55, the three dimensions of the wedge are w (width of the footing), p (maximum soil pressure), and x (limiting dimension of the stressed portion of the cracked section). In this situation the footing width is known, so the definition of the wedge requires only the determination of p and x.

For the rectangular section the centroid is at the third point of the triangle. Defining this distance from the edge as a, as shown in Figure 3.55, then x is equal to three times a. It may be observed that a is equal to half the footing width minus the eccentricity e. Thus, once we compute the eccentricity, we can determine the values of a and x.

The volume of the stress wedge can be expressed in terms of its three dimensions as

$$V = \frac{wpx}{2}$$

With w and x determined, the remaining dimension of the wedge can be established by transforming the equation of the volume to

$$p = \frac{2V}{wx}$$

or, because the volume is equal to the force N:

$$p = \frac{2N}{wx}$$

All four cases of combined stress in Figure 3.53 will cause some tilting of the footing due to deformation of the compressible soil. The extent of this tilting and its effect on the structure supported by the footing must be carefully considered in the design of the footing. It is usually desired that the soil pressure be evenly distributed for anything other than very short time loadings such as those caused by wind or seismic effects.

Example 26. Find the value of maximum soil pressure for a square footing subjected to a load of 100 kips and a moment of 100 kip-ft. Find values for footing widths of (a) 8 ft, (b) 6 ft, (c) 5 ft.

Solution: The first step is to find the equivalent eccentricity and compare it to the kern limit for the footing to establish which of the cases shown in Figure 3.53 applies. For all parts:

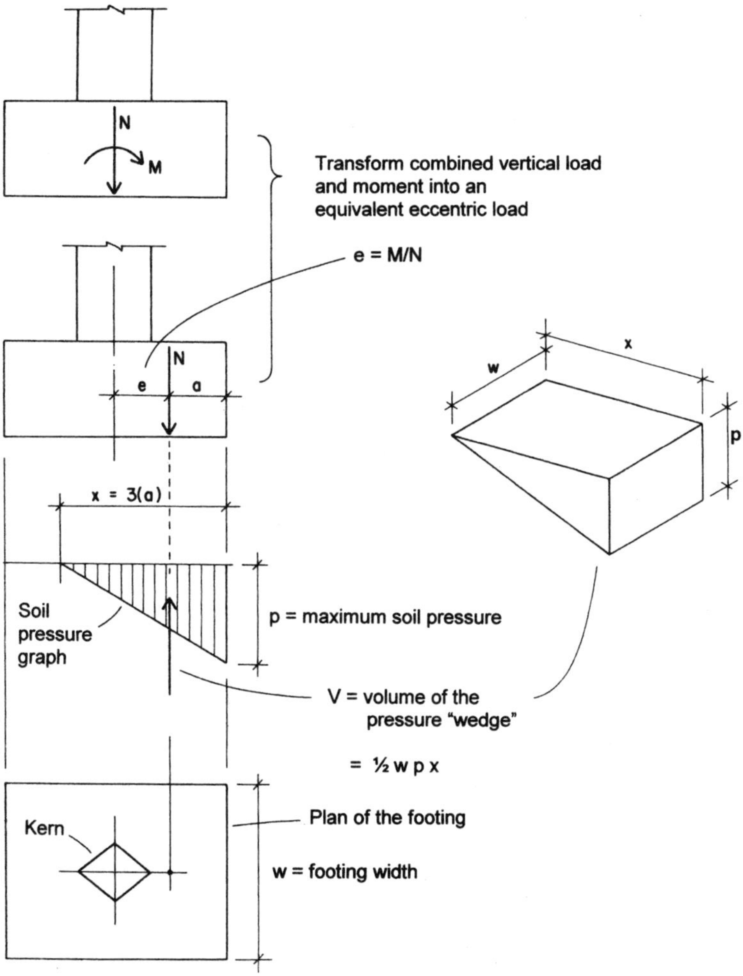

Figure 3.55 Investigation of combined stress on a "cracked" section by the pressure wedge method.

$$e = \frac{M}{N} = \frac{100}{100} = 1 \text{ ft}$$

For (a) the 8 ft wide footing has a kern limit of 8/6 = 1.33 ft; thus, Case 1 applies. For computations the properties of the 8-ft square footing are

$$A = 8 \times 8 = 64 \text{ ft}^2$$

$$I = \frac{bd^3}{12} = \frac{(8)(8)^3}{12} = 341.3 \text{ ft}^4$$

and the maximum soil pressure is

$$p = \frac{N}{A} + \frac{Mc}{I} = \frac{100}{64} + \frac{(100)(4)}{341.3} = 1.56 + 1.17 = 2.73 \text{ ksf}$$

For (b) the 6-ft-wide footing has a kern limit of 1 ft, the same as the eccentricity. Thus, the situation is Case 2 in Figure 3.53 with $N/A = Mc/I$ and

$$p = 2\left(\frac{N}{A}\right) = 2\left(\frac{100}{36}\right) = 5.56 \text{ ksf}$$

For (c) the eccentricity exceeds the kern limit and the investigation must be done as illustrated in Figure 3.55. Thus:

$$a = \frac{5}{2} - e = 2.5 - 1 = 1.5 \text{ ft}$$

$$x = 3a = 3(1.5) = 4.5 \text{ ft}$$

$$p = \frac{2N}{wx} = \frac{2(100)}{5(4.5)} = 8.89 \text{ ksf}$$

Problem 3.11.A. A square footing sustains a force of 40 kips [178kN] and a bending moment of 30 kip-ft [40.7 kN-m]. Find the maximum soil pressure for widths of: (*a*) 5 ft [1.5 m]; (*b*) 4 ft [1.2 m].

Problem 3.11.B. A square footing sustains a force of 60 kips [267 kN] and a bending moment of 60 kip-ft [81.4 kN-m]. Find the maximum soil pressure for widths of: (*a*) 7 ft [2.13 m]; (*b*) 5 ft [1.5 m].

3.12 RIGID FRAMES

Frames in which two or more of the members are attached to each other with connections that are capable of transmitting bending between the ends of the members are called *rigid frames*. The connections used to achieve such a frame are called *moment connections* or *moment-resisting connections.* Most rigid frame structures are statically indeterminate and do not yield to investigation by consideration of static equilibrium alone. The rigid-frame structure occurs quite frequently as a multiple-level, multiple-span bent, constituting part of the structure for a multistory building. In most cases, such a bent is used as a lateral bracing element; although once it is formed as a moment-resistive framework, it will respond as such for all types of loads. The computational examples presented in this section are all rigid frames that have conditions that make them statically determinate and thus capable of being fully investigated by methods developed in this book.

Cantilever Frames

Consider the frame shown in Figure 3.56*a*, consisting of two members rigidly joined at their intersection. The vertical member is fixed at its base, providing the necessary support condition for stability of the frame. The horizontal member is loaded with a uniformly distributed loading and functions as a simple cantilever beam. The frame is described as a cantilever frame because of the single fixed support. The five sets of figures shown in Figure 3.56*b* through *f* are useful elements for the investigation of the behavior of the frame. They consist of the following:

1. The free-body diagram of the entire frame, showing the loads and the components of the reactions (Figure 3.56*b*). Studying this diagram will help in establishing the nature of the reactions and in the determination of the conditions necessary for stability of the frame as a whole.
2. The free-body diagrams of the individual elements (Figure 3.56*c*). These are of great value in visualizing the interaction of the parts of the frame. They are also useful in the computations for the internal forces in the frame.
3. The shear diagrams of the individual elements (Figure 3.56*d*). These are sometimes useful for visualizing, or for actually computing, the variations of moment in the individual elements. No particular sign convention is necessary unless in conformity with the sign used for moment.
4. The moment diagrams for the individual elements (Figure 3.56*e*). These are very useful, especially in determination of the deformation of the frame. The sign convention used is that of plotting the moment on the compression (concave) side of the flexed element.

5. The deformed shape of the loaded frame (Figure 3.56*f*). This is the exaggerated profile of the bent frame, usually superimposed on an outline of the unloaded frame for reference. This is very useful for the general visualization of the frame behavior. It is particularly useful for determining the character of the external reactions and the form of interaction between the parts of the frame. Correlation between the deformed shape and the form of the moment diagram is a useful check.

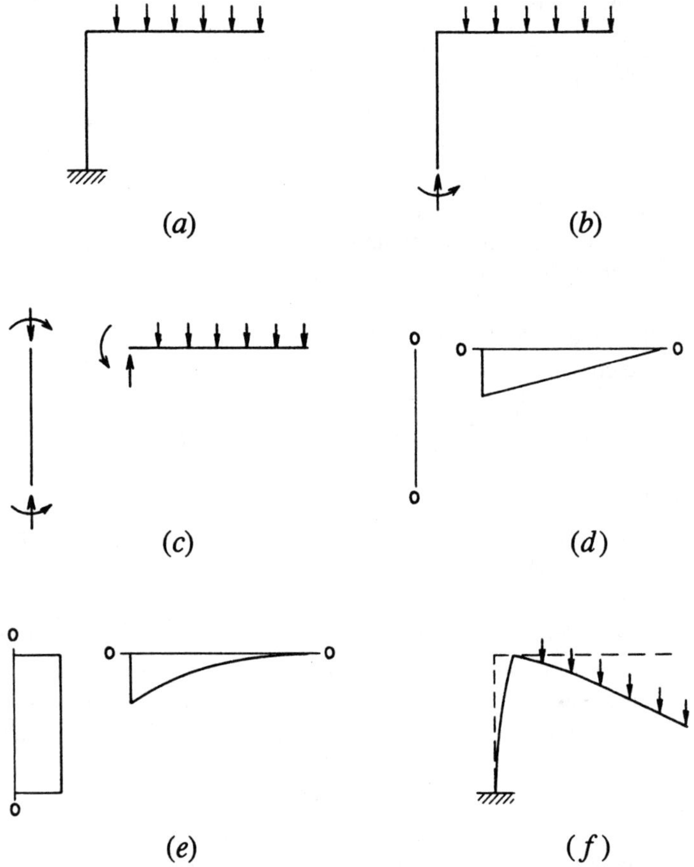

Figure 3.56 Diagrams for investigation of the rigid frame.

When we are performing investigations, these elements are not usually produced in the sequence just described. In fact, it is generally recommended that the deformed shape be sketched first so that its correlation with other factors in the investigation may be used as a check on the work. The following examples illustrate the process of investigation for simple cantilever frames.

Example 27. Find the components of the reactions and draw the free-body diagrams, shear and moment diagrams, and the deformed shape of the frame shown in Figure 3.57*a*.

Solution: The first step is to determine the reactions. Considering the free-body diagram of the whole frame (Figure 3.57*b*):

$$\Sigma F = 0 = +8 - R_v, \qquad R_v = 8 \text{ kips} \quad \text{(up)}$$

and with respect to the support:

$$\Sigma M = 0 = M_R - (8 \times 4), \qquad M_R = 32 \text{ kip-ft} \quad \text{(clockwise)}$$

Note that the sense, or sign, of the reaction components is visualized from the logical development of the free-body diagram.

Consideration of the free-body diagrams of the individual members will yield the actions required to be transmitted by the moment connection. These may be computed by applying the conditions for equilibrium for either of the members of the frame. Note that the sense of the force and moment is opposite for the two members, simply indicating that what one does to the other is the opposite of what is done to it.

In this example, there is no shear in the vertical member. As a result, there is no variation in the moment from the top to the bottom of the member. The free-body diagram of the member, the shear and moment diagrams, and the deformed shape should all corroborate this fact. The shear and moment diagrams for the horizontal member are simply those for a cantilever beam.

With this example, as with many simple frames, it is possible to visualize the nature of the deformed shape without recourse to any mathematical computations. It is advisable to attempt to do so as a first step in investigation, and to check continually during the work that individual computations are logical with regard to the nature of the deformed structure.

Example 28. Find the components of the reactions and draw the shear and moment diagrams and the deformed shape of the frame in Figure 3.58*a*.

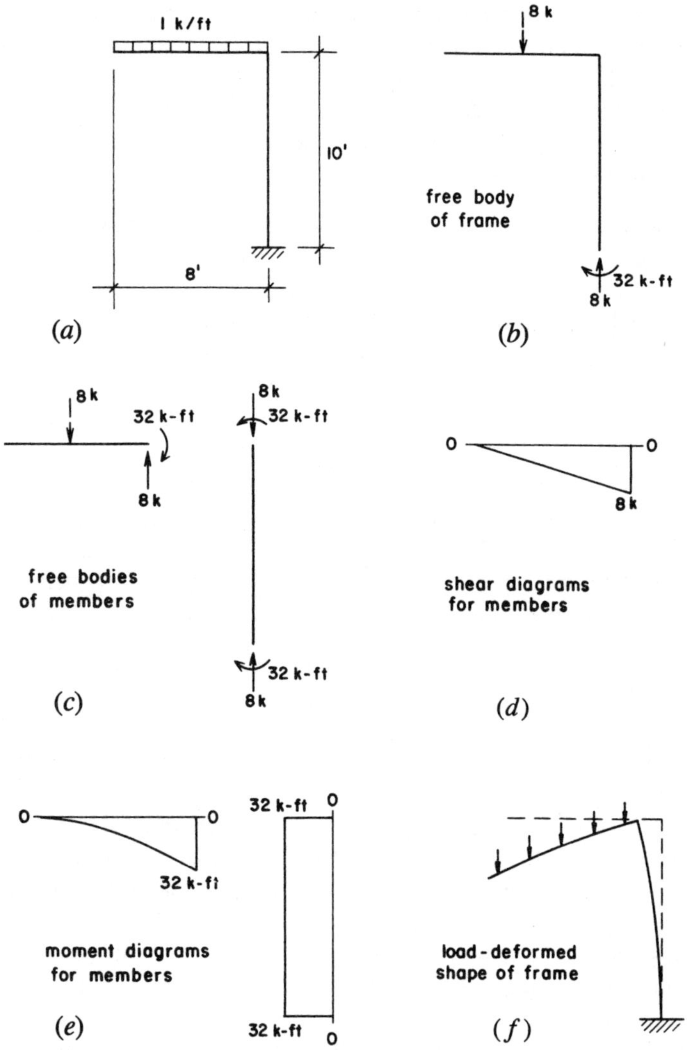

Figure 3.57 Reference for Example 27.

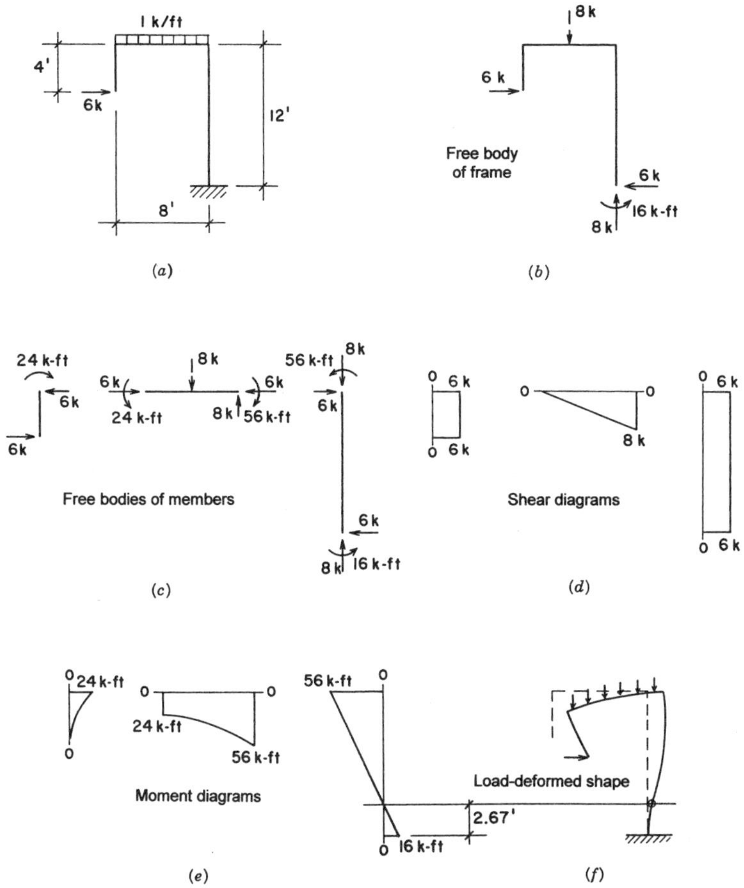

Figure 3.58 Reference for Example 28.

Solution: In this frame there are three reaction components required for stability because the loads and reactions constitute a general coplanar force system. Using the free-body diagram of the whole frame (Figure 3.58*b*), we use the three conditions for equilibrium for a coplanar system to find the horizontal and vertical reaction components and the moment component. If necessary, we can combine the reaction force components into a single-force vector, although this is seldom required for design purposes.

Note that the inflection occurs in the larger vertical member because the moment of the horizontal load about the support is greater than that of the vertical load. In this case, this computation must be done before the deformed shape can be accurately drawn.

The reader should verify that the free-body diagrams of the individual members are truly in equilibrium and that there is the required correlation between all the diagrams.

Problems 3.12.A–C. For the frames shown in Figure 3.59*a–c*, find the components of the reactions, draw the free-body diagrams of the whole frame and the individual members, draw the shear and moment diagrams for the individual members, and sketch the deformed shape of the loaded structure.

Single-Span Frames

Single-span rigid frames with two supports are ordinarily statically indeterminate, and as such, they are discussed in the next section. The following example illustrates the case of a statically determinate, single-span frame, made so by the particular conditions of its support and internal construction. In fact, these conditions are technically achievable, but a little weird for practical use. The example is offered here as an exercise for readers that is within the scope of the work in this section.

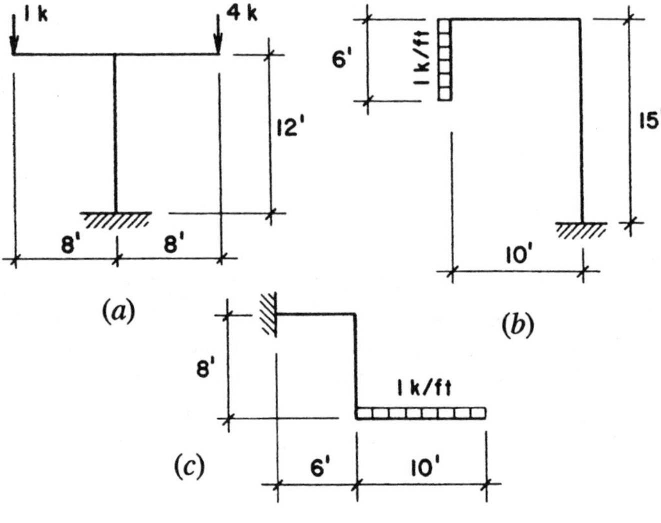

Figure 3.59 Reference for Problem 3.12, part 1.

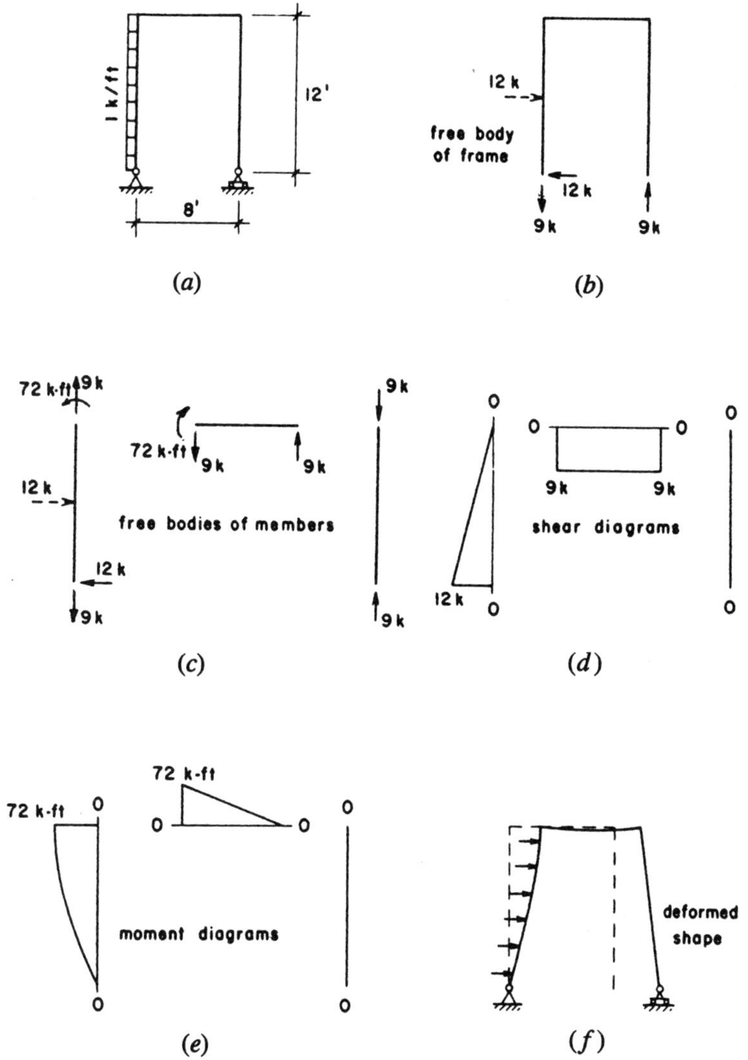

Figure 3.60 Reference for Example 29.

Example 29. Investigate the frame shown in Figure 3.60 for the reactions and internal conditions. Note that the right-hand support allows for an upward vertical reaction only, whereas the left-hand support allows for both vertical and horizontal components. Neither support provides moment resistance.

Solution: The typical elements of investigation, as illustrated for the preceding examples, are shown in Figure 3.60. The suggested procedure for the work is as follows:

1. Sketch the deflected shape (a little tricky in this case, but a good exercise). Consider the equilibrium of the free-body diagram for the whole frame to find the reactions.
2. Consider the equilibrium of the left-hand vertical member to find the internal actions at its top.
3. Proceed to the equilibrium of the horizontal member.
4. Finally, consider the equilibrium of the right-hand vertical member.
5. Draw the shear and moment diagrams and check for correlation of all work.

Before attempting the exercise problems, the reader is advised to attempt to produce the results shown in Figure 3.60 independently.

Problems 3.12.D–E. Investigate the frames shown in Figure 3.61*d* and *e* for reactions and internal conditions, using the procedure shown for the preceding examples.

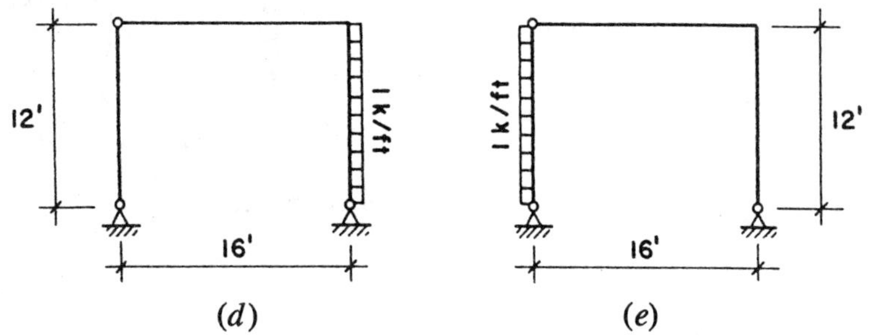

Figure 3.61 Reference for Problem 3.12, part 2.

3.13 BUCKLING OF BEAMS

Buckling of beams, in one form or another, is mostly a problem with beams that are relatively weak on their transverse axes, that is, the axis of the beam cross section at right angles to the axis of bending. This is not a frequent condition in concrete beams, but it is a common one with beams of wood or steel or with trusses that perform beam functions. The cross sections shown in Figure 3.62 illustrate members that are relatively susceptible to buckling in beam action.

When buckling is a problem, one solution is to redesign the beam for more resistance to lateral movement. Another possibility is to analyze for the lateral buckling effect and reduce the usable bending capacity as appropriate. However, the solution most often used is to brace the beam against the movement developed by the buckling effect. To visualize where and how such bracing should be done, we must first consider the various possibilities for buckling. The three main forms of beam buckling are shown in Figure 3.63.

Figure 3.63*b* shows the response described as *lateral* (that is *sideways) buckling*. This action is caused by the compressive stresses in the top of the beam that make it act like a long column, which is thus subject to a sideways movement as with any slender column. Bracing the beam for this action means simply preventing its sideways movement at

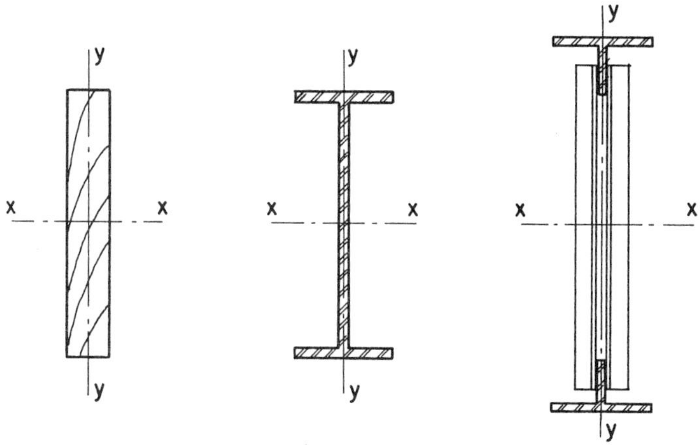

Figure 3.62 Beam shapes with low resistance to lateral bending and buckling.

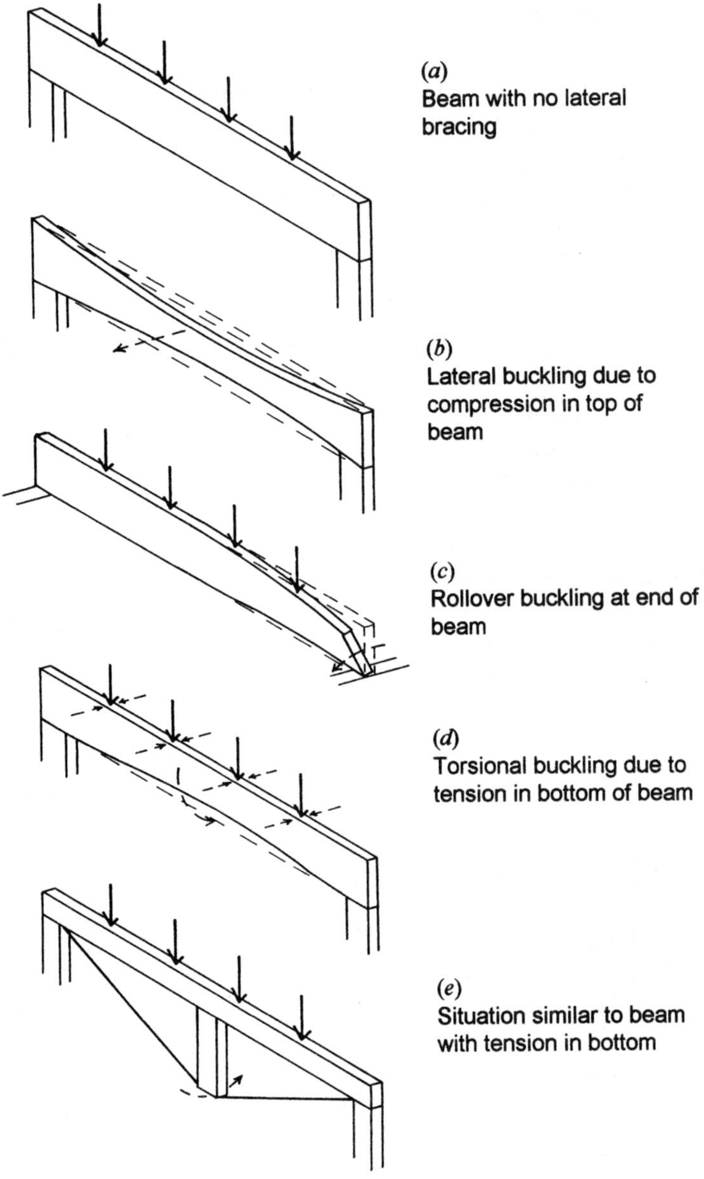

Figure 3.63 Forms of buckling of beams.

the beam edge where compression exists. For simple span beams this edge is the top of the beam. For beams, joists, rafters, or trusses that directly support roof or floor decks, the supported deck may provide this bracing if it is adequately attached to the supporting members. For beams that support other beams in a framing system, the supported beams at right angles to the supporting member may provide lateral bracing. In the latter case, the unsupported length of the buckling member becomes the distance between the supported beams, rather than its entire span length.

Another form of buckling for beams is that described as *torsional buckling*, as shown in Figure 3.63*d*. This action may be caused by tension stress, resulting in a rotational, or twisting, effect. This action can occur even when the top of the beam is braced against lateral movement and is often due to a lack of alignment of the plane of the loading and the vertical axis of the beam. Thus, a beam that is slightly tilted is predisposed to a torsional response. An analogy for this is shown in Figure 3.63*e*, which shows a trussed beam with a vertical post at the center of the span. Unless this post is perfectly vertical, a sideways motion at the bottom end of the post is highly likely.

To prevent both lateral and torsional buckling, it is necessary to brace the beam sideways at both its top and bottom. If the roof or floor deck is capable of bracing the top of the beam, the only extra bracing required

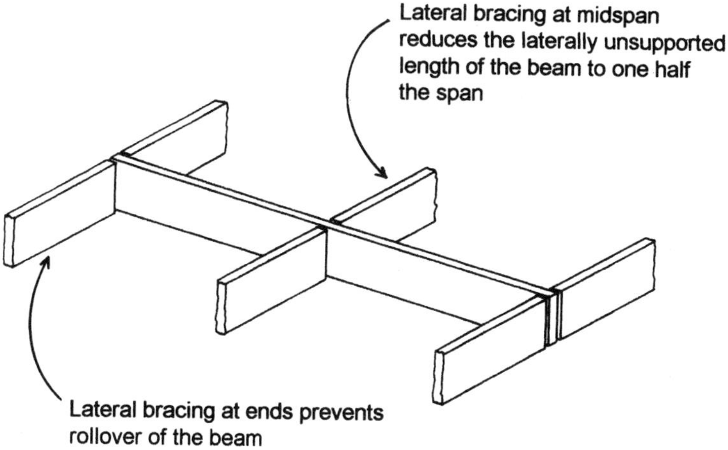

Figure 3.64 Lateral bracing for beams.

is that for the bottom. For closely spaced trusses, this bracing is usually provided by simple horizontal ties between adjacent trusses. For beams in wood or steel framing systems, lateral bracing may be provided as shown in Figure 3.64. The beam shown is braced for both lateral and torsional buckling. Other forms of bracing are described in Part II for wood structures and in Part III for steel structures.

4

STRUCTURAL DESIGN LOADS AND METHODS

This chapter treats the issues of determination of loads on the structure and selection of the basic structural design method to be used.

4.1 LOAD SOURCES

Structural tasks are defined primarily in terms of the loading conditions imposed on the structure. There are many potential sources of load for building structures. Designers must consider all the potential sources and the logical combinations with which they may occur. Building codes currently stipulate both the load sources and the form of combinations to be used for design. The following loads are listed in the 2003 edition of the *ASCE Minimum Design Loads for Buildings and Other Structures* (Ref. 1), hereinafter referred to as ASCE 2003.

D = Dead load

E = Earthquake-induced force

L = Live load, except roof load
L_r = Roof live load
S = Snow load.
W = Load due to wind pressure.

Additional special loads are listed, but these are the commonly occurring loads. This chapter contains descriptions of some of these loads.

Dead Loads

Dead load consists of the weight of the materials of which the building is constructed such as walls, partitions, columns, framing, floors, roofs, and ceilings. In the design of a beam or column, the dead load used must include an allowance for the weight of the structural member itself. Table 4.1, which lists the weights of many construction materials, may be used in the computation of dead loads. Dead loads are due to gravity and result in downward vertical forces.

Dead load is generally a permanent load, once the building construction is completed, unless remodeling or rearrangement of the construction occurs. Because of this permanent, long-time character, the dead load requires certain considerations in design, such as the following:

1. Dead load is always included in design loading combinations, except for investigations of singular effects, such as deflections due to only live load.
2. Its long-time character has some special effects that cause sag and require reduction of design stresses in wood structures, development of long-term, continuing settlements in some soils, and production of creep effects in concrete structures.
3. Dead load contributes some unique responses, such as the stabilizing effects that resist uplift and overturn caused by wind forces.

Although we can determine weights of materials with reasonable accuracy, the complexity of most building construction makes the computation of dead loads possible only on an approximate basis. This adds to other factors to make design for structural behaviors a very approximate science. As in other cases, this should not be used as an excuse for sloppiness in the computational work, but it should be recognized as a fact to temper concern for high accuracy in design computations.

TABLE 4.1 Weight of Building Construction

	psf[a]	kPa[a]
Roofs		
3-ply ready roofing (roll, composition)	1	0.05
3-ply felt and gravel	5.5	0.26
5-ply felt and gravel	6.5	0.31
Shingles: Wood	2	0.10
Asphalt	2–3	0.10–0.15
Clay tile	9–12	0.43–0.58
Concrete tile	6–10	0.29–0.48
Slate, 3 in.	10	0.48
Insulation: Fiber glass batts	0.5	0.025
Foam plastic, rigid panels	1.5	0.075
Foamed concrete, mineral aggregate	2.5/in.	0.0047/mm
Wood rafters: 2 × 6 at 24 in.	1.0	0.05
2 × 8 at 24 in.	1.4	0.07
2 × 10 at 24 in.	1.7	0.08
2 × 12 at 24 in.	2.1	0.10
Steel deck, painted: 22 gage	1.6	0.08
20 gage	2.0	0.10
Skylights: Steel frame with glass	6–10	0.29–0.48
Aluminum frame with plastic	3–6	0.15–0.29
Plywood or softwood board sheathing	3.0/in.	0.0057/mm
Ceilings		
Suspended steel channels	1	0.05
Lath: Steel mesh	0.5	0.025
Gypsum board, 1/2 in.	2	0.10
Fiber tile	1	0.05
Drywall, gyspum board, 1/2 in.	2.5	0.12
Plaster: Gypsum	5	0.24
Cement	8.5	0.41
Suspended lighting and HVAC, average	3	0.15
Floors		
Hardwood, 1/2 in.	2.5	0.12
Vinyl tile	1.5	0.07
Ceramic tile: 3/4 in.	10	0.48
Thin-set	5	0.24
Fiberboard underlay, 0.625 in.	3	0.15
Carpet and pad, average	3	0.15
Timber deck	2.5/in.	0.0047/mm
Steel deck, stone concrete fill, average	35–40	1.68–1.92
Concrete slab deck, stone aggregate	12.5/in.	0.024/mm
Lightweight concrete fill	8.0/in.	0.015/mm

(*continued*)

TABLE 4.1 (*Continued*)

	psf[a]	kPa[a]
Floors (Continued)		
Wood joists: 2 × 8 at 16 in.	2.1	0.10
2 × 10 at 16 in.	2.6	0.13
2 × 12 at 16 in.	3.2	0.16
Walls		
2 × 4 studs at 16 in., average	2	0.10
Steel studs at 16 in., average	4	0.20
Lath. plaster—see *Ceilings*		
Drywall, gypsum board, 1/2 in.	2.5	0.10
Stucco, on paper and wire backup	10	0.48
Windows, average, frame + glazing:		
Small pane, wood or metal frame	5	0.24
Large pane, wood or metal frame	8	0.38
Increase for double glazing	2–3	0.10–0.15
Curtain wall, manufactured units	10–15	0.48–0.72
Brick veneer, 4 in., mortar joints	40	1.92
1/2 in., mastic-adhered	10	0.48
Concrete block:		
Lightweight, unreinforced, 4 in.	20	0.96
6 in.	25	1.20
8 in.	30	1.44
Heavy, reinforced, grouted, 6 in.	45	2.15
8 in.	60	2.87
12 in.	85	4.07

[a]Average weight per square foot of surface, except as noted.

Values given as in. or mm are to be multiplied by actual thickness of material.

Building Code Requirements

Structural design of buildings is most directly controlled by building codes, which are the general basis for the granting of building permits—the legal permission required for construction. Building codes (and the permit-granting process) are administered by some unit of government: city, county, or state. Most building codes, however, are based on some model code

Model codes are more similar than different and are in turn largely derived from the same basic data and standard reference sources, including many industry standards. In the several model codes and many city, county, and state codes, however, there are some items that reflect particular regional concerns. With respect to control of structures, all codes have materials (all essentially the same) that relate to the following issues:

1. *Minimum required live loads.* All building codes have tables that provide required values to be used for live loads. Tables 4.2 and 4.3 contain some loads as specified in ASCE 2003 (Ref. 1).
2. *Wind loads.* These are highly regional in character with respect to concern for local windstorm conditions. Model codes provide data with variability on the basis of geographic zones.
3. *Seismic (earthquake) effects.* These are also regional, with predominant concerns in the western states. This data, including recommended investigations, is subject to quite frequent modification, as the area of study responds to ongoing research and experience.

TABLE 4.2 Minimum Floor Live Loads

Building Occupancy or Use	Uniformly Distributed Load (psf)	Concentrated Load (lb)
Apartments and Hotels		
Private rooms and corridors serving them	40	
Public rooms and corridors serving them	100	
Dwellings, One and Two-Family		
Uninhabitable attics without storage	10	
Uninhabitable attics with storage	20	
Habitable attics and sleeping rooms	30	
All other areas except stairs and balconies	40	
Office Buildings		
Offices	50	2000
Lobbies and first-floor corridors	100	2000
Corridors above first floor	80	2000
Stores		
Retail		
First floor	100	1000
Upper floors	75	1000
Wholesale, all floors	125	1000

Source: ASCE 2003 (Ref. 1), used with permission of the publishers, American Society of Civil Engineers.

TABLE 4.3 Live Load Element Factor, K_{LL}

Element	K_{LL}
Interior columns	4
Exterior columns without cantilever slabs	4
Edge columns with cantilever slabs	3
Corner columns with cantilever slabs	2
Edge beams without cantilever slabs	2
Interior beams	2
All other members not identified above	1

Source: ASCE 2003 (Ref. 1), used with permission of the publishers, American Society of Civil Engineers.

4. *Load duration.* Loads or design stresses are often modified on the basis of the time span of the load, varying from the life of the structure for dead load to a few seconds for a wind gust or a single major seismic shock. Safety factors are frequently adjusted on this basis. Some applications are illustrated in the work in the design examples.
5. *Load combinations.* Formerly mostly left to the discretion of designers, these are now quite commonly stipulated in codes, mostly because of the increasing use of ultimate strength design and the use of factored loads.
6. *Design data for types of structures.* These deal with basic materials (wood, steel, concrete, masonry, etc.), specific structures (rigid frames, towers, balconies, pole structures, etc.), and special problems (foundations, retaining walls, stairs, etc.). Industry-wide standards and common practices are generally recognized, but local codes may reflect particular local experience or attitudes. Minimal structural safety is the general basis, and some specified limits may result in questionably adequate performances (bouncy floors, cracked plaster, etc.).
7. *Fire resistance.* For the structure, there are two basic concerns, both of which produce limits for the construction. The first concern is for structural collapse or significant structural loss. The second concern is for containment of the fire to control its spread. These concerns produce limits on the choice of materials (e.g., combustible or noncombustible) and some details of the construction (cover on reinforcement in concrete, fire insulation for steel beams, etc.).

The work in the design examples in Chapters 18 through 20 is based largely on criteria from ASCE 2003 (Ref. 1).

Live Loads

Live loads technically include all the nonpermanent loadings that can occur, in addition to the dead loads. However, the term as commonly used usually refers only to the vertical gravity loadings on roof and floor surfaces. These loads occur in combination with the dead loads but are generally random in character and must be dealt with as potential contributors to various loading combinations, as discussed in Section 4.2.

Roof Loads

In addition to the dead loads they support, roofs are designed for a uniformly distributed live load. The minimum specified live load accounts for general loadings that occur during construction and maintenance of the roof. For special conditions, such as heavy snowfalls, additional loadings are specified.

The minimum roof live load in psf is specified in ASCE 2003 (Ref. 1) in the form of an equation, as follows:

$$L_r = 20R_1R_2 \text{ in which } 12 \leq L_r \leq 20$$

In the equation, R_1 is a reduction factor based on the tributary area supported by the structural member being designed (designated as A_t and quantified in ft^2) and is determined as follows:

$$\begin{aligned} R_1 &= 1 \qquad \text{for } A_t \leq 200 \text{ ft}^2 \\ &= 1.2 - 0.001A_t \qquad \text{for } 200 \text{ ft}^2 < A_t < 600 \text{ ft}^2 \\ &= 0.6 \qquad \text{for } A_t \geq 600 \text{ ft}^2 \end{aligned}$$

Reduction factor R_2 accounts for the slope of a pitched roof and is determined as follows:

$$\begin{aligned} R_2 &= 1 \text{ for} \qquad F \leq 4 \\ &= 1.2 - 0.05\,F \qquad \text{for } 4 < F < 12 \\ &= 0.6 \qquad \text{for } F \geq 12 \end{aligned}$$

The quantity F in the equations for R_2 is the number of inches of rise per ft for a pitched roof (for example: $F = 12$ indicates a rise of 12 in.12 or an angle of 45°).

The design standard also provides data for roof surfaces that are arched or domed and for special loadings for snow or water accumulation. Roof surfaces must also be designed for wind pressures on the roof surface, both upward and downward. A special situation that must be considered is that of a roof with a low dead load and a significant wind load that exceeds the dead load.

Although the term *flat roof* is often used, there is generally no such thing; all roofs must be designed for some water drainage. The minimum required pitch is usually ¼ in./ft, or a slope of approximately 1:50. With roof surfaces that are close to flat, a potential problem is that of *ponding*, a phenomenon in which the weight of the water on the surface causes deflection of the supporting structure, which in turn allows for more water accumulation (in a pond), causing more deflection, and so on, resulting in an accelerated collapse condition.

Floor Live Loads

The live load on a floor represents the probable effects created by the occupancy. It includes the weights of human occupants, furniture, equipment, stored materials, and so on. All building codes provide minimum live loads to be used in the design of buildings for various occupancies. Because there is a lack of uniformity among different codes in specifying live loads, the local code should always be used. Table 4 .2 contains a sample of values for floor live loads as given in ASCE 2003 (Ref. 1) and commonly specified by building codes.

Although expressed as uniform loads, code-required values are usually established large enough to account for ordinary concentrations that occur. For offices, parking garages, and some other occupancies, codes often require the consideration of a specified concentrated load as well as the distributed loading. This required concentrated load is listed in Table 4.2 for the appropriate occupancies. Where buildings are to contain heavy machinery, stored materials, or other contents of unusual weight, these must be provided for individually in the design of the structure.

When structural framing members support large areas, most codes allow some reduction in the total live load to be used for design. These reductions, in the case of roof loads, are incorporated in the formulas for roof loads given previously. The following is the method given in ASCE

2003 (Ref. 1) for determining the reduction permitted for beams, trusses, or columns that support large floor areas.

The design live load on a member may be reduced in accordance with the formula

$$L = L_0\left(0.25 + \frac{15}{\sqrt{K_{LL}A_T}}\right)$$

where L = reduced design live load per square foot of area supported by the member

L_0 = unreduced live load supported by the member

K_{LL} = live load element factor (see Table 4.3)

A_T = tributary area supported by the member

L shall not be less than $0.50L_0$ for members supporting one floor, and L shall not be less than $0.40L_0$ for members supporting two or more floors.

In office buildings and certain other building types, partitions may not be permanently fixed in location but may be erected or moved from one position to another in accordance with the requirements of the occupants. In order to provide for this flexibility, it is customary to require an allowance of 15 to 20 psf, which is usually added to other dead loads.

Lateral Loads (Wind and Earthquake)

As used in building design, the term *lateral load* is usually applied to the effects of wind and earthquakes, as they induce horizontal forces on stationary structures. From experience and research, design criteria and methods in this area are continuously refined, with recommended practices being presented through the various model building codes.

Space limitations do not permit a complete discussion of the topic of lateral loads and design for their resistance. The following discussion summarizes some of the criteria for design in ASCE 2003 (Ref. 1). Examples of application of these criteria are given in the design examples of building structural design in Chapters 18 through 20. For a more extensive discussion the reader is referred to *Simplified Building Design for Wind and Earthquake Forces* (Ref. 11).

Wind. Where wind is a regional problem, local codes are often developed in response to local conditions. Complete design for wind effects on buildings includes a large number of both architectural and structural concerns. The following is a discussion of some of the requirements from ASCE 2003 (Ref. 1).

Basic Wind Speed. This is the maximum wind speed (or velocity) to be used for specific locations. It is based on recorded wind histories and adjusted for some statistical likelihood of occurrence. For the United States, recommended minimum wind speeds are taken from maps provided in the ASCE standard. As a reference point, the speeds are those recorded at the standard measuring position of 10 m (approximately 33 ft) above the ground surface.

Wind Exposure. This refers to the conditions of the terrain surrounding the building site. The ASCE standard uses three categories, labeled B, C, and D. Qualifications for categories are based on the form and size of wind-shielding objects within specified distances around the building.

Simplified Design Wind Pressure (p_s). This is the basic reference equivalent static pressure based on the critical wind speed and is determined as follows

$$p_s = \lambda I p_{S30}$$

where λ = adjustment factor for building height and exposure

I = importance factor

p_{S30} = simplified design wind pressure for exposure B, at height of 30 ft, and for $I = 1.0$

The importance factor for ordinary circumstances of building occupancy is 1.0. For other buildings, factors are given for facilities that involve hazard to a large number of people, for facilities considered to be essential during emergencies (such as windstorms), and for buildings with hazardous contents.

The design wind pressure may be positive (inward) or negative (outward, suction) on any given surface. Both the sign and the value for the pressure are given in the design standard. Individual building surfaces, or parts thereof, must be designed for these pressures.

Design Methods. Two methods are described in the Code for the application of wind pressures.

Method 1 (simplified procedure). This method is permitted to be used for relatively small, low-rise buildings of simple symmetrical shape. It is the method described here and used for the examples in Part V.

Method 2 (analytical procedure). This method is much more complex and is prescribed to be used for buildings that do not fit the limitations described for Method 1

Uplift. Uplift may occur as a general effect, involving the entire roof or even the whole building. It may also occur as a local phenomenon such as that generated by the overturning moment on a single shear wall

Overturning Moment. Most codes require that the ratio of the dead load resisting moment (called the restoring moment, stabilizing moment, etc.) to the overturning moment be 1.5 or greater. When this is not the case, uplift effects must be resisted by anchorage capable of developing the excess overturning moment. Overturning may be a critical problem for the whole building, as in the case of relatively tall and slender tower structures. For buildings braced by individual shear walls, trussed bents, and rigid-frame bents, overturning is investigated for the individual bracing units

Drift. Drift refers to the horizontal deflection of the structure caused by lateral loads. Code criteria for drift are usually limited to requirements for the drift of a single story (horizontal movement of one level with respect to the next above or below). As in other situations involving structural deformations, effects on the building construction must be considered; thus, the detailing of curtain walls or interior partitions may affect limits on drift.

Special Problems. The general design criteria given in most codes are applicable to ordinary buildings. More thorough investigation is recommended (and sometimes required) for special circumstances such as the following:

Tall buildings. These are critical with regard to their height dimension as well as the overall size and number of occupants inferred. Local wind speeds and unusual wind phenomena at upper elevations must be considered.

Flexible structures. These may be affected in a variety of ways, including vibration or flutter as well as simple magnitude of movements.

Unusual shapes. Open structures, structures with large overhangs or other projections, and any building with a complex shape should be careful studied for the special wind effects that may occur. Wind-tunnel testing may be advised or even required by some codes.

Earthquakes. During an earthquake, a building is shaken up and down and back and forth. The back-and-forth (horizontal) movements are typically more violent and tend to produce major destabilizing effects on buildings; thus, structural design for earthquakes is mostly done in terms of considerations for horizontal (called lateral) forces. The lateral forces are actually generated by the weight of the building— or, more specifically, by the mass of the building that represents both an inertial resistance to movement and the source for kinetic energy once the building is actually in motion. In the simplified procedures of the equivalent static force method, the building structure is considered to be loaded by a set of horizontal forces consisting of some fraction of the building weight. An analogy would be to visualize the building as being rotated vertically 90° to form a cantilever beam, with the ground as the fixed end and with a load consisting of the building weight.

In general, design for the horizontal force effects of earthquakes is quite similar to design for the horizontal force effects of wind. The same basic types of lateral bracing (shear walls, trussed bents, rigid frames, etc.) are used to resist both force effects. There are indeed some significant differences, but in the main, a system of bracing that is developed for wind bracing will most likely serve reasonably well for earthquake resistance as well.

Because of its considerably more complex criteria and procedures, we have chosen not to illustrate the design for earthquake effects in the examples in this book. Nevertheless, the development of elements and systems for the lateral bracing of the building in the design examples here is quite applicable in general to situations where earthquakes are a predominant concern. For structural investigation, the principal difference is in the determination of the loads and their distribution in the building. Another major difference is in the true dynamic effects, critical wind force being usually represented by a single, major, one-direction punch from a gust, whereas earthquakes represent rapid back-and-forth, reversing-direction actions. However, once the dynamic effects are translated into equivalent static forces, design concerns for the bracing systems are very similar, involving considerations for shear, overturning, horizontal sliding, and so on.

For a detailed explanation of earthquake effects and illustrations of the investigation by the equivalent static force method, refer to *Simplified Building Design for Wind and Earthquake Forces* (Ref. 9).

4.2 LOAD COMBINATIONS

The various types of load sources, as described in the preceding section, must be individually considered for quantification. However, for design work the possible combination of loads must be also be considered. Using the appropriate combinations, we must determine the design load for individual structural elements. The first step in finding the design load is to establish the critical combinations of load for the individual element. Using ASCE 2003 (Ref. 1) as a reference, consider the following combinations. Because this process is different for the two basic methods of design, they are presented separately.

Allowable Stress Method

For this method, the individual loads are used directly for the following possible combinations:

Dead load only
Dead load + live load.
Dead load + roof load
Dead load + 0.75(live load) + 0.75(roof load)
Dead load + wind load or 0.7(earthquake load)
Dead load + 0.75(live load) + 0.75(roof load) + 0.75(wind load) or 0.7(earthquake load)
0.6(dead load) + wind load
0.6(dead load) + 0.7(earthquake load)

The combination that produces the critical design situation for individual structural elements depends on the load magnitudes and the loading condition for the elements. A demonstration of examples of the use of these combinations is given in the building design cases in Part V.

Strength Design Method

Some adjustment of the percentage of loads (called *factoring*) is done with the allowable stress method. However, factoring is done with all the loads for the strength method. The need here is to produce a load higher than the true anticipated load (called the *service load*)—the difference representing a margin of safety. The structural elements will be designed at their failure limits with the design load, and they really shouldn't fail with the actual expected loads.

For the strength method, the following combinations are considered:

1.4(dead load)
1.2(dead load) + 1.6(live load) + 0.5(roof load)
1.2(dead load) + 1.6(roof load) + live load or 0.8(wind load)
1.2(dead load) + 1.6(wind load) + (live load) + 0.5(roof load)
1.2(dead load) + 1.0(earthquake load) + live load + 0.2(snow load)
0.9(dead load) + 1.0(earthquake load) or 1.6(wind load)

Use of these load combinations is demonstrated in the building design cases in Part V.

4.3 DETERMINATION OF DESIGN LOADS

The following example demonstrates the process of determination of loading for individual structural elements. Additional examples are presented in the building design cases in Part V.

Figure 4.1 shows the plan layout for the framed structure of a multistory building. The vertical structure consists of columns and the horizontal floor structure of a deck and beam system. The repeating plan unit of 24 × 32 ft is called a column bay. Assuming lateral bracing of the building to be achieved by other structural elements, the columns and beams shown here will be designed for dead load and live load only.

The load to be carried by each element of the structure is defined by the unit loads for dead load and live load and the *load periphery* for the individual elements. The load periphery for an element is established by the layout and dimensions of the framing system. Referring to the labeled elements in Figure 4.1, the load peripheries are as follows.

Beam A: $8 \times 24 = 192$ ft^2
Beam B: $4 \times 24 = 96$ ft^2
Beam C: $24 \times 24 = 576$ ft^2 (Note that beam C carries only three of the four beams per bay of the system, the fourth being carried directly by the columns.)
Column 1: $24 \times 32 = 768$ ft^2
Column 2: $12 \times 32 = 384$ ft^2
Column 3: $16 \times 24 = 384$ ft^2
Column 4: $12 \times 16 = 192$ ft^2

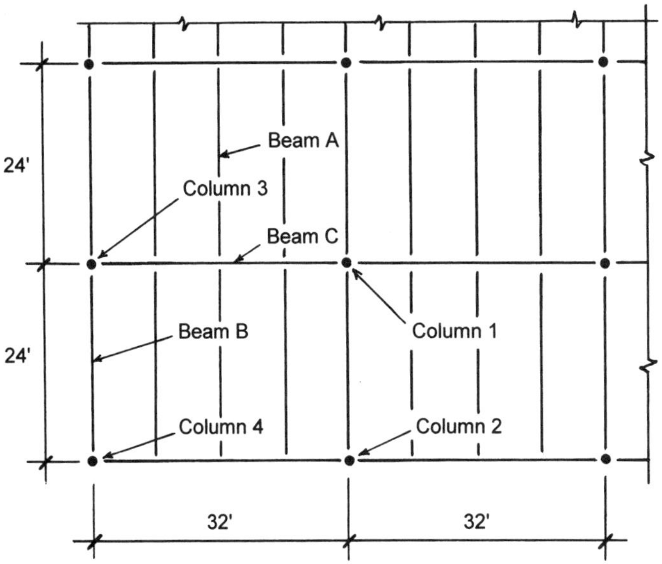

Figure 4.1 Reference for determination of distributed loads.

For each of these elements, the unit dead load and unit live load from the floor is multiplied by the floor areas computed for the individual elements. Any possible live load reduction (as described in Section 4.1) is made for the individual elements based on their load periphery area.

Additional dead load for the elements consists of the dead weight of the elements themselves. For the columns and beams at the building edge, another additional dead load consists of the portion of the exterior wall construction supported by the elements. Thus, column 2 carries an area of the exterior wall defined by the multiple of the story height times 32 ft. Column 3 carries 24 ft of wall, and Column 4 carries 28 ft of wall (12 + 16).

The column loads are determined by the indicated supported floor, to which is added the weight of the columns. For an individual story column, this would be added to loads supported above this level—from the roof and any upper levels of floor.

The loads as described are used in the defined combinations described in Section 4.2. If any of these elements are involved in the development of the lateral bracing structure, the appropriate wind or earthquake loads are also added.

Floor live loads may be reduced by the method described in Sec. 4.1. Reductions are based on the tributary area supported and the number of levels supported by members. Computations of design loads using the process described here are given for the building design cases in Part V.

4.4 DESIGN METHODS

Use of allowable stress as a design condition relates to the classic method of structural design once known as the *working stress method* and now called the *allowable stress design* method (ASD). The loads used for this method are generally those described as *service loads*; that is, they are related to the service (use) of the structure. Deformation limits are also related to service loads.

Even from the earliest times of use of stress methods, it was known that for most materials and structures the true ultimate capacity was not predictable by use of elastic stress methods. Compensating for this with the working stress method was mostly accomplished by considerations for the establishing of the limiting design stresses. For more accurate predictions of true failure limits, however, it was necessary to abandon elastic methods and to use true ultimate strength behaviors. This lead eventually to the so-called *strength method* for design, presently described as the LRFD method, or *load and resistance factor design* method.

The procedures of the stress method are still applicable in many cases—especially for design for deformation limitations. However, the LRFD methods are now very closely related to more accurate use of test data and risk analysis, and purport to be more realistic with regard to true structural safety.

4.5 THE ALLOWABLE STRESS DESIGN METHOD (ASD)

The allowable stress method generally consists of the following:

1. The service (working) load conditions are visualized and quantified as intelligently as possible. Adjustments may be made here by the determination of various statistically likely load combinations (dead load plus live load plus wind load, etc.), by consideration of load duration, and so on.
2. Stress, stability, and deformation limits are set by standards for the various responses of the structure to the loads: in tension, bending, shear, buckling, deflection, uplift, overturning, and so on.
3. The structure is then evaluated (investigated) for its adequacy or is proposed (designed) for an adequate response.

An advantage obtained in working with the stress method is that the real usage condition (or at least an intelligent guess about it) is kept continuously in mind. The principal disadvantage comes from its detached nature regarding real failure conditions, because most structures develop much different forms of stress and strain as they approach their failure limits.

4.6 THE STRENGTH DESIGN METHOD (LRFD)

In essence, the allowable stress design method consists of designing a structure to *work* at some established appropriate percentage of its total capacity. The strength method consists of designing a structure to *fail*, but at a load condition well beyond what it should have to experience in use. A major reason for favoring of strength methods is that the failure of a structure is relatively easily demonstrated by physical testing. What is truly appropriate as a working condition, however, is pretty much a theoretical speculation. The strength method is now largely preferred in professional design work. It was first largely developed for design of concrete structures but has now generally taken over all areas of structural design.

Nevertheless, it is considered necessary to study the classic theories of elastic behavior as a basis for visualization of the general ways that structures work. Ultimate responses are usually some form of variant from the classic responses (because of inelastic materials, secondary effects, multimode responses, etc.). In other words, the usual study procedure is to first consider a classic, elastic response, and then to observe (or speculate about) what happens as failure limits are approached.

For the strength method, the process is as follows:

1. The service loads are quantified as in Step 1 for the stress method and then are multiplied by an adjustment factor (essentially a safety factor) to produce the *factored load.*
2. The form of response of the structure is visualized and its ultimate (maximum, failure) resistance is quantified in appropriate terms (resistance to compression, to buckling, to bending, etc.). This quantified resistance is also subject to an adjustment factor called the *resistance factor*. Use of resistance factors is discussed in Part III for steel structures and in Part IV for concrete structures.
3. The usable resistance of the structure is then compared to the ultimate resistance required (an investigation procedure), or a structure with an appropriate resistance is proposed (a design procedure).

When the design process using the strength method employs both load and resistance factors, it is now called *load and resistance factor design* (abbreviated LRFD).

4.7 CHOICE OF DESIGN METHOD

Applications of design procedures in the stress method tend to be simpler and more direct-appearing than in the strength methods. For example, the design of a beam may amount to the simple inversion of a few stress or strain equations to derive some required properties (section modulus for bending, area for shear, moment of inertia for deflection, etc.). Applications of strength methods tend to be more obscure, simply because the mathematical formulations for describing failure conditions are more complex than the refined forms of the classic elastic methods.

As strength methods are increasingly used, however, the same kinds of shortcuts, approximations, and round-number rules of thumb will emerge to ease the work of designers. And, of course, use of the computer combined with design experience will permit designers to utilize highly complex formulas and massive databases with ease—all the while hopefully keeping some sense of the reality of it all.

Arguments for the use of the stress method or the strength method are essentially academic. An advantage of the stress method may be a closer association with the in-use working conditions of the structure. On the other hand, strength design has a tighter grip on true safety through its focus on failure modes and mechanisms. The successful structural designer, however, needs both forms of consciousness and will develop them through the work of design, whatever methods are employed for the task.

Deformations of structures (such as deflection of beams) that are of concern for design will occur at the working stress level. Visualization and computation of these deformations require the use of basic techniques developed for the ASD method, regardless of whether ASD or LRFD is used for the design work in general.

Work in this book demonstrates the use of both the ASD and LRFD methods. The work for wood in Part II and in the wood structures in Part V is done with the ASD method. The work for steel in Part III and for concrete in Part IV and the steel and concrete structures in Part V is done with the LRFD method. Either method can be used for wood, steel, or concrete structures. However, professional structural design for steel and concrete is now done almost exclusively with the LRFD method. For wood, design work is still done with both methods, although the trend is steadily toward the use of the LRFD method.

II

WOOD CONSTRUCTION

Wood has long been the structural material of choice in the United States whenever conditions permit its use. For small buildings—where fire codes permit—it is extensively used. As with other building products, elements of wood used for building structures are produced in a highly refined industrialized production system, and quality of the materials and products is considerably controlled. In addition to the many building codes, there are several organizations in the United States that provide standards for design of wood products. The work in this part is based primarily on one of these standards, the *National Design Specification for Wood Construction*, published by the American Forest and Paper Association (Ref. 3), herein referred to as the *NDS*.

5

WOOD SPANNING ELEMENTS

This chapter deals with applications of wood products for the development of spanning structures for building roofs and floors. Spanning systems used for roofs and floors commonly employ a variety of wood products. The solid wood material, cut directly from logs (here called *solid-sawn)*, is used as standard-sized structural lumber, such as the all-purpose 2 by 4 (here called 2 × 4). Solid pieces can be mechanically connected and assembled to form various structures and can be glued together to form glued-laminated products.

A widely used product is the plywood panel, formed by gluing together three or more very thin plies of wood and used extensively for wall sheathing or for roof and floor decking. This represents a moderate reconstitution of the basic wood material, essentially retaining the grain character both structurally and for surface appearance. A more extensive reconstitution involves the use of the basic wood fiber reduced to small pieces and adhered with a binding matrix to form fiber products such as paper, cardboard, and particleboard. Because of the major commercial use of paper products, the fiber industry is strongly established and application of wood fiber products for building construction advances steadily. Wood fiber paneling is slowly replacing plywood and wood boards in many applications.

5.1 STRUCTURAL LUMBER

Structural lumber consists of solid-sawn, standard-sized elements produced for various construction applications. Individual pieces of lumber are marked for identification as to wood species (type of tree of origin), grade (quality), size, usage classification, and grading authority. On the basis of this identity, various structural properties are established for engineering design.

Aside from the natural properties of the species (tree), the most important factors that influence structural grading are density (unit weight), basic grain pattern (as sawn from the log), natural defects (knots, checks, splits, pitch pockets, etc.), and moisture content. Because the relative effects of natural defects vary with the size of sawn pieces and the usage application, structural lumber is classified with respect to its size and use. Incorporating these and other considerations, the four major classifications follow. (Note that *nominal dimensions*, as explained in the appendix, are used here.)

1. *Dimension lumber.* Sections with thickness of 2 to 4 in. and width of 2 in. or more (includes most studs, rafters, joists, and planks).
2. *Beams and stringers.* Rectangular sections 5 in. or more in thickness, with width 2 in. or more greater than thickness, graded for strength in bending when loaded on the narrow face.
3. *Posts and timbers.* Square or nearly square sections, 5×5 or larger, with width not more than 2 in. greater than thickness, graded primarily for use as compression elements with bending strength not especially important.
4. *Decking.* Lumber from 2 to 4 in. thick, tongued and grooved or splined on the narrow face, graded for flat application (mostly as plank deck).

Design data and standards for structural lumber is provided in the publications of the wood industry, such as the *NDS*.

A broad grouping of trees identifies them as *softwoods* or *hardwoods*. Softwoods, such as pine, cypress, and redwood, mostly come from trees that are coniferous or cone-bearing, whereas hardwoods come mostly from trees that have broad leaves, as exemplified by oaks and maples. Two species of trees used extensively for structural lumber in the United States are Douglas fir and Southern pine, both of which are classified among softwoods.

Dimensions

As discussed in Chapter 4, structural lumber is described in terms of a nominal size, which is slightly larger than the true dimensions of pieces. However, properties for structural computations, as given in Table A.8 in the appendix, are based on the true dimensions, which are also listed in the table.

For sake of brevity, we have omitted metric units from the text and the tabular data. However, example computations and exercise problems are presented with data in both U.S. and metric units.

Design Value Tables

We must consider many factors in determining the unit stresses to be used for design for wood structures. Extensive testing has produced values known as *clear wood strength values.* To obtain values for design work, the clear wood values are modified by factors that take into account the loss of strength from defects, the size and position of knots, the size of members, the degree of density of the wood, and the condition of seasoning or specific value of moisture content of the lumber at the time of use. For specific design applications, other modifications may be made for the type of loads and the particular structural usage.

Table 5.1 gives design values to be used for ordinary allowable stress design. It is adapted from the *NDS* (Ref. 2) and gives data for one popular wood species: Douglas fir-larch. To obtain values from the table, the following is determined:

1. *Species.* The *NDS* publication lists values for 27 different species, only one of which is included in Table 5.1.
2. *Moisture condition at time of use.* The moisture condition corresponding to the table values is given with the species designation in the table. Adjustments for other conditions are described in the table footnotes or in various specifications in the *NDS.*
3. *Grade.* This is indicated in the first column of the table and is based on visual grading standards.

4. *Size and use.* The second column of the table identifies size ranges or usages of the lumber.
5. *Structural function.* Individual columns in the table yield values for various stress conditions. The last column yields the material modulus of elasticity.

In the reference document, there are extensive footnotes for this table. Data from Table 5.1 is used in various example computations in this book, and some issues treated in the document footnotes are explained. In many situations, there are modifications to the design values, as will be explained later.

Bearing Stress

There are various situations in which a wood member may develop a contact bearing stress—essentially a surface compression stress. Some examples are the following:

1. At the base of a wood column supported in direct bearing. This is a case of bearing stress that is in a direction parallel to the grain.
2. At the end of a beam that is supported by bearing on a support. This is a case of bearing stress that is perpendicular to the grain.
3. Within a bolted connection at the contact surface between the bolt and the wood at the edge of the bolt hole.
4. In a timber truss where a compression force is developed by direct bearing between the two members. This is frequently a situation involving bearing stress that is at some angle to the grain other than parallel or perpendicular.

For connections, the bearing condition is usually incorporated into the general assessment of the unit value of connecting devices. Limiting stress values for direct bearing are based only on the wood species and relative density. Ordinarily, a single value is given for dense grades and another single value for other (ordinary or not dense) grades. The value for compression perpendicular to the grain (at beam ends, for example) is given in Table 5.1.

The situation of stress at an angle to the grain requires the determination of a compromise value somewhere between the allowable values for the two limiting stress conditions for stresses parallel and perpendicular to the wood grain. This is obtained from a formula in the *NDS*.

TABLE 5.1 Design Values for Visually-Graded Lumber of Douglas Fir-Larch[a] (Values in psi)

Species and Commercial Grade	Size and Use Classification	Extreme Fiber in Bending F_b: Single Member Uses	Extreme Fiber in Bending F_b: Repetitive Member Uses	Tension Parallel to Grain, F_t	Horizontal Shear, F_v	Compression Perpendicular to Grain, $F_{c\perp}$	Compression Parallel to Grain, F_c	Modulus of Elasticity, E
Dimension Lumber 2 to 4 in. thick								
Select structural	2 in. & wider	1500	1725	1000	180	625	1700	1,900,000
No. 1 and better		1200	1380	800	180	625	1550	1,800,000
No. 1		1000	1150	675	180	625	1500	1,700,000
No. 2		900	1035	575	180	625	1350	1,600,000
No. 3		525	603	325	180	625	775	1,400,000
Stud	2–6 in. wide	700	805	450	180	625	850	1,400,000
Timbers								
Dense sel. struc.	Beams and stringers	1900	—	1100	170	730	1300	1,700,000
Select structural		1600	—	950	170	625	1100	1,600,000
Dense no. 1		1550	—	775	170	730	1100	1,700,000
No. 1		1350	—	675	170	625	925	1,600,000
No. 2		875	—	425	170	625	600	1,300,000
Dense sel. struc.	Posts and timbers	1750	—	1150	170	730	1350	1,700,000
Select structural		1500	—	1000	170	625	1150	1,600,000
Dense no. 1		1400	—	950	170	730	1200	1,700,000
No. 1		1200	—	825	170	625	1000	1,600,000
No. 2		750	—	475	170	625	700	1,300,000
Decking								
Select dex	Decking	1750	2000	—	—	625	—	1,800,000
Commercial dex		1450	1650	—	—	625	—	1,700,000

Source: Data adapted from *National Design Specification for Wood Construction,* 2001 edition (Ref. 3), with permission of the publishers, American Forest & Paper Association. The table in the reference document lists several other species and has extensive footnotes.

[a] Values listed are for normal duration loading and dry service conditions. See Table 5.1A for size adjustment factors for dimension lumber (2–4 in. thick). Values for design are subject to various other adjustment factors in the *NDS* specification.

TABLE 5.1A Size Adjustment Factors for Dimension Lumber and Decking

		Thickness (breadth), F_b			
Grades	Width (depth)	2-in. & 3-in.	4-in.	F_t	F_c
	2, 3 and 4-in.	1.5	1.5	1.5	1.15
Select structural,	5-in.	1.4	1.4	1.4	1.1
No. 1 and better,	6-in.	1.3	1.3	1.3	1.1
No. 1, No. 2, No. 3	8-in.	1.2	1.3	1.2	1.05
	10-in.	1.1	1.2	1.1	1.0
	12-in.	1.0	1.1	1.0	1.0
	14-in. and wider	0.9	1.0	0.9	0.9
Stud	2, 3, and 4-in.	1.1	1.1	1.1	1.05
	5 and 6-in.	1.0	1.0	1.0	1.0
	8-in. and wider	Use no. 3 grade tabulated design values and size factors			
Decking		2-in.	3-in.		
		1.10	1.04		

Modifications of Design Values

The values given in Table 5.1 are basic references for establishing the allowable values to be used for design. The table values are based on some defined norms, and in many cases, the design values will be modified for actual use in structural computations. In some cases, the form of the modification is a simple increase or decrease achieved by a percentage factor. The following are some common types of modifications.

1. *Moisture.* The table or footnotes define a specific assumed moisture content on which the table values are based. Increases may be allowed for wood that is specially cured to a lower moisture content. If exposed to weather or other high moisture conditions, a reduction may be required.
2. *Load duration.* The table values are based on so-called *normal* duration loading, which is actually somewhat meaningless. Increases are permitted for very short duration loading, such as wind and earthquakes. A decrease is required when the critical design loading is long in duration (such as a major dead load). Table 5.2 gives a summary of the *NDS* requirements for modifications for load duration.
3. *Temperature.* Where prolonged exposure to temperatures over 150°F exists, some design values must be reduced.

TABLE 5.2 Modification Factors for Design Values for Structural Lumber for Load Duration[a]

Load Duration	Multiply Design Values by:	Typical Design Loads
Permanent	0.9	Dead load
Ten years	1.0	Occupancy live load
Two months	1.15	Snow load
Seven days	1.25	Construction load
Ten minutes	1.6	Wind or earthquake load
Impact[b]	2.00	Impact load

Source: Adapted from the *National Design Specification for Wood Construction*, 2001 edition (Ref. 3), with permission of the publishers, American Forest & Paper Association.

[a] Load duration factors shall not apply to modulus of elasticity, *E*, nor to compression perpendicular to grain design values, F_{cz}, based on a deformation limit.

[b] Load duration factors greater than 1.6 shall not apply to structural members pressure-treated with water-borne preservatives, or fire retardant chemicals. The impact load duration factor shall not apply to connections.

4. *Chemical treatments.* Impregnation with chemicals for resistance to fire, rot, vermin, and insects may require reductions in some values.
5. *Size.* Effectiveness in flexure is reduced in beams exceeding 12 in. in depth. This is described in the next section.
6. *Buckling.* Slender columns or beams may have reduced capacities. Bracing is the best solution; otherwise, stresses must be reduced.
7. *Load orientation to the wood grain.* This mostly affects design of connections.

Each design situation must be analyzed to determine necessary modifications.

5.2 DESIGN FOR BENDING

The design of a wood beam for strength in bending is accomplished by using the flexure formula (Section 3.7). The form of this equation used in design is

$$S = \frac{M}{F_b}$$

where M = maximum bending moment

F_b = allowable extreme fiber (bending) stress

S = required beam section modulus

Beams must be considered for shear, deflection, end bearing, and lateral buckling, as well as for bending stress. However, a common procedure is to first find the beam size required for bending and then to investigate for other conditions. Such a procedure is as follows:

1. Determine the maximum bending moment.
2. Select the wood species and grade of lumber to be used.
3. From Table 5.1, determine the basic allowable bending stress.
4. Consider appropriate modifications for the design stress value to be used.
5. Using the allowable bending stress in the flexure formula, find the required section modulus
6. Select a beam size from Table A.8.

Example A simple beam has a span of 16 ft [4.88 m] and supports a total uniformly distributed load, including its own weight, of 6500 lb [28.9 kN]. Using Douglas fir-larch, select structural grade, determine the size of the beam with the least cross-sectional area on the basis of limiting bending stress.

Solution: The maximum bending moment for this condition is

$$M = \frac{WL}{8} = \frac{6500 \times 16}{8} = 13{,}000 \text{ ft-lb } [17.63 \text{ kN-m}]$$

The next step is to use the flexure formula with the allowable stress to determine the required section modulus. A problem with this is that there are two different size/use groups in Table 5.1, yielding two different values for the allowable bending stress. Assuming single member use, the part listed under "Dimension Lumber" yields a stress of 1500 psi for the chosen grade, while the part under "Timber, Beams, and Stringers" yields a stress of 1600 psi. Using the latter category, the required value for the section modulus is

$$S = \frac{M}{F_b} = \frac{13{,}000 \times 12}{1600} = 97.5 \text{ in.}^3 \ [1.60 \times 10^6 \text{ mm}^3]$$

whereas the value for $F_b = 1500$ psi may be determined by proportion as

$$S = \frac{1600}{1500} \times 97.5 = 104 \text{ in.}^3$$

From Table A.8, the smallest members in these two size categories are

$$4 \times 16, \quad S = 135.661 \text{ in.}^3, \quad A = 53.375 \text{ in.}^2$$

$$6 \times 12, \quad S = 121.229 \text{ in.}^3, \quad A = 63.25 \text{ in.}^2$$

For the 4 × 16 the allowable stress is not changed by Table 5.1.A as the factor from the table is 1.0. Thus, the 4 × 16 is the choice for the least cross-sectional area, in spite of having the lower value for bending stress.

Size Factors for Beams

Beams greater than 12 in. in depth, with a thickness of 5 in. or more, have reduced values for the maximum allowable bending stress. This reduction is achieved with a reduction factor determined as

$$C_F = \left(\frac{12}{d}\right)^{1/9}$$

Values for this factor for standard lumber sizes are given in Table 5.3. For the preceding example, neither section qualifies for size reduction modification.

Repetitive Member Use

Table 5.1 yields two values for allowable bending stress for dimension lumber. The first value is given for "Single Member Uses," which refers

TABLE 5.3 Size Factors for Wood Beams

Beam Depth (in.)	Bending Moment Capacity Reduction Factor C_F
13.5	0.987
15.5	0.972
17.5	0.959
19.5	0.947
21.5	0.937
23.5	0.928

to the case of individual beams. The second value under "Repetitive Member Uses" refers to closely spaced joists and rafters, for which some load sharing occurs; thus prompting a 15 percent increase in stress value for design.

Using the value for repetitive member use requires that there be at least three members in a group and that they not be spaced farther than 24 in. on center.

The following example illustrates the application of the beam design procedure for the case of a roof rafter.

Example 2. Rafters of Douglas fir-larch, No. 2 grade, are to be used at 16 in. spacing for a span of 20 ft. Live load without snow is 20 psf, and the total dead load, including the rafters, is 15 psf. Find the minimum size for the rafters, based only on bending stress.

Solution: At this spacing, the rafters qualify for the increased bending stress described as "Repetitive Member Uses" in Table 5.1. Thus, for the No. 2 grade rafters, $F_b = 1035$ psi. The loading condition as described qualifies the situation with regard to load duration for an allowable stress increase factor of 1.25 (see Table 5.2). For the rafters at 16 in. spacing, the maximum bending moment is

$$M = \frac{wL^2}{8} = \frac{\left(\frac{16}{12}\right)(20 + 15)(20)^2}{8} = 2333 \text{ ft-lb}$$

and the required section modulus is

$$S = \frac{M}{F_b} = \frac{2333 \times 12}{1.25 \times 1035} = 21.64 \text{ in.}^3$$

From Table A.8, the smallest section with this property is a 2 × 12, with an S of 31.64 in.^{3}. Note that the allowable stress is not changed by Table 5.1.A, as the table factor is 1.0.

Lateral Bracing

Design specifications provide for the adjustment of bending capacity or allowable bending stress when a member is vulnerable to a compression buckling failure. To reduce this effect, thin beams (mostly joists and rafters) are often provided with bracing that is adequate to prevent both lateral (sideways) buckling and torsional (rollover) buckling. The NDS requirements for bracing are given in Table 5.4. If bracing is not provided, a reduced bending capacity must be determined from rules given in the specifications.

Common forms of bracing consist of bridging and blocking. Bridging consists of crisscrossed wood or metal members in rows. Blocking consists of solid, short pieces of lumber the same size as the framing; these are fit tightly between the members in rows.

Problem 5.2.A. No. 1 grade of Douglas fir-larch is to be used for a series of floor beams 6 ft [1.83 m] on center, spanning 14 ft [4.27 m]. If the total uniformly distributed load on each beam, including the beam weight, is 3200 lb [14.23 kN], select the section with the least cross-sectional area based on bending stress.

Problem 5.2.B. A simple beam of Douglas fir-larch, select structural grade, has a span of 18 ft [5.49 m] with two concentrated loads of 4 kips [13.34 kN] each placed at the third points of the span. Neglecting its own weight, determine the size of the beam with the least cross-sectional area based on bending stress.

Problem 5.2.C. Rafters are to be used on 24-in. centers for a roof span of 16 ft. Live load is 20 psf (without snow), and the dead load is 15 psf, including the weight of the rafters. Find the rafter size required for Douglas fir-larch of (a) No. 1 grade and (b) No. 2 grade, based on bending stress.

TABLE 5.4 Lateral Support Requirements for Wood Beams

Ratio of Depth to Thickness[a]	Required Conditions
2:1 or less (2 × 4)	No support required
3:1, 4:1 (2 × 6,8)	Ends held in position to resist lateral rotation
5:1 (2 × 10)	One edge held in position for entire span
6:1 (2 × 12)	Bridging or blocking at maximum spacing of 8 ft; or both edges held in position for entire span; or one edge held in position for entire span (compression edge) and ends held against lateral rotation
7:1 (2 × 14)	Both edges held against lateral rotation for entire span

Source: Adapted from data in *National Design Specification for Wood Construction,* 2001 ed. (Ref. 3), with permission of the publishers, American Forest and Paper Association.

[a]Ratio of nominal dimensions for standard elements of structural lumber.

5.3 BEAM SHEAR

As discussed in Section 3.7, the maximum beam shear stress for the rectangular sections ordinarily used for wood beams is expressed as

$$f_v = \frac{1.5V}{A}$$

where f_v = maximum unit horizontal shear stress in pounds per square inch

V = total vertical shear force at the section, in pounds

A = the cross-sectional area of the beam

Wood is relatively weak in shear resistance, with the typical failure producing a horizontal splitting of the beam ends. This is most frequently only a problem with heavily loaded beams of short span, for which bending moment may be low but the shear force is high. Because the failure is one of horizontal splitting, it is common to describe this stress as horizontal shear in wood design; which is how the allowable shear stress is labeled in Table 5.1.

Example 3. A 6 × 10 beam of Douglas fir-larch, No. 2 dense grade, has a total horizontally distributed load of 6000 lb [26.7 kN]. Investigate for shear stress.

Solution: For this loading condition, the maximum shear at the beam end is one half of the total load, or 3000 lb. Using the true dimensions of the section from Table A.8, the maximum stress is

$$f_v = \frac{1.5V}{A} = \frac{1.5 \times 3000}{52.25} = 86.1 \text{ psi } [0.594 \text{ MPa}]$$

Referring to Table 5.1, under the classification "Beams and Stringers," the allowable stress is 170 psi. The beam is therefore adequate for this loading condition.

For uniformly loaded beams that are supported by end bearing, the code permits a reduction in the design shear force to that which occurs at a distance from the support equal to the depth of the beam.

Note: In the following problems, use Douglas fir-larch and neglect the beam weight.

Problem 5.3.A. A 10 × 10 beam of select structural grade supports a single concentrated load of 10 kips [44.5 kN] at the center of the span. Investigate the beam for shear.

Problem 5.3.B. A 10 × 14 beam of dense select structural grade is loaded symmetrically with three concentrated loads of 4300 lb [19.13 kN], each placed at the quarter points of the span. Is the beam safe for shear?

Problem 5.3.C. A 10 × 12 beam of No. 2 dense grade is 8 ft [2.44 m] long and has a concentrated load of 8 kips [35.58 kN] located 3 ft [0.914 m] from one end. Investigate the beam for shear.

Problem 5.3.D. What should be the nominal cross-sectional dimensions for the beam of least weight that supports a total uniformly distributed load of 12 kips [53.4 kN] on a simple span and consists of No. 1 grade? Consider only the limiting shear stress.

5.4 BEARING

Bearing occurs at beam ends when a beam sits on a support, or when a concentrated load is placed on top of a beam within the span. The stress developed at the bearing contact area is compression perpendicular to the grain, for which an allowable value ($F_{c\perp}$) is given in Table 5.1.

Although the design values given in the table may be safely used, when the bearing length is quite short, the maximum permitted level of stress may produce some indentation in the edge of the wood member. If the appearance of such a condition is objectionable, a reduced stress is recommended. Excessive deformation may also produce some significant vertical movement, which may be a problem for the construction.

Example 4. An 8 × 14 beam of Douglas fir-larch, No. 1 grade, has an end bearing length of 6 in [152 mm]. If the end reaction is 7400 lb [32.9 kN], is the beam safe for bearing?

Solution: The developed bearing stress is equal to the end reaction divided by the product of the beam width and the length of bearing. Thus:

$$f_c = \frac{\text{Bearing force}}{\text{Contact area}} = \frac{7400}{7.5 \times 6} = 164 \text{ psi } [1.13 \text{ MPa}]$$

This is compared to the allowable stress of 625 psi from Table 5.1, which shows the beam to be quite safe.

Example 5. A 2 × 10 rafter cantilevers over and is supported by the 2 × 4 top plate of a stud wall. The load from the rafter is 800 lb [3.56 kN]. If both the rafter and the plate are No. 2 grade, is the situation adequate for bearing?

Solution: The bearing stress is determined as

$$f = \frac{800}{1.5 \times 3.5} = 152 \text{ psi } [1.05 \text{ MPa}]$$

This is considerably less than the allowable stress of 625 psi, so the bearing is safe.

Example 5. A two-span 3 × 12 beam of Douglas fir-larch, No. 1 grade, bears on a 3 × 14 beam at its center support. If the reaction force is 4200 lb [18.7 kN], is this safe for bearing?

Solution: Assuming the bearing to be at right angles, the stress is

$$f = \frac{4200}{2.5 \times 2.5} = 672 \text{ psi } [4.63 \text{ MPa}]$$

This is slightly in excess of the allowable stress of 625 psi.

Problem 5.4.A. A 6 × 12 beam of Douglas fir-larch, No. 1 grade, has 3 in. of end bearing to develop a reaction force of 5000 lb [22.2 kN]. Is the situation adequate for bearing?

Problem 5.4.B. A 3 × 16 rafter cantilevers over a 3 × 16 support beam. If both members are of Douglas fir-larch, No. 1 grade, is the situation adequate for bearing? The rafter load on the support beam is 3000 lb [13.3 kN].

5.5 DEFLECTION

Deflections in wood structures tend to be most critical for rafters and joists, where span-to-depth ratios are often pushed to the limit. However, long-term high levels of bending stress can also produce sag, which may be visually objectionable or cause problems with the construction. In general, it is wise to be conservative with deflections of wood structures. Push the limits, and you will surely get sagging floors and roofs and possibly very bouncy floors. This may in some cases make a strong argument for use of glued-laminated beams or even steel beams.

For the common uniformly loaded beam, the deflection takes the form of the equation

$$\Delta = \frac{5WL^3}{384EI}$$

Substitutions of relations between W, M, and flexural stress in this equation can result in the form

$$\Delta = \frac{5L^2 f_b}{24Ed}$$

Using average values of 1500 psi for f_b and 1500 ksi for E, the expression reduces to

$$\Delta = \frac{0.03L^2}{d}$$

where Δ = deflection in inches
L = span in feet
d = beam depth in inches

Figure 5.1 is a plot of this expression with curves for nominal dimensions of depth for standard lumber. For reference, the lines on the graph corresponding to ratios of deflection of $L/180$, $L/240$, and $L/360$ are shown. These are commonly used design limitations for total load and live load deflections, respectively. Also shown for reference is the limiting span-to-depth ratio of 25 to 1, which is commonly considered to be a practical span limit for general purposes. For beams with other values for bending stress and modulus of elasticity, true deflections can be obtained as follows:

$$\text{True } \Delta = \frac{\text{True } f_b}{1500} \times \frac{1{,}500{,}000}{\text{True } E} \times \Delta \text{ from graph}$$

The following examples illustrate problems involving deflection. Douglas fir-larch is used for these examples and for the problems that follow them.

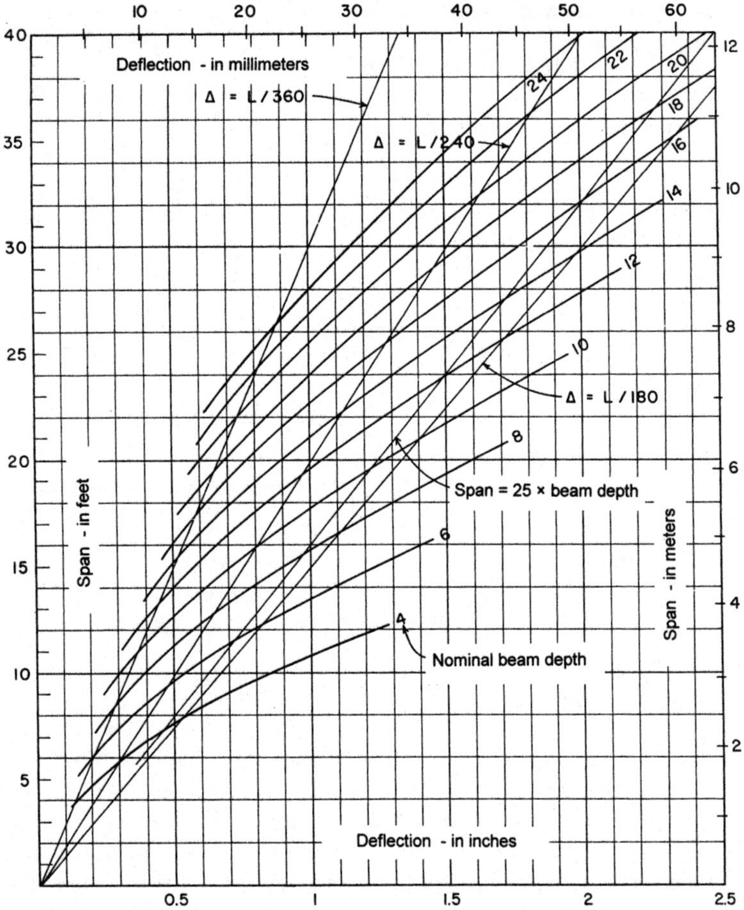

Figure 5.1 Deflection of wood beams. Assumed conditions: maximum bending stress of 1500 psi, modulus of elasticity of 1,500,000 psi.

Example 7. An 8 × 12 wood beam with $E = 1{,}600{,}000$ psi is used to carry a total uniformly distributed load of 10 kips on a simple span of 16 ft. Find the maximum deflection of the beam.

Solution: From Table A.8, find the value of $I = 950$ in.^{4} for the 8 × 12 section. Then, using the deflection formula for this loading:

$$\Delta = \frac{5WL^3}{384EI} = \frac{5 \times 10{,}000 \times (16 \times 12)^3}{384 \times 1{,}600{,}000 \times 950} = 0.61 \text{ in.}$$

Or, using the graph in Figure 5.1:

$$M = \frac{WL}{8} = \frac{10{,}000 \times 16}{8} = 20{,}000 \text{ ft-lb}$$

$$f_b = \frac{M}{S} = \frac{20{,}000 \times 12}{165} = 1455 \text{ psi}$$

From Figure 5.1, Δ = approximately 0.66 in. Then:

$$\Delta = \frac{1455}{1500} \times \frac{1{,}500{,}000}{1{,}600{,}000} \times 0.66 = 0.60 \text{ in.}$$

which shows reasonable agreement with the computed value.

Example 8. A beam consisting of a 6 × 10 section with E = 1,400,000 psi spans 18 ft and carries two concentrated loads. One load is 1800 lb and is placed at 3 ft from one end of the beam, and the other load is 1200 lb, placed at 6 ft from the opposite end of the beam. Find the maximum deflection due only to the concentrated loads.

Solution: For an approximate computation, use the equivalent uniform load method, consisting of finding the hypothetical total uniform load that will produce a moment equal to the actual maximum moment in the beam. Then the deflection for uniformly distributed load may be used with this hypothetical (equivalent uniform) load. Thus:

$$\text{If:} \quad M = \frac{WL}{8}, \quad \text{then:} \quad W = \frac{8M}{L}$$

For this loading the maximum bending moment is 6600 ft-lb (the reader should verify this by the usual procedures), and the equivalent uniform load is thus

$$W = \frac{8M}{L} = \frac{8 \times 6600}{18} = 2933 \text{ lb}$$

and the approximate deflection is

$$\Delta = \frac{5WL^3}{384EI} = \frac{5 \times 2933 \times (18 \times 12)^3}{384 \times 1{,}400{,}000 \times 393} = 0.70 \text{ in.}$$

As in the previous example, the deflection could also be found by using Figure 5.1, with adjustments made for the true maximum bending stress and the true modulus of elasticity.

Note: For the following problems, neglect the beam weight and consider deflection to be limited to 1/240 of the beam span. Wood is Douglas fir-larch.

Problem 5.5.A. A 6 × 14 beam of No. 1 grade is 16 ft [4.88 m] long and supports a total uniformly distributed load of 6000 lb [26.7 kN]. Investigate the deflection.

Problem 5.5.B. An 8 × 12 beam of dense No. 1 grade is 12 ft [3.66 m] in length and has a concentrated load of 5 kips [22.2 kN] at the center of the span. Investigate the deflection.

Problem 5.5.C. Two concentrated loads of 3500 lb [15.6 kN] each are located at the third points of a 15-ft [4.57 m] beam. The 10 × 14 beam is of select structural grade. Investigate the deflection.

Problem 5.5.D. An 8 × 14 beam of select structural grade has a span of 16 ft [4.88 m] and a total uniformly distributed load of 8 kips [35.6 kN]. Investigate the deflection.

Problem 5.5.E. Find the least weight section that can be used for a simple span of 18 ft [5.49 m] with a total uniformly distributed load of 10 kips [44.5 kN] based on deflection. Wood is No. 1 grade.

5.6 JOISTS AND RAFTERS

Joists and rafters are closely spaced beams that support structural floor or roof decks. These may consist of solid-sawn lumber, light trusses, or composite construction with combined elements of solid wood, laminated pieces, plywood, or particleboard. The discussion in this section deals only with solid-sawn lumber, typically in the class called *dimension lumber*, and having nominal thicknesses of 2 to 3 in. The sizes most commonly used are 2 × 6, 2 × 8, 2 × 10, and 2 × 12. Although the strength of the structural deck is a factor, the dimension used for the spacing of joists (center-to-center) is typically determined by the dimensions of the panel materials used for decking and for the development of ceilings. Nailed edges of panels must fall at the centers of the joists. The most used panel size is 48 in. × 96 in., making desired spacings some even division of these dimensions. Most used are spacings of 24, 16, and 12 in.

Floor Joists

A common form of wood floor construction is shown in Figure 5.2. The structural deck shown is plywood, which produces a top not generally usable for a finished surface. Thus, some finish must be used, such as the hardwood flooring shown here. More common now for most interiors is carpet or thin tile, both of which require some smoother surface than the plywood, resulting in some panel material for *underlay* (usually particleboard) or a thin fill of concrete.

A drywall panel finish is shown here for a ceiling directly attached to the underside of the joists. If a suspended ceiling is required, a second structural frame must be developed beneath the joists.

Lateral bracing of joists is achieved with bridging in Figure 5.2. The need for this is a function of the slenderness of the joist cross-section, which determines requirements for lateral bracing (see Table 5.4). A problem to be considered with this construction is the lack of support for the edges of the surfacing materials in a direction perpendicular to the joists. Use of structural deck with tongue-and-groove edges may be a solution for the deck, but the edges of ceiling panels must still be supported.

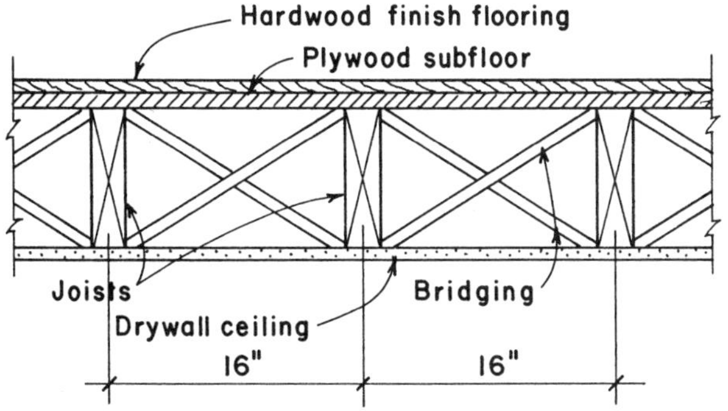

Figure 5.2 Typical joist floor construction.

One solution is to use solid blocking (short pieces of joist lumber) fit tightly between the joists in rows matched to the surfacing material panel size. The blocking also performs the lateral bracing function. Another factor in this decision may be the need for the development of horizontal diaphragm action for lateral forces (see Chapter 18).

Solid blocking is also used under any supported walls that are perpendicular to the joists or under walls parallel to the joists but not falling directly over one joist. Any loading on the general joist system, other than the typical floor construction, should be considered. A simple way to give local increase in strength to the system is to double up joists, which is commonly done at edges of openings, such as those for stairs.

Bridging or blocking provides a tying together of adjacent joists, which permits some load sharing by groups of joists. This is the basis for the category of allowable bending stress in Table 5.1 described as "Repetitive Member Uses" and described in Section 5.1.

The design of joists consists of determining the load to be supported and then applying the procedures for beam design as explained in the preceding sections. However, to facilitate the selection of joists carrying uniformly distributed loads (by far the most common loading), many tables have been prepared that give maximum safe spans for joists of various common sizes and spacings. Table 5.5 is representative of such tables and has been reproduced from the 1997 edition of the *Uniform Building Code* (UBC). Examining the table, we note that spans are computed on the basis of modulus of elasticity, with the required bending stress F_b listed at the bottom of the table for the various joist spacings. The modulus of elasticity of the wood is an index of its stiffness and thus its resistance to deflection, whereas the limiting bending stress relates to maximum bending that can be developed; these are the two considerations incorporated in the table. Shear is ignored as not critical, which is the usual condition for joists with light uniformly distributed loading—required sizes are based strictly on bending and deflection limitations.

Joist spacings in the table are based on dividing the usual length of panels (96 in.) by 4 to get 24, by 5 to get 19.2, by 6 to get 16, or by 8 to get 12. Live load deflection in this table is limited to 1/360 of the joist span. The use of Table 5.5 is illustrated in the following example.

Example 9. Using Table 5.5, select joists to carry a live load of 40 psf on a span of 15 ft 6 in. if the spacing is 16 in. on center.

Solution: Referring to Table 5.5, we find that 2×10 joists with an E of 1,400,000 psi and F_b of at least 1148 psi may be used for up to a span of

TABLE 5.5 Allowable Spans in Feet and Inches for Floor Joists

DESIGN CRITERIA:
Deflection — For 40 psf (1.92 kN/m^2) live load.
Limited to span in inches (mm) divided by 360.
Strength — Live load of 40 psf (1.92 kN/m^2) plus dead load of 10 psf (0.48 kN/m^2) determines the required bending design value.

Joist Size (in)	Spacing (in)	Modulus of Elasticity, *E*, in 1,000,000 psi × 0.00689 for N/mm^2																
× 25.4 for mm		0.8	0.9	1.0	1.1	1.2	1.3	1.4	1.5	1.6	1.7	1.8	1.9	2.0	2.1	2.2	2.3	2.4
2 × 6	12.0	8-6	8-10	9-2	9-6	9-9	10-0	10-3	10-6	10-9	10-11	11-2	11-4	11-7	11-9	11-11	12-1	12-3
	16.0	7-9	8-0	8-4	8-7	8-10	9-1	9-4	9-6	9-9	9-11	10-2	10-4	10-6	10-8	10-10	11-0	11-2
	19.2	7-3	7-7	7-10	8-1	8-4	8-7	8-9	9-0	9-2	9-4	9-6	9-8	9-10	10-0	10-2	10-4	10-6
	24.0	6-9	7-0	7-3	7-6	7-9	7-11	8-2	8-4	8-6	8-8	8-10	9-0	9-2	9-4	9-6	9-7	9-9
2 × 8	12.0	11-3	11-8	12-1	12-6	12-10	13-2	13-6	13-10	14-2	14-5	14-8	15-0	15-3	15-6	15-9	15-11	16-2
	16.0	10-2	10-7	11-0	11-4	11-8	12-0	12-3	12-7	12-10	13-1	13-4	13-7	13-10	14-1	14-3	14-6	14-8
	19.2	9-7	10-0	10-4	10-8	11-0	11-3	11-7	11-10	12-1	12-4	12-7	12-10	13-0	13-3	13-5	13-8	13-10
	24.0	8-11	9-3	9-7	9-11	10-2	10-6	10-9	11-0	11-3	11-5	11-8	11-11	12-1	12-3	12-6	12-8	12-10
2 × 10	12.0	14-4	14-11	15-5	15-11	16-5	16-10	17-3	17-8	18-0	18-5	18-9	19-1	19-5	19-9	20-1	20-4	20-8
	16.0	13-0	13-6	14-0	14-6	14-11	15-3	15-8	16-0	16-5	16-9	17-0	17-4	17-8	17-11	18-3	18-6	18-9
	19.2	12-3	12-9	13-2	13-7	14-0	14-5	14-9	15-1	15-5	15-9	16-0	16-4	16-7	16-11	17-2	17-5	17-8
	24.0	11-4	11-10	12-3	12-8	13-0	13-4	13-8	14-0	14-4	14-7	14-11	15-2	15-5	15-8	15-11	16-2	16-5
2 × 12	12.0	17-5	18-1	18-9	19-4	19-11	20-6	21-0	21-6	21-11	22-5	22-10	23-3	23-7	24-0	24-5	24-9	25-1
	16.0	15-10	16-5	17-0	17-7	18-1	18-7	19-1	19-6	19-11	20-4	20-9	21-1	21-6	21-10	22-2	22-6	22-10
	19.2	14-11	15-6	16-0	16-7	17-0	17-6	17-11	18-4	18-9	19-2	19-6	19-10	20-2	20-6	20-10	21-2	21-6
	24.0	13-10	14-4	14-11	15-4	15-10	16-3	16-8	17-0	17-5	17-9	18-1	18-5	18-9	19-1	19-4	19-8	19-11
F_b	12.0	718	777	833	888	941	993	1,043	1,092	1,140	1,187	1,233	1,278	1,323	1,367	1,410	1,452	1,494
	16.0	790	855	917	977	1,036	1,093	1,148	1,202	1,255	1,306	1,357	1,407	1,456	1,504	1,551	1,598	1,644
	19.2	840	909	975	1,039	1,101	1,161	1,220	1,277	1,333	1,388	1,442	1,495	1,547	1,598	1,649	1,698	1,747
	24.0	905	979	1,050	1,119	1,186	1,251	1,314	1,376	1,436	1,496	1,554	1,611	1,667	1,722	1,776	1,829	1,882

NOTE: The required bending design value, F_b, in pounds per square inch (× 0.00689 for N/mm^2) is shown at the bottom of this table and is applicable to all lumber sizes shown. Spans are shown in feet-inches (1 foot = 304.8 mm, 1 inch = 25.4 mm) and are limited to 26 feet (7925 mm) and less.

Source: Reproduced from the *Uniform Building Code,* 1997 ed. (Ref. 2), with permission of the publishers, International Conference of Building Officials.

15 ft 8 in. From Table 5.1 it may be determined that a stress grade of No. 1 may be used for Douglas fir-larch joists (category "Dimension Lumber" allowable stress of 1150 psi).

Note that the joists in Table 5.5 are designed for a dead load of 10 psf, which relates to common conditions for light wood construction, such as that shown in Figure 5.2. Other tables may be used for different loadings—both live and dead—or the usual procedures for beam design can always be resorted to.

Rafters

Rafters are used for roof decks in a manner similar to floor joists. Whereas floor joists are typically installed dead flat, rafters are commonly sloped to achieve roof drainage. For structural design, it is common to consider the rafter span to be the horizontal projection, as indicated in Figure 5.3.

As with floor joists, rafter design is frequently accomplished with the use of safe load tables. Table 5.6 is representative of such tables and has been reproduced from the 1997 edition of the UBC. Organization of the table is similar to that for Table 5.5, except that the columns are arranged under maximum bending stress and modulus of elasticity is given at the bottom—opposite to the arrangement in Table 5.5. This generally reflects the common situation that deflection is more critical for floors (to reduce bounce) and less critical for roofs.

The following example illustrates the use of the data in Table 5.6.

Example 10. Rafters are to be used on 24-in. centers for a roof span of 16 ft. Live load is 20 psf; total dead load is 15 psf; live load deflection is

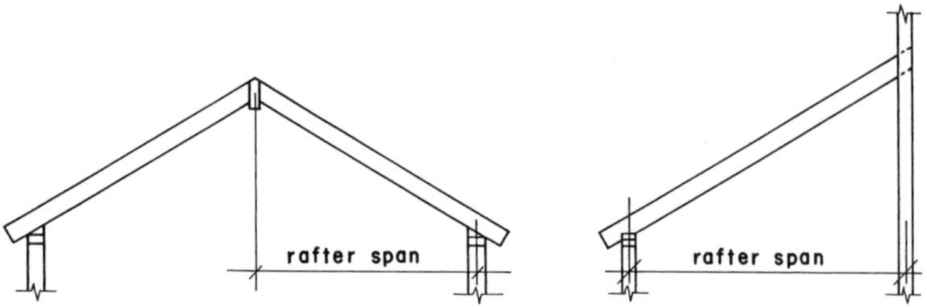

Figure 5.3 Span of sloping rafters.

limited to 1/240 of the span. Find the rafter size required for Douglas fir-larch of (1) No. 1 grade and (2) No. 2 grade.

Solution: (1) From Table 5.1, we find the design values for No. 1 grade to be $E = 1{,}700{,}000$ psi and $F_b = 1150$ psi. Although Table 5.6 does not have a column for $F_b = 1150$ psi, it is apparent that the size choice is for a 2×12 rafter. It is also apparent that E is not a critical concern.

(2) From Table 5.1, we find the design values for No. 2 grade to be $E = 1{,}600{,}000$ psi and $F_b = 1035$ psi. It should again be apparent that the 2×12 is the choice for the rafters at 24 in. centers.

It should be noted that the 1997 edition of the UBC that is used as a general reference for this book, and from which Tables 5.5 and 5.6 are taken, uses the 1991 edition of the NDS, not the 2001 edition (Ref. 3) that is generally used here. The adjustments given in Table 5.1.A have therefore not been used in deriving the materials in Tables 5.5 and 5.6, and computations based on the 2001 NDS may produce some minor discrepancies with the table entries.

Problems 5.6.A, B, C, D. Using Douglas fir-larch, No. 2 grade, pick the joist size required from Table 5.5 for the stated conditions. Live load is 40 psf; dead load is 10 psf; deflection is limited to 1/360 of the span under live load only.

	Joist Spacing (in.)	Joist Span (ft)
A	16	14
B	12	14
C	16	18
D	12	22

Problems 5.6.E, F, G, H. Using Douglas fir-larch, No. 2 grade, pick the rafter size required from Table 5.6 for the stated conditions. Live load is 20 psf; dead load is 15 psf; deflection is limited to 1/240 of the span under live load only.

	Rafter Spacing (in.)	Rafter Span (ft)
E	16	12
F	24	12
G	16	18
H	24	18

TABLE 5.6 Allowable Spans in Feet and Inches for Low- or High-Slope Rafters

DESIGN CRITERIA:
Strength — Live load of 20 psf (0.96 kN/m^2) plus dead load of 15 psf (0.72 kN/m^2) determines the required bending design value.
Deflection — For 20 psf (0.96 kN/m^2) live load.
Limited to span in inches (mm) divided by 240.

Rafter Size (in)	Spacing (in)	Bending Design Value, F_b (psi) × 0.00689 for N/mm^2												
× 25.4 for mm		300	400	500	600	700	800	900	1000	1100	1200	1300	1400	1500
2 × 6	12.0	6-7	7-7	8-6	9-4	10-0	10-9	11-5	12-0	12-7	13-2	13-8	14-2	14-8
	16.0	5-8	6-7	7-4	8-1	8-8	9-4	9-10	10-5	10-11	11-5	11-10	12-4	12-9
	19.2	5-2	6-0	6-9	7-4	7-11	8-6	9-0	9-6	9-11	10-5	10-10	11-3	11-7
	24.0	4-8	5-4	6-0	6-7	7-1	7-7	8-1	8-6	8-11	9-4	9-8	10-0	10-5
2 × 8	12.0	8-8	10-0	11-2	12-3	13-3	14-2	15-0	15-10	16-7	17-4	18-0	18-9	19-5
	16.0	7-6	8-8	9-8	10-7	11-6	12-3	13-0	13-8	14-4	15-0	15-7	16-3	16-9
	19.2	6-10	7-11	8-10	9-8	10-6	11-2	11-10	12-6	13-1	13-8	14-3	14-10	15-4
	24.0	6-2	7-1	7-11	8-8	9-4	10-0	10-7	11-2	11-9	12-3	12-9	13-3	13-8
2 × 10	12.0	11-1	12-9	14-3	15-8	16-11	18-1	19-2	20-2	21-2	22-1	23-0	23-11	24-9
	16.0	9-7	11-1	12-4	13-6	14-8	15-8	16-7	17-6	18-4	19-2	19-11	20-8	21-5
	19.2	8-9	10-1	11-3	12-4	13-4	14-3	15-2	15-11	16-9	17-6	18-2	18-11	19-7
	24.0	7-10	9-0	10-1	11-1	11-11	12-9	13-6	14-3	15-0	15-8	16-3	16-11	17-6
2 × 12	12.0	13-5	15-6	17-4	19-0	20-6	21-11	23-3	24-7	25-9				
	16.0	11-8	13-5	15-0	16-6	17-9	19-0	20-2	21-3	22-4	23-3	24-3	25-2	26-0
	19.2	10-8	12-3	13-9	15-0	16-3	17-4	18-5	19-5	20-4	21-3	22-2	23-0	23-9
	24.0	9-6	11-0	12-3	13-5	14-6	15-6	16-6	17-4	18-2	19-0	19-10	20-6	21-3
E	12.0	0.12	0.19	0.26	0.35	0.44	0.54	0.64	0.75	0.86	0.98	1.11	1.24	1.37
	16.0	0.11	0.16	0.23	0.30	0.38	0.46	0.55	0.65	0.75	0.85	0.96	1.07	1.19
	19.2	0.10	0.15	0.21	0.27	0.35	0.42	0.51	0.59	0.68	0.78	0.88	0.98	1.09
	24.0	0.09	0.13	0.19	0.25	0.31	0.38	0.45	0.53	0.61	0.70	0.78	0.88	0.97

NOTE: The required modulus of elasticity, E, in 1,000,000 pounds per square inch (psi) (× 0.00689 for N/mm^2) is shown at the bottom of this table, is limited to 2.6 million psi (17 914 N/mm^2) and less, and is applicable to all lumber sizes shown. Spans are shown in feet-inches (1 foot = 304.8 mm, 1 inch = 25.4 mm) and are limited to 26 feet (7925 mm) and less.

Source: Reproduced from the *Uniform Building Code,* 1997 ed. (Ref. 2), with permission of the publishers, International Conference of Building Officials.

5.7 DECKING FOR ROOFS AND FLOORS

Materials used to produce roof and floor surfaces include the following:

1. Boards of nominal 1-in.-thick solid-sawn wood, typically with tongue-and-groove edges
2. Solid-sawn wood elements thicker than 1 inch nominal dimension (usually called planks or planking) with tongue-and-groove or other edge development to prevent vertical slipping between adjacent units
3. Plywood of appropriate thickness for the span and the construction
4. Other panel materials, including those of compressed wood fibers or particles

Plank deck is especially popular for roof decks that are exposed to view from below. A variety of forms of products used for this construction is shown in Figure 5.4. Widely used is a nominal 2-in.-thick unit, which may be of solid-sawn form (Figure 5.6*a*) but is now more likely to be of glue-laminated form (Figure 5.6*c*). Thicker units can be obtained for considerable spans between supporting members, but the thinner plank units are most popular.

Plank decks and other special decks are fabricated products produced by individual manufacturers. Information about their properties should be obtained from suppliers or the manufacturers. Plywood decks are

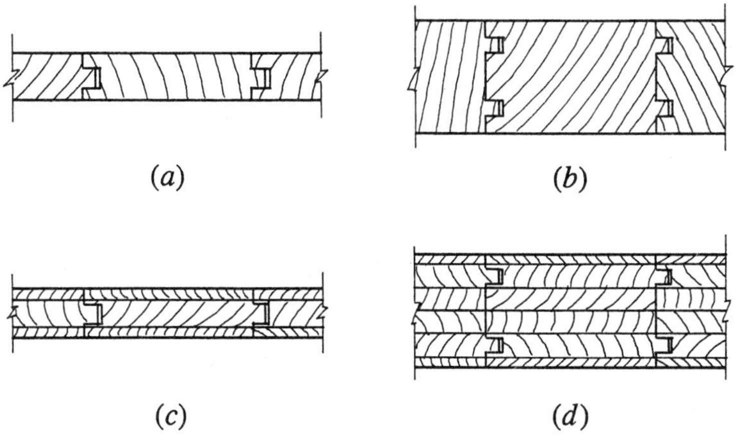

(*a*) (*b*) (*c*) (*d*)

Figure 5.4 Units for board and plank decks.

widely used where their structural properties are critical. Plywood is an immensely variable material, although a few selected types are commonly used for structural purposes. The remainder of this section deals with plywood used for roof and floor decks.

Plywood Decks

Structural plywood consists primarily of that made with all plies of softwood such as Douglas fir. Other than panel thickness, principal distinctions are the following:

1. *Glue type.* Panels are identified for exterior (exposed to weather) or interior use, based on the type of glue used.
2. *Grade of plies.* Individual plies are rated on the basis of presence of flaws, common ratings being A, B, C, and D, with A the best. Quality of face plies is important for construction use, but middle plies are also important for structural applications.
3. *Structural classification of panels.* Individual panels are identified with markings for specific construction uses.
4. *Panel identification index.* This is an industry-standard method for identification consisting of an indelible stamp on the panel face. Data and symbols in the stamp indicate the intended usage and the strength of panels for deck applications.

Plywood decks and wall sheathing are frequently used to develop resistance to wind and earthquake forces on structures. In this case, the objective is to turn the entire sheathed surface into a continuous rigid panel (called a *shear wall* or a *diaphragm*). Structural quality of the plywood is important for this application, but of equal concern is the nailing of plywood edges to achieve the necessary attachment to develop the diaphragm continuity through the many joints. Use of plywood diaphragms is discussed in Chapter 18.

For gravity load spanning functions, plywood is strongest when the face ply grain direction is perpendicular to the parallel supports (usually rafters or joists). However, for various reasons, it is sometimes desired to turn the panels the other way, which somewhat limits the spanning capability of the panels. This latter weakness is most critical for panels of fewer plies (that is, for thinner plywood).

Depending on the details of the construction, there may be some concern for plywood panel edges that do not fall over supports (edges perpendicular to the rafters or joists). Blocking (solid short pieces of the rafter or joist members fitted between spanning members) may provide support, usually while doubling as lateral bracing for the spanning members. Other remedies include tongue-and-groove edges for thicker panels or metal clips (H-shaped) that fit between panels.

The all-American panel size is, of course, 48 × 96 inches. If ordered in large batches, however, other sizes are obtainable. The most common panel thicknesses used for structural decks and wall sheathing are ⅜, ½, ⅝, and ¾ in. (Or slightly modified thicknesses relating to metric units).

Plywood deck load capacities and span limits are rated in industry standards and building codes. The following tables from the *Uniform Building Code* (Ref. 1) are presented here.

1. UBC Table 23-II-E-1, giving load capacities or span limits for gravity-load spanning decks for roofs or floors with the face grain direction perpendicular to supports (Table 5.7)
2. UBC Table 23-II-E-2, giving load capacities or span limits for roof decks with the face grain direction parallel to supports (Table 5.8)
3. UBC Table 23-II-H, giving shear capacities for horizontal plywood diaphragms (roof or floor decks) - (Table 18.1)
4. UBC Table 23-II-I-1, giving shear capacities for plywood sheathing for shear wall actions (Table 18.2)

5.8 GLUED LAMINATED PRODUCTS

In addition to plywood panels, we use a number of other products for wood construction that are fabricated by gluing together pieces of wood into solid form. Girders, framed bents, and arch ribs of large size are produced by assembling standard 2-in. nominal lumber (2 × 6. etc.). The resulting thickness of such elements is essentially the width of the standard lumber used, with a small dimensional loss due to finishing. The depth is a multiple of the lumber thickness of 1.5 in.

Availability of large glued-laminated products should be investigated on a regional basis, because shipping to job sites is a major cost factor. Information about these products can be obtained from local suppliers or from the product manufacturers in the region. As with other widely used products, there are industry standards and usually some building code data for design.

TABLE 5.7 Data for Spanning Plywood Decks—Face Grain Perpendicular to Supports[1–9]

Panel Span Rating	Panel Thickness (inches)	Maximum Span (inches) × 25.4 for mm		Load[5] (pounds per square foot) × 0.0479 for kN/m²		Maximum Span (inches)
Roof/Floor Span	× 25.4 for mm	With Edge Support[6]	Without Edge Support	Total Load	Live Load	× 25.4 for mm
12/0	5/16	12	12	40	30	0
16/0	5/16, 3/8	16	16	40	30	0
20/0	5/16, 3/8	20	20	40	30	0
24/0	3/8, 7/16, 1/2	24	20[7]	40	30	0
24/16	7/16, 1/2	24	24	50	40	16
32/16	15/32, 1/2, 5/8	32	28	40	30	16[8]
40/20	19/32, 5/8, 3/4, 7/8	40	32	40	30	20[8,9]
48/24	23/32, 3/4, 7/8	48	36	45	35	24
54/32	7/8, 1	54	40	45	35	32
60/48	7/8, 1, 1 1/8	60	48	45	35	48
SINGLE-FLOOR GRADES		ROOF[3]				FLOOR[4]
Panel Span Rating (inches)	Panel Thickness (inches)	Maximum Span (inches) × 25.4 for mm		Load[5] (pounds per square foot) × 0.0479 for kN/m²		Maximum Span (inches)
× 25.4 for mm		With Edge Support[6]	Without Edge Support	Total Load	Live Load	× 25.4 for mm
16 oc	1/2, 19/32, 5/8	24	24	50	40	16[8]
20 oc	19/32, 5/8, 3/4	32	32	40	30	20[8,9]
24 oc	23/32, 3/4	48	36	35	25	24
32 oc	7/8, 1	48	40	50	40	32
48 oc	1 3/32, 1 1/8	60	48	50	50	48

[1]Applies to panels 24 inches (610 mm) or wider.
[2]Floor and roof sheathing conforming with this table shall be deemed to meet the design criteria of Section 2312.
[3]Uniform load deflection limitations 1/180 of span under live load plus dead load, 1/240 under live load only.
[4]Panel edges shall have approved tongue-and-groove joints or shall be supported with blocking unless 1/4-inch (6.4 mm) minimum thickness underlayment or 1 1/2 inches (38 mm) of approved cellular or lightweight concrete is placed over the subfloor, or finish floor is 3/4-inch (19 mm) wood strip. Allowable uniform load based on deflection of 1/360 of span is 100 pounds per square foot (psf) (4.79 kN/m²) except the span rating of 48 inches on center is based on a total load of 65 psf (3.11 kN/m).
[5]Allowable load at maximum span.
[6]Tongue-and-groove edges, panel edge clips [one midway between each support, except two equally spaced between supports 48 inches (1219 mm) on center], lumber blocking, or other. Only lumber blocking shall satisfy blocked diaphgrams requirements.
[7]For 1/2-inch (12.7 mm) panel, maximum span shall be 24 inches (610 mm).
[8]May be 24 inches (610 mm) on center where 3/4-inch (19 mm) wood strip flooring is installed at right angles to joist.
[9]May be 24 inches (610 mm) on center for floors where 1 1/2 inches (38 mm) of cellular or lightweight concrete is applied over the panels.

Source: Table 23-II-E-1 from the *Uniform Building Code*, 1997 ed. (Ref. 2), reproduced with permission of the publishers, International Conference of Building Officials.

TABLE 5.8 Data for Spanning Plywood Decks—Face Grain Parallel to Supports[1–3]

PANEL GRADE	THICKNESS (inch)	MAXIMUM SPAN (inches)	LOAD AT MAXIMUM SPAN (psf) × 0.0479 for kN/m²	
	× 25.4 for mm		Live	Total
Structural I	$^{7}/_{16}$	24	20	30
	$^{15}/_{32}$	24	35[3]	45[3]
	$^{1}/_{2}$	24	40[3]	50[3]
	$^{19}/_{32}$, $^{5}/_{8}$	24	70	80
	$^{23}/_{32}$, $^{3}/_{4}$	24	90	100
Other grades covered in UBC Standard 23-2 or 23-3	$^{7}/_{16}$	16	40	50
	$^{15}/_{32}$	24	20	25
	$^{1}/_{2}$	24	25	30
	$^{19}/_{32}$	24	40[3]	50[3]
	$^{5}/_{8}$	24	45[3]	55[3]
	$^{23}/_{32}$, $^{3}/_{4}$	24	60[3]	65[3]

[1]Roof sheathing conforming with this table shall be deemed to meet the design criteria of Section 2312.

[2]Uniform load deflection limitations: $^{1}/_{180}$ of span under live load plus dead load, $^{1}/_{240}$ under live load only. Edges shall be blocked with lumber or other approved type of edge supports.

[3]For composite and four-ply plywood structural panel, load shall be reduced by 15 pounds per square foot (0.72 kN/m²).

Source: Table 23-II-E-2 from the *Uniform Building Code*, 1997 ed. (Ref. 2), reproduced with permission of the publishers, International Conference of Building Officials.

5.9 WOOD FIBER PRODUCTS

Various products are produced with wood that is reduced to fiber form from the logs of trees. Major considerations are those for the size and shape of the wood fiber elements and their arrangement in the finished products. For paper, cardboard, and some fine hardboard products, the wood is reduced to very fine particles and generally randomly placed in the mass of the products. This results in little orientation of the material, other than that produced by the manufacturing process of the particular products.

For structural products, we use somewhat larger wood particle elements and obtain some degree of orientation. Two types of products with this character are the following:

Wafer board or flake board. These are panel products produced with wood chips in wafer form. The wafers are laid randomly on top of each other, producing a panel with a two-way, fiber-oriented nature that simulates the character of plywood panels. Applications include wall sheathing and some structural decks.

Strip or strand elements. Produced from long strands that are shredded from the logs, these products are bundled with the strands all in the same direction to produce elements that have something approaching the character of the linear orientation in solid sawn wood. Applications include studs, rafters, joists, and small beams.

For decking or wall sheathing, these products are generally used in thicknesses greater than that of plywood for the same spans. Other construction issues must be considered, such as nail holding for materials attached to the deck. We must also consider the type and magnitude of loads, the type of finished flooring for floor decks, and the need for diaphragm action for lateral loads. Code approval is an important issue and must be determined on a local basis.

Information about these products should be obtained from the manufacturers or suppliers of particular proprietary products. Some data is now included in general references, such as the UBC, but particular products are competitively produced by individual companies.

This is definitely a growth area as plywood becomes increasingly expensive and logs for producing plywood are harder to find. Resources for fiber products include small trees, smaller sections from large trees, and even some recycled wood. A general trend to use composite materials certainly indicates the likelihood of more types of products for future applications.

5.10 MISCELLANEOUS WOOD STRUCTURAL PRODUCTS

Various types of structural components can be produced with assembled combinations of plywood, fiber panels, laminated products, and solid-sawn lumber. Figure 5.5 shows some commonly used elements that can serve as structural components for buildings.

The unit shown in Figure 5.5*a* consists of two panels of plywood attached to a frame of solid-sawn lumber elements. This is generally described as a *sandwich panel*; however, when used for structural purposes, it is called a *stressed-skin panel*. For spanning actions, the plywood

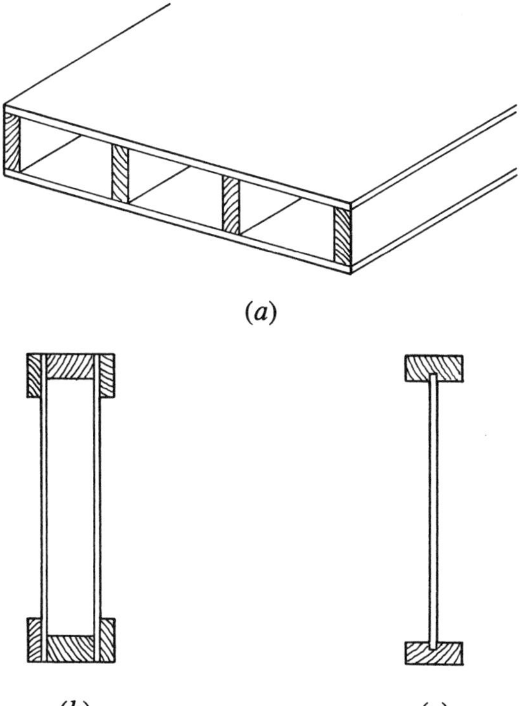

(*a*) (*b*) (*c*)

Figure 5.5 Composite, built-up components with elements of solid-sawn lumber and panels of plywood or wood fiber.

panels serve as bending, stress-resisting flanges, and the lumber elements as beam webs for shear development.

Another common type of product takes the form of the box beam (Figure 5.5*b*) or the built-up I-beam (Figure 5.5*c*). In this case, the roles defined for the sandwich panel are reversed, with the solid-sawn elements serving as flanges and the panel material as the web. These elements are highly variable, using both plywood and fiber products for the panels, and solid-sawn lumber or glued-laminated products for the flange elements. It is also possible to produce various profiles, with a flat chord opposed to a sloped or curved one on the opposite side. Using these elements allows for production of relatively large components from small trees, resulting in a saving of large solid-sawn lumber and old growth forests.

The box beam shown in Figure 5.5*b* can be assembled with attachments of ordinary nails or screws. The I-beam uses glued joints to attach the web and flanges. Box beams may be custom-assembled at the building site, but the I-beams are produced in highly controlled factory conditions.

I-beam products have become highly popular for use in the range of spans just beyond the feasibility for solid-sawn lumber joists and rafters, that is, over about 15 ft for joists and about 20 ft for rafters.

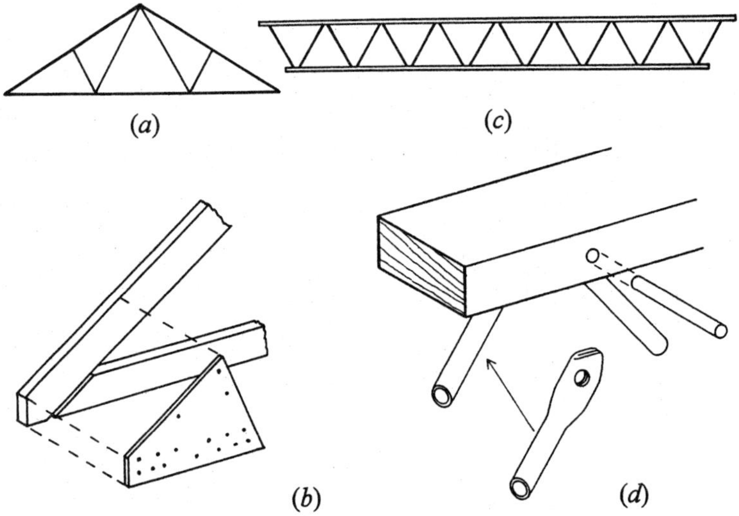

Figure 5.6 Light wood trusses.

Two types of light wood trusses are widely used. The W-truss, shown in Figure 5.6*a*, is widely used for short-span gable-form roofs. Achieved with a single layer of 2× lumber members, and with simple gusset-plated joints (Figure 5.6*b*), this has been the form of the roof structure for small wood-framed buildings for many years. Gussets may consist of pieces of plywood, attached with nails, but are now mostly factory-assembled with metal connector plates.

For flat spanning structures—both roofs and floors—the truss shown in Figure 5.6*c* is used, mostly for spans just beyond the spanning length feasible for solid-sawn wood rafters or joists. One possible assembly is shown in Figure 5.6*d*, using steel tubes with flattened ends, connected to the chords with pins driven through drilled holes. Chords may be simple solid-sawn lumber elements but are also made of proprietary laminated elements that permit virtually unlimited length for single-piece members.

6

WOOD COLUMNS

The wood column that is used most frequently is the *solid-sawn section* consisting of a single piece of wood that is square or oblong in cross section. Single-piece round columns are also used as building columns or foundation piles. This chapter deals with these common elements and some other special forms used as compression members in building construction.

6.1 SOLID-SAWN COLUMNS

For all columns, a fundamental consideration is the column slenderness. For the solid-sawn wood column, slenderness is established as the ratio of the laterally unbraced length to the least side dimension, or L/d (Figure 6.1*a*). The unbraced length (height) is typically the overall vertical length of the column. It takes very little force to brace a column from moving sideways (buckling under compression), so that where construction constrains a column, there may be a shorter unbraced length on one or both axes (Figure 6.1*b*).

Readers are referred to the discussions of relative slenderness in and to the relationship established between the slenderness ratio and the axial compression capacity of linear compression members in Section 3.11. Three general zones of behavior are identified, with the boundaries of the range of slenderness being the very short, stocky compression member and the extremely long, slender member. The very short member fails essentially in compressive stress action (crushing), whereas the very slender member buckles (bends sideways) under a small load.

An important point to make here is that the short compression member is limited by stress resistance, whereas the very slender member is limited essentially by its stiffness –(that is, by the resistance of the member to lateral deflection). Deflection resistance is measured in terms of the stiffness (modulus of elasticity) of the material of the column and the geometric property of its cross-section (moment of inertia). Therefore, *stress* establishes the limit at the low range of relative stiffness and *stiffness* (modulus of elasticity, slenderness ratio) that establishes the limit at extreme values of relative stiffness.

Most building columns, however, fall in a range of stiffness that is transitional between these extremes (Zone 2, as described in Section 3.11).

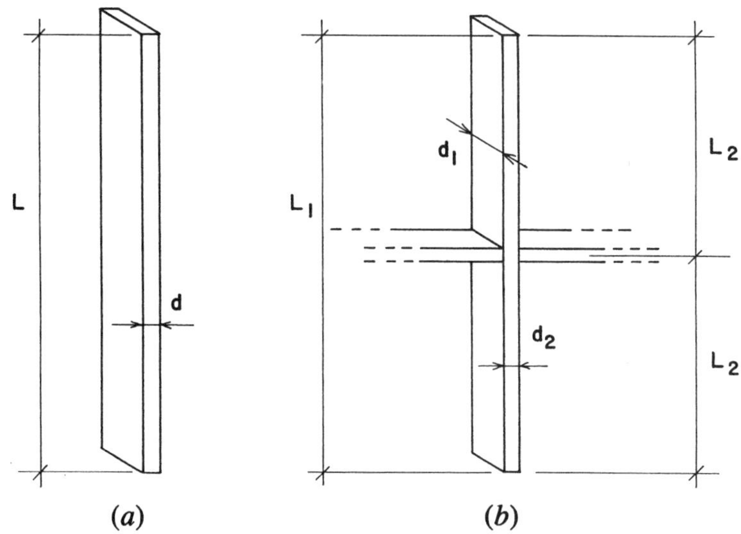

Figure 6.1 Determination of unbraced height for a column, as related to the critical column thickness dimension.

It becomes necessary, therefore, to establish some means for determination of the axial capacity of columns that treats the complete range—from very short to very tall—and all points between. Current column design standards establish complex formulas to describe a single curve that makes the full transition of column behavior related to slenderness. It is important to understand the effect of the variables in these formulas, although for practical design work, use is generally made of one or more design aids that permit shortcuts to pragmatic answers.

Excessively slender building columns are neither safe nor practical. In fact, the point of separation between Zones 2 and 3 in Figure 3.49 generally represents a practical limit for maximum slenderness for columns. Some codes specify a limit, but this degree of slenderness is a general guide for designers. For wood columns, a limit used in the past was a slenderness ratio of 1/50.

Column Load Capacity

The following discussion presents materials from the NDS (Ref. 2) for design of axially loaded columns. The basic formula for determination of the capacity of a wood column, based on the working stress method, is

$$P = (F_c{}^*)(C_p)(A)$$

where A = area of the column cross section

$F_c{}^*$ = the allowable design value for compression parallel to the grain, as modified by applicable factors, except C_p

C_p = the column stability factor

P = the allowable column axial compression load

$$C_p = \frac{1 + (F_{cE}/F_c{}^*)}{2c} - \sqrt{\left[\frac{1 + F_{cE}/F_c{}^*}{2c}\right]^2 - \frac{F_{cE}/F_c{}^*}{c}}$$

where F_{cE} = the Euler buckling stress, as determined by the following formula

c = 0.8 for sawn lumber, 0.85 for round poles, 0.9 for glued-laminated timbers

For the buckling stress:

$$F_{cE} = \frac{K_{cE}E}{(L_c/d)^2}$$

where K_{cE} = 0.3 for visually graded lumber and machine evaluated lumber, 0.418 for machine stress rated lumber and glued-laminated timber

E = modulus of elasticity for the wood species and grade

L_e = the effective length (unbraced height as modified by any factors for support conditions) of the column

d = the column cross section dimension (column width) measured in the direction that buckling occurs

The values to be used for the effective column length and the corresponding column width should be considered as discussed for the conditions displayed in Figure 6.1. For a basic reference, the buckling phenomenon typically uses a member that is pinned at both ends and prevented from lateral movement only at the ends, for which no modification for support conditions is made; this is a common condition for wood columns. The NDS presents methods for modified buckling lengths that are essentially similar to those used for steel design (see Section 10.2). These will be illustrated for steel columns, but not here.

The following examples illustrate the use of the NDS formulas for columns.

Example 1. A wood column consists of a 6 × 6 of Douglas fir-larch, No. 1 grade. Find the safe axial compression load for unbraced lengths of: (1) 2 ft, (2) 8 ft, (3) 16 ft.

Solution: From Table 5.1, find values of F_c = 1000 psi and E = 1,600,000 psi. With no basis for adjustment given, the F_c value is used directly as the F_c value in the column formulas.

For (1), $L/d = 2(12)/5.5 = 4.36$. Then

$$F_{cE} = \frac{K_{cE}E}{(L_c/d)^2} = \frac{0.3 \times 1{,}600{,}000}{(4.36)^2} = 25{,}250 \text{ psi}$$

$$\frac{F_{cE}}{F_c^*} = \frac{25{,}250}{1000} = 25.25$$

$$C_p = \frac{1 + 25.25}{1.6} - \sqrt{\left[\frac{1 + 25.25}{1.6}\right]^2 - \frac{25.25}{0.8}} = 0.992$$

And the allowable compression load is

$$P = (F_c^*)(C_p)(A) = (1000)(0.992)(5.5)^2 = 30{,}008 \text{ lb}$$

For (2), $L/d = 8(12)/5.5 = 17.45$ for which $F_{cE} = 1576$ psi, $F_{cE}/F_c^* = 1.576$, and $C_p = 0.821$; thus:

$$P = (1000)(0.821)(5.5)^2 = 24{,}835 \text{ lb}$$

For (3): $L/d = 16(12)/5.5 = 34.9$ for which $F_{cE} = 394$ psi, $F_{cE}/F_c^* = 0.394$, and $C_p = 0.355$; thus:

$$P = (1000)(0.355)(5.5)^2 = 10{,}739 \text{ lb}$$

Example 2. Wood 2 × 4 elements are to be used as vertical compression members to form a wall (ordinary stud construction). If the wood is Douglas fir-larch, stud grade, and the wall is 8.5 ft high, what is the column load capacity of a single stud?

Solution: It is assumed that the wall has a covering attached to the studs or blocking between the studs to brace them on their weak (1.5-in. dimension) axis. Otherwise, the practical limit for the height of the wall is 50 × 1.5 = 75 in. Therefore, using the larger dimension:

$$\frac{L}{d} = \frac{8.5 \times 12}{3.5} = 29.14$$

From Table 5.1 $F_c = 850$ psi, $E = 1{,}400{,}000$ psi, with the value for F_c adjusted to $1.05(850) = 892.5$ psi. Then

$$\frac{F_{cE}}{F^*} = \frac{495}{892.5} = 0.555, \text{ then } F_{cE} = \frac{0.3 \times 1{,}400{,}000}{(29.14)^2} = 495 \text{ psi}$$

$$C_p = \frac{1.555}{1.6} - \sqrt{\left[\frac{1.555}{1.6}\right]^2 - \frac{0.555}{0.8}} = 0.471$$

$$P = (F_c^*)(C_p)(A) = (892.5)(0.471)(1.5 \times 3.5) = 2207 \text{ lb}$$

Problem 6.1.A–D. Find the allowable axial compression load for the following wood columns of Douglas fir-larch, No. 2 grade.

	Nominal Size (in.)	Unbraced Length (ft)	(m)
A	4 × 4	8	2.44
B	6 × 6	10	3.05
C	8 × 8	18	5.49
D	10 × 1	14	4.27

6.2 DESIGN OF WOOD COLUMNS

The design of columns is complicated by the relationships in the column formulas. The allowable stress for the column is dependent upon the actual column dimensions, which are not known at the beginning of the design process. This does not allow for simply inverting the column formulas to derive required properties for the column. A trial-and-error process is therefore indicated. For this reason, designers typically use various design aids: graphs, tables, or computer-aided processes.

Because of the large number of wood species, resulting in many different values for allowable stress and modulus of elasticity, precisely tabulated capacities become impractical. Nevertheless, aids using average values are available and simple to use for design. Figure 6.2 is a graph on which the axial compression load capacity of some square column sections of a single species and grade are plotted. Table 6.1 yields the capacity for a range of columns. Note that the smaller-sized column sections fall into the classification in Table 5.1 for "Dimension Lumber," rather than for "Timbers." This makes for one more complication in the column design process.

Problem 6.2.A, B, C, D. Select square column sections of Douglas fir-larch, No. 1 grade from Table 6.1, for the following data.

	Required Axial Load (kips)	(kN)	Unbraced Length (ft)	(m)
A	20	89	8	2.44
B	50	222	12	3.66
C	50	222	20	6.10
D	100	445	16	4.88

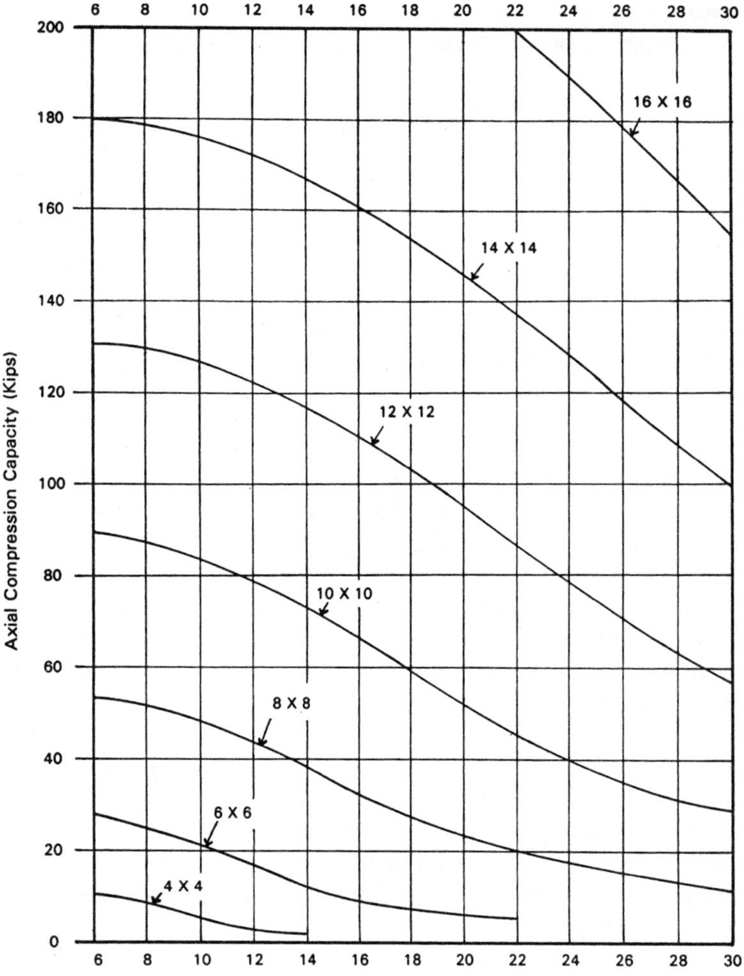

Figure 6.2 Axial compression load capacity for wood members of square cross section. Derived from NDS requirements for Douglas fir-larch, No. 1 grade.

6.3 WOOD STUD CONSTRUCTION

Stud wall construction is often used as part of a general system described as *light wood frame construction.* The joist and rafter construction discussed in Chapter 5, together with the stud wall construction discussed here, are the primary structural elements of this system. In most applica-

TABLE 6.1 Safe Loads for Wood Columns[a]

Column Section		Unbraced Length (ft)										
Nominal Size	Area (in.^{2})	6	8	10	12	14	16	18	20	22	24	26
4 × 4	12.25	11.1	7.28	4.94	3.50	2.63						
4 × 6	19.25	17.4	11.4	7.76	5.51	4.14						
4 × 8	25.375	22.9	15.1	10.2	7.26	6.46						
6 × 6	30.25	27.6	24.8	20.9	16.9	13.4	10.7	8.71	7.17	6.53		
6 × 8	41.25	37.6	33.9	28.5	23.1	18.3	14.6	11.9	9.78	8.91		
6 × 10	52.25	47.6	43.0	36.1	29.2	23.1	18.5	15.0	13.4	11.3		
8 × 8	56.25	54.0	51.5	48.1	43.5	38.0	32.3	27.4	23.1	19.7	16.9	14.6
8 × 10	71.25	68.4	65.3	61.0	55.1	48.1	41.0	34.7	29.3	24.9	21.4	18.4
8 × 12	86.25	82.8	79.0	73.8	66.7	58.2	49.6	42.0	35.4	30.2	26.0	22.3
10 × 10	90.25	88.4	85.9	83.0	79.0	73.6	67.0	60.0	52.9	46.4	40.4	35.5
10 × 12	109.25	107	104	100	95.6	89.1	81.2	72.6	64.0	56.1	48.9	42.9
10 × 14	128.25	126	122	118	112	105	95.3	85.3	75.1	65.9	57.5	50.4
12 × 12	132.25	130	128	125	122	117	111	104	95.6	86.9	78.3	70.2
14 × 14	182.25	180	178	176	172	168	163	156	148	139	129	119
16 × 16	240.25	238	236	234	230	226	222	216	208	200	190	179

[a] Load capacity in kips for solid-sawn sections of No. 1 grade Douglas fir-larch with no adjustment for moisture or load duration conditions.

tions, the system is almost entirely developed with 2-in. nominal thickness lumber. Timber elements are sometimes used for freestanding columns or beams for long spans.

Studs are most commonly of 2-in. nominal thickness (actually 1.5-in.) and must be braced against buckling on their weak axis. Just about any wall covering attached to the studs will perform this function, but if no covering exists, blocking between studs must be provided as shown in Figure 6.3.

Stud spacing is typically related to the dimension of panels of covering material (plywood, gypsum drywall, etc.), with the common size being 48 × 96 in. This yields the same situation as discussed for spacing of rafters and joists in Section 5.6, with 16-in. spacing being the most common.

Studs may be of relatively low grade wood; a special stud grade is available for some species. Straightness is probably the single most desirable quality, in order to obtain flat wall surfaces.

Exterior walls are subject to wind loadings, producing a combination of lateral bending plus vertical compression, a situation described in the

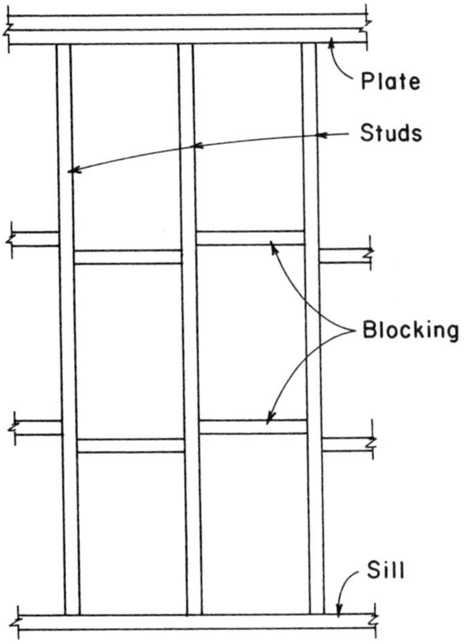

Figure 6.3 Stud wall construction with blocking.

next section. Studs for heavily loaded walls, tall walls, and walls with exceptionally high lateral loading need to be investigated as columns using the procedure given in Section 6.2.

6.4 COLUMNS WITH BENDING

In wood structures, columns with bending occur most frequently as shown in Figure 6.4. Studs in exterior walls represent the situation shown in Figure 6.4*a*, with a loading consisting of vertical gravity plus horizontal wind loads. Due to use of common construction details, columns carrying only vertical loads may sometimes be loaded eccentrically, as shown in Figure 6.4*b*.

The general case of columns subjected to combined compression and bending is discussed in Section 3.11. Current criteria for design of wood columns begins with the straight-line interaction relationship and then adds considerations for buckling due to bending, p-delta effects, and so on. For solid-sawn columns, the NDS provides the following formula for investigation:

$$\left(\frac{f_c}{F'_c}\right)^2 + \frac{f_b}{F_b\left(1 - \dfrac{f_c}{F_{cE}}\right)} \leq 1$$

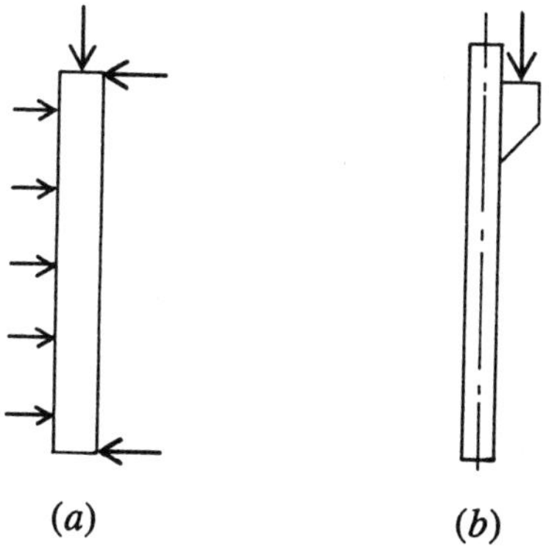

Figure 6.4 Common cases involving combined axial compression and bending in columns: (a) exterior stud or truss chord; (b) column with bracketed support for spanning member.

where f_c = computed compressive stress due to column load

F'_c = tabulated design value for compressive stress, adjusted by all modification factors

f_b = computed bending stress due to bending moment

F_b = tabulated design stress for bending

F_{cE} = the value determined for solid-sawn columns as discussed in Section 6.2

The following examples demonstrate some applications for the procedure.

Example 3. An exterior wall stud of Douglas fir-larch, stud grade, is loaded as shown in Figure 6.5*a*. Investigate the stud for the combined loading. (*Note:* This is the wall stud from the building example in Chapter 18.)

Solution: From Table 5.1, F_b = 805 psi (repetitive stress use), F_c = 850 psi, and E = 1,400,000 psi. Note that the allowable stresses are not changed by Table 5.1.A, because the table factors are 1.0. With inclusion of the wind loading, the stress values (but not E) may be increased by a factor of 1.6 (see Table 5.2).

Assume that wall surfacing braces the 2 × 6 studs adequately on their weak axis (d = 1.5 in.), so d = 5.5 in. Thus:

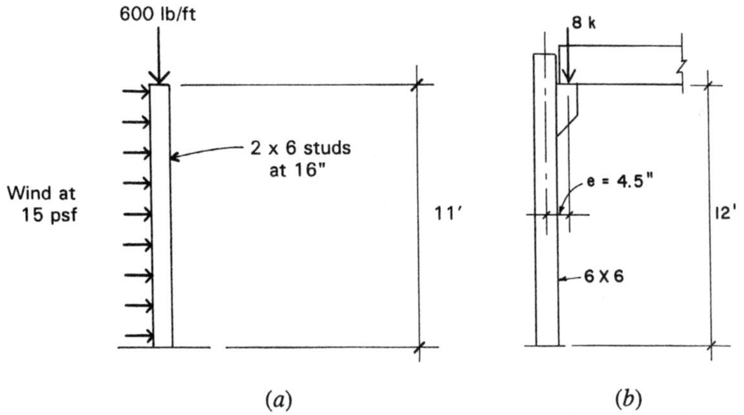

Figure 6.5 Reference for Examples 3 and 4.

$$\frac{L}{d} = \frac{11 \times 12}{5.5} = 24$$

$$F_{cE} = \frac{K_{cE}E}{\left(\frac{L}{d}\right)^2} = \frac{0.3 \times 1{,}400{,}000}{(24)^2} = 729 \text{ psi}$$

$$\frac{F_{cE}}{F_c^*} = \frac{729}{1.6 \times 850} = 0.536$$

$$C_p = \frac{1 + (F_{cE}/F_c^*)}{2c} - \sqrt{\left(\frac{1 + (F_{cE}/F_c^*)}{2c}\right)^2 - \frac{F_{cE}/F_c^*}{c}}$$

$$= \frac{1 + 0.536}{1.6} - \sqrt{\left(\frac{1 + 0.536}{1.6}\right)^2 - \frac{0.536}{0.8}}$$

$$= 0.458$$

The first investigation involves the gravity load without the wind, for which the stress increase factor of 1.6 is omitted. Thus:

$$P = (F_c^*)(C_p)(A) = (850)(0.458)(8.25) = 3212 \text{ lb}$$

Compared to the given load for the 16-in. stud spacing, this is

$$P = (16/12)(600) = 800 \text{ lb}$$

which demonstrates that this is not a critical concern.

Proceeding with consideration for the combined loading, we determine that

$$f_c = \frac{P}{A} = \frac{800}{8.25} = 97 \text{ psi}$$

$$F'_c = C_p F_c^* = 0.458 \times 1.6 \times 850 = 623 \text{ psi}$$

$$M = \left(\frac{16}{12}\right)\left(\frac{wL^2}{8}\right) = \left(\frac{16}{12}\right)\left[\frac{15(11)^2}{8}\right] = 302.5 \text{ lb-ft}$$

$$f_b = \frac{M}{S} = \frac{302.5 \times 12}{7.563} = 480 \text{ psi}$$

$$\frac{f_c}{F_{cE}} = \frac{97}{729} = 0.133$$

Then, using the code formula for the interaction:

$$\left(\frac{97}{623}\right)^2 + \frac{480}{1.6 \times 805(1 - 0.133)} = 0.024 + 0.430 = 0.454$$

Because the result is less than 1, the stud is adequate.

Example 4. The column shown in Figure 6.5*b* is of Douglas fir-larch, dense No. 1 grade. Investigate the column for combined column action and bending.

Solution: From Table 5.1, $F_b = 1400$ psi, $F_c = 1200$ psi, $E = 1{,}700{,}000$ psi. From Table A.8, $A = 30.25$ in.^{2}, and $S = 27.7$ in.^{3}. Then

$$\frac{L}{d} = \frac{12 \times 12}{5.5} = 26.18$$

$$F_{cE} = \frac{0.3 \times 1{,}700{,}000}{(5.5)^2} = 16{,}860 \text{ psi}$$

$$\frac{F_{cE}}{F_c} = \frac{16{,}860}{1200} = 14.05$$

$$C_p = \left(\frac{1 + 14.05}{1.6}\right) - \sqrt{\left(\frac{1 + 14.05}{1.6}\right)^2 - \left(\frac{14.05}{0.8}\right)} = 0.985$$

$$f_c = \frac{8000}{30.25} = 264 \text{ psi}$$

$$F'_c = C_p F_c = (0.985)(1200) = 1182 \text{ psi}$$

$$\frac{f_c}{F_{cE}} = \frac{264}{16{,}860} = 0.016$$

$$f_b = \frac{M}{S} = \frac{8000 \times 4.5}{27.7} = 1300 \text{ psi}$$

And for the column interaction:

$$\left(\frac{264}{1182}\right)^2 + \frac{1300}{1400(1 - 0.016)} = 0.050 + 0.944 = 0.994$$

Because this is less than 1, the column is adequate, but just barely.

Problem 6.4.A. Nine-ft-high 2 × 4 studs of Douglas fir-larch, No. 1 grade, are used in an exterior wall. Wind load is 17 psf on the wall surface; studs are 24 in. on center; the gravity load on the wall is 500 lb/ft of wall length. Investigate the studs for combined action of compression plus bending.

Problem 6.4.B. Ten-ft-high 2 × 4 studs of Douglas fir-larch, No. 1 grade, are used in an exterior wall. Wind load is 25 psf on the wall surface; studs are 16 in. on center; the gravity load on the wall is 500 lb/ft of wall length. Investigate the studs for combined action of compression plus bending.

Problem 6.4.C. A 10 × 10 column of Douglas fir-larch, No. 1 grade, is 9 ft high and carries a compression load of 20 kips that is 7.5 in. eccentric from the column axis. Investigate the column for combined compression and bending.

Problem 6.4.D. A 12 × 12 column of Douglas fir-larch, No. 1 grade, is 12 ft high and carries a compression load of 24 kips that is 9.5 in. eccentric from the column axis. Investigate the column for combined compression plus bending.

7

CONNECTIONS FOR WOOD STRUCTURES

Structures of wood typically consist of large numbers of separate pieces that must be joined together. Fastening of pieces is rarely achieved directly, as in the fitted and glued joints of furniture, except for the production of glued products, such as plywood. For assemblage of building construction, fastening is most often achieved by using some steel device, common ones being nails, screws, bolts, and sheet metal fasteners. A major portion of the NDS (Ref. 3) is devoted to concerns for structural fastenings for wood. This chapter presents a highly condensed treatment of the simple cases for some very common fasteners.

7.1 BOLTED JOINTS

When steel bolts are used to connect wood members, there are several design concerns. Some of the principle concerns are the following:

1. *Net cross section in member.* Holes made for the placing of bolts reduce the wood member cross section. For this investigation, the hole diameter is assumed to be $\frac{1}{16}$ in. larger than that of the bolt. Common situations are shown in Figure 7.1. When bolts in multiple rows are staggered, it may be necessary to make two investigations, as shown in the illustration.
2. *Bearing of the bolt on the wood.* This compressive stress limit varies with the angle of the wood grain to the load direction.

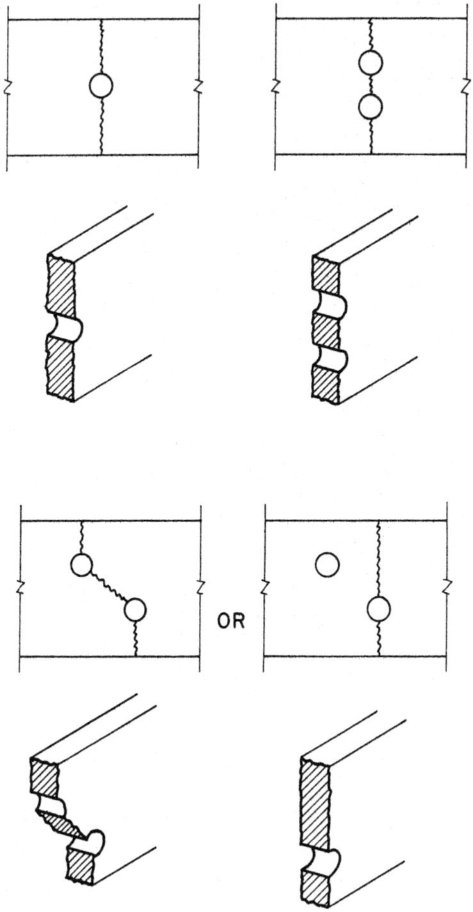

Figure 7.1 Effect of bolt holes on reduction of cross section for tension members.

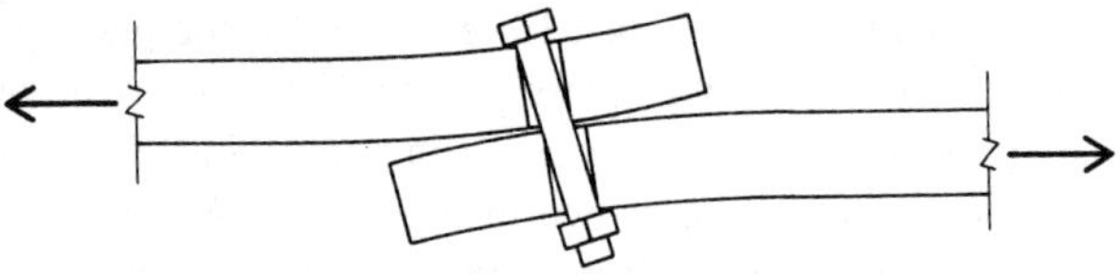

Figure 7.2 Twisting in the two-member bolted joint.

3. *Bending of the bolt.* Long, thin bolts in thick wood members will bend considerably, causing a concentration of bearing at the edge of the hole.
4. *Number of members bolted at a single joint.* The worst case, as shown in Figure 7.2, is that of the two-member joint. In this case, the lack of symmetry in the joint produces considerable twisting. This situation is referred to as *single shear* because the bolt is subjected to shear on a single cross section of the bolt. With more members in the joint, twisting may be eliminated and the bolt is sheared at multiple cross sections.
5. *Ripping out the bolt when too close to an edge.* This problem, together with that of the minimum spacing of bolts, is dealt with by using the criteria given in the NDS.

The NDS presents materials for design of bolted joints in layered stages. The first stage involves general concerns for the basic wood construction—notably, adjustments for load duration, moisture condition, and the particular species and grade of the wood.

The second layer of concern has to do with some considerations for structural fastenings in general. Some of the general concerns are for the form of connections, arrangements of multiple fasteners, and the eccentricity of forces in connections.

The third layer of concern involves considerations for the particular type of fastener. All in all, the number of separate concerns is considerable. The reader is referred to Ref. 4 or other engineering texts for the complete design procedure for bolted joints in wood.

7.2 NAILED JOINTS

A great variety of nails are used in building construction. For structural fastening, the nail most commonly used is called, appropriately, the *common wire nail*. As shown in Figure 7.3, the critical concerns for such nails are the following:

1. *Nail size.* Critical dimensions are the diameter and length (see Figure 7.3*a*). Sizes are specified in pennyweight units, designated as 4d, 6d, and so on, and referred to as four penny, six penny, and so on.
2. *Load direction.* Pullout loading in the direction of the nail shaft is called *withdrawal*; shear loading perpendicular to the nail shaft is called *lateral load.*
3. *Penetration.* Nailing is typically done through one element and into another, and the load capacity is essentially limited by the amount of the length of embedment of the nail in the second member (see Figure 7.3*b*). The length of this embedment is called the penetration.
4. *Species and grade of wood.* The heavier the wood (indicating generally harder, tougher material), the more the load resistance capability.

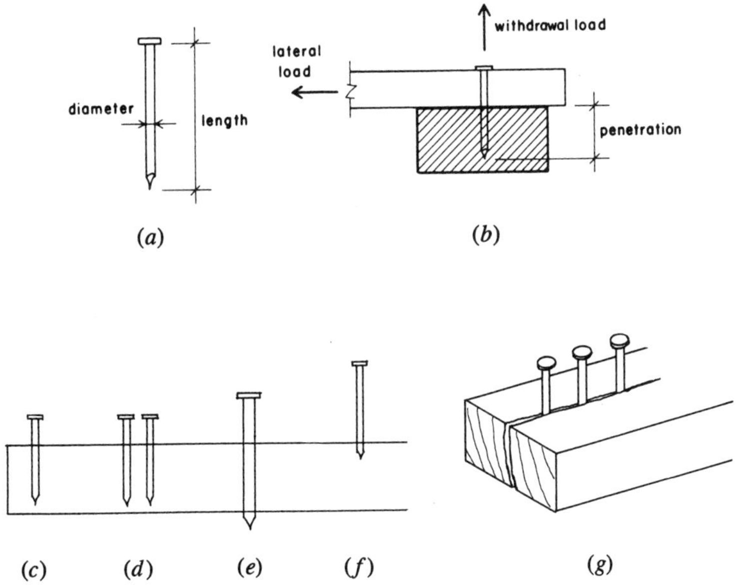

Figure 7.3 Use of common wire nails. (a) Critical dimensions. (b) Loading considerations. (c) through (g)—Poor nailing practices: (c) Too close to edge; (d) nails too close together; (e) nail too large for wood piece; (f) too little penetration of nail into holding wood piece; (g) too many closely-spaced nails in a single row parallel to wood grain and/or nails too close to member end.

Design of good nailed joints requires a little engineering and a lot of good carpentry. Some obvious situations to avoid are those shown in Figures 7.3c through *g*. A little actual carpentry experience is highly desirable for anyone who designs nailed joints.

Withdrawal load capacities of nails are given in units of force per inch of nail penetration length. This unit load is multiplied by the actual penetration length to obtain the total force capacity of the nail. For structural connections, withdrawal resistance is relied on only when the nails are perpendicular to the wood grain direction.

TABLE 7.1 Lateral Load Capacity of Common Wire Nails (lb/nail)

Side Member Thickness, t_s (in.)	Nail Length, L (in.)	Nail Diameter, D (in.)	Pennyweight	Load per Nail for Douglas Fir-Larch G = 0.50, Z (lb)
Structural Plywood Side Members				
3/8	2	0.113	6d	48
	2½	0.131	8d	63
	3	0.148	10d	76
½	2	0.113	6d	50
	2½	0.131	8d	65
	3	0.148	10d	78
	3½	0.162	16d	92
¾	2	0.113	6d	58
	2½	0.131	8d	73
	3	0.148	10d	86
	3½	0.162	16d	100
Solid-Sawn Lumber Side Members				
¾	2½	0.131	8d	90
	3	0.148	10d	105
	3½	0.162	16d	121
	4	0.192	20d	138
1½	3	0.148	10d	118
	3½	0.162	16d	141
	4	0.192	20d	170
	4½	0.207	30d	186
	5	0.225	40d	205
	5½	0.244	50d	211

Source: Adapted from *National Design Specification for Wood Construction*, 2001 edition (Ref. 3), with permission of the publisher, American Forest & Paper Association.

Lateral load capacities for common wire nails are given in Table 7.1 for joints with both plywood and lumber side pieces. The NDS contains very extensive tables for many wood types as well as metal side pieces. The following example illustrates the design of a nailed joint using the table data.

Example 1. A structural joint is formed as shown in Figure 7.4, with the wood members connected by 16d common wire nails. Wood is Douglas fir-larch. What is the maximum value for the compression force in the two side members?

Solution: From Table 7.1, we read a value of 141 lb per nail. (Side member thickness of 1.5-in., 16d nails). As shown in the illustration, there are five nails on each side or a total of 10 nails in the joint. The total joint load capacity is thus

$$C = (10)(141) = 1410 \text{ lb}$$

No adjustment is made for direction of load to the grain. However, the basic form of nailing assumed here is so-called side grain nailing, in which the nail is inserted at 90 degrees to the grain direction and the load is perpendicular (lateral) to the nails.

Minimum adequate penetration of the nails into the supporting member is also a necessity, but use of the combinations given in Table 7.1 ensures adequate penetration if the nails are fully buried in the members.

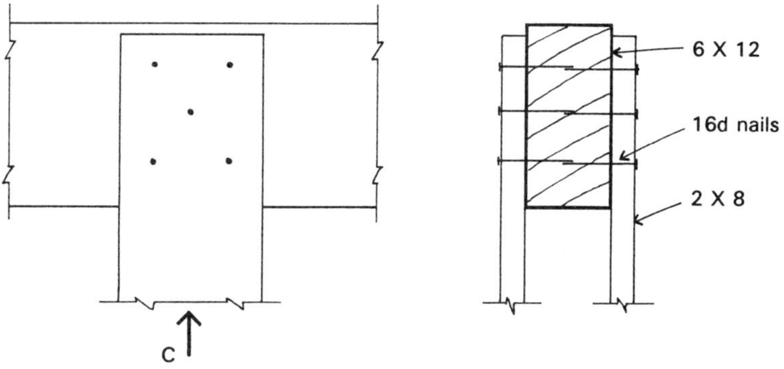

Figure 7.4 Reference for Example 4.

The following example treats a joint for a light wood truss using lumber members for the truss and connecting panels (called gusset plates) of structural plywood.

Example 2. The truss heel joint shown in Figure 7.5 is made with 2-in. nominal thickness lumber and gusset plates of ½-in.-thick plywood. Nails are 6d common wire with the nail layout shown occurring in both sides of the joint. Find the tension load capacity for the bottom chord member (load 3 in the figure).

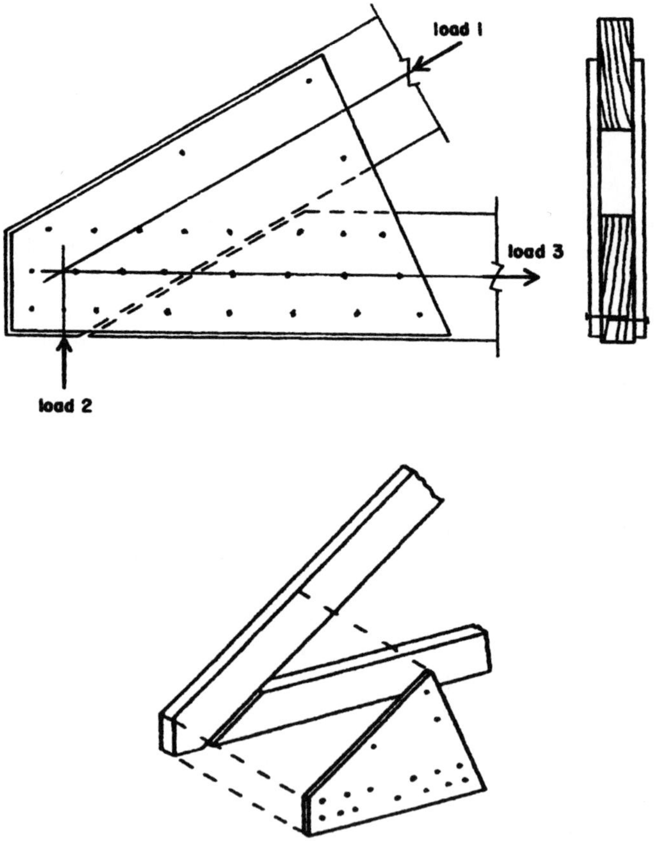

Figure 7.5 Reference for Example 2.

Solution. From Table 7.1 the capacity of one nail is 50 lb. With 12 nails on each side of the joint, the total capacity of the joint is thus

$$T = 24 \times 50 = 1200 \text{ lb}$$

Problem 7.2.A. A joint similar to that in Figure 7.4 is formed with outer members of 1- inch nominal thickness (¾-in. actual thickness) and 10d common wire nails. Find the compression force that can be transferred to the two side members.

Problem 7.2.B. Same as Problem 7.2.A, except outer members are 2 × 10, middle member is 4 × 10, nails are 20d.

Problem 7.2.C. A truss heel joint similar to that in Figure 7.5 is made with gusset plates of ½-in. plywood and 8d nails. Find the tension force limit for the bottom chord.

Problem 7.2.D. A truss heel joint similar to that in Figure 7.5 is made with ¾-in.-thick gusset plates and 10d nails. Find the tension force limit for the bottom chord.

III

STEEL CONSTRUCTION

Steel is used in a wide variety of forms for many tasks in building construction. Wood, concrete, and masonry structures require many steel objects. This part of the book, however, deals with steel as a material for the production of components and systems for steel structures. This usage includes some common ones and an endless range of special possibilities. The concentration here is on ordinary uses.

8

STEEL STRUCTURAL PRODUCTS

For the assemblage of building structures, components of steel consist mostly of standard forms of industrially produced products. The most common and widely used of these products are in forms that have been developed and produced for a long time. Modifications of production and assemblage methods are made continuously, but the basic forms of most steel structures are pretty much as they have been for many years.

8.1 DESIGN METHODS FOR STEEL STRUCTURES

Presently two fundamentally different methods for structural investigation and design are in use. The first of these, traditionally used for many years by designers and researchers, is referred to as the *working stress method* or the *allowable stress method.* At present, this method is called the *allowable stress design* method, designated ASD. The second method, first adopted in 1986, is called the *ultimate strength method* or simply the *strength method.* At present, this method is called the *load and resistance factor design* method, designated LRFD.

In general, the techniques and operational procedures of the ASD method are simpler to use and to demonstrate. They are based largely on direct use of classical analytical formulas for stress and strain and a direct use of the actual working loads (called *service loads*) assumed for the structure. In fact, it is often necessary to explain the analytical methods of the LRFD method by comparing them to those used for the ASD method. The advantage of the LRFD method is a more uniform reliability of structures subjected to many types of loading and, typically, a selection of more economical sections.

The ASD method for steel design is similar to the method for wood design demonstrated earlier in this book, where the stress experienced by the structure under maximum service loading needs to be less or equal to an allowable stress. For steel structures the allowable stress is equal to the yield stress of the member divided by a factor of safety. The factors of safety vary for different types of steel members and are based largely on years of practical design experience.

The LRFD method for steel design is based upon the theory that when a structural member is loaded, it responds by resisting that load. Failure of the member occurs when either the load is larger than what was designed for or the resistance of the member is smaller than that anticipated. Both the loading and resistance is variable and difficult to accurately predict; therefore, the LRFD method uses probability-based rationale to help predict both the loading upon and resistance of the members. The design strength of the member must be equal to or greater than the resistance required due to the actual loading. This is expressed mathematically as:

$$\phi R_n \geq \Sigma \delta_i Q_i$$

where ϕ = resistance factor ($\phi < 1$)
R_n = nominal resistance of the member
δ_i = load factor ($\delta_i > 1$)
Q_i = different load effects

The chief source of information for design of steel structures—the American Institute of Steel Construction (AISC)—presently publishes two separate and complete references for the purposes of supporting both the ASD and LRFD methods. These references are used extensively throughout this book, with considerable presentation of sample materials from the LRFD reference (Ref. 54). AISC is publishing a new specification in 2005 that will combine the two references. The new specification will cover both the ASD and the LRFD methods.

This situation—with alternative methods—is understandably confusing but is a product of our times that readers should at least become familiar with. For practical purposes, the ASD method should be viewed as a somewhat simpler involvement for learning of basic issues, and the LRFD method should be considered the one that produces more economical steel sections. The LRFD method is becoming the industry standard for steel design and therefore is the method that will largely be demonstrated in this book.

8.2 MATERIALS FOR STEEL PRODUCTS

The strength, hardness, corrosion resistance, and some other properties of steel can be varied through a considerable range by changes in the production processes. Literally hundreds of different steels are produced, although only a few standard products are used for the majority of the elements of building structures. Working and forming processes, such as rolling, drawing, machining, and forging, may also alter some properties. However, certain properties, such as density (unit weight), stiffness (modulus of elasticity), thermal expansion, and fire resistance, tend to remain constant for all steels.

For various applications, other properties will be significant. Hardness effects the ease with which cutting, drilling, planing, and other working can be done. For welded connections, the weldability of the base material must be considered. Resistance to rusting is normally low but can be enhanced by various materials added to the steel, producing various types of special steels, such as stainless steel and the so-called "rusting steels" that rust at a very slow rate.

These various properties of steel must be considered when working with the material and when designing for its use. However, in this book, we are most concerned with the unique structural nature of steel.

Structural Properties of Steel

Basic structural properties, such as strength, stiffness, ductility, and brittleness, can be interpreted from laboratory load tests on specimens of the material. Figure 8.1 displays characteristic forms of curves that are ob-

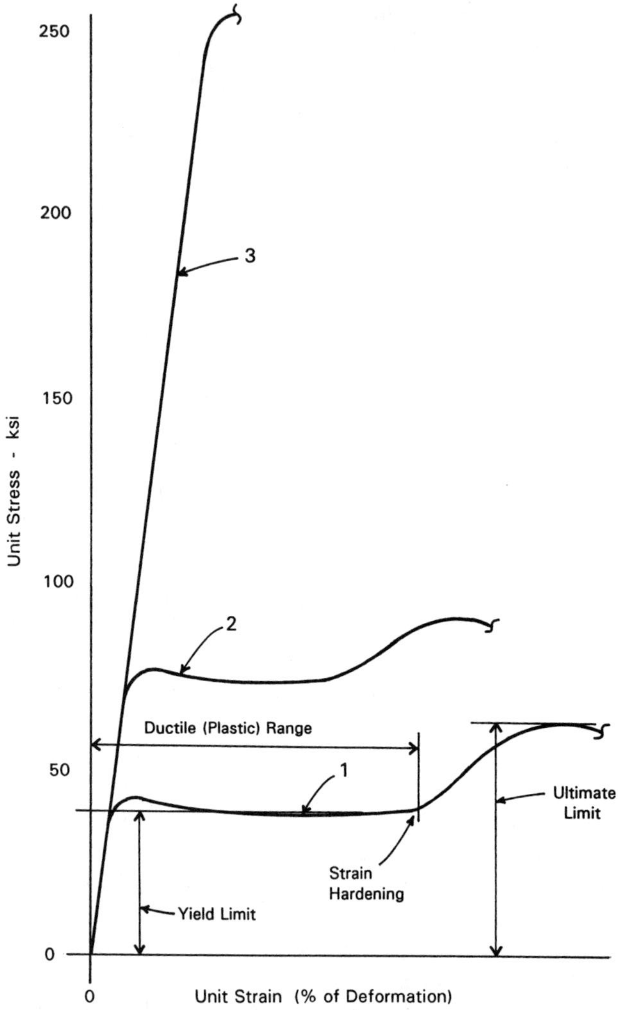

Figure 8.1 Stress/strain response of steel: 1) ordinary structural steel; 2) high-strength steel for rolled shapes; 3) super-strength steel—mostly in wire form.

tained by plotting stress and strain values from such tests. An important property of many structural steels is the plastic deformation (yield) phenomenon. This is demonstrated by curve 1 in Figure 8.1. For steels with this character, there are two different stress values of significance: the yield limit and the ultimate failure limit.

Generally, the higher the yield limit, the less the degree of ductility. The extent of ductility is measured as the ratio of the plastic deformation between first yield and strain hardening (see Figure 8.1) to the elastic deformation at the point of yield. Curve 1 in Figure 8.1 is representative of ordinary structural steel (ASTM A36), and curve 2 indicates the typical effect as the yield strength is raised a significant amount. Eventually, the significance of the yield phenomenon becomes virtually negligible when the yield strength approaches as much as three times the yield of ordinary steel (36 ksi for ASTM A36 steel).

Some of the highest-strength steels are produced only in thin-sheet or drawn-wire forms. Bridge strand is made from wire with strength as high as 300,000 psi. At this level, yield is almost nonexistent and the wires approach the brittle nature of glass rods.

For economical use of the expensive material, steel structures are generally composed of elements with relatively thin parts. This results in many situations in which the ultimate limiting strength of elements in bending, compression, and shear is determined by buckling, rather than the stress limits of the material. Because buckling is a function of stiffness (modulus of elasticity) of the material, and because this property remains the same for all steels, there is limited opportunity to make effective use of higher-strength steels in many situations. The grades of steel most commonly used are to some extent ones that have the optimal effective strength for most typical tasks.

Because many structural elements are produced as some manufacturer's product line, choices of basic materials are often mostly out of the hands of individual building designers. The proper steel for the task—on the basis of many properties—is determined as part of the product design, although a range of grades may be obtained for some products.

Steel that meets the requirements of the American Society for Testing and Materials (ASTM) Specification A36 is the grade of structural steel most commonly used to produce rolled steel elements for building construction. It must have an ultimate tensile strength of 58 to 80 ksi and a minimum yield point of 36 ksi. It may be used for bolted, riveted, or welded fabrication. This is the steel used for much of the work in this book, and it is referred to simply as A36 steel.

High-strength steel members are gaining popularity in today's construction market. Steel is designated high-strength when its yield stress is greater than 50 ksi, and because of this, it is sometimes referred to as 50-ksi steel. Traditionally, high-strength steel was used when the savings of weight or added durability was important. Today it is available mainly because steel produced by electric arc furnaces is made largely of recycled steel and one of the outcomes of using recycled steel is that the steel's yield stress increases. Electric arc furnaces produce the steel for all domestically made wide flange shapes.

Other Uses of Steel

Steel used for other purposes than the production of rolled products generally conforms to standards developed for the specific product. This is especially true for steel connectors, wire, cast and forged elements, and very high strength steels produced in sheet, bar, and rod form for fabricated products. The properties and design stresses for some of these product applications are discussed in other sections of this book. Standards used typically conform to those established by industry-wide organizations, such as the Steel Joist Institute (SJI) and the Steel Deck Institute (SDI). In some cases, larger fabricated products make use of ordinary rolled products, produced from A36 steel or other grades of steel from which hot-rolled products can be obtained.

8.3 TYPES OF STEEL STRUCTURAL PRODUCTS

Steel itself is formless, used basically for production as a molten material or a heat- softened lump. The structural products produced derive their basic forms from the general potentialities and limitations of the industrial processes of forming and fabricating. A major process used for structural products is that of *hot-rolling,* which is used to produce the familiar cross-sectional forms (called *shapes*) of I, H, L, T, U, C, and Z, as well as flat plates and round or square bars. Other processes include drawing (used for wire), extrusion, casting, and forging.

Raw stock can be assembled by various means into objects of multiple parts, such as a manufactured truss or a prefabricated wall panel, or a whole building framework. In the assemblage process, various connecting means or devices may be employed. Learning to design with steel begins with acquiring some familiarity with the standard industrial products and with the processes of reforming them and attaching them to other elements in structural assemblages.

Rolled Structural Shapes

The products of the steel rolling mills used as beams, columns, and other structural members are designated as *sections* or *shapes*, and their usages are related to the profiles of their cross sections. American standard I-beams (Figure 8.2*a*) were the first beam sections rolled in the United

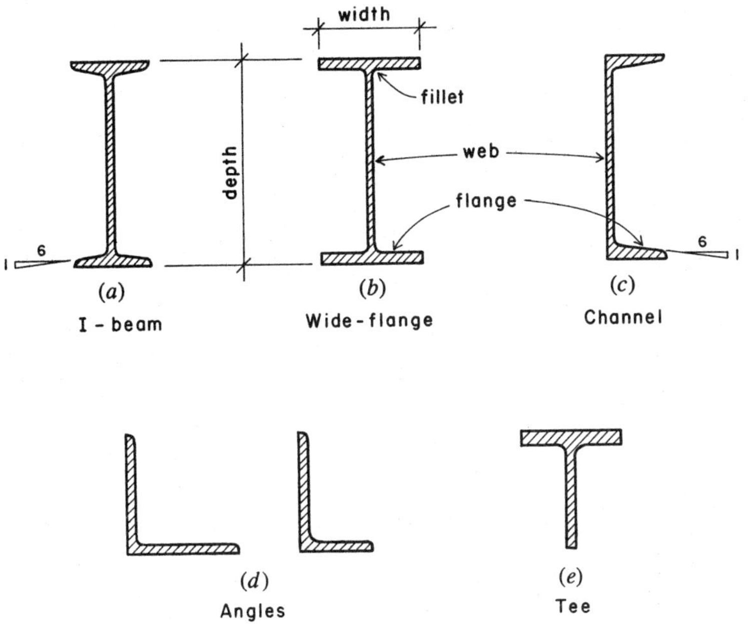

Figure 8.2 Shapes of typical hot-rolled products.

States and are currently produced in sizes of 3 to 24 in. in depth. The W shapes (Figure 8.2*b*, originally called wide-flange shapes) are a modification of the I cross section and are characterized by parallel flange surfaces as contrasted with the tapered inside flange surfaces of standard I-beams; they are available in depths of 4 to 44 in. In addition to the standard I and W sections, the structural steel shapes most commonly used in building construction are channels, angles, tees, plates, and bars. The tables in the appendix list the dimensions, weights, and various properties of some of these shapes. Complete tables of structural shapes are given in the AISC Manual (Ref.5).

W Shapes

In general, W shapes have greater flange widths and relatively thinner webs than standard I-beams; and, as noted earlier, the inner faces of the flanges are parallel to the outer faces. These sections are identified by the alphabetical symbol W, followed by the *nominal* depth in inches and the weight in pounds per linear foot. Thus, the designation W 12 × 26 indicates a wide-flange shape of nominal-12-in. depth, weighing 26 lb per linear foot (plf).

The actual depths of wide-flange shapes vary within the nominal depth groupings. From Table A.3, we know that a W 12 × 26 has an actual depth of 12.22 in., whereas the depth of a W 12 × 30 is 12.34 in. This is a result of the rolling process during manufacture in which the cross-sectional areas of wide-flange shapes are increased by spreading the rollers both vertically and horizontally. The additional material is thereby added to the cross section by increasing flange and web thickness as well as flange width (Figure 8.2*b*). The resulting higher percentage of material in the flanges makes wide-flange shapes more efficient for bending resistance than standard I-beams. A wide variety of weights is available within each nominal depth group.

In addition to shapes with profiles similar to the W 12 × 26, which has a flange width of 6.490 in., many wide-flange shapes are rolled with flange widths approximately equal to their depths. The resulting H configurations of these cross sections are much more suitable for use as columns than the I profiles. From Table A.3, we know that the following shapes, among others, fall into this category: W 14 × 120, W 12 × 65, and W 10 × 49.

Cold-Formed Steel Products

Sheet steel can be bent, punched, or rolled into a variety of forms. Structural elements so formed are called *cold-formed* or *light-gage* steel products. Steel decks and small framing elements are produced in this manner. These products are described in Chapter 12.

Fabricated Structural Components

A number of special products are formed of both hot-rolled and cold-formed elements for use as structural members in buildings. Open-web steel joists consist of prefabricated, light steel trusses. For short spans and light loads, a common design is that shown in Figure 8.3*a* in which the web consists of a single, continuous bent steel rod and the chords of steel rods or cold-formed elements. For larger spans or heavier loads, the forms more closely resemble those for ordinary light steel trusses; single angles, double angles, and structural tees constitute the truss members. Open-web joists for floor framing are discussed in Section 9.10.

Another type of fabricated joist is shown in Figure 8.3*b*. This member is formed from standard rolled shapes by cutting the web in a zigzag fashion. The resulting product has a greatly reduced weight-to-depth ratio when compared with the lightest of the rolled shapes.

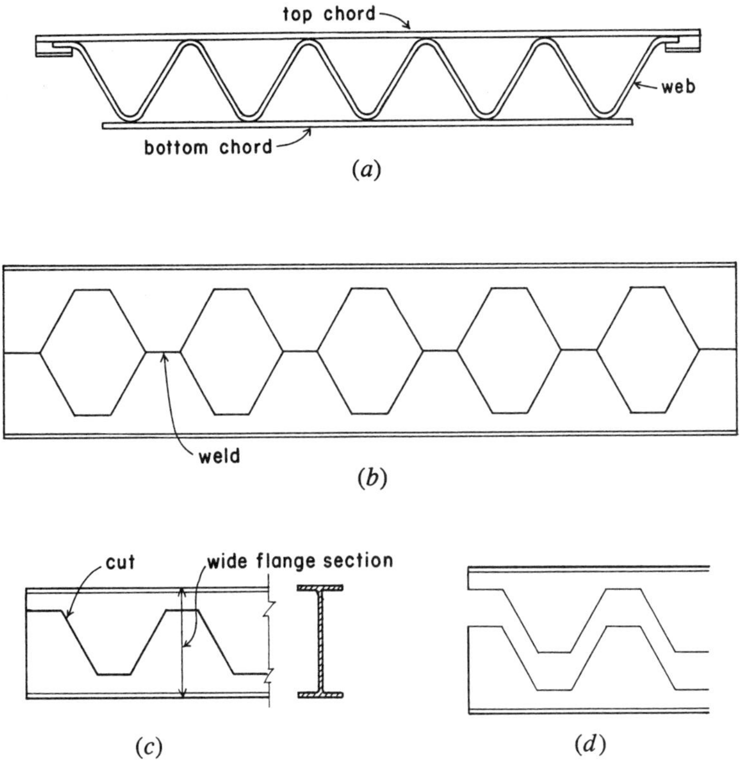

Figure 8.3 Fabricated products formed from steel elements.

Other fabricated steel products range from those used to produce whole building systems to individual elements for construction of windows, doors, curtain wall systems, and the framing for interior partition walls. Many components and systems are produced as proprietary items by a single manufacturer, although some are developed under controls of industry-wide standards, such as those published by the Steel Joist Institute and Steel Deck Institute. Some structural products in this category are discussed and illustrated in the development of the building system design examples in Part V.

Development of Structural Systems

Structural systems that comprise entire roof, floor, or wall constructions—or even entire buildings—are typically assembled from many individual elements. These individual elements may be of some variety, as in the case of the typical floor using rolled steel shapes for beams and a formed sheet steel deck. Selection of individual elements is often done with some data from structural investigations but is also often largely a matter of practical development of the form of the construction.

It is common for a building to incorporate more than a single material for its entire structural system. Various combinations, including a wood deck on steel beams, or masonry bearing walls for support of a steel spanning floor or roof structure, are possible. This part of the book deals primarily with structures of steel, but some of these mixed material situations are very common and are discussed in other parts of this book.

Connection Methods

Connection of structural steel members that consist of rolled elements is typically achieved by direct welding or by steel rivets or bolts. Riveting for building structures has become generally obsolete in favor of high-strength bolts. However, the forms of structural members, and even of connections, that were developed for riveted construction are still used with little change for bolted construction. The design of bolted connections is discussed in Chapter 11. In general, welding is preferred for shop (factory) fabrication and bolting for field (construction site) connections.

Thin elements of cold-formed steel may be attached by welding, bolting, or using sheet metal screws. Thin deck and wall paneling elements are sometimes attached to one another by simple interlocking at their abutting edges; the interlocked parts can be folded or crimped to give further security to the connection.

Adhesives or sealants may be used to seal joints or to bond thin sheet materials in laminated fabrications. Some elements used in connections may be attached to connected parts by adhesion to facilitate the work of fabrication and erection, but adhesive connection is not used for major structural joints.

A frequent structural design problem is that of the connections of columns and beams in heavy frames for multistory buildings. For rigid frame action to resist lateral loads, these connections are achieved with very large welds, generally developed to transfer the full strength of the connected members. Design of these connections is beyond the scope of this book, although lighter framing connections of various form are discussed in Chapter 11.

9

STEEL BEAMS AND FRAMING ELEMENTS

There are many steel elements that can be used for the basic functions of spanning, including rolled shapes, cold-formed shapes, and fabricated beams and trusses. This chapter deals with fundamental considerations for these elements, with an emphasis on rolled shapes. For simplicity, it is assumed that all the rolled shapes used for the work in this chapter are of steel with F_y = 36 ksi [250 MPa] or F_y = 50 ksi [350 MPa].

9.1 FACTORS IN BEAM DESIGN

Various rolled shapes may serve beam functions, although the most widely used shape is the *wide flange shape*; that is, the member with an I-shaped cross section that bears the standard designation of W shape. Except for those members of the W series that approach a square in cross section (flange width approximately equal to nominal depth), the proportions of members in this series are developed for optimal use in flexure about their major axis (designated as x-x). Design for beam use may involve any combination of the following considerations:

Flexural stress. Flexural stresses generated by bending moments are the primary stress concern in beams. There are several failure modes for steel beams that define our approach to designing with them, but the general equation for the design of bending members is

$$\phi_b M_n \geq M_u$$

where ϕ_b = 0.9 for rolled sections
M_n = Nominal moment capacity of the member
M_u = Maximum moment due to the factored loading

Shear stress. Whereas shear stress is quite critical in wood and concrete beams, it is less often a problem in steel beams, except for situations where buckling of the thin beam web may be part of a general buckling failure of the cross section. For beam shear alone, the actual critical stress is the diagonal compression, which is the direct cause of buckling in beam action. The general equation for the design for shear is

$$\phi_v V_n \geq V_u$$

where ϕ_v = 0.9 for rolled sections
V_n = Nominal shear capacity of the member
V_u = Maximum shear due to the factored loading

Buckling. In general, beams that are not adequately braced may be subject to various forms of buckling. Especially critical are beams with very thin webs or narrow flanges or with cross sections especially weak in the lateral direction (on the minor, or y-y axis). Buckling controls the failure mechanism in inadequately braced members and greatly reduces bending capacity. The most effective solution is to provide adequate bracing to eliminate this mode of failure.

Deflection. Although steel is the stiffest material used for ordinary structures, steel structures tend to be quite flexible; thus, vertical deflection of beams must be carefully investigated. A significant value to monitor is the span-to-depth ratio of beams; if this is kept within certain limits, deflection is much less likely to be critical.

Connections and supports. Framed structures contain many joints between separate pieces, and the details of the connections and supports must be developed for proper construction as well as for the transfer of necessary structural forces through the joints.

Individual beams are often parts of a system in which they play an interactive role. Besides their basic beam functions, there are often design considerations that derive from the overall system actions and interactions. Discussions in this chapter focus mostly on individual beam actions, but discussions in many other chapters treat the usage and overall incorporation of beams in structural systems.

There are several hundred different W shapes for which properties are listed in the AISC manuals (Table A.3). In addition, there are several other shapes that frequently serve beam functions in special circumstances. Selection of the optimal shape for a given situation involves many considerations; an overriding consideration is often the choice of the most economical shape for the task. In general, the least costly shape is usually the one that weighs the least—other things being equal, because steel is priced by unit weight. In most design cases, therefore, the *least weight* selection is typically considered the most economical. Just as a beam may be asked also to develop other actions, such as tension, compression, or torsion, other structural elements may also develop beam actions. Walls may span for bending against wind pressures, columns may receive bending moments as well as compression loads, and truss chords may span as beams as well as function for basic truss actions. The basic beam functions described in this chapter may thus be part of the design work for various structural elements besides the real, singular-purpose beam.

9.2 INELASTIC VERSUS ELASTIC BEHAVIOR

As discussed in Chapters 4 and 8, there are two competing methods of design for steel, allowable stress design (ASD) and load and resistance factor design (LRFD). ASD in steel is very similar to ASD for wood discussed in Part II of this text. This text has adopted LRFD for steel because it has become the standard for the construction industry. The basic difference between these two methods is rooted in the distinction between elastic or inelastic theory of member failure. ASD is rooted in elastic theory, and LRFD is rooted in inelastic theory. The purpose of this section is to compare these two theories so one can better understand inelastic theory and, therefore, the LRFD method of design.

The maximum resisting moment by elastic theory is predicted to occur when the stress at the extreme fiber reaches the elastic yield value, F_y, and it may be expressed as

$$M_y = F_y \times S$$

Beyond this condition the resisting moment can no longer be expressed by elastic theory equations because an inelastic, or *plastic*, stress condition will start to develop on the beam cross section.

Figure 9.1 represents an idealized form of a load-test response for a specimen of ductile steel. The graph shows that up to the yield point the deformations are proportional to the applied stress and that beyond the yield point there is a deformation without an increase in stress. For A36 steel, this additional deformation, called the *plastic range*, is approximately 15 times that produced just before yield occurs. This relative magnitude of the plastic range is the basis for qualification of the material as significantly ductile.

Note that beyond the plastic range the material once again stiffens, called the *strain hardening* effect, which indicates a loss of the ductility and the onset of a second range in which additional deformation is produced only by additional increase in stress. The end of this range establishes the *ultimate stress* limit for the material.

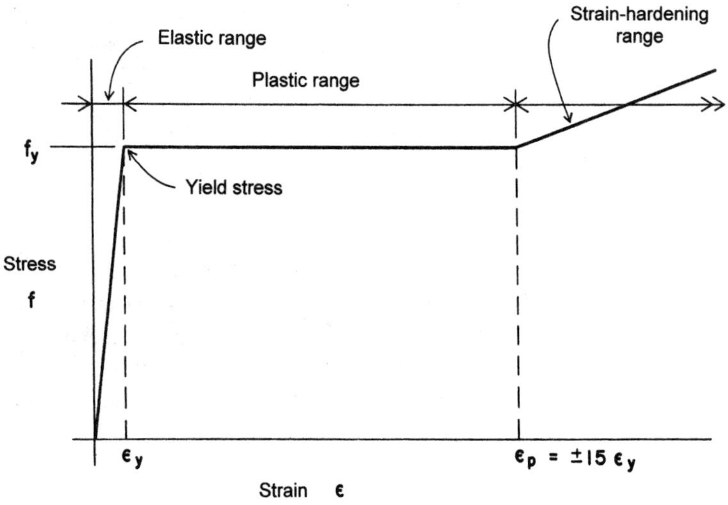

Figure 9.1 Idealized form of the stress-strain response of ductile steel.

The following example illustrates the application of the elastic theory and will be used for comparison with an analysis of plastic behavior.

Example 1. A simple beam has a span of 16 ft [4.88 m] and supports a single concentrated load of 18 kips [80 kN] at its center. If the beam is a W 12 × 30, compute the maximum flexural stress.

Solution: See Figure 9.2. For the maximum value of the bending moment

$$M = \frac{PL}{4} = \frac{18 \times 16}{4} = 72 \text{ kip-ft [98 kNm]}$$

In Table A.3 find the value of S for the shape as 38.6 in.^{3} [632 × 10^3 mm^3]. Thus, the maximum stress is

$$f = \frac{M}{S} = \frac{72 \times 12}{38.6} = 22.4 \text{ ksi [154 MPa]}$$

and it occurs as shown in Figure 9.2*d*. Note that this stress condition occurs only at the beam section at midspan. Figure 9.2*e* shows the form of the deformations that accompany the stress condition. This stress level is well below the elastic stress limit (yield point).

The limiting moment for elastic stress is that which occurs when the maximum flexural stress reaches the yield limit, as stated before in the expression for M_y. This condition is illustrated by the stress diagram in Figure 9.3*a*.

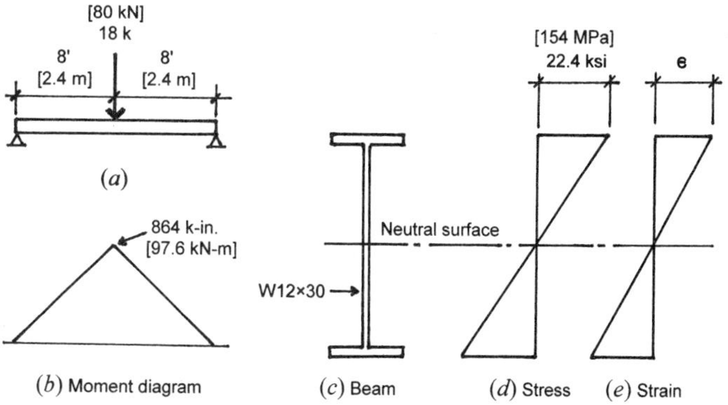

Figure 9.2 Elastic behavior of the beam.

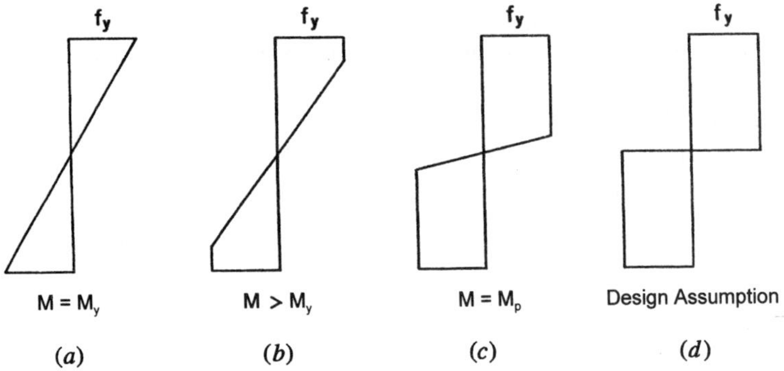

Figure 9.3 Progression of development of flexural stress, from the elastic to the plastic range.

If the loading (and the resulting bending moment) that causes the yield limit flexural stress is increased, a stress condition like that illustrated in Figure 9.3*b* begins to develop as the ductile material deforms plastically. This spread of the higher stress level over the beam cross section indicates the development of a resisting moment in excess of M_y. With a high level of ductility, a limit for this condition takes a form as shown in Figure 9.3*c*, and the limiting resisting moment is described as the *plastic moment*, designated M_p. Although a small percentage of the cross section near the beam's neutral axis remains in an elastic stress condition, its effect on the development of the resisting moment is quite negligible. Thus, it is assumed that the full plastic limit is developed by the condition shown in Figure 9.3*d*.

Attempts to increase the bending moment beyond the value of M_p will result in large rotational deformation, with the beam acting as though it were hinged (pinned) at this location. For practical purposes, therefore, the resisting moment capacity of the ductile beam is considered to be exhausted with the attaining of the plastic moment; additional loading will merely cause a free rotation at the location of the plastic moment. This location is thus described as a *plastic hinge* (see Figure 9.4), and its effect on beams and frames will be discussed further.

In a manner similar to that for elastic stress conditions, the value of the resisting plastic moment is expressed as

$$M_p = F_y \times Z$$

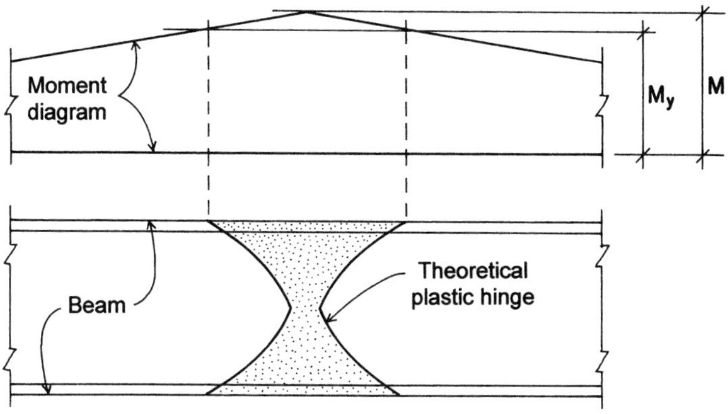

Figure 9.4 Development of the plastic hinge.

The term Z is called the *plastic section modulus*, and its value is determined as follows:

Refer to Figure 9.5, which shows a W shape subjected to a level of flexural stress corresponding to the fully plastic section (Figure 9.3*d* and 9.4), and note the following:

A_u = upper area of the cross section, above the neutral axis
y_u = distance of the centroid of A_u from the neutral axis
A_l = lower area of the cross section, below the neutral axis
y_l = distance of the centroid of A_l from the neutral axis

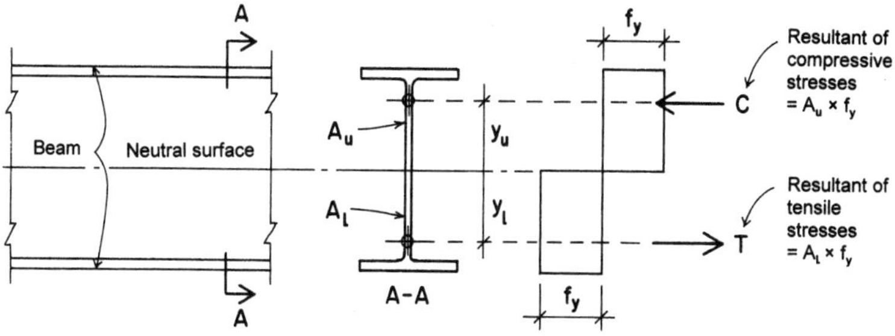

Figure 9.5 Development of the plastic resisting moment.

For equilibrium of the internal forces on the cross section (the resulting forces C and T developed by the flexural stresses), the condition can be expressed as

$$\Sigma F_h = 0$$

or

$$[A_u \times (+f_y)] + [A_l \times (-f_y)] = 0$$

and, thus:

$$A_u = A_l$$

This shows that the plastic stress neutral axis divides the cross section into equal areas, which is apparent for symmetrical sections, but it applies to unsymmetrical sections as well. The resisting moment equals the sum of the moments of the stresses; thus, the value for M_p may be expressed as

$$M_p = C \times y_u + T \times y_l$$

or

$$M_p = (A_u \times f_y \times y_u) + (A_l \times f_y \times y_l)$$

or

$$M_p = f_y \, [(A_u \times y_u) + (A_l \times y_l)]$$

or

$$M_p = f_y \times Z$$

and the quantity $[(A_u \times y_u) + (A_l \times y_l)]$ is the property of the cross section defined as the plastic section modulus, designated Z.

Using the expression for Z just derived, its value for any cross section can be computed. However, values of Z are tabulated in the AISC Manual (Ref. 5) for all rolled sections used as beams. See Table A.3 in the appendix.

Comparison of the values for S_x and Z_x for the same W shape will show that the values for Z are larger. This presents an opportunity to compare the fully plastic resisting moment to the yield stress limiting moment by elastic stress—that is, the advantage of using plastic analysis.

Example 2. A simple beam consisting of a W 21 × 57 is subjected to bending. Find the limiting moments (a) based on elastic stress conditions and a limiting stress of $F_y = 36$ ksi, and (b) based on full development of the plastic moment.

Solution: For (a) the limiting moment is expressed as

$$M_y = F_y \times S_x$$

From Table A.3, for the W 21 × 57, S_x is 111 in.^{3}, so the limiting moment is

$$M_y = (36) \times (111) = 3996 \text{ kip-in.} \quad \text{or} \quad \frac{3996}{12} = 333 \text{ kip-ft } [452 \text{ kN-m}]$$

For (b) the limiting plastic moment, using the value of $Z_x = 129$ in.^{3} from Table A.3, is

$$M_p = (36) \times (129) = 4644 \text{ kip-in.} \quad \text{or} \quad \frac{4644}{12} = 387 \text{ kip-ft } [525 \text{ kN-m}]$$

The increase in moment resistance represented by the plastic moment indicates an increase of 387 – 333 = 54 kip-ft, or a percentage gain of (54/333)(100) = 16.2 percent.

Problem 9.2.A. A simple-span, uniformly-loaded beam consists of a W 18 × 50 with $F_y = 36$ ksi. Find the percentage of gain in the limiting bending moment if a fully plastic condition is assumed, instead of a condition limited by elastic stress.

Problem 9.2.B. A simple-span, uniformly loaded beam consists of a W 16 × 45 with $F_y = 36$ ksi. Find the percentage of gain in the limiting bending moment if a fully plastic condition is assumed, instead of a condition limited by elastic stress.

9.3 NOMINAL MOMENT CAPACITY OF STEEL BEAMS

The first step in design for bending using LRFD is the determination of the bending capacity (M_n) of a steel section. The capacity of a steel section is based upon the cross-sectional properties, its yield stress, and bracing of the member from out of plane buckling. Each of these parameters affect how the beam will ultimately fail and thus how much capacity it will have for bending.

Ideally, the failure mode that every beam will be controlled by is the inelastic failure described in Section 9.2. If a member section is capable of failing in a plastic hinge, it is considered a "compact" cross section. A compact shape is one that meets the following criteria:

$$\frac{b_f}{2t_f} \leq \frac{65}{\sqrt{F_y}} \quad \text{and} \quad \frac{h_c}{t_w} \leq \frac{640}{\sqrt{F_y}}$$

where b_f = flange width in inches
t_f = flange thickness in inches
F_y = minimum yield stress in ksi
h_c = height of the web in inches
t_w = web thickness in inches

For A36 steel, this translates to

$$\frac{b_f}{2t_f} \leq 10.8 \quad \text{and} \quad \frac{h_c}{t_w} \leq 107$$

and for steel with a yield stress of 50 ksi it translates to

$$\frac{b_f}{2t_f} \leq 9.19 \quad \text{and} \quad \frac{h_c}{t_w} \leq 90.5$$

In the AISC manual (Ref. 5) and in Table 9.1, the ratios needed to determine compactness are computed for each structural shape, and in many of the AISC tables, the non-compact sections are clearly labeled. It should be noted that most shapes used for beams are compact by design.

Compact beams that are adequately laterally supported will fail with a plastic hinge, and, therefore, the moment capacity is the yield moment for the section.

$$M_n = M_p = F_y \times Z$$

Example 3. Determine the moment capacity of an A36 W 24 × 76 steel beam that is adequately laterally supported.

Solution: First, we check to make sure that it is a compact section. Compact section criteria is taken from Table 9.1.

$$\frac{b_f}{2t_f} \le \frac{65}{\sqrt{F_y}} \qquad \text{and} \qquad \frac{h}{t_w} \le \frac{640}{\sqrt{F_y}}$$

$$6.61 \le 10.8 \qquad\qquad 49 \le 107$$

The test for compactness was correct; therefore, the moment capacity will be equal to the plastic moment.

$$M_n = M_y = F_y \times Z = 36 \text{ ksi} \times 200 \text{ in.}^3 = 7200 \text{ kip-in.} = \frac{7200}{12}$$

$$= 600 \text{ kip-ft } [814 \text{ kN-m}]$$

To ensure that a compact section fails plastically, the maximum spacing between lateral supports of the beam (L_b) must be less than a limiting laterally unbraced length for full plastic flexural strength (L_p), which is defined as

$$L_p = \frac{300 \times r_y}{\sqrt{F_y}}$$

where r_y = radius of gyration about the y-axis in inches
F_y = minimum yield stress in ksi

If $L_b > L_p$, the beam will fail because of the buckling of the compression side of the beam similar to buckling in columns. The buckling added to the load on the beam causes a torisional failure of the beam (Figure 3.63). Plastic deformation continues on the cross section prior to buckling as long as the unbraced lateral length (L_b) does not exceed a limiting laterally unbraced length for inelastic lateral-torsional buckling (L_r). When $L_b > L_r$, the cross-section will fail in buckling with the section in the elastic range. The limiting laterally unbraced length for inelastic lateral-torsional buckling (L_r) is defined as

$$L_r = \left(\frac{r_y \times X_1}{F_y - F_r}\right) \times \sqrt{1 + \sqrt{1 + X_2 \times (F_y - F_r)^2}}$$

where r_y = radius of gyration about the y-axis in inches
X_1 = beam buckling factor
F_y = minimum yield stress in ksi
F_r = compressive residual stress in section = 10 ksi (for rolled sections)
X_2 = beam buckling factor

TABLE 9.1 Load Factor Resistance Design Selection for Shapes Used as Beams

Designation	Z_x in.^{3}	$F_y = 36$ ksi L_p ft	L_r ft	M_p kip-ft	M_r kip-ft	$F_y = 50$ ksi L_p ft	L_r ft	M_p kip-ft	M_r kip-ft	r_y in.	$b_f/2t_f$	h/t_w	X_1 ksi	$X_2 \times 10^6$ $(1/\text{ksi})^2$
W 33 × 141	**514**	**10.1**	**30.1**	**1,542**	**971**	**8.59**	**23.1**	**2,142**	**1,493**	**2.43**	**6.01**	**49.6**	**1,800**	**17,800**
W 30 × 148	500	9.50	30.6	1,500	945	8.06	22.8	2,083	1,453	2.28	4.44	41.6	2,310	6,270
W 24 × 162	468	12.7	45.2	1,404	897	10.8	32.4	1,950	1,380	3.05	5.31	30.6	2,870	2,260
W 24 × 146	418	12.5	42.0	1,254	804	10.6	30.6	1,742	1,237	3.01	5.92	33.2	2,590	3,420
W 33 × 118	**415**	**9.67**	**27.8**	**1,245**	**778**	**8.20**	**21.7**	**1,729**	**1,197**	**2.32**	**7.76**	**54.5**	**1,510**	**37,700**
W 30 × 124	408	9.29	28.2	1,224	769	7.88	21.5	1,700	1,183	2.23	5.65	46.2	1,930	13,500
W 21 × 147	373	12.3	46.4	1,119	713	10.4	32.8	1,554	1,097	2.95	5.44	26.1	3,140	1,590
W 24 × 131	370	12.4	39.3	1,110	713	10.5	29.1	1,542	1,097	2.97	6.70	35.6	2,330	5,290
W 18 × 158	356	11.4	56.5	1,068	672	9.69	38.0	1,483	1,033	2.74	3.92	19.8	4,410	403
W 30 × 108	**346**	**8.96**	**26.3**	**1,038**	**648**	**7.60**	**20.3**	**1,442**	**997**	**2.15**	**6.89**	**49.6**	**1,680**	**24,200**
W 27 × 114	343	9.08	28.2	1,029	648	7.71	21.3	1,429	997	2.18	5.41	42.5	2,100	9,220
W 24 × 117	327	12.3	37.1	981	631	10.4	27.9	1,363	970	2.94	7.53	39.2	2,090	8,190
W 21 × 122	307	12.2	41.0	921	592	10.3	29.8	1,279	910	2.92	6.45	31.3	2,630	3,160
W 18 × 130	290	11.3	47.7	870	555	9.55	32.8	1,208	853	2.7	4.65	23.9	3,680	810
W 30 × 90	**283**	**8.71**	**24.8**	**849**	**531**	**7.39**	**19.4**	**1,179**	**817**	**2.09**	**8.52**	**57.5**	**1,410**	**49,600**
W 24 × 103	280	8.29	27.0	840	531	7.04	20.0	1,167	817	1.99	4.59	39.2	2,390	5,310
W 27 × 94	278	8.83	25.9	834	527	7.50	19.9	1,158	810	2.12	6.70	49.5	1,740	19,900
W 14 × 145	260	16.6	81.6	780	503	14.1	54.7	1,083	773	3.98	7.11	16.8	4,400	348
W 24 × 94	254	8.25	25.9	762	481	7.00	19.4	1,058	740	1.98	5.18	41.9	2,180	7,800

W 21 × 101	253	12.0	37.1	759	492	10.2	27.6	1,054	757	2.89	7.68	37.5	2,200	6,400
W 12 × 152	243	13.3	94.8	729	453	11.3	62.1	1,013	697	3.19	4.46	11.2	6,510	79
W 18 × 106	230	11.1	40.4	690	442	9.40	28.7	958	680	2.66	5.96	27.2	2,990	1,880
W 14 × 120	212	15.6	67.9	636	412	13.2	46.2	883	633	3.74	7.80	19.3	3,830	601
W 24 × 76	**200**	**8.00**	**23.4**	**600**	**381**	**6.79**	**18.0**	**833**	**587**	**1.92**	**6.61**	**49.0**	**1,760**	**18,600**
W 16 × 100	200	10.4	42.7	600	384	8.84	29.6	833	590	2.5	5.29	23.2	3,530	947
W 21 × 83	196	7.63	24.9	588	371	6.47	18.5	817	570	1.83	5.00	36.4	2,400	5,250
W 18 × 86	186	11.0	35.5	558	360	9.30	26.1	775	553	2.63	7.20	33.4	2,460	4,060
W 12 × 120	186	13.0	75.5	558	353	11.1	50.0	775	543	3.13	5.57	13.7	5,240	184
W 21 × 68	**160**	**7.50**	**22.8**	**480**	**303**	**6.36**	**17.3**	**667**	**467**	**1.8**	**6.04**	**43.6**	**2,000**	**10,900**
W 24 × 62	**154**	**5.71**	**17.2**	**462**	**286**	**4.84**	**13.3**	**642**	**440**	**1.37**	**5.97**	**49.7**	**1,730**	**23,800**
W 16 × 77	152	10.3	35.4	456	295	8.70	25.5	633	453	2.46	6.77	29.9	2,770	2,460
W 12 × 96	147	12.9	61.4	441	284	10.9	41.4	613	437	3.09	6.76	17.7	4,250	407
W 10 × 112	147	11.2	86.4	441	273	9.48	56.5	613	420	2.68	4.17	10.4	7,080	57
W 18 × 71	146	7.08	24.5	438	275	6.01	17.8	608	423	1.7	4.71	32.4	2,690	3,290
W 14 × 82	139	10.3	42.8	417	267	8.77	29.5	579	410	2.48	5.92	22.4	3,590	849
W 24 × 55	**135**	**5.58**	**16.6**	**405**	**249**	**4.74**	**12.9**	**563**	**383**	**1.34**	**6.94**	**54.1**	**1,570**	**36,500**
W 21 × 57	129	5.63	17.3	387	241	4.77	13.1	538	370	1.35	5.04	46.3	1,960	13,100
W 18 × 60	123	7.00	22.3	369	234	5.94	16.6	513	360	1.68	5.44	38.7	2,290	6,080
W 12 × 79	119	12.7	51.8	357	232	10.8	35.7	496	357	3.05	8.22	20.7	3,530	839
W 14 × 68	115	10.3	37.3	345	223	8.70	26.4	479	343	2.46	6.97	27.5	3,020	1,660
W 10 × 88	113	11.0	68.4	339	213	9.30	45.1	471	328	2.63	5.18	13.0	5,680	132

(continued)

TABLE 9.1 (*Continued*)

Designation	Z_x in.^{3}	$F_y = 36$ ksi L_p ft	L_r ft	M_p kip-ft	M_r kip-ft	$F_y = 50$ ksi L_p ft	L_r ft	M_p kip-ft	M_r kip-ft	r_y in.	$b_f/2t_f$	h/t_w	X_1 ksi	$X_2 \times 10^6$ $(1/\text{ksi})^2$
W 21 × 50	**110**	**5.42**	**16.2**	**330**	**205**	**4.60**	**12.5**	**458**	**315**	**1.3**	**6.10**	**49.4**	**1,730**	**22,600**
W 16 × 57	105	6.67	22.8	315	200	5.66	16.6	438	307	1.6	4.98	33.0	2,650	3,400
W 18 × 50	**101**	**6.88**	**20.5**	**303**	**193**	**5.83**	**15.6**	**421**	**296**	**1.65**	**6.57**	**45.2**	**1,920**	**12,400**
W 21 × 44	**95.4**	**5.25**	**15.4**	**286**	**177**	**4.45**	**12.0**	**398**	**272**	**1.26**	**7.22**	**53.6**	**1,550**	**36,600**
W 18 × 46	90.7	5.38	16.6	272	171	4.56	12.6	378	263	1.29	5.01	44.6	2,060	10,100
W 14 × 53	87.1	8.00	28.0	261	169	6.79	20.1	363	259	1.92	6.11	30.9	2,830	2,250
W 10 × 68	85.3	10.8	53.7	256	164	9.16	36.0	355	252	2.59	6.58	16.7	4,460	334
W 16 × 45	82.3	6.54	20.2	247	158	5.55	15.2	343	242	1.57	6.23	41.1	2,120	8,280
W 18 × 40	**78.4**	**5.29**	**15.7**	**235**	**148**	**4.49**	**12.1**	**327**	**228**	**1.27**	**5.73**	**50.9**	**1,810**	**17,200**
W 12 × 53	77.9	10.3	35.8	234	153	8.77	25.6	325	235	2.48	8.69	28.1	2,820	2,100
W 14 × 43	69.6	7.88	24.7	209	136	6.68	18.3	290	209	1.89	7.54	37.4	2,330	4,880
W 10 × 54	66.6	10.7	43.9	200	130	9.05	30.2	278	200	2.56	8.15	21.2	3,580	778
W 12 × 45	64.2	8.13	28.3	193	125	6.89	20.3	268	192	1.95	7.00	29.6	2,820	2,210

W 16 × 36	**64.0**	**6.33**	**18.2**	**192**	**122**	**5.37**	**14.0**	**267**	**188**	**1.52**	**8.12**	**48.1**	**1,700**	**20,400**
W 10 × 45	54.9	8.38	35.1	165	106	7.11	24.1	229	164	2.01	6.47	22.5	3,650	758
W 14 × 34	**54.6**	**6.38**	**19.0**	**164**	**105**	**5.41**	**14.4**	**228**	**162**	**1.53**	**7.41**	**43.1**	**1,970**	**10,600**
W 12 × 35	51.2	6.42	20.7	154	99	5.44	15.2	213	152	1.54	6.31	36.2	2,430	4,330
W 16 × 26	**44.2**	**4.67**	**13.4**	**133**	**83**	**3.96**	**10.4**	**184**	**128**	**1.12**	**7.97**	**56.8**	**1,480**	**40,300**
W 14 × 26	**40.2**	**4.50**	**13.4**	**121**	**76**	**3.82**	**10.2**	**168**	**118**	**1.08**	**5.98**	**48.1**	**1,880**	**14,100**
W 10 × 33	38.8	8.08	27.5	116	76	6.86	19.8	162	117	1.94	9.15	27.1	2,720	2,480
W 12 × 26	**37.2**	**6.29**	**18.1**	**112**	**72**	**5.34**	**13.8**	**155**	**111**	**1.51**	**8.54**	**47.2**	**1,820**	**13,900**
W 10 × 26	**31.3**	**5.67**	**18.6**	**94**	**60**	**4.81**	**13.6**	**130**	**93**	**1.36**	**6.56**	**34.0**	**2,510**	**3,760**
W 12 × 22	**29.3**	**3.53**	**11.2**	**88**	**55**	**3.00**	**8.41**	**122**	**85**	**0.848**	**4.74**	**41.8**	**2,170**	**8,460**
W 10 × 19	**21.6**	**3.64**	**12.0**	**65**	**41**	**3.09**	**8.89**	**90**	**63**	**0.874**	**5.09**	**35.4**	**2,440**	**5,030**

Source: Compiled from data in the *Manual of Steel Construction* with permission of the publishers, American Institute of Steel Construction.

Example 4. Determine the limiting lateral lengths L_p and L_r for an A36 W 24 × 76 steel beam.

Solution: From Table 9.1 for a W 24 × 76:

$$r_y = 1.92 \text{ in.}$$
$$X_1 = 1760 \text{ ksi}$$
$$X_2 = 0.0186\ (1/\text{ksi})^2$$

$$L_p = \frac{300 \times r_y}{\sqrt{F_y}} = \frac{300 \times 1.92 \text{ in.}}{\sqrt{36 \text{ ksi}}} = 96 \text{ in.} = 8 \text{ ft } [2.44 \text{ m}]$$

$$L_r = \left(\frac{r_x \times X_1}{F_y - F_r}\right) \times \sqrt{1 + \sqrt{1 + X_2 \times (F_y - F_r)^2}}$$
$$= \left(\frac{1.92 \text{ in.} \times 1760 \text{ ksi}}{36 \text{ ksi} - 10 \text{ ksi}}\right)$$
$$\times \sqrt{1 + \sqrt{1 + 0.0186\ (1/\text{ksi})^2 \times (36 \text{ ksi} - 10 \text{ ksi})^2}}$$
$$= 281 \text{ in.} = 23.4 \text{ ft } [7.14 \text{ m}]$$

Knowing the relationship between the unbraced length (L_b) and the limiting lateral unbraced lengths (L_p and L_r), we can determine the nominal moment capacity (M_n) for any beam. Figure 9.6 shows the form of the relation between M_n and L_b using an example of a W 18 ×50 with F_y = 50 ksi. The AISC Manual (Ref. 3) contains a series of such graphs for shapes commonly used as beams.

If $L_b \leq L_p$:

$$M_n = M_p = F_y \times Z$$

If $L_p < L_b \leq L_r$:

$$M_n = M_p - (M_p - M_r) \times \left(\frac{L_b - L_p}{L_r - L_p}\right)$$

where $M_r = (F_y - F_r)\, S_x$
F_r = compressive residual stress = 10 ksi for rolled shapes
S_x = section modulus about x-axis in in.^{3}

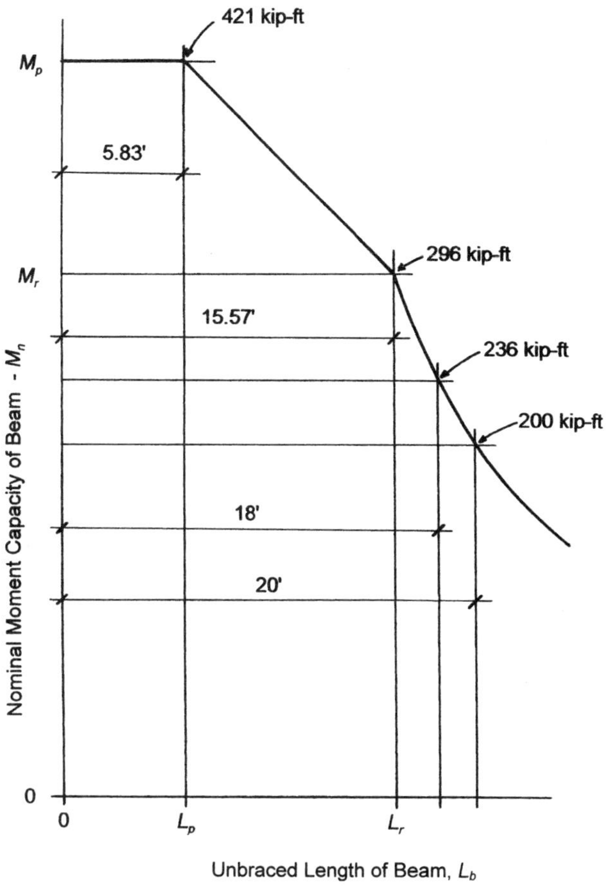

Figure 9.6 Relation between nominal moment capacity M_n and unbraced length L_b for a W 18 × 50 steel beam with F_y = 50 ksi.

If $L_b > L_r$

$$M_n = \left(\frac{S_x \times X_1 \times \sqrt{2}}{L_b/r_y}\right) \times \sqrt{1 + \frac{(X_1)^2 \times X_2}{2 \times (L_b/r_y)^2}}$$

where X_1 and X_2 = beam buckling factors

r_y = radius of gyration about the y axis

Example 5. Determine the moment capacity of an A36 W 24 × 76 steel beam that is laterally supported every 10 ft.

Solution: $(L_p = 8\text{ ft}) < (L_b = 10\text{ ft}) < (L_r = 23.4\text{ ft})$ (from Example 4)

Determine M_r:

$$M_r = (F_y - F_r) \times S_x = (36\text{ ksi} - 10\text{ ksi}) \times 176\text{ in.}^3$$
$$= 4576\text{ kip-in.} = 381\text{ kip-ft [571 kN-m]}$$

Determine M_n:

$$M_n = M_p - (M_p - M_r) \times \left(\frac{L_b - L_p}{L_r - L_p}\right)$$
$$= 600\text{ kip-ft} - (600\text{ kip-ft} - 381\text{ kip-ft}) \times \left(\frac{10\text{ ft} - 8\text{ ft}}{23.4\text{ ft} - 10\text{ ft}}\right)$$
$$= 567\text{ kip-ft [769 kN-m]}$$

Example 6. Determine the moment capacity of a 25-ft [7.63 m] A36 W 24 × 76 steel beam that is laterally unsupported.

Solution: $L_b = 25\text{ ft} > L_r = 23.1\text{ ft}$ (from Example 4). Next determine M_n:

$$M_n = \left(\frac{S_x \times X_1 \times \sqrt{2}}{L_b/r_y}\right) \times \sqrt{1 + \frac{(X_1)^2 \times X_2}{2 \times (L_b/r_y)}}$$
$$= \left(\frac{176\text{ in.}^3 \times 1760\text{ ksi} \times \sqrt{2}}{\left[25\text{ ft} \times \left(\frac{12\text{ in.}}{1\text{ ft}}\right)\right] \Big/ 1.92\text{ in.}}\right)$$
$$\times \sqrt{1 + \frac{(1760\text{ ksi})^2 \times (0.0186\text{ ksi})}{2 \times \left\{\left[25\text{ ft} \times \left(\frac{12\text{ in.}}{1\text{ ft}}\right)\right] \Big/ 1.92\text{ in.}\right\}^2}}$$
$$= \left(\frac{438{,}000\text{ kip-in.}}{156}\right) \times \sqrt{1 + \frac{57{,}600}{48{,}800}}$$
$$= 2810\text{ kip-in.} \times 1.48 = 4160\text{ kip-in.} = 347\text{ kip-ft [471 kN-m]}$$

Problem 9.3.A. Determine the nominal moment capacity (M_n) for a W 30 × 90 made of A36 steel and the following unbraced lengths: (1) 5 ft, (2) 15 ft, (3) 30 ft.

Problem 9.3.B. Determine the nominal moment capacity (M_n) for a W 16 × 36 made of steel with a yield stress of 50 ksi and the following unbraced lengths: (1) 5 ft, (2) 10 ft, and (3) 30 ft.

9.4 DESIGN FOR BENDING

Design for bending usually involves the determination of the ultimate bending moment (M_u) that the beam must resist and the use of formulas derived in Section 9.3 for definition of bending resistance of the member (M_n). The basic formulation of this equation is

$$M_u = \phi_b M_n$$

where $\phi_b = 0.9$.

Design for Plastic Failure Mode

The formula $F_y = M_u/Z_x$ or $Z_x = M_u/F_y$ is used to determine the minimum plastic section modulus required. Because weight is determined by area, not plastic section modulus, the beam chosen may have more plastic section modulus than required and still be the most economical choice. The following example illustrates the basic procedure.

Example 7. Design a simply supported floor beam to carry a superimposed load of 2 kips per ft [29.2 kN/m] over a span of 24 ft [7.3 m]. (The term *superimposed load* is used to denote any load other than the weight of a structural member itself.) The superimposed load is 25 percent dead load and 75 percent live load. The yield stress is 36 ksi [250 MPa]. The floor beam is continuously supported along its length against lateral buckling.

Solution: The load must first be factored for the load combination to determine the maximum factored load. The combined load factors are given in Section 4.2 under "Strength Design Method." The applicable factors are

$$\begin{aligned} w_u &= 1.4 \times (\text{dead load}) \\ &= 1.4 \times (0.5 \text{ kip/ft}) = 0.7 \text{ kip/ft } [10.2 \text{ kN/m}] \end{aligned}$$

or

$$\begin{aligned} w_u &= 1.2 \times (\text{dead load}) + 1.6 \times (\text{live load}) \\ &= 1.2 \times (0.5 \text{ kip/ft}) + 1.6 \times (1.5 \text{ kip/ft}) = 3.0 \text{ kip/ft } [43.8 \text{ kN/m}] \end{aligned}$$

The bending moment due to the maximum factored superimposed load is

$$M_u = \frac{wL^2}{8} = \frac{3 \times (24)^2}{8} = 216 \text{ kip-ft } [293 \text{ kN-m}]$$

The required bending resistance of the member is

$$M_u = \phi_b \times M_n$$

$$M_n = \frac{M_u}{\phi_b} = \frac{216 \text{ kip-ft}}{0.9} = 240 \text{ kip-ft } [325 \text{ kN-m}]$$

The required plastic section modulus for this moment is

$$Z_x = \frac{M_n}{F_y} = \frac{240 \text{ kip-ft} \times (12 \text{ in./1 ft})}{36 \text{ ksi}} = 80.0 \text{ in.}^3\ [1.31 \times 10^6 \text{ mm}^3]$$

Table 9.1 lists a number of shapes commonly used as beams in descending order of the value of their plastic modulus. Also listed in Table 9.1 are values for L_p, L_r, M_p, and M_r. From Table 9.1, a W 16 × 45 is found with a plastic section modulus of 82.3 in.^{3} [1192 ×10^3 mm^3]. Further scanning of the table reveals a W 14 × 53 with a Z_x of 87.1 in.^{3} [1328 × 10^3 mm^3] and a W 18 × 46 with a Z_x of 90.7 in.^{3} [1291 × 10^3 mm^3]. In the absence of any known restriction on the beam depth, try the lightest section (W 16 × 45). The weight of the beam, which was not taken into consideration in the earlier calculations, must now be considered. First, the weight of the member must be factored:

$$\begin{aligned} w_u &= 1.2 \times (\text{dead load}) + 1.6 \times (\text{live load}) \\ &= 1.2 \times (45 \text{ lb/ft}) + 1.6 \times (0 \text{ lb/ft}) = 54.0 \text{ lb/ft } [788 \text{ N/m}] \end{aligned}$$

The bending moment at the center of the span caused by the beam weight is

$$\begin{aligned} M_u &= \frac{wL^2}{8} = \frac{54 \text{ lb/fit} \times (24 \text{ ft})^2}{8} \\ &= 3890 \text{ ft-lb or } 3.89 \text{ kip-ft } [5.27 \text{ kN-m}] \end{aligned}$$

The required bending capacity due to the beam weight is

$$M_n = \frac{M_u}{\phi_b} = \frac{3.89 \text{ kip-ft}}{0.9} = 4.32 \text{ kip-ft } [5.86 \text{ kN-m}]$$

Thus, the total bending moment at midspan is

$$M_n = 240 + 4.32 = 244 \text{ kip-ft } [331 \text{ kN-m}]$$

The plastic section modulus required for this moment is

$$Z_x = \frac{M_n}{F_y} = \frac{244 \text{ kip-ft} \times (12 \text{ in./1 ft})}{36 \text{ ksi}} = 81.3 \text{ in.}^3 \ [1.33 \times 10^6 \text{ mm}^3]$$

Because this required value is less than that of the W 16 × 45, this section is still acceptable.

Use of Plastic Section Modulus Tables

Selection of rolled shapes on the basis of required plastic section modulus may be achieved by the use of tables in the AISC Manual (Ref. 3) in which beam shapes are listed in descending order of their section modulus values. Table 9.1 is similar to these tables and presents a small sample of the reference table data. Note that certain shapes have their designations listed in boldface type. These are sections that have an especially efficient bending moment resistance, indicated by the fact that there are other sections of greater weight but the same or smaller section modulus. Thus, for a savings of material cost, these *least-weight* sections offer an advantage. Consideration of other beam design factors, however, may sometimes make this a less important concern.

Data are also supplied in Table 9.1 for the consideration of lateral support for beams. For consideration of lateral support, the values are given for the two limiting lengths L_p and L_r. If a calculation has been made by assuming the minimum yield stress of 36 ksi [250 MPa], the required plastic section modulus obtained will be proper only for beams in which the lateral unsupported length is equal to or less than L_p under the column labeled "36 ksi." Similary, if the required plastic section modulus was obtained using a minimum yield stress of 50 ksi [350 Mpa], the lateral unsupported length needs to be equal to or less than L_p under the "50 ksi" column.

A second method of using Table 9.1 for beams omits the calculation of a required plastic section modulus and refers directly to the listed values for the plastic bending moment of the sections, given as M_p in the tables. It is an appropriate section if $L_b < L_p$.

Example 8. Rework the problem in the preceding example in this section by using Table 9.1.

Solution: As before, the bending moment caused by the superimposed loading is found to be 240 kip-ft [325 kN-m]. Noting that some additional bending capacity will be required because of the beam's own weight, scan the tables for shapes with an M_p of more than 240 kip-ft [325 kN-m]. Thus:

Shape	M_p (kip-ft)	M_p (kN-m)
W 21 × 44	286	388
W 18 × 46	272	369
W 14 × 53	261	354
W 10 × 68	256	347
W 16 × 45	247	335

Although the W 21 × 44 is the least-weight section, other design considerations, such as restricted depth or lateral support, may make any of the other shapes the appropriate choice. In the original problem, the lateral support was said to be continuous along the beam. The reality is that continuous support is rare and that it is usually given at various points along the beam. The W 21 × 44 would remain the most appropriate section if the maximum distance between lateral support (L_b) was less than or equal to 5.25 ft (L_p). If the distance was greater, other members should be chosen. An example of this is as follows:

Lateral Supports at:	Maximum Unbraced Length	Most Economical Section
Quarter points	6 ft [1.83 m]	W 16 × 45
Third points	8 ft [2.44 m]	W 14 × 53
Midpoint	12 ft [3.66 m]	W 12 × 79
Ends of beam only	24 ft [7.32 m]	No section qualifies

It should be noted that not all of the available W shapes listed in Table A.3 are included in Table 9.1. Specifically excluded are the shapes that are approximately square (depth equal to flange width) and are ordinarily used for columns rather than beams.

The following problems involve design for bending under plastic failure mode only. Use A36 steel and assume that least-weight members are desired for each case.

Problem 9.4.A. Design for flexure a simple beam 14 ft [4.3 m] in length and having a total uniformly distributed dead load of 13.2 kips [59 kN] and a total uniformly distributed live load of 26.4 kips [108 kN].

Problem 9.4.B. Design for flexure a beam having a span of 16 ft [4.9 m] with a concentrated live load of 40 kips [178 kN] at the center of the span.

Problem 9.4.C. A beam of 15 ft [4.6 m] in length has three concentrated live loads of 6 kips, 7.5 kips, and 9 kips at 4 ft, 10 ft, and 12 ft [26.7 kN, 33.4 kN, and 40.0 kN at 1.2 m, 3 m, and 3.6 m], respectively, from the left-hand support. Design the beam for flexure.

Problem 9.4.D. A beam 30 ft [9 m] long has concentrated live loads of 9 kips [40 kN] each at the third points and also a total uniformly distributed dead load of 20 kips [89 kN] and a total uniformly distributed live load of 10 kips [44 kN]. Design the beam for flexure.

Problem 9.4.E. Design for flexure a beam 12 ft [3.6 m] in length, having a uniformly distributed dead load of 1 kip/ft [14.6 kN/m], a total uniformly distributed live load of 1 kip/ft [14.6 kN/m], and a concentrated dead load of 8.4 kips [37.4 kN] a distance of 5 ft [1.5 m] from one support.

Problem 9.4.F. A beam of 19 ft [5.8 m] in length has concentrated live loads of 6 kips [26.7 kN] and 9 kips [40 kN] at 5 ft [1.5 m] and 13 ft [4 m], respectively, from the left-hand support. In addition, there is a uniformly distributed dead load of 1.2 kip/ft [17.5 kN/m] beginning 5 ft [1.5 m] from the left support and continuing to the right support. Design the beam for flexure.

Problem 9.4.G. A steel beam 16 ft [4.9 m] long has a uniformly distributed dead load of 100 lb/ft [1.46 kN/m] extending over the entire span and a uniformly distributed live load of 100 lb/ft [1.46 kN/m] extending 10 ft [3 m] from the left support. In addition, there is a concentrated live load of 8 kips [35.6 kN] at 10 ft [3 m] from the left support. Design the beam for flexure.

Problem 9.4.H. Design for flexure a simple beam 21 ft [6.4 m] in length, having two concentrated loads of 20 kips [89 kN] each, one 7 ft [2.13 m] from the left end and the other 7 ft [2.13 m] from the right end. The concentrated loads are each made up of equal parts dead load and live load.

Problem 9.4.I. A cantilever beam 12 ft [3.6 m] long has a uniformly distributed dead load of 600 lb/ft [8.8 kN/m] and a uniformly distributed dead load of 1000 lb/ft [14.6 kN/m]. Design the beam for flexure.

Problem 9.4.J. A cantilever beam 6 ft [1.8 m] long has a concentrated live load of 12.3 kips [54.7 kN] at its unsupported end. Design the beam for flexure.

9.5 DESIGN OF BEAMS FOR BUCKLING FAILURE

Although it is preferable to design beams to fail under the plastic hinge mode discussed in Section 9.4, it is not always possible to do so. This is commonly caused by excessive unbraced lengths for lateral support of the beam (L_b). The simplest solution is to decrease the maximum unbraced length to make it less than the plastic limit length (L_p). If this is not possible, the solution is to accept that the beam failure mode is buckling and use the appropriate equations to determine the nominal moment capacity (M_n) of the beam as described in Section 9.3.

If $L_p < L_b \leq L_r$:

$$M_n = M_p - (M_p - M_r) \times \left(\frac{L_b - L_p}{L_r - L_p} \right)$$

If $L_b > L_r$:

$$M_n = \left(\frac{S_x \times X_1 \times \sqrt{2}}{L_b/r_y} \right) \times \sqrt{1 + \frac{(X_1)^2 \times X_2}{2 \times (L_b/r_y)^2}}$$

Example 9. A 14 ft [4.7 m] long simply supported beam has a uniform live load of 3 kip/ft [43.3 kN/m] and a dead load of 2 kip/ft [29.2 kN/m]. It is laterally supported only at its ends. Determine the most economical W shape available.

Solution: First, we need to determine the appropriate load combination and maximum factored moment:

$$w_u = 1.4 \times (\text{dead load})$$
$$= 1.4 \times (2 \text{ kip/ft}) = 2.8 \text{ kip/ft } [40.9 \text{ kN/m}]$$

or

$$w_u = 1.2 \times (\text{dead load}) + 1.6 \times (\text{live load})$$
$$= 1.2 \times (2 \text{ kip/ft}) + 1.6 \times (3 \text{ kip/ft}) = 7.2 \text{ kip/ft } [105 \text{ N/m}]$$

The bending moment caused by the maximum factored superimposed load is

$$M_u = \frac{wL^2}{8} = \frac{7.2 \times (14)^2}{8} = 176 \text{ kip-ft } [239 \text{ kNm}]$$

The required bending resistance of the member is

$$M_u = \phi_b \times M_n$$

$$M_n = \frac{M_u}{\phi_b} = \frac{176 \text{ kip-ft}}{0.9} = 196 \text{ kip-ft } [266 \text{ kNm}]$$

The required plastic section modulus for this moment is

$$Z_x = \frac{M_n}{F_y} = \frac{196 \text{ kip-ft} \times (12 \text{ in./1 ft})}{36 \text{ ksi}} = 65.3 \text{ in.}^3 \ [4213 \text{ mm}^3]$$

From Table 9.1, we find that the most economical cross section not taking into account unbraced length is a W 18 × 40, but this section has a plastic limit on length (L_p) of 5.29 ft; therefore, it does not work for this problem. It should be investigated to determine if it is possible to support this beam at its third points. If it can be supported, a W 18 × 40 would be the most economical shape available. If it cannot be supported, we need to look in Table 9.1 to see if there are any beams that have $Z_x >$ 65.3 in.^{3} and $L_p \geq 14$ ft. There are two sections that match this criteria, a W 14 × 120 and a W 14 × 145. The only problem with these sections is that they require three times the amount of steel to do the work than the amount required by the plastic section modulus Z_x. Next, we look for a section that has $Z_x > 65.3$ in.$_3$ and $L_p < 14$ ft $< L_r$ and whose moment capacity $M_n > 196$ kip-ft.

We begin with the W 18 × 40 we chose before because it would be the most economical if it works. We take the values for M_p, M_r, L_p, and L_r from Table 9.1.

$$M_n = M_p - (M_p - M_r) \times \left(\frac{L_b - L_p}{L_r - L_p}\right)$$

$$= 235 \text{ kip-ft} - (235 \text{ kip-ft} - 148 \text{ kip-ft}) \times \left(\frac{14 \text{ ft} - 5.29 \text{ ft}}{15.7 \text{ ft} - 5.29 \text{ ft}}\right)$$

$$= 162 \text{ kip-ft} < 196 \text{ kip-ft}$$

Next, we choose the W 14 × 43 (the next most economical section) and check to see if it works:

$$M_n = 209 \text{ kip-ft} - (209 \text{ kip-ft} - 136 \text{ kip-ft}) \times \left(\frac{14 \text{ ft} - 7.88 \text{ ft}}{24.7 \text{ ft} - 7.88 \text{ ft}}\right)$$

$$= 182 \text{ kip-ft} < 196 \text{ kip-ft}$$

Trying a W 21 × 44, we find

$$M_n = 286 \text{ kip-ft} - (286 \text{ kip-ft} - 177 \text{ kip-ft}) \times \left(\frac{14 \text{ ft} - 5.25 \text{ ft}}{15.4 \text{ ft} - 5.25 \text{ ft}}\right)$$

$$= 192 \text{ kip-ft} < 196 \text{ kip-ft}$$

Trying a W 16 × 45, we find

$$M_n = 247 \text{ kip-ft} - (247 \text{ kip-ft} - 158 \text{ kip-ft}) \times \left(\frac{14 \text{ ft} - 6.54 \text{ ft}}{20.2 \text{ ft} - 6.54 \text{ ft}}\right)$$

$$= 198 \text{ kip-ft} > 196 \text{ kip-ft}$$

A W 16 × 45 appears to work, but we need to check if the weight of the beam increases the required bending capacity. The weight of the beam is 45 lb/ft.

$$w_u = 1.2 \times (\text{dead load}) + 1.6 \times (\text{live load})$$
$$= 1.2 \times (45 \text{ lb/ft}) + 1.6 \times (0 \text{ lb/ft}) = 54 \text{ lb/ft } [0.79 \text{ kN/m}]$$

The bending moment caused by the maximum factored self weight of the beam is

$$M_u = \frac{wL^2}{8} = \frac{54 \text{ lb/ft} \times (14 \text{ ft})^2}{8} = 1323 \text{ lb-ft} = 1.32 \text{ kip-ft } [1.79 \text{ kNm}]$$

The total bending moment caused by the maximum factored self weight of the beam and the factored superimposed load is

$$M_u = 1.32 \text{ kip-ft} + 176 \text{ kip-ft} = 177 \text{ kip-ft } [240 \text{ kNm}]$$

The required bending resistance of the member is

$$M_u = \phi_b \times M_n$$

$$M_n = \frac{M_u}{\phi_b} = \frac{177 \text{ kip-ft}}{0.9} = 197 \text{ kip-ft } [267 \text{ kNm}]$$

Because the bending capacity of the selected beam is still greater than the required bending capacity, a W 16 × 45 is selected as the most economical section.

The following problems involve use of Table 9.1 to choose the most economical beams when lateral bracing is a concern. All beams use A36 steel.

Problem 9.5.A. A W shape steel is to be used for a uniformly loaded simple beam carrying a total dead load of 27 kip [120 kN] and a total live load of 50 kip [222 kN] on a 45-ft [13.7 m] span. Select the lightest weight shape for unbraced lengths of (a) 10 ft [3.05 m]; (b) 15 ft [4.57 m]; (c) 22.5 ft [6.90 m].

Problem 9.5.B. A W shape is to be used for a uniformly loaded simple beam carrying a total dead load of 30 kip [133 kN] and a total live load of 40 kip [178 kN] on a 24-ft [7.32 m] span. Select the lightest weight shape for unbraced lengths of (a) 6 ft [1.83 m]; (b) 8 ft [2.44 m]; (c) 12 ft [3.66 m].

Problem 9.5.C. A W shape is to be used for a uniformly loaded simple beam carrying a total dead load of 22 kip [98 kN] and a total live load of 50 kip [222 kN] on a 30-ft [9.15 m] span. Select the lightest weight shape for unbraced lengths of (a) 6 ft [1.83 m]; (b) 10 ft [3.05 m]; (c) 15 ft [4.57 m].

Problem 9.5.D. A W shape is to be used for a uniformly loaded simple beam carrying a total dead load of 26 kip [116 kN] and a total live load of 26 kip [116 kN] on a 36-ft [11 m] span. Select the lightest weight shape for unbraced lengths of (a) 9 ft [2.74 m]; (b) 12 ft [3.66 m]; (c) 18 ft [5.49 m].

9.6 SHEAR IN STEEL BEAMS

Investigation and design for shear forces in the LRFD method is similar to that for bending moment in that the maximum factored shear force must be less than the factored shear capacity of the beam chosen. This is expressed as

$$\phi_v V_n \geq V_u$$

where ϕ_v = 0.90

V_n = nominal shear capacity of the beam

V_u = required (factored) shear strength of the beam.

Shear in beams consists of the vertical slicing effect produced by the opposition of the vertical loads on the beams (downward) and the reactive forces at the beam supports (upward). The internal shear force mechanism is visualized in the form of the shear diagram for the beam. With a uniformly distributed load on a simply supported beam, this diagram takes the form of that shown in Figure 9.7*a*.

As the shear diagram for the uniformly loaded beam shows, this load condition results in an internal shear force that peaks to a maximum value at the beam supports and steadily decreases in magnitude to zero at the center of the beam span. With a beam having a constant cross section throughout the span, the critical location for shear is thus at the supports, and if conditions there are adequate, there is no concern for shear at other locations along the beam. Because this is the common condition of loading for many beams, it is therefore necessary only to investigate the support conditions for such beams.

Figure 9.7*b* shows another loading condition, that of a major concentrated load within the beam span. Framing arrangements for roof and floor systems frequently employ beams that carry the end reactions of other beams, so this is also a common condition. In this case, a major internal shear force is generated over some length of the beam. If the con-

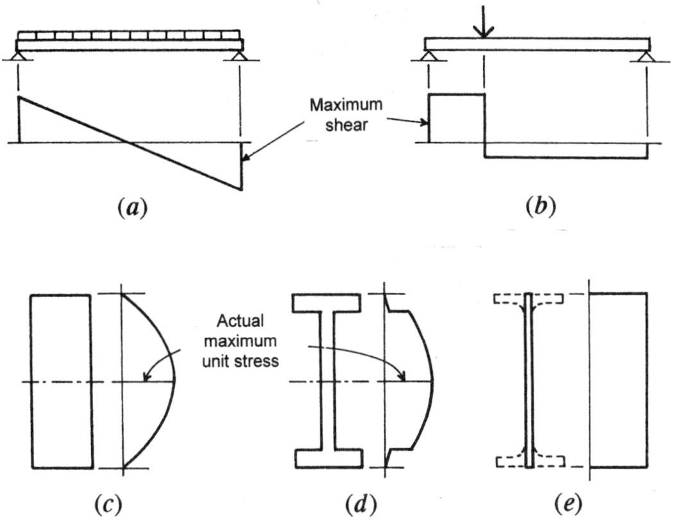

Figure 9.7 Development of shear in beams.

centrated load is close to one support, a critical internal shear force is created in the shorter portion of the beam length between the load and the closer support.

Internal shear force develops shear stresses in a beam (see Section 3.8). The distribution of these stresses over the cross section depends on the geometric properties of the beam cross section and significantly depends on the general form of the cross section. For a simple rectangular cross section, such as that of a wood beam, the distribution of beam shear stress is as shown in Figure 9.7*c*, taking the form of a parabola with a maximum shear stress value at the beam neutral axis and a decrease to zero stress at the extreme fiber distances (top and bottom edges).

For the I-shaped cross section of the typical W-shape rolled steel beam, the beam shear stress distribution takes the form shown in Figure 9.7*d* (referred to as the "derby hat" form). Again, the shear stress is a maximum at the beam neutral axis, but the falloff is less rapid between the neutral axis and the inside of the beam flanges. Although the flanges indeed take some shear force, the sudden increase in beam width results in an abrupt drop in the beam unit shear stress. A traditional shear stress investigation for the W shape, therefore, is based on ignoring the flanges and assuming the shear-resisting portion of the beam to be an equivalent vertical plate (Figure 9.7*e*) with a width equal to the beam web thickness and a height equal to the full beam depth. For the ASD method, an allowable value is established for a unit shear stress on this basis, and the computation is performed as

$$f_v = \frac{V}{t_w d_b} = \frac{V}{A_w}$$

where f_v = the average unit shear stress, based on an assumed distribution as shown in Figure 9.7*e*

V = the value for the internal shear force at the cross section

t_w = the beam web thickness

d_b = the overall beam depth

A_w = area of the web

Shear forces internal to a W shape beam may be problematic in terms of compression forces on the section's web. Figure 9.8*d* illustrates this behavior at a beam support consisting of a bearing of the beam end on top of the support, usually in this case the top of a wall or a wall ledge. The potential problem here has more to do with vertical compression force, which results in a squeezing of the beam end and a columnlike action in the thin beam web. This may actually produce a columnlike form of failure. As with a column, the range of possibilities for this form of failure relate to the relative slenderness of the web. These three cases are defined by the slenderness ratio of the web (h/t_w) and control the shear capacity (V_n) of a beam. The three distinct cases are as follows:

1. A very stiff (thick) web, that may actually reach something close to the full yield stress limit of the material where

$$h/t_w \leq 418/\sqrt{F_y}$$

$$V_n = (0.6 \times F_y) \times A_w$$

2. A somewhat slender web that responds with some combined yield stress and buckling effect (called an *inelastic buckling* response) where

$$418/\sqrt{F_y} \leq h/t_w \leq 523/\sqrt{F_y}$$

$$V_n = (0.6 \times F_y) \times A_w \times \left(\frac{418/\sqrt{F_y}}{h/t_w}\right)$$

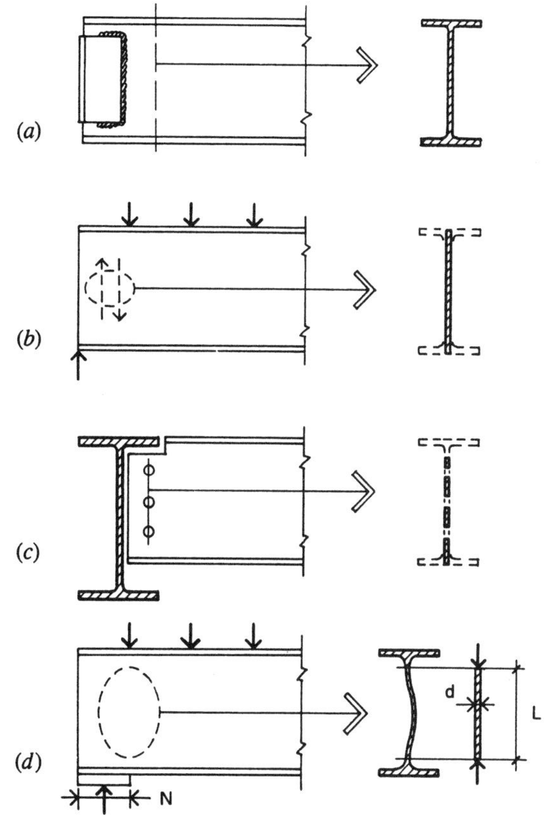

Figure 9.8 Considerations for end support in rolled steel beams: (*a*) Development of shear by the full beam section with a framed connection. (*b*) Design assumption for resistance of the web only for shear. (*c*) Development of shear across a reduced section. (*d*) Development of compression in the beam web at a bearing support.

3. A very slender web that fails essentially in *elastic buckling* in the classic Euler formula manner; basically, a deflection failure rather than a stress failure where

$$h/t_w > 523/\sqrt{F_y}$$

$$V_n = 132{,}000 \times \frac{A_w}{(h/t_w)^2}$$

Example 10. A simple beam of A36 steel is 6 ft [1.83 m] long and has a concentrated live load of 36 kips [160 kN] applied 1 ft [0.3 m] from one end. It is found that a W 10 × 19 is adequate for the bending moment. Investigate the beam to determine if the shear capacity is adequate for the required shear.

Solution: The load must first be factored.

$$\begin{aligned} P_u &= 1.2 \times (\text{Dead load}) + 1.6\ (\text{Live load}) \\ &= 1.2(0\ \text{kip}) + 1.6(36\ \text{kips}) \\ &= 57.6\ \text{kips}\ [256\ \text{kN}] \end{aligned}$$

The two reactions for this loading are 48 kips [214 kN] and 9.6 kips [43.2 kN]. The maximum shear (V_u) in the beam is equal to the larger reaction force.

From Table A.3, for the given shape, $d = 10.24$ in., $t_w = 0.250$ in. and $h/t_w = 37.8$. Then

$$418/\sqrt{F_y} = 418/\sqrt{36\ \text{ksi}} = 69.7 > h/t_w = 37.8$$

Therefore, the shear capacity will be determined using the equation associated with the full yield stress limit of the material:

$$\begin{aligned} A_w &= d \times t_w = 10.24\ \text{in.} \times 0.250\ \text{in.} = 2.56\ \text{in.}^2\ [1651\ \text{mm}^2] \\ V_n &= (0.6 \times F_y) \times A_w \\ V_n &= (0.6 \times 36\ \text{ksi}) \times 2.56\ \text{in.}^2 = 55.3\ \text{kips}\ [246\ \text{kN}] \\ \phi_v V_n &= 0.9 \times 55.3\ \text{kips} = 49.8\ \text{kips} > V_u = 48\ \text{kips}\ [214\ \text{kN}] \end{aligned}$$

Because the capacity of the beam is greater than the factored shear, the shape is acceptable.

The net effect of investigations of all the situations so far described, relating to end shear in beams, may be to influence a choice of beam shape with a web that is sufficient. However, other criteria for selection (flexure, deflection, framing details, and so on) may indicate an ideal choice that has a vulnerable web. In the latter case, it is sometimes decided to *reinforce* the web; the usual means is to insert vertical plates on either side of the web and to fasten them to the web as well as to the beam flanges. These plates then both brace the slender web (column) and absorb some of the vertical compression stress in the beam.

For practical design purposes, the beam end shear and end support limitations of unreduced webs can be handled by data supplied in AISC tables.

Problems 9.6.A, B, C. Compute the shear capacity ($\phi_v V_n$) for the following beams of A36 steel: A, W 24 × 84; B, W 12 × 45; C, W 10 × 33.

9.7 DEFLECTION OF BEAMS

Deformations of structures must often be controlled for various reasons. These reasons may relate to the proper functioning of the structure, but more often they relate to effects on the supported construction or to the overall purpose of the structure.

To steel's advantage is the relative stiffness of the material itself. With a modulus of elasticity of 29,000 ksi, it is 8 to 10 times as stiff as average structural concrete and 15 to 20 times as stiff as structural lumber. However, it is often the overall deformation of whole structural assemblages that must be controlled; in this regard, steel structures are frequently quite deformable and flexible. Because of its cost, steel is usually formed into elements with thin parts (beam flanges and webs, for example), and because of its high strength, it is frequently formed into relatively slender elements (beams and columns, for example).

For a beam in a horizontal position, the critical deformation is usually the maximum sag, called the beam's *deflection*. For most beams, this deflection will be too small in magnitude to be detected by eye. However, any load on a beam, such as that in Figure 9.9, will cause some amount of deflection, beginning with the beam's own weight. In the case of a simply supported, symmetrical, single-span beam, the maximum deflection will occur at midspan, and it usually is the only deformation value of

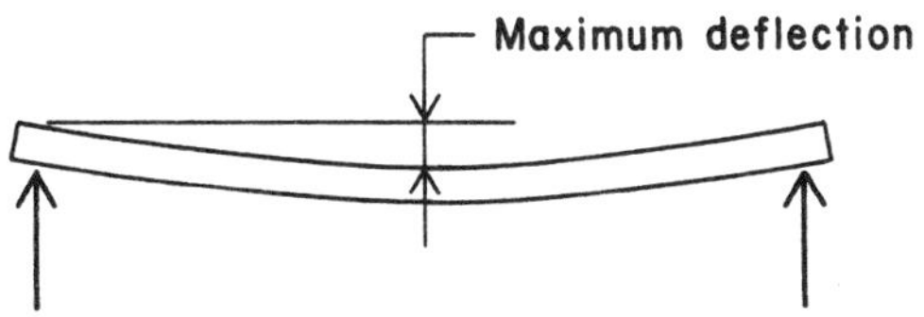

Figure 9.9 Deflection of a simple beam under symmetrical loading.

concern for design. However, as the beam deflects, its ends rotate unless restrained, and this twisting deformation may also be of concern in some situations.

If deflection is determined to be excessive, the usual remedy is to select a deeper beam. Actually, the critical geometric property of the beam cross section is its *moment of inertia* (I) about its major axis (I_x for a W shape), which is typically affected significantly by increases in depth of the beam. Formulas for deflection of beams take a typical form that involves variables as follows:

$$\Delta = C\frac{WL^3}{EI}$$

In the formula:

Δ = the deflection, measured vertically (usually in inches or millimeters)
C = a constant related to the form of the load and support conditions
W = the total load on the beam
L = the span of the beam
E = the modulus of elasticity of the material of the beam
I = the moment of inertia of the beam cross section for the axis about which bending occurs

Note that the magnitude of the deflection is directly proportional to the magnitude of the load; that is, if you double the load, you will double the deflection. However, the deflection is proportional to the third power of the span; double the span and you get 2^3 or eight times the deflection. For resistance to deflection, increases in either the material's stiffness or the beam's geometric form, I, will cause direct proportional reduction of the deflection. Because E is constant for all steel, design modulation of deflections must deal only with the beam's shape.

Excessive deflection may cause various problems. For roofs, an excessive sag may disrupt the intended drainage patterns for generally flat surfaces. For floors, a common problem is the development of some perceivable bounciness. The form of the beam and its supports may also be a consideration. For the simple-span beam in Figure 9.9, the usual concern is simply for the maximum sag at the beam midspan. For a beam with a projected (cantilevered) end, however, a problem may be created at the unsupported cantilevered end; depending on the extent of the cantilever, this may involve downward deflection (as shown in Figure 9.10*a*) or upward deflection (as shown in Figure 9.10*b*).

With continuous beams, a potential problem derives from the fact that load in any span causes some deflection in all the spans. This is most critical when loads vary in different spans or the length of the spans differ significantly (see Figure 9.10*c*).

Most deflection problems in buildings stem from effects of the structural deformation on adjacent or supported elements of the building. When beams are supported by other beams (usually referred to as girders), excessive rotation caused by deflection occurring at the ends of the supported beams can result in cracking or other separation of the floor deck that is continuous over the girders, as shown in Figure 9.10*d*. For such a system, there is also an accumulative deflection caused by the independent deflections of the deck, the beams, and the girders, which can

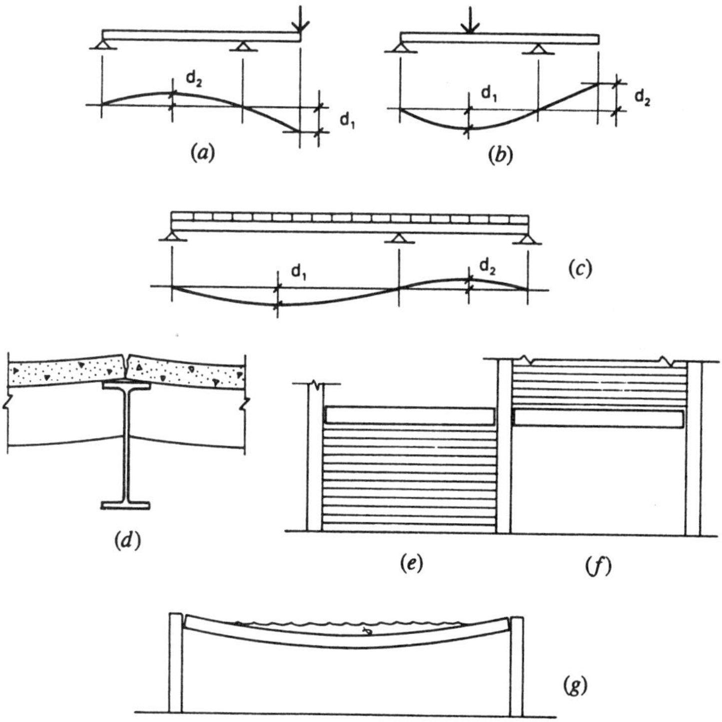

Figure 9.10 Considerations for deflection of beams.

cause problems for maintaining a flat floor surface or a desired roof surface profile for drainage.

An especially difficult problem related to deflections is the effect of beam deflections on nonstructural elements of the construction. Figure 9.10*e* shows the case of a beam occurring directly over a solid wall. If the wall is made to fit tightly beneath the beam, any deflection of the beam will cause it to bear on top of the wall—not an acceptable situation if the wall is relatively fragile (a metal and glass curtain wall, for example). A different sort of problem occurs when relatively rigid walls (plastered, for example) are supported by spanning beams, as shown in Figure 9.10*f*. In this case, the wall is relatively intolerant of *any* deformation, so anything significant in the form of sag of the beam is really critical.

For longspan structures (an ambiguous class, usually meaning 100 ft or more span), A special problem is that of the relatively flat roof surface. In spite of provisions for code-mandated minimum drainage, heavy rain will run slowly off of the surface and linger to cause some considerable loading. Because this results in deflection of the spanning structure, the sag may form a depressed area that the rain can swiftly turn into a pond (see Figure 9.10*g*). The pond then constitutes an additional load that causes more sag—and a resulting deeper pond. This progression can quickly pyramid into a failure of the structure, so that codes (including the AISC Specification) now provide design requirements for investigation of potential ponding.

Standard Equations for Deflection

Determining deflection on a beam is usually done using a series of standard equations. These equations are listed in the AISC Manual (Ref. 5) and in Figure 3.25. The purpose of these equations is to determine the actual deflection of a beam for a given loading, and therefore when using them in steel design, we use non-factored loading. It is important to stay consistent with your units. In steel this usually means working in kips and inches. The most used of these equations for simply supported beams are listed in the following table.

Loading Condition	Maximum Deflection
Uniform load over entire span	$\Delta = \dfrac{5wL^4}{384EI}$
Point load at midspan	$\Delta = \dfrac{PL^3}{48EI}$

Example 11. A simple beam has a span of 20 ft [6.10 m] with a uniformly distributed load of 1.95 kip/ft [28.5 kN/M]. The beam is a steel W 14 × 34. Find the maximum deflection.

Solution: We can use the standard equation for a uniform load over the entire span. Moment of inertia for the W 14 × 34 is 340 in.^{4}.

$$\Delta = \frac{5wL^4}{384EI}$$

$$= \frac{5\left[1.95 \text{ kip/ft} \times \left(\frac{1 \text{ ft}}{12 \text{ in.}}\right)\right]\left[20 \text{ ft} \times \left(\frac{12 \text{ in.}}{1 \text{ ft.}}\right)\right]^4}{384(29{,}000 \text{ ksi})(340 \text{ in.}^4)}$$

$$= 0.712 \text{ in. } [18.1 \text{ mm}]$$

Allowable Deflections

What is permissible for beam deflection is mostly a matter of judgment by experienced designers. It is difficult to provide any useful guidance for specific limitations to avoid the various problems described in Figure 9.10. Each situation must be investigated individually, and the designers of the structure and those who develop the rest of the building construction must make some cooperative decisions about the necessary design controls.

For spanning beams in ordinary situations, some rules of thumb have been derived over many years of experience. These usually consist of establishing some maximum degree of beam curvature described in the form of a limiting ratio of the deflection to the beam span (L), expressed as a fraction of the span. These are sometimes, although not always, specified in design codes or legally enacted building codes. Some typical limitations recognized by designers are the following:

To avoid visible sag under total load on short to medium spans, $L/150$

For total load deflection of a roof structure, $L/180$

For deflection under live load only for a roof structure, $L/240$

For total load deflection of a floor structure, $L/240$

For deflection under live load only for a floor structure, $L/360$

Deflection of Uniformly Loaded Simple Beams

The most frequently used beam in flat roof and floor systems is the uniformly loaded beam with a single, simple span (no end restraint). This situation is shown in Figure 3.25 as Case 2. For this case, the following values may be obtained for the beam behavior:

Maximum bending moment:

$$M = \frac{wL^2}{8}$$

Maximum stress on the beam cross section:

$$f = \frac{Mc}{I}$$

Maximum midspan deflection:

$$\Delta = \frac{5}{384} \times \frac{wL^4}{EI}$$

Using these relationships, together with the case of a known modulus of elasticity ($E = 29{,}000$ ksi for steel), we can derive a convenient formula for deflection. Noting that the dimension c in the bending stress formula is $d/2$ for symmetrical shapes, and substituting the expression for M, we can say

$$f = \frac{Mc}{I} = \left(\frac{wL^2}{8}\right)\left(\frac{d/2}{I}\right) = \frac{wL^2d}{16I}$$

Then

$$\Delta = \frac{5wL^4}{384EI} = \left(\frac{wL^2d}{16I}\right)\left(\frac{5L^2}{24Ed}\right) = (f)\left(\frac{5L^2}{24Ed}\right) = \frac{5fL^2}{24Ed}$$

This is a basic formula for any beam symmetrical about its bending axis. For a shorter version, we will use 29,000 ksi for E. Because these deflection equations are only appropriate within the elastic range, we will set the bending stress (f) to the allowable bending stress in allowable stress design ($F_b = 0.67 \times F_y$). Also, for convenience, spans are usually measured in feet, not inches, so a factor of 12 is added. Thus, for A36 steel:

$$\Delta = \frac{5fL^2}{24Ed} = \left(\frac{5}{24}\right)\left[\frac{(0.67 \times 36)}{29{,}000}\right]\left(\frac{(12L)^2}{d}\right) = \frac{0.02483L^2}{d}$$

In metric units, with $f = 165$ MPa, $E = 200$ GPa, and the span in meters:

$$\Delta = \frac{0.0001719L^2}{d}$$

For $F_y = 50$ ksi:

$$\Delta = \frac{5fL^2}{24Ed} = \left(\frac{5}{24}\right)\left[\frac{(0.67 \times 50)}{29{,}000}\right] = \left(\frac{(12L)^2}{d}\right) = \frac{0.03449L^2}{d}$$

In metric units, with $f = 230$ MPa, $E = 200$ GPa, and the span in meters:

$$\Delta = \frac{0.0002396L^2}{d}$$

The derived deflection formula involving only span and beam depth can be used to plot a graph that displays the deflection of a beam of a constant depth for a variety of spans. Figure 9.11 consists of a series of such graphs for beams from 6 to 36 in. in depth and a yield stress of 36 ksi. Figure 9.12 is for steel with a yield stress of 50 ksi. Use of these graphs presents yet another means for determining beam deflections. An answer within about 5 percent should be considered reasonable from the graphs.

The real value of the graphs in Figure. 9.11 and 9.12, however, is in the design process. Once the span is known, it may be initially determined from the graphs what beam depth is required for a given deflection. The limiting deflection may be given in an actual dimension, or more commonly, as a limiting percentage of the span (1/240, 1/360, etc.), as previously discussed. To aid in the latter situation, lines are drawn on the graph representing the usual percentage limits of 1/360, 1/240, and 1/180. Thus, if a beam is to be used for a span of 36 ft, and the total load

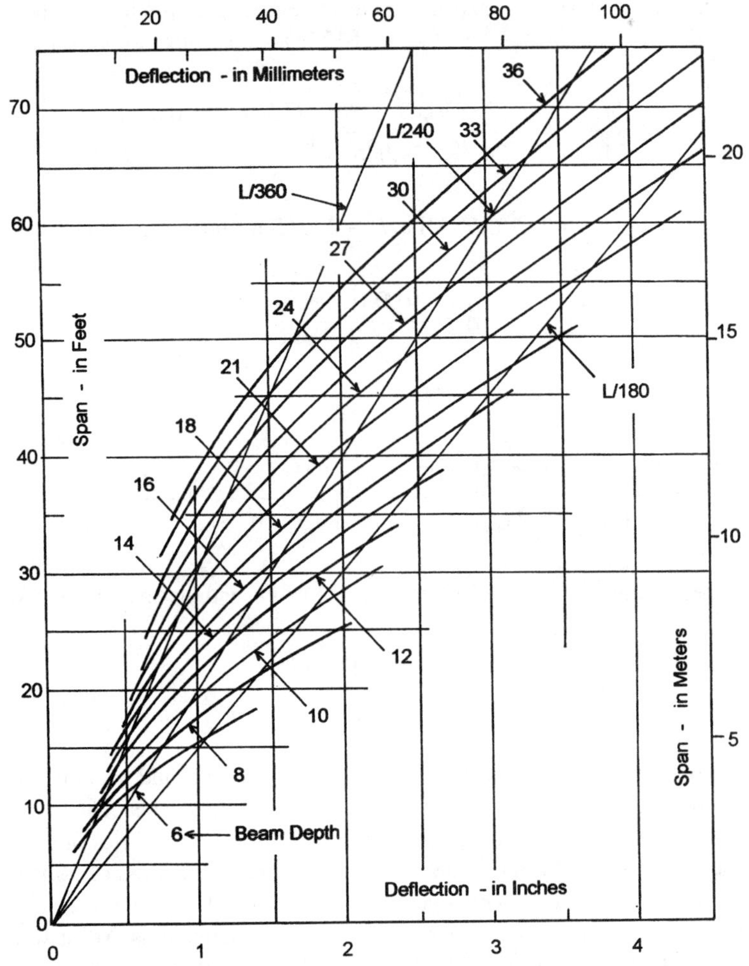

Figure 9.11 Deflection of steel beams with yield stress of 36 ksi [250 Mpa] under a constant bending stress of 24 ksi [165 MPa].

deflection limit is *L*/240, it may be observed on Figure 9.11 that the lines for a span of 36 ft and a ratio of 1/240 intersect almost precisely on the curve for an 18-in.-deep beam. This means that an 18-in.-deep beam will deflect almost precisely 1/240th of the span if stressed in bending to 24 ksi. Thus, any beam chosen with less depth will be inadequate for de-

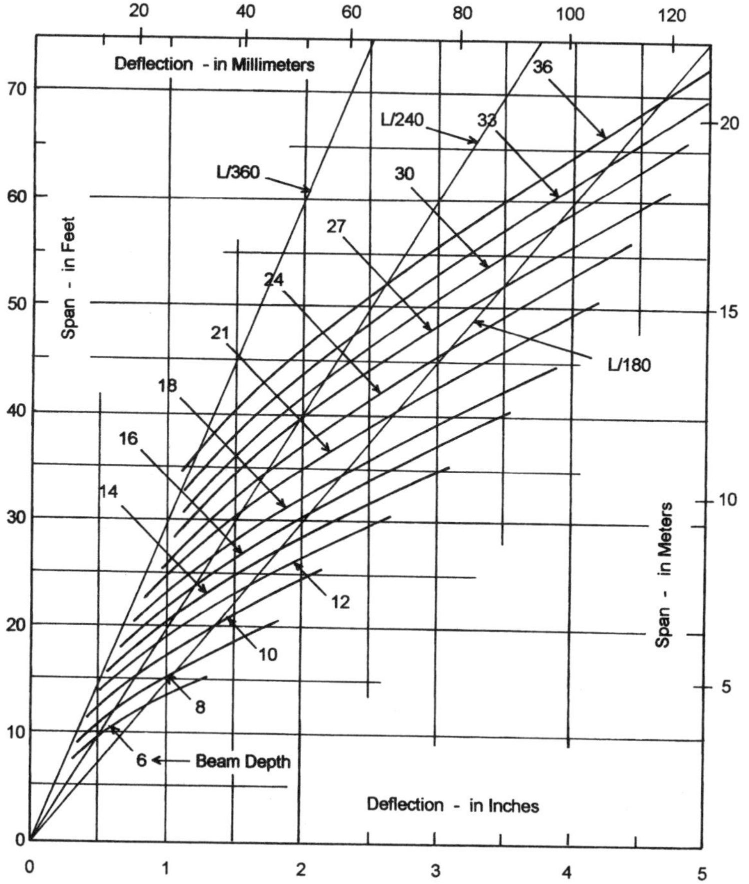

Figure 9.12 Deflection of steel beams with yield stress of 50 ksi [345 Mpa] under a constant bending stress of 33 ksi [228 MPa].

flection, and any beam greater in depth will be conservative in regard to deflection.

The minimum depth of a beam required by deflection criteria can be approximated from these equations. They are derived by placing the allowable deflection criteria into the equation derived earlier. It is important to remember in these equations that the beam length (L) is in feet, whereas the beam depth (d) is in inches. The equations are shown in the table that follows.

Yield Stress	$\frac{L}{180}$	$\frac{L}{240}$
36 ksi	$d_{min} = 0.372 \times L \approx \frac{L}{3}$	$d_{min} = 0.497 \times L \approx \frac{L}{2}$
50 ksi	$d_{min} = 0.517 \times L \approx \frac{L}{2}$	$d_{min} = 0.690 \times L \approx \frac{2L}{3}$

Problems 9.7.A, B, C, D. Find the maximum deflection in inches for the following simple beams of A36 steel with uniformly distributed load. Find the values using (a) the equation for deflection of a uniformly distributed load, and (b) the curves in Figure 9.11.

(A) W 10 × 33, span = 18 ft, total service load = 1.67 kips/ft [5.5 m, 24.2 kN/m]

(B) W 16 × 36, span = 20 ft, total service load = 2.5 kips/ft [6 m, 37 kN/m]

(C) W 18 × 46, span = 24 ft, total service load = 2.29 kips/ft [7.3 m, 33.6 kN/m]

(D) W 21 × 57, span = 27 ft, total service load = 2.5 kips/ft [8.2 m, 36.5 kN/m]

9.8 SAFE LOAD TABLES

The simple beam with uniformly distributed load occurs so frequently that it is useful to have a rapid design method for quick selection of shapes based on knowing only the beam load and span. The AISC Manual (Ref. 3) provides a series of such tables with data for the W, M, S, and C shapes most often used as beams.

Tables 9.2 and 9.3 present data for selected beams for $F_y = 36$ ksi and $F_y = 50$ ksi, respectively. Table values are the total factored load for a simple span beam with uniformly distributed loading and lateral bracing at points not further apart than the limiting dimension of L_p (see discussion in Section 9.5). The table values are determined from the maximum bending moment capacity of the shapes ($M_p = Z_x \times F_y$).

For very short spans, loads are often limited by beam shear or end support conditions rather than by bending or deflection limits. For this reason, table values are not shown for spans of less than 12 times the beam depth.

For long spans, loads are often limited by deflection rather than by bending. Thus, table values are not shown for spans exceeding a limit of 24 times the beam depth.

When the distance between points of lateral support exceeds the limiting value of L_p, the values in Tables 9.2 and 9.3 should not be used. See the discussion regarding buckling in Section 9.5.

The self-weight of the beam is also included in the weight given in these tables. It has not been removed from consideration because the factoring will vary depending on the given loading condition. Once a beam is selected, the self-weight of the beam must be factored and added to the total load to determine if the beam is still acceptable.

The following examples illustrate the use of Table 9.2 for some common design situations.

Example 12. Design a simply supported beam to carry a uniformly distributed live load of 1.33 kips/ft [19.4 kN/m] and a superimposed uniformly distributed dead load of 0.66 kips/ft [9.6 kN/m]on a span of 24 ft [7.32 m]. The yield stress is 36 ksi [250 MPa]. Find (1) the lightest shape permitted and (2) the shallowest (least-depth) shape permitted.

Solution: First, the loading must be factored and totaled for the entire beam.

$$w_u = 1.4 \times \text{(Dead load)}$$

$$= 1.4 \times (0.66 \text{ kip/ft}) = 0.924 \text{ kip/ft } [13.5 \text{ kN/m}]$$

or

$$w_u = 1.2 \times \text{(Dead load)} + 1.6 \times \text{(Live load)}$$

$$= 1.2 \times (0.66 \text{ kip/ft}) + 1.6 \times (1.33 \text{ kip/ft}) = 2.92 \text{ kip/ft } [42.6 \text{ kN/m}]$$

$$\text{Total superimposed load} = 2.92 \text{ kip/ft} \times 24 \text{ ft} = 70.1 \text{ kips } [312 \text{ kN}]$$

From Table 9.2, we find the following:

For Shape	Allowable Load
W 12 × 53	70.1 kips [312 kN]
W 18 × 40	70.6 kips [314 kN]
W 16 × 45	74.1 kips [330 kN]
W 14 × 53	78.4 kips [349 kN]
W 18 × 46	81.6 kips [363 kN]
W 21 × 44	85.9 kips [382 kN]
W 12 × 79	107 kips [476 kN]

TABLE 9.2 Factored Load-Span Values for 36 ksi Beams

	F_y = 36 ksi				Span (ft)									
Designation	L_p ft	L_r ft	M_p kip-ft	M_r kip-ft	12	14	16	18	20	22	24	26	28	30
W 10 × 17	3.52	11.1	56	35	33.7	28.9	25.2	22.4	20.2					
W 12 × 16	3.22	9.58	60	37	36.2	31.0	27.1	24.1	21.7	19.7	18.1			
W 10 × 19	3.64	12.0	65	41	38.9	33.3	29.2	25.9	23.3					
W 12 × 22	3.53	11.2	88	55	52.7	45.2	39.6	35.2	31.6	28.8	26.4			
W 10 × 26	5.67	18.6	94	60	56.3	48.3	42.3	37.6	33.8					
W 12 × 26	6.29	18.1	112	72	67.0	57.4	50.2	44.6	40.2	36.5	33.5			
W 10 × 33	8.08	27.5	116	76	69.8	59.9	52.4	46.6	41.9					
W 14 × 26	4.50	13.4	121	76		62.0	54.3	48.2	43.4	39.5	36.2	33.4	31.0	
W 16 × 26	4.67	13.4	133	83			59.7	53.0	47.7	43.4	39.8	36.7	34.1	31.8
W 12 × 35	6.42	20.7	154	99	92.2	79.0	69.1	61.4	55.3	50.3	46.1			
W 14 × 34	6.38	19.0	164	105		84.2	73.7	65.5	59.0	53.6	49.1	45.4	42.1	
W 10 × 45	8.38	35.1	165	106	98.8	84.7	74.1	65.9	59.3					
W 16 × 36	6.33	18.2	192	122			86.4	76.8	69.1	62.8	57.6	53.2	49.4	46.1
W 12 × 45	8.13	28.3	193	125	116	99.1	86.7	77.0	69.3	63.0	57.8			
W 10 × 54	10.7	43.9	200	130	120	103	89.9	79.9	71.9					
W 14 × 43	7.88	24.7	209	136		107	94.0	83.5	75.2	68.3	62.6	57.8	53.7	
W 12 × 53	10.3	35.8	234	153	140	120	105	93.5	84.1	76.5	70.1			

W 18 × 40	5.29	15.7	235	148				94.1	84.7	77.0	70.6	65.1	60.5	56.4
W 16 × 45	6.54	20.2	247	158			111	98.8	88.9	80.8	74.1	68.4	63.5	59.3
W 10 × 68	10.8	53.7	256	164	154	132	115	102	92.1					
W 14 × 53	8.00	28.0	261	169		134	118	105	94.1	85.5	78.4	72.4	67.2	
W 18 × 46	5.38	16.6	272	171				109	98.0	89.1	81.6	75.4	70.0	65.3
W 18 × 50	6.88	20.5	303	193				121	109	99.2	90.9	83.9	77.9	72.7
W 16 × 57	6.67	22.8	315	200			142	126	113	103	94.5	87.2	81.0	75.6
W 10 × 88	11.0	68.4	339	213	203	174	153	136	122					
W 14 × 68	10.3	37.3	345	223		177	155	138	124	113	104	95.5	88.7	
W 12 × 79	12.7	51.8	357	232	214	184	161	143	129	117	107			
W 18 × 60	7.00	22.3	369	234				148	133	121	111	102	94.9	88.6
W 14 × 82	10.3	42.8	417	267		214	188	167	150	136	125	115	107	
W 18 × 71	7.08	24.5	438	275				175	158	143	131	121	113	105
W 10 × 112	11.2	86.4	441	273	265	227	198	176	159					
W 12 × 96	12.9	61.4	441	284	265	227	198	176	159	144	132			
W 16 × 77	10.3	35.4	456	295			205	182	164	149	137	126	117	109
W 12 × 120	13.0	75.5	558	353	335	287	251	223	201	183	167			
W 16 × 100	10.4	42.7	600	384			270	240	216	196	180	166	154	144
W 14 × 120	15.6	67.9	636	412		327	286	254	229	208	191	176	164	
W 12 × 152	13.3	94.8	729	453	437	375	328	292	262	239	219			
W 14 × 145	16.6	81.6	780	503		401	351	312	281	255	234	216	201	

(continued)

TABLE 9.2 (*Continued*)

	$F_y = 36$ ksi				Span (ft)									
Designation	L_p ft	L_r ft	M_p kip-ft	M_r kip-ft	24	26	28	30	32	34	36	38	40	42
W 21 × 44	5.25	15.4	286	177	85.9	79.3	73.6	68.7	64.4	60.6	57.2	54.2	51.5	49.1
W 21 × 50	5.42	16.2	330	205	99.0	91.4	84.9	79.2	74.3	69.9	66.0	62.5	59.4	56.6
W 21 × 57	5.63	17.3	387	241	116	107	99.5	92.9	87.1	82.0	77.4	73.3	69.7	66.3
W 24 × 55	5.58	16.6	405	249	122	112	104	97.2	91.1	85.8	81.0	76.7	72.9	69.4
W 24 × 62	5.71	17.2	462	286	139	128	119	111	104	97.8	92.4	87.5	83.2	79.2
W 21 × 68	7.50	22.8	480	303	144	133	123	115	108	101.6	96.0	90.9	86.4	82.3
W 18 × 86	11.0	35.5	558	360	167	155	143	134	126	118	112			
W 21 × 83	7.63	24.9	588	371	176	163	151	141	132	125	118	111	106	101
W 24 × 76	8.00	23.4	600	381	180	166	154	144	135	127	120	114	108	103
W 18 × 106	11.1	40.4	690	442	207	191	177	166	155	146	138			
W 21 × 101	12.0	37.1	759	492	228	210	195	182	171	161	152	144	137	130
W 24 × 94	8.25	25.9	762	481	229	211	196	183	171	161	152	144	137	131
W 27 × 94	8.83	25.9	834	527			214	200	188	177	167	158	150	143

W 24 × 103	8.29	27.0	840	531	252	233	216	202	189	178	168	159	151	144
W 30 × 90	8.71	24.8	849	531				204	191	180	170	161	153	146
W 24 × 104	12.1	35.2	867	559	260	240	223	208	195	184	173	164	156	149
W 18 × 130	11.3	47.7	870	555	261	241	224	209	196	184	174			
W 21 × 122	12.2	41.0	921	592	276	255	237	221	207	195	184	175	166	158
W 24 × 117	12.3	37.1	981	631	294	272	252	235	221	208	196	186	177	168
W 27 × 114	9.08	28.2	1,029	648			265	247	232	218	206	195	185	176
W 30 × 108	8.96	26.3	1,038	648				249	234	220	208	197	187	178
W 18 × 158	11.4	56.5	1,068	672	320	296	275	256	240	226	214			
W 24 × 131	12.4	39.3	1,110	713	333	307	285	266	250	235	222	210	200	190
W 21 × 147	12.3	46.4	1,119	713	336	310	288	269	252	237	224	212	201	192
W 30 × 124	9.29	28.2	1,224	769				294	275	259	245	232	220	210
W 33 × 118	9.67	27.8	1,245	778						264	249	236	224	213
W 24 × 146	12.5	42.0	1,254	804	376	347	322	301	282	266	251	238	226	215
W 24 × 162	12.7	45.2	1,404	897	421	389	361	337	316	297	281	266	253	241
W 30 × 148	9.50	30.6	1,500	945				360	338	318	300	284	270	257
W 33 × 141	10.1	30.1	1,542	971						327	308	292	278	264

Source: Compiled from data in the *Manual of Steel Construction* with permission of the publishers, American Institute of Steel Construction.

TABLE 9.3 Factored Load-Span Values for 50 ksi Beams

	F_y = 50 ksi				Span (ft)									
Designation	L_p (ft)	L_r ft	M_p kip-ft	M_r kip-ft	12	14	16	18	20	22	24	26	28	30
W 10 × 17	2.99	8.37	78	54	46.8	40.1	35.1	31.2	28.1					
W 12 × 16	2.73	7.44	84	57	50.3	43.1	37.7	33.5	30.2	27.4	25.1			
W 10 × 19	3.09	8.89	90	63	54.0	46.3	40.5	36.0	32.4					
W 12 × 22	3.00	8.41	122	85	73.3	62.8	54.9	48.8	44.0	40.0	36.6			
W 10 × 26	4.81	13.6	130	93	78.3	67.1	58.7	52.2	47.0					
W 12 × 26	5.34	13.8	155	111	93.0	79.7	69.8	62.0	55.8	50.7	46.5			
W 10 × 33	6.86	19.8	162	117	97.0	83.1	72.8	64.7	58.2					
W 14 × 26	3.82	10.2	168	118		86.1	75.4	67.0	60.3	54.8	50.3	46.4	43.1	
W 16 × 26	3.96	10.4	184	128			82.9	73.7	66.3	60.3	55.3	51.0	47.4	44.2
W 12 × 35	5.44	15.2	213	152	128	110	96.0	85.3	76.8	69.8	64.0			
W 14 × 34	5.41	14.4	228	162		117	102	91.0	81.9	74.5	68.3	63.0	58.5	
W 10 × 45	7.11	24.1	229	164	137	118	103	91.5	82.4					
W 16 × 36	5.37	14.0	267	188			120	107	96.0	87.3	80.0	73.8	68.6	64.0
W 12 × 45	6.89	20.3	268	192	161	138	120	107	96.3	87.5	80.3			
W 10 × 54	9.05	30.2	278	200	167	143	125	111	100					
W 14 × 43	6.68	18.3	290	209		149	131	116	104	94.9	87.0	80.3	74.6	
W 12 × 53	8.77	25.6	325	235	195	167	146	130	117	106	97.4			

W 18 × 40	4.49	12.1	327	228				131	118	107	98.0	90.5	84.0	78.4
W 16 × 45	5.55	15.2	343	242			154	137	123	112	103	95.0	88.2	82.3
W 10 × 68	9.16	36.0	355	252	213	183	160	142	128					
W 14 × 53	6.79	20.1	363	259		187	163	145	131	119	109	101	93.3	
W 18 × 46	4.56	12.6	378	263				151	136	124	113	105	97.2	90.7
W 18 × 50	5.83	15.6	421	296				168	152	138	126	117	108	101
W 16 × 57	5.66	16.6	438	307			197	175	158	143	131	121	113	105
W 10 × 88	9.30	45.1	471	328	283	242	212	188	170					
W 14 × 68	8.70	26.4	479	343		246	216	192	173	157	144	133	123	
W 12 × 79	10.8	35.7	496	357	298	255	223	198	179	162	149			
W 18 × 60	5.94	16.6	513	360				205	185	168	154	142	132	123
W 14 × 82	8.77	29.5	579	410		298	261	232	209	190	174	160	149	
W 18 × 71	6.01	17.8	608	423				243	219	199	183	168	156	146
W 10 × 112	9.48	56.5	613	420	368	315	276	245	221					
W 12 × 96	10.9	41.4	613	437	368	315	276	245	221	200	184			
W 16 × 77	8.70	25.5	633	453			285	253	228	207	190	175	163	152
W 12 × 120	11.1	50.0	775	543	465	399	349	310	279	254	233			
W 16 × 100	8.84	29.6	833	590			375	333	300	273	250	231	214	200
W 14 × 120	13.2	46.2	883	633		454	398	353	318	289	265	245	227	
W 12 × 152	11.3	62.1	1,013	697	608	521	456	405	365	331	304			
W 14 × 145	14.1	54.7	1,083	773		557	488	433	390	355	325	300	279	

(continued)

TABLE 9.3 (*Continued*)

	$F_y = 50$ ksi				Span (ft)									
Designation	L_p (ft)	L_r ft	M_p kip-ft	M_r kip-ft	24	26	28	30	32	34	36	38	40	42
W 21 × 44	4.45	12.0	398	272	119	110	102	95.4	89.4	84.2	79.5	75.3	71.6	68.1
W 21 × 50	4.60	12.5	458	315	138	127	118	110	103	97.1	91.7	86.8	82.5	78.6
W 21 × 57	4.77	13.1	538	370	161	149	138	129	121	114	108	102	96.8	92.1
W 24 × 55	4.74	12.9	563	383	169	156	145	135	127	119	113	107	101	96.4
W 24 × 62	4.84	13.3	642	440	193	178	165	154	144	136	128	122	116	110
W 21 × 68	6.36	17.3	667	467	200	185	171	160	150	141	133	126	120	114
W 18 × 86	9.30	26.1	775	553	233	215	199	186	174	164	155			
W 21 × 83	6.47	18.5	817	570	245	226	210	196	184	173	163	155	147	140
W 24 × 76	6.79	18.0	833	587	250	231	214	200	188	176	167	158	150	143
W 18 × 106	9.40	28.7	958	680	288	265	246	230	216	203	192			
W 21 × 101	10.2	27.6	1,054	757	316	292	271	253	237	223	211	200	190	181
W 24 × 94	7.00	19.4	1,058	740	318	293	272	254	238	224	212	201	191	181
W 27 × 94	7.50	19.9	1,158	810			298	278	261	245	232	219	209	199
W 24 × 103	7.04	20.0	1,167	817	350	323	300	280	263	247	233	221	210	200
W 30 × 90	7.39	19.4	1,179	817				283	265	250	236	223	212	202

W 24 × 104	10.3	26.8	1,204	860	361	333	310	289	271	255	241	228	217	206
W 18 × 130	9.55	32.8	1,208	853	363	335	311	290	272	256	242			
W 21 × 122	10.3	29.8	1,279	910	384	354	329	307	288	271	256	242	230	219
W 24 × 117	10.4	27.9	1,363	970	409	377	350	327	307	289	273	258	245	234
W 27 × 114	7.71	21.3	1,429	997			368	343	322	303	286	271	257	245
W 30 × 108	7.60	20.3	1,442	997				346	324	305	288	273	260	247
W 18 × 158	9.69	38.0	1,483	1,033	445	411	381	356	334	314	297			
W 24 × 131	10.5	29.1	1,542	1,097	463	427	396	370	347	326	308	292	278	264
W 21 × 147	10.4	32.8	1,554	1,097	466	430	400	373	350	329	311	294	280	266
W 30 × 124	7.88	21.5	1,700	1,183				408	383	360	340	322	306	291
W 33 × 118	8.20	21.7	1,729	1,197						366	346	328	311	296
W 24 × 146	10.6	30.6	1,742	1,237	523	482	448	418	392	369	348	330	314	299
W 24 × 162	10.8	32.4	1,950	1,380	585	540	501	468	439	413	390	369	351	334
W 30 × 148	8.06	22.8	2,083	1,453				500	469	441	417	395	375	357
W 33 × 141	8.59	23.1	2,142	1,493						454	428	406	386	367

Source: Compiled from data in the *Manual of Steel Construction* with permission of the publishers, American Institute of Steel Construction.

The W 12 × 53 should be checked to see if it is still appropriate after the self-weight of the beam is added to the superimposed factored load.

$$1.2 \times (\text{Dead load}) + 1.6 \times (\text{Live load})$$

$$= 1.2 \times (53 \text{ lb/ft}) + 1.6 \times (0 \text{ lb/ft}) = 63.6 \text{ lb/ft } [0.93 \text{ kN/m}]$$

$$\text{Total self-weight} = 63.6 \text{ lb/ft} \times 24 \text{ ft} \times \left(\frac{1 \text{ kip}}{1000 \text{ lb}}\right) = 1.53 \text{ kips } [6.8 \text{ kN}]$$

$$\text{Total load} = 1.53 \text{ kips} + 70.1 \text{ kips} = 71.6 \text{ kips } [318 \text{ kN}]$$

Thus, the W 12 × 53 is not viable for this given loading. Similarly, the W 18 × 40 does not work. The lightest shape is the W 21 × 44, and the shallowest is the W 12 × 79. It should be noted that although the W 12 × 79 is the shallowest section, it comes at a cost of considerably more steel. If a W 14 × 53 were substituted for the W 12 x 79, it would come with a 45 percent weight savings.

Problems 9.8.A, B, C, D, E, F, G, H. For each of the following conditions, find (a) the lightest permitted shape and (b) the shallowest permitted shape of A36 steel:

	Span	Live Load	Superimposed Dead Load
A	16 ft	3 kips/ft [43.8 kN/m]	3 kips/ft [43.8 kN/m]
B	20 ft	1 kips/ft [14.6 kN/m]	0.5 kips/ft [7.30 kN/m]
C	36 ft	1 kips/ft [14.6 kN/m]	0.5 kips/ft [7.30 kN/m]
D	40 ft	1.25 kips/ft [18.2 kN/m]	1.25 kips/ft [18.2 kN/m]
E	18 ft	0.333 kips/ft [4.86 kN/m]	0.625 kips/ft [9.12 kN/m]
F	32 ft	1.167 kips/ft [17.0 kN/m]	3.5 kips/ft [51.0 kN/m]
G	42 ft	1 kips/ft [14.6 kN/m]	0.238 kips/ft [3.47 kN/m]
H	28 ft	0.5 kips/ft [7.3 kN/m]	0.5 kips/ft [7.3 kN/m]

Equivalent Load Techniques

The safe service loads in Table 9.2 are uniformly distributed loads on simple beams. Actually, the table values are determined on the basis of bending moment and limiting bending stress so that it is possible to use the tables for other loading conditions for some purposes. Because framing systems always include some beams with other than simple uniformly distributed loadings, this is sometimes a useful process for design.

Consider the following situation: a beam with a load consisting of two equal concentrated loads placed at the beam third points—in other words, Case 3 in Figure 3.25. For this condition, the figure yields a maximum moment value expressed as $PL/3$. By equating this to the moment value for a uniformly distributed load, a relationship between the two loads can be derived. Thus:

$$\frac{WL}{8} = \frac{PL}{3} \qquad \text{or} \qquad W = 2.67P$$

which shows that if the value of one of the concentrated loads in Case 3 of Figure 3.25 is multiplied by 2.67, the result would be an *equivalent uniform load* or *equivalent tabular load* (called EUL or ETL) that would produce the same magnitude of loading as the true loading condition.

Although the expression "equivalent uniform load" is the general name for this converted loading, when derived to facilitate the use of tabular materials, it is also referred to as the *equivalent tabular load* (ETL), which is the designation used in this book. Figure 3.25 yields the ETL factors for several common loading conditions.

It is important to remember that the EUL or ETL is based only on consideration of flexure (that is, on limiting bending stress) so that investigations for deflection, shear, or bearing must use the true loading conditions for the beam.

This method may also be used for any loading condition, not just the simple, symmetrical conditions shown in Figure 3.25. The process consists of first finding the true maximum moment due to the actual loading; then this is equated to the expression for the maximum moment for a theoretical uniform load, and the EUL is determined. Thus:

$$M = \frac{WL}{8} \qquad \text{or} \qquad W = \frac{8M}{L}$$

The expression $W = 8M/L$ is the general expression for an equivalent uniform load (or ETL) for any loading condition.

9.9 STEEL TRUSSES

When iron, and then steel, emerged as major industrial materials in the eighteenth and nineteenth centuries, one of the earliest applications to spanning structures was in the development of trusses and trussed forms of arches and bents. One reason for this was the early limit on the size of members that could be produced. In order to create a reasonably large structure, therefore, it was necessary to link together a large number of small parts.

Another major technical problem for such assemblages was the achievement of the many joints. Thus, the creation of steel structural assemblages involves design of many joints, which must be both economical and practical for formation. Various connecting devices or methods have been employed; a major one used for building structures in earlier times was hot-driven rivets. The process for this consisted of matching up of holes in members to be connected, placing a heat-softened steel pin in the hole, and then beating the heck out of the protruding ends of the pin to form a rivet.

Basic forms developed for early connections are still widely used. Today, however, joints are mostly achieved by welding or with highly tightened bolts in place of rivets. Welding is mostly employed for connections made in the fabricating shop (called *shop connections*), whereas bolting is preferred for connections made at the erection site (called *field connections*). Riveting and bolting are often achieved with an intermediate connecting device (gusset plate, connecting angles, etc.), whereas welding is frequently achieved directly between attached members.

Truss forms relate to the particular structural application (bridge, gable-form roof, arch, flat-span floor, etc.), to the magnitude of the span, and to the materials and methods of construction.

Some typical forms for individual trusses used in steel construction are shown in Figure 9.13. The forms in widest use are the parallel-chorded types, shown in Figure 9.13*a* and *b*; these often are produced as manufactured, proprietary products by individual companies. (See discussion in Section 9.10.) Variations of the gable-form truss (Figure 9.13*c*) can be produced for a wide range of spans.

As planar elements, trusses are quite unstable in a lateral direction (perpendicular to the plane of the truss). Structural systems employing trusses require considerable attention for development of bracing. The delta truss (Figure 9.13*d*) is a unique self-stabilizing form that is frequently used for towers and columns, but it may also be used for a spanning truss that does not require additional bracing.

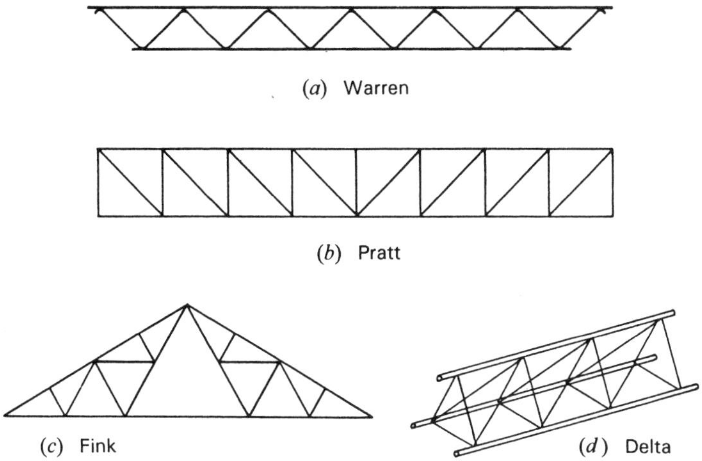

Figure 9.13 Common forms of light steel trusses.

9.10 MANUFACTURED TRUSSES FOR FLAT SPANS

Factory-fabricated, parallel-chord trusses are produced in a wide range of sizes by a number of manufacturers. Most producers comply with the regulations of industry-wide organizations; for light steel trusses, the principal such organization is the Steel Joist Institute, called the SJI. Publications of the SJI are a chief source of general information (see Ref. 8), although the products of individual manufacturers vary, so that much valuable design information is available directly from the suppliers of a specific product. Final design and development of construction details for a particular project must be done in cooperation with the supplier of the products.

Light steel parallel-chord trusses, called *open web joists*, have been in use for many years. Early versions used all-steel bars for the chords and the continuously bent web members (see Figure. 9.14), so that they were also referred to as *bar joists*. Although other elements are now used for the chords, the bent steel rod is still used for the diagonal web members for some of the smaller-sized joists. The range of size of this basic element has now been stretched considerably, resulting in members as long as 150 ft [46 m] and depths of 7 ft [2.14 m] and more. At the larger size range, the members are usually more common forms for steel trusses—double angles, structural tees, and so on. Still, a considerable usage is made of the smaller sizes for both floor joists and roof rafters.

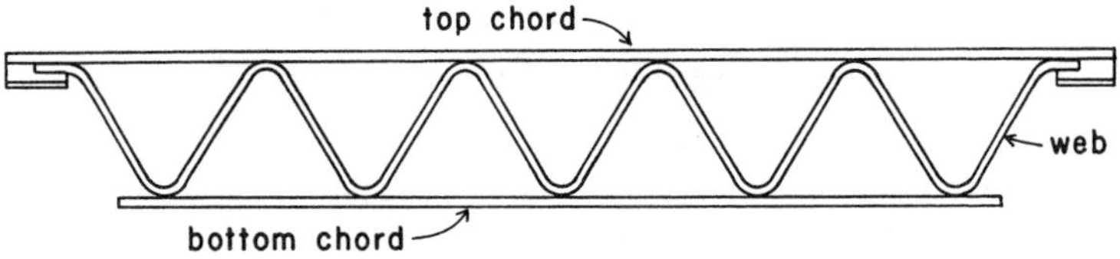

Figure 9.14 Form of a short span open web steel joist.

Table 9.4 is adapted from a standard table in a publication of the SJI (Ref. 6). This table lists a number of sizes available in the K series, which is the lightest group of joists. Joists are identified by a three-unit designation. The first number indicates the overall nominal depth of the joist, the letter indicates the series, and the second number indicates the class of size of the members—the higher the number, the heavier and stronger the joist.

Table 9.4 can be used to select the proper joist for a determined load and span situation. Figure 9.15 shows the basis for determination of the span for a joist. There are two entries in the table for each span: The first number represents the total factored load capacity of the joist in pounds per ft of the joist length (lb/ft), and the number in parentheses is the load that will produce a deflection of 1/360 of the span. The following examples illustrate the use of the table data for some common design situations. For the purpose of illustration, the examples use data from Table 9.4. However, more joists sizes are available, and their capacities are given in the reference for Table 9.4.

Example 13. Open web steel joists are to be used to support a roof with a unit live load of 20 psf and a unit dead load of 15 psf (not including the weight of the joists) on a span of 40 ft. Joists are spaced at 6 ft center to center, Select the lightest joist if deflection under live load is limited to 1/360 of the span.

Solution: The first step is to determine the unit load per ft on the joists, thus:

Live load: 6(20) = 120 lb/ft [1.8 kN/m]

Dead load: 6(15) = 90 lb/ft (not including joist weight) [1.3 kN/m]

Total factored load: 1.2(90) + 1.6(120) = 108 + 192 = 300 lb/ft [4.4 kN/m]

TABLE 9.4 Safe Factored Loads for K-Series Open Web Joists[a]

Joist Designation:	12K1	12K3	12K5	14K1	14K3	14K6	16K2	16K4	16K6	18K3	18K5	18K7	20K3	20K5	20K7
Weight (lb/ft):	5.0	5.7	7.1	5.2	6.0	7.7	5.5	7.0	8.1	6.6	7.7	9.0	6.7	8.2	9.3
Span (ft)															
20	357	448	607	421	528	729	545	732	816	687	816	816	767	816	816
	(142)	(177)	(230)	(197)	(246)	(347)	(297)	(386)	(426)	(423)	(490)	(490)	(517)	(550)	(550)
22	295	369	500	347	435	641	449	602	739	567	769	816	632	816	816
	(106)	(132)	(172)	(147)	(184)	(259)	(222)	(289)	(351)	(316)	(414)	(438)	(393)	(490)	(490)
24	246	308	418	291	363	537	377	504	620	475	644	781	530	720	816
	(81)	(101)	(132)	(113)	(141)	(199)	(170)	(221)	(269)	(242)	(318)	(382)	(302)	(396)	(448)
26				246	310	457	320	429	527	403	547	665	451	611	742
				(88)	(110)	(156)	(133)	(173)	(211)	(190)	(249)	(299)	(236)	(310)	(373)
28				212	267	393	276	369	454	347	472	571	387	527	638
				(70)	(88)	(124)	(106)	(138)	(168)	(151)	(199)	(239)	(189)	(248)	(298)
30							239	320	395	301	409	497	337	457	555
							(86)	(112)	(137)	(123)	(161)	(194)	(153)	(201)	(242)
32							210	282	346	264	359	436	295	402	487
							(71)	(92)	(112)	(101)	(132)	(159)	(126)	(165)	(199)
36										209	283	344	233	316	384
										(70)	(92)	(111)	(88)	(115)	(139)
40													188	255	310
													(64)	(84)	(101)

(continued)

TABLE 9.4 (*Continued*)

Joist Designation:	22K4	22K6	22K9	24K4	24K6	24K9	26K5	26K7	26K9	28K6	28K8	28K10	30K7	30K9	30K12
Weight (lb/ft):	8.0	8.8	11.3	8.4	9.7	12.0	9.8	10.9	12.2	11.4	12.7	14.3	12.3	13.4	17.6
Span (ft)															
28	516	634	816	565	693	816	692	816	816	813	816	816			
	(270)	(328)	(413)	(323)	(393)	(456)	(427)	(501)	(501)	(541)	(543)	(543)			
30	448	550	738	491	602	807	601	730	816	708	816	816	816	816	816
	(219)	(266)	(349)	(262)	(319)	(419)	(346)	(417)	(459)	(439)	(500)	(500)	(543)	(543)	(543)
32	393	484	647	430	530	709	528	641	770	620	764	815	743	815	815
	(180)	(219)	(287)	(215)	(262)	(344)	(285)	(343)	(407)	(361)	(438)	(463)	(461)	(500)	(500)
36	310	381	510	340	417	559	415	504	607	490	602	723	586	705	723
	(126)	(153)	(201)	(150)	(183)	(241)	(199)	(240)	(284)	(252)	(306)	(366)	(323)	(383)	(392)
40	250	307	412	274	337	451	337	408	491	395	487	629	473	570	650
	(91)	(111)	(146)	(109)	(133)	(175)	(145)	(174)	(207)	(183)	(222)	(284)	(234)	(278)	(315)
44	206	253	340	227	277	372	277	337	405	326	402	519	390	470	591
	(68)	(83)	(109	(82)	(100)	(131)	(108)	(131)	(155)	(137)	(167)	(212)	(176)	(208)	(258)
48				190	233	313	233	282	340	273	337	436	328	395	542
				(63)	(77)	(101)	(83)	(100)	(119)	(105)	(128)	(163)	(135)	(160)	(216)
52							197	240	289	233	286	371	279	335	498
							(65)	(79)	(102)	(83)	(100)	(128)	(106)	(126)	(184)
56										200	246	319	240	289	446
										(66)	(80)	(102)	(84)	(100)	(153)
60													209	250	389
													(69)	(81)	(124)

Source: Data adapted from more extensive tables in the *Guide for Specifying Steel Joists with LRFD, 2000* (Ref. 7), with permission of the publishers, Steel Joist Institute. The Steel Joist Institute publishes both specifications and load tables; each of these contains standards that are to be used in conjunction with one another.

[a]Loads in pounds per ft of joist span. First entry represents the total factored joist capacity; entry in parentheses is the load that produces a deflection of 1/360 of the span. See Fig. 9.16 for definition of span.

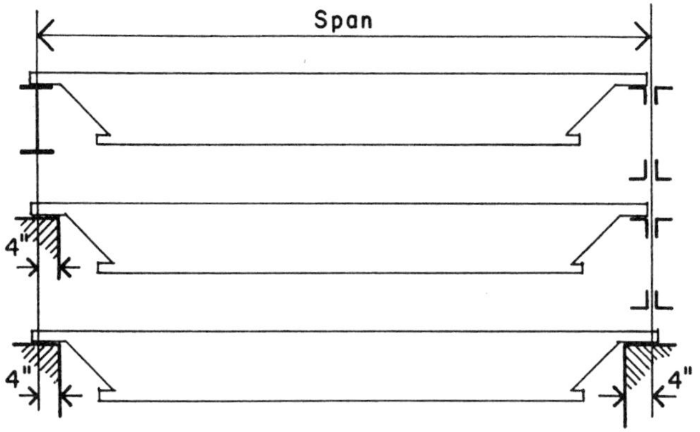

Figure 9.15 Definition of span for open web steel joists, as given in Ref. 6. Reprinted with permission of the Steel Joist Institute.

This yields the two numbers (total factored load and live load only) that can be used to scan the data for the given span in Table 9.4. Note that the joist weight—so far excluded in the computation—is included in the total load entries in the table. Once a joist is selected, therefore, the actual joist weight (given in the table) must be deducted from the table entry for comparison with the computed values. We thus note from the table the possible choices listed in Table 9.5. Although the joists weights are all very close, the 24K6 is the lightest choice.

Example 14. Open web steel joists are to be used for a floor with a unit live load of 75 psf [3.59 kN/m^2 and a unit dead load of 40 psf [1.91 kN/m^2 (not including the joist weight) on a span of 30 ft [9.15 m]. Joists

TABLE 9.5 Possible Choices for the Roof Joist

Load Condition	Required Capacity (lb/ft)	Capacity of the Indicated Joists (lb/ft) 22K9	24K6	26K5
Factored total capacity		412	337	337
Joist weight from Table 9.4		11.3	9.7	9.8
Factored joist weight		14	12	12
Net usable capacity	300	398	325	325
Load for deflection of 1/360	120	146	133	157

are 2 ft [0.61 m] on center, and deflection is limited to 1/240 of the span under total load and 1/360 of the span under live load only. Determine the lightest possible joist and the lightest joist of least depth possible.

Solution: As in the previous example, the unit loads are first determined, thus:

Live load: 2(75) = 150 lb/ft (for limiting deflection of L/360) [2.2 kN/m]

Dead load: 2(40) = 80 lb/ft (not including joist weight) [1.2 kN/m]

Total service load: 150 + 80 = 230 lb/ft [3.36 kN/m]

Total factored load: 1.2(80) + 1.6(150) = 96 + 240 = 336 lb/ft [4.9 kN/m]

To satisfy the deflection criteria for total load, the limiting value for deflection in parentheses in the table should be not less than (240/360) (230) = 153 lb/ft 2.2 kN/m]. Because this is slightly larger than the live load, it becomes the value to look for in the table. Possible choices obtained from Table 9.4 are listed in Table 9.6, from which the following may be observed:

The lightest joist is the 22K4.

The shallowest depth joist is the 18K5.

In some situations, it may be desirable to select a deeper joist, even though its load capacity may be somewhat redundant. Total sag, rather than an abstract curvature limit, may be of more significance for a flat roof structure. For example, for the 40 ft [12.2 m] span in Example 13, a sag of 1/360 of the span = (1/360)(40 × 12) = 1.33 in. [338 mm].

TABLE 9.6 Possible Choices for the Floor Joist

		Capacity of the Indicated Joists (lb/ft)		
Load Condition	Required Capacity (lb/ft)	18K5	20K5	22K4
Factored total capacity		409	457	448
Joist weight from Table 9.4		7.7	8.2	8.0
Factored joist weight		10	10	10
Net usable capacity	300	399	447	438
Load for deflection	153	161	201	219

The actual effect of this dimension on roof drainage or in relation to interior partition walls must be considered. For floors, a major concern is for bounciness, and this very light structure is highly vulnerable in this regard. Designers therefore sometimes deliberately choose the deepest feasible joist for floor structures in order to get all the help possible to reduce deflection as a means of stiffening the structure in general against bouncing effects.

As mentioned previously, joists are available in other series for heavier loads and longer spans. The SJI, as well as individual suppliers, also have considerably more information regarding installation details, suggested specifications, bracing, and safety during erection for these products.

Stability is a major concern for these elements because they have very little lateral or torsional resistance. Other construction elements, such as decks and ceiling framing, may help, but the whole bracing situation must be carefully studied. Lateral bracing in the form of x-braces or horizontal ties is generally required for all steel joist construction, and the reference source for Table 9.4 (Ref. 6) has considerable information on this topic.

One means of assisting stability has to do with the typical end support detail, as shown in Figure 9.14. The common method of support consists of hanging the trusses by the ends of their top chords, which is a general means of avoiding the rollover type of rotational buckling at the supports that is illustrated in Figure 3.63*c*. For construction detailing, however, this adds a dimension to the overall depth of the construction, in comparison to an all-beam system with the joist/beams and supporting girders all having their tops level. This added dimension (the depth of the end of the joist) is typically 2.5 in. [63.5 mm] for small joists and 4 in. [101 mm] for larger joists.

For development of a complete truss system, a special type of prefabricated truss available is that described as a *joist girder*. This truss is specifically designed to carry the regularly spaced, concentrated loads consisting of the end support reactions of joists. A common form of joist girder is shown in Figure 9.16. Also shown in the figure is the form of standard designation for a joist girder, which includes indications of the nominal girder depth, the number of spaces between joists (called the girder *panel unit*), and the end reaction force from the joists—which is the unit concentrated load on the girder.

Predesigned joist girders (that is, girders actually designed for fabrication by the suppliers) may be selected from catalogs in a manner similar to that for open web joists. The procedure is usually as follows:

1. The designer determines the joist spacing, joist load, and girder span. (The joist spacing should be a full number division of the girder span.)
2. The designer uses this information to specify the girder by the standard designation.
3. The designer chooses the girder from a particular manufacturer's catalog or simply specifies it for the supplier.

Illustrations of the use of joists and complete truss systems are given in the building design examples in Part V.

Problem 9.10.A. Open web steel joists are to be used for a roof with a live load of 25 psf [1.2 kN/m2] and a dead load of 20 psf [957 N/m^2] (not including the joist weight) on a span of 48 ft [14.6 m]. Joists are 4 ft [1.22 m] on center, and deflection under live load is limited to 1/360 of the span. Select the lightest joist.

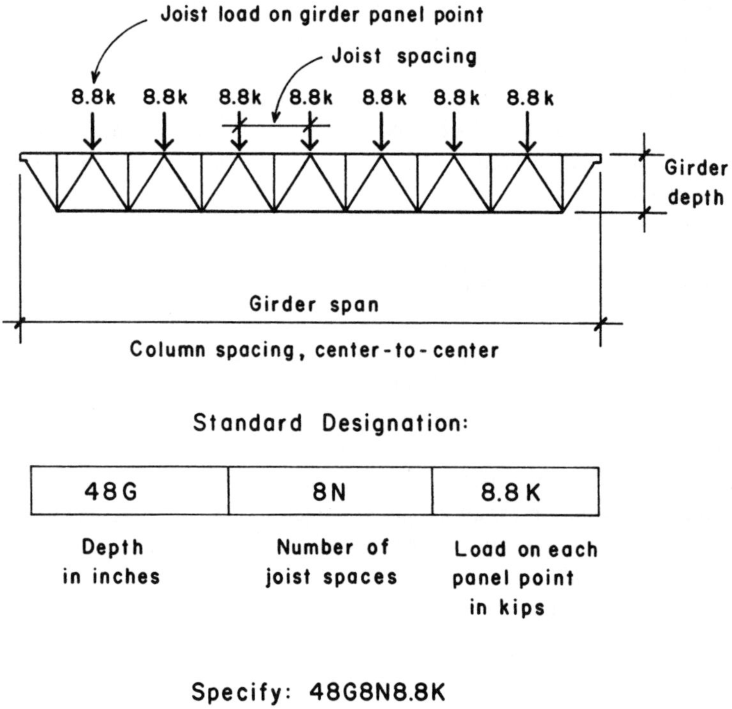

Figure 9.16 Considerations for layout and designation of joist girders.

Problem 9.10.B. Open web steel joists are to be used for a roof with a live load of 30 psf [1.44 kN/m^2] and a dead load of 18 psf [862 N/m^2] (not including the joist weight) on a span of 44 ft [13.42 m]. Joists are 5 ft [1.53 m] on center, and deflection is limited to 1/360 of the span. Select the lightest joist.

Problem 9.10.C. Open web steel joists are to be used for a floor with a live load of 50 psf [2.39 kN/m^2] and a dead load of 45 psf [2.15 kN/m^2] (not including the joist weight) on a span of 36 ft [11 m]. Joists are 2 ft [0.61 m] on center, and deflection is limited to 1/360 of the span under live load only and to 1/240 of the span under total load. Select (a) the lightest possible joist and (b) the shallowest depth possible joist.

Problem 9.10.D. Repeat Problem 9.10.C, except that the live load is 100 psf [4.79 kN/m^2], the dead load is 35 psf [1.67 kN/m^2], and the span is 26 ft [7.93 m].

9.11 DECKS WITH STEEL FRAMING

Figure 9.17 shows four possibilities for a floor deck used in conjunction with a framing system of rolled steel beams. When a wood deck is used (Figure 9.17*a*), it is usually supported by and nailed to a series of wood joists, which are in turn supported by the steel beams. However, in some cases the deck may be nailed to wood members that are bolted to the tops of the steel beams, as shown in the figure. For floor construction, it is now also common to use a concrete fill on top of the wood deck, for added stiffness, fire protection, and improved acoustic behavior.

A site-cast concrete deck (Figure 9.17*b*) is typically formed with plywood panels placed against the bottoms of the top flanges of the beams. This helps to lock the slab and beams together for lateral effects, although steel lugs are also typically welded to the tops of the beams for composite construction.

Concrete may also be used in the form of precast deck units. In this case, steel elements are imbedded in the ends of the precast units and are welded to the beams. A site-poured concrete fill is typically used to provide a smooth top surface and is bonded to the precast units for added structural performance.

Formed sheet steel units may be used in one of three ways: as the primary structure, as strictly forming for the concrete deck, or as a composite element in conjunction with the concrete. Attachment of this type of deck to the steel beams is usually achieved by welding the steel units to the beams before the concrete is placed.

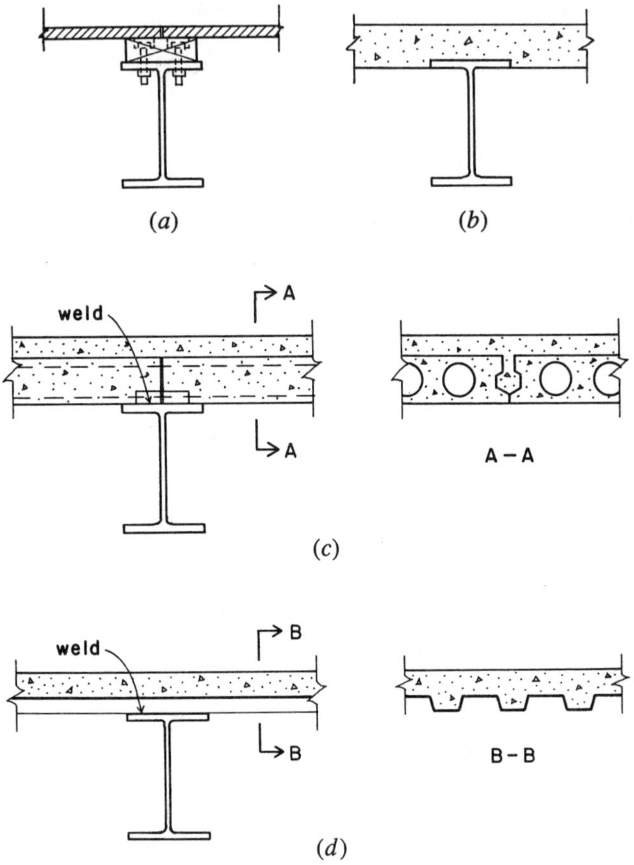

Figure 9.17 Typical forms of floor deck construction used with steel framing.

Three possibilities for roof decks using steel elements are shown in Figure 9.18. Roof loads are typically lighter than floor loads, and bounciness of the deck is usually not a major concern. (A possible exception is where suspended elements may be hung from the deck and can create a problem with vertical movements during an earthquake.)

A fourth possibility for the roof is the plywood deck shown in Figure 9.17*a*. This is, in fact, probably a wider use of this form of construction.

Decks using formed sheet steel elements are discussed more fully in Chapter 12, together with other structural products and systems developed from formed sheet stock.

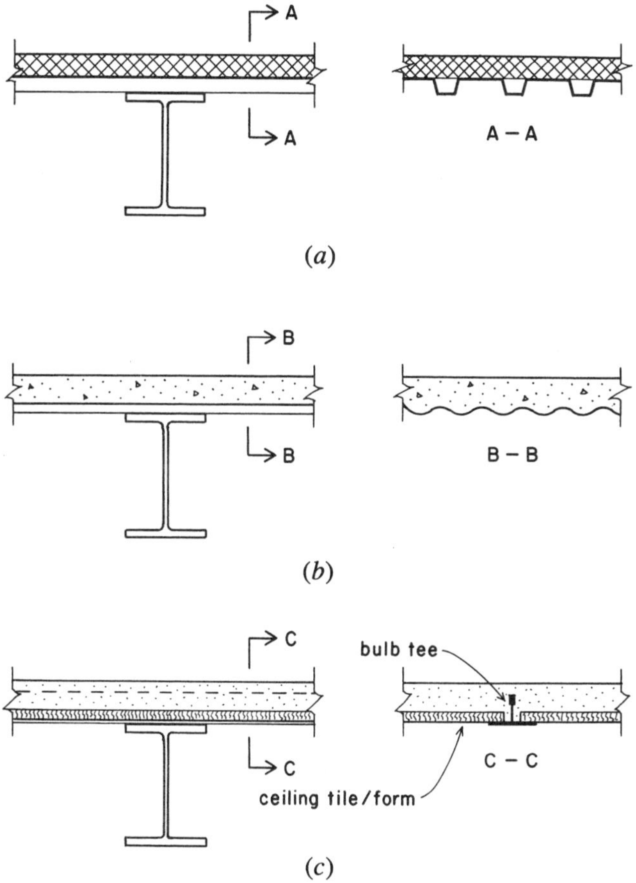

Figure 9.18 Typical forms of roof deck construction used with steel framing.

9.12 CONCENTRATED LOAD EFFECTS ON BEAMS

An excessive bearing reaction on a beam, or an excessive concentrated load at some point in the beam span, may cause either localized yielding or *web crippling* (buckling of the thin beam web). The AISC Specification requires that beam webs be investigated for these effects and that web stiffeners be used if the concentrated load exceeds limiting values.

The three common situations for this effect occur as shown in Figure 9.19. Figure 9.19*a* shows the beam end bearing on a support (commonly

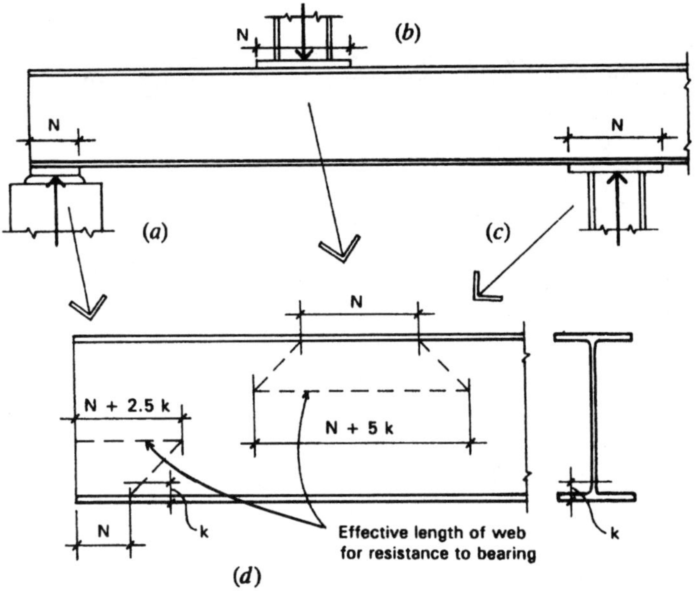

Figure 9.19 Considerations for bearing in steel beams with thin webs as related to web crippling (buckling of the thin web in diagonal compression).

a masonry or concrete wall), with the reaction force transferred to the beam bottom flange through a steel bearing plate. Figure 9.19*b* shows a column load applied to the top of the beam at some point within the beam span. Figure 9.19*c* shows what may be the most frequent occurrence of this condition: a beam supported in bearing on top of a column with the beam continuous through the joint.

Figure 9.19*d* shows the development of the effective portion of the web length (along the beam span) that is assumed to resist bearing forces. For yield resistance, the maximum end reaction and the maximum load within the beam span are defined as follows (see Figure 9.19*d*):

$$\text{Maximum end reaction} = (0.66F_y)(t_w)[N + 2.5(k)]$$

$$\text{Maximum interior load} = (0.66F_y)(t_w)[N + 5(k)]$$

where t_w = thickness of the beam web
N = length of the bearing

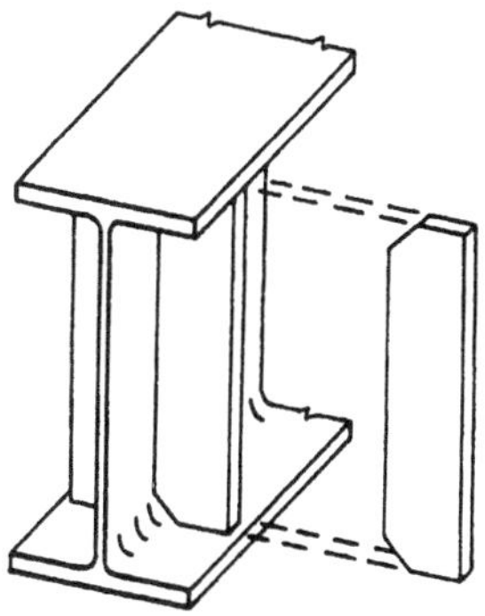

Figure 9.20 Use of stiffeners to prevent buckling of a thin web in a beam.

k = distance from the outer face of the beam flange to the web toe of the fillet (radius) of the corner between the web and the flange

For W shapes, the dimensions t_w and k are provided in the AISC Manual (Ref. 5) tables of properties for rolled shapes.

When these values are exceeded, it is recommended that web stiffeners—usually consisting of steel plates welded into the channel-shaped sides of the beam as shown in Figure 9.20—be provided at the location of the concentrated load. These stiffeners may indeed add to bearing resistance, but the usual reason for using them is to alleviate the other potential form of failure: web crippling due to an excessively slender web; hence, their usual description as *web stiffeners*.

The AISC provides additional formulas for computation of limiting loads due to web crippling. However, the AISC Manual (Ref. 5) also provides data for shortcut methods.

10

STEEL COLUMNS AND FRAMES

Steel compression members range from small, single-piece columns and truss members to huge, built-up sections for high-rise buildings and large tower structures. The basic column function is one of simple compressive force resistance, but it is often complicated by the effects of buckling and the possible presence of bending actions. This chapter deals with various issues relating to the design of individual compression members and with the development of building structural frameworks.

10.1 COLUMN SHAPES

For modest load conditions, the most frequently used shapes are the round pipe, the rectangular tube, and the W shapes with wide flanges (see Figure 10.1). Accommodation of beams for framing is most easily achieved with W shapes of 10 in. or larger nominal depth. For various reasons, it is sometimes necessary to make up a column section by assembling two or more individual steel elements. Figure 10.2 shows

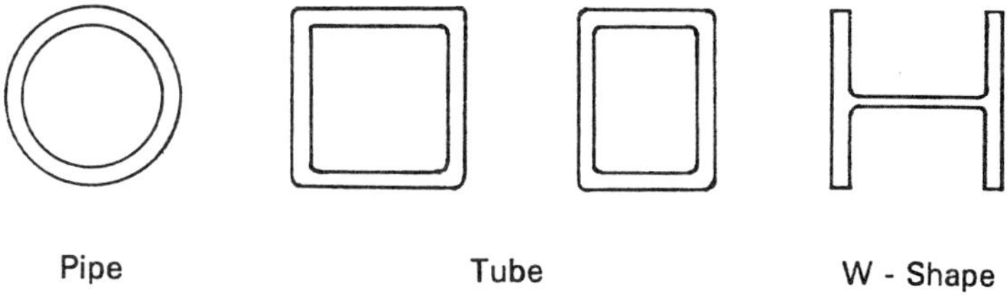

Figure 10.1 Common cross-sectional shapes for steel columns.

some such shapes that are used for special purposes. The customized assemblage of built-up sections is usually costly, so a single piece is typically favored if one is available.

One widely used built-up section is the double angle, shown in Figure 10.2*f*. This occurs most often as a member of a truss or as a bracing member in a frame; the general stability of the paired members is much better than that of a single angle. This section is seldom used as a building column, however.

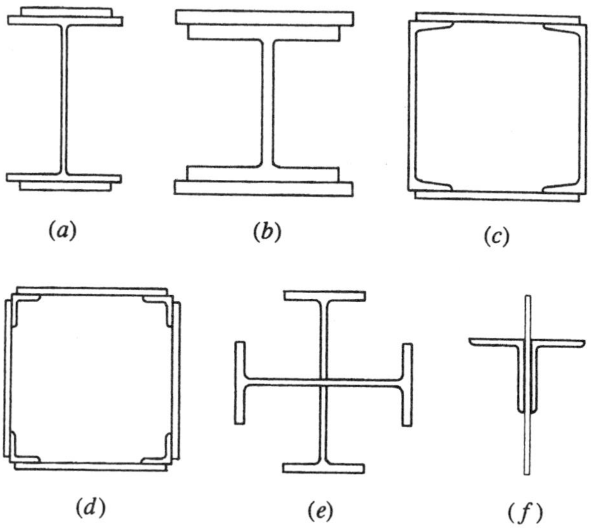

Figure 10.2 Various forms of combined, built-up column shapes for steel columns.

10.2 SLENDERNESS AND END CONDITIONS

The general effect of slenderness on the axial compression load capacity of columns is discussed in Section 3.11. For steel columns, the value of the critical stress (F_c) in compression is determined from formulas in the AISC Specification; it includes variables of the steel yield stress and modulus of elasticity, the relative slenderness of the column, and special considerations for the restraint at the column ends.

Column slenderness is determined as the ratio of the column unbraced length to the radius of gyration of the column section: L/r. Effects of end restraint are considered by use of a modifying factor (K). (See Figure 10.3.) The modified slenderness is thus expressed as KL/r.

Figure 10.4 is a graph of the critical axial compressive stress for a column for two grades of steel with F_y of 36 ksi and 50 ksi. Values for full number increments of KL/r are also given in Table 10.1.

	(a)	(b)	(c)	(d)	(e)	(f)
Buckled shape of column is shown by dashed line	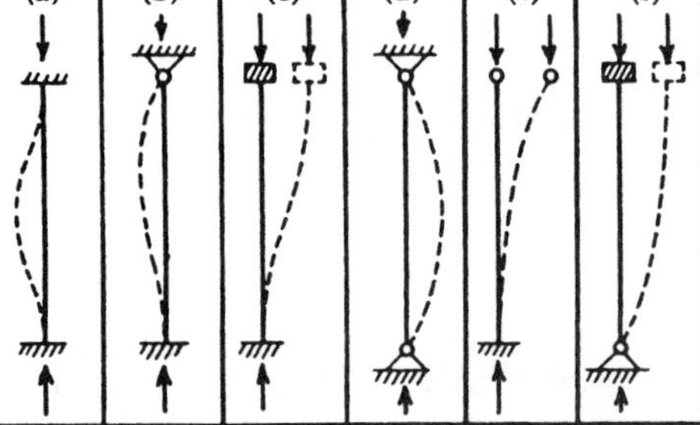					
Theoretical *K* value	0.5	0.7	1.0	1.0	2.0	2.0
Recommended design value when ideal conditions are approximated	0.65	0.80	1.2	1.0	2.10	2.0
End condition code		**Rotation fixed and translation fixed** **Rotation free and translation fixed** **Rotation fixed and translation free** **Rotation free and translation free**				

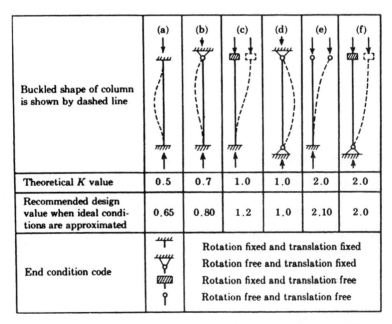

Figure 10.3 Determination of effective column length for buckling. Reprinted from *Manual of Steel Construction,* with permission of the publishers, the American Institute of Steel Construction.

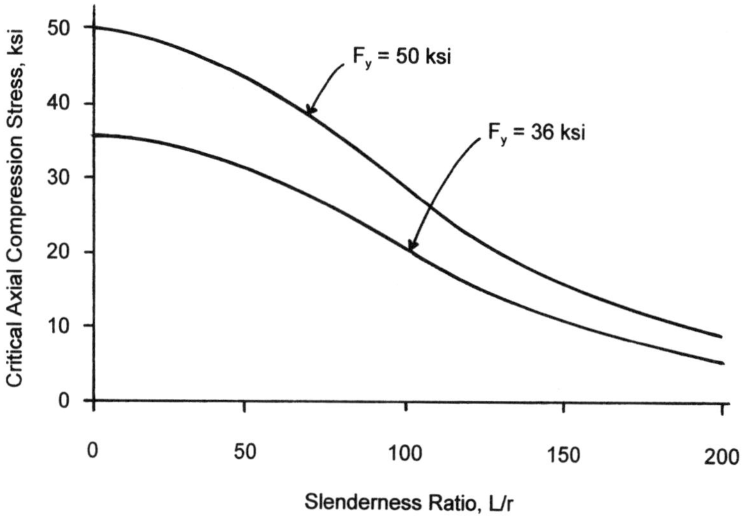

Figure 10.4 Allowable axial compression stress for steel columns as a function of yield limit stress and column slenderness.

For practical reasons, most building columns tend to have relative slenderness between about 50 and 100, with only very heavily loaded columns falling below this. Most designers avoid using extremely slender columns. The AISC specification for steel discourages use of any compression member with a slenderness ratio greater than 200.

10.3 SAFE AXIAL LOADS FOR STEEL COLUMNS

The design strength in axial compression for a column is computed by multiplying the design stress ($\phi_c F_c$) by the cross-sectional area of the column where $\phi_c = 0.85$. Thus:

$$P_u = \phi_c \times P_n = \phi_c \times F_c \times A$$

where P_u = the maximum factored load

ϕ_c = the resistance factor, 0.85 for columns

P_n = the nominal load resistance (unfactored) of the column

F_c = the critical compressive stress for the column, based on KL/r

A = the area of the column cross section

The following examples demonstrate the process. For single-piece columns, a more direct process consists of using column load tables. For built-up sections, however, it is necessary to compute the properties of the section.

Example 1. A W 12 × 53 of A36 steel is used as a column with an unbraced length of 16 ft [4.88 m]. Compute the maximum factored load (P_u).

Solution: Referring to Table A.3, note that $A = 15.6$ in.^{2}, $r_x = 5.23$ in., and $r_y = 2.48$ in. If the column is unbraced on both axes, it is limited by the lower r value for the weak axis. With no stated end conditions, we assume Case (d) in Figure 10.3, for which $K = 1.0$; that is, no modification is made. (This is the unmodified condition.) Thus, the relative stiffness is computed as

$$\frac{KL}{r} = \frac{1 \times 16 \times 12}{2.48} = 77.4$$

In design work, it is usually considered acceptable to round the slenderness ratio off to the nearest whole number. Thus, with a KL/r value of 77, Table 10.1 yields a value for F_c of 26.3 ksi. The maximum factored load for the column is then

$$P_u = \phi_c \times F_c \times A = 0.85 \times 26.3 \text{ ksi} \times 15.6 \text{ in.}^2$$
$$= 349 \text{ kips } [1.55 \times 10^3 \text{ kN}]$$

Example 2. Compute the maximum factored load for the column in Example 1 if the top is pinned but prevented from lateral movement and the bottom is totally fixed.

Solution: Referring to Figure 10.3, note that the case for this is (b) and the modifying factor is 0.8. Then

$$\frac{KL}{r} = \frac{0.8 \times 16 \times 12}{2.48} = 62$$

From Table 10.1, $F_c = 29.4$ ksi.

$$P_u = \phi_c \times F_c \times A = 0.85 \times 29.4 \text{ ksi} \times 15.6 \text{ in.}^2$$
$$= 390 \text{ kips } [1.73 \times 10^3 \text{ kN}]$$

The following example illustrates the situation where a W shape is braced differently on its two axes.

Example 3. Figure 10.5*a* shows an elevation of the steel framing at the location of an exterior wall. The column is laterally restrained but rotationally free at the top and bottom in both directions. (The end condition is as shown for Case (d) in Figure 10.3.) With respect to the x-axis of the section, the column is laterally unbraced for its full height. However, the existence of the horizontal framing in the wall plane provides lateral bracing with respect to the y-axis of the section; thus, the buckling of the column in this direction takes the form shown in Figure 10.5*b*. If the column is a W 12 × 53 of A36 steel, L_1 is 30 ft [9.15 m], and L_2 is 18 ft [5.49 m], what is the maximum factored compression load?

Solution: The basic procedure here is to investigate both axes separately and to use the highest value for relative slenderness obtained to find the critical stress. (*Note:* This is the same section used in Example 1, for which properties were previously obtained from Table A.3.) For the x-axis, the situation is Case (d) from Figure 10.3. Thus:

$$x\text{-axis: } \frac{KL}{r} = \frac{1 \times 30 \times 12}{5.23} = 68.8, \text{ say } 69$$

For the y-axis, the situation is also assumed to be Case (d) from Figure 10.3, except that the deformation occurs in two parts (see Figure 10.5*b*). The lower part is used as it has the greater unbraced length. Thus:

$$y\text{-axis: } \frac{KL}{r} = \frac{1 \times 18 \times 12}{2.48} = 87.1, \text{ say } 87$$

Despite the bracing, the column is still critical on its weak axis. From Table 10.1 the value for F_c is 24.2 ksi, and the allowable load is thus

$$P_u = \phi_c \times F_c \times A = 0.85 \times 24.2 \text{ ksi} \times 15.6 \text{ in.}^2$$
$$= 321 \text{ kips } [1.43 \times 10^3 \text{ kN}]$$

For the following problems use A36 steel with $F_y = 36$ ksi.

Problem 10.3.A. Determine the maximum factored axial compression load (P_u) for a W 10 × 49 column with an unbraced height of 15 ft [4.57 m]. Assume $K = 1.0$.

TABLE 10.1 Critical Unfactored Compressive Stress for Columns, F_c [a]

	Critical Stress, F_c			Critical Stress, F_c			Critical Stress, F_c	
KL/r	F_y = 36 ksi	F_y = 50 ksi	*KL/r*	F_y = 36 ksi	F_y = 50 ksi	*KL/r*	F_y = 36 ksi	F_y = 50 ksi
1	36.0	50.0	41	33.0	45.8	81	25.5	35.4
2	36.0	50.0	42	32.8	45.6	82	25.3	35.1
3	36.0	50.0	43	32.7	45.4	83	25.0	34.8
4	36.0	50.0	44	32.5	45.2	84	24.8	34.5
5	36.0	49.9	45	32.4	44.9	85	24.6	34.2
6	35.9	49.9	46	32.2	44.7	86	24.4	33.9
7	35.9	49.9	47	32.0	44.5	87	24.2	33.6
8	35.9	49.8	48	31.9	44.3	88	23.9	33.3
9	35.8	49.8	49	31.7	44.1	89	23.7	33.0
10	35.8	49.7	50	31.6	43.8	90	23.5	32.6
11	35.8	49.7	51	31.4	43.6	91	23.3	32.3
12	35.7	49.6	52	31.2	43.4	92	23.1	32.0
13	35.7	49.6	53	31.1	43.1	93	22.8	31.7
14	35.6	49.5	54	30.9	42.9	94	22.6	31.4
15	35.6	49.4	55	30.7	42.6	95	22.4	31.1
16	35.5	49.3	56	30.5	42.4	96	22.2	30.8
17	35.5	49.2	57	30.3	42.1	97	21.9	30.5
18	35.4	49.2	58	30.2	41.9	98	21.7	30.2
19	35.3	49.1	59	30.0	41.6	99	21.5	29.8

20	35.2	49.0	60	29.8	41.4	100	21.3	29.5
21	35.2	48.9	61	29.6	41.1	101	21.0	29.2
22	35.1	48.7	62	29.4	40.8	102	20.8	28.9
23	35.0	48.6	63	29.2	40.6	103	20.6	28.6
24	34.9	48.5	64	29.0	40.3	104	20.4	28.3
25	34.8	48.4	65	28.8	40.0	105	20.1	28.0
26	34.7	48.3	66	28.6	39.8	106	19.9	27.7
27	34.6	48.1	67	28.4	39.5	107	19.7	27.4
28	34.5	48.0	68	28.2	39.2	108	19.5	27.1
29	34.4	47.8	69	28.0	38.9	109	19.3	26.8
30	34.3	47.7	70	27.8	38.6	110	19.0	26.4
31	34.2	47.5	71	27.6	38.3	111	18.8	26.1
32	34.1	47.4	72	27.4	38.1	112	18.6	25.8
33	34.0	47.2	73	27.2	37.8	113	18.4	25.5
34	33.9	47.0	74	27.0	37.5	114	18.2	25.2
35	33.8	46.9	75	26.8	37.2	115	17.9	24.9
36	33.6	46.7	76	26.6	36.9	116	17.7	24.6
37	33.5	46.5	77	26.3	36.6	117	17.5	24.3
38	33.4	46.3	78	26.1	36.3	118	17.3	24.0
39	33.2	46.2	79	25.9	36.0	119	17.1	23.7
40	33.1	46.0	80	25.7	35.7	120	16.9	23.4

[a] Usable design stress limit in ksi for obtaining of nominal strength (unfactored) of steel columns.

*Source:*Developed from data in the *Manual of Steel Construction* with permission of the publishers, American Institute of Steel Construction.

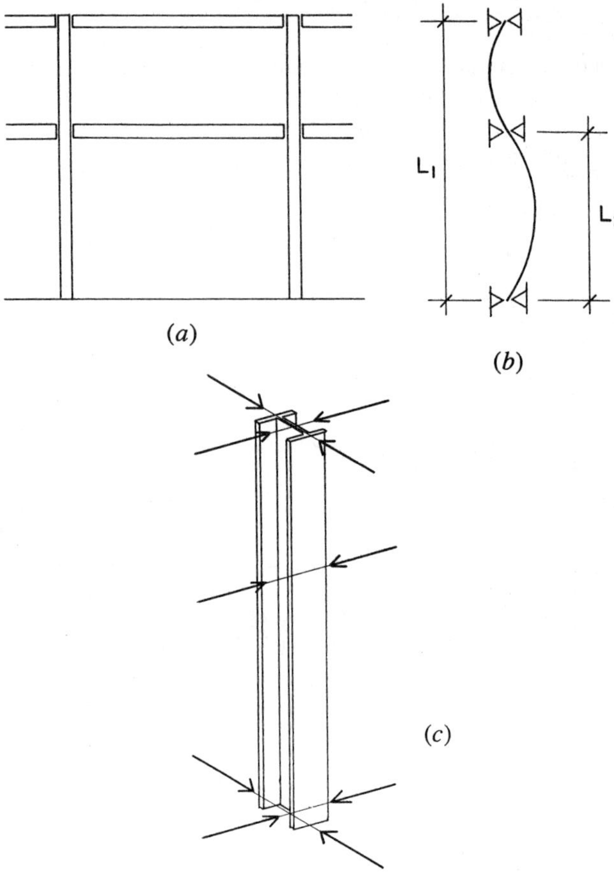

Figure 10.5 Biaxial bracing for steel columns.

Problem 10.3.B. Determine the maximum factored axial compression load (P_u) for a W 12 × 120 column with an unbraced height of 22 ft [6.71 m], if both ends are fixed against rotation and horizontal movement.

Problem 10.3.C. Determine the maximum factored axial compression load (P_u) in Problem 10.3.A if the conditions are as shown in Figure 10.5 with L_1 = 15 ft [4.6 m] and L_2 = 8 ft [2.44 m].

Problem 10.3.D. Determine the maximum factored axial compression load (P_u) in Problem 10.3.B if the conditions are as shown in Figure 10.5 with L_1 = 40 ft [12 m] and L_2 = 22 ft [6.7 m].

10.4 DESIGN OF STEEL COLUMNS

Unless a computer-supported procedure is used, design of steel columns is mostly accomplished through the use of tabulated data. The following discussions in this section consider the latter process, using materials from the AISC Manual (Ref. 5). Design of a column without using load tables is hampered by the fact that the critical stress (F_c) is not able to be precisely determined until *after* the column shape is selected. That is, KL/r and the associated value for F_c cannot be determined until the critical r value for the column is known. This leads to a trial-and-error approach, which can become laborious in even simple circumstances. This process is unavoidable for built-up sections, so their design is unavoidably tedious.

The real value of the safe load tables for single rolled shapes is the ability to use them directly, once only the factored load, column height, and K factor are determined. The AISC Manual for LRFD (Ref. 5) provides load tables for W shapes for F_y equal to 50 ksi.

The work that follows in this section provides examples of the use of the AISC Manual tables for selection of columns of a variety of shapes. For the sake of brevity, the examples are limited to a single steel grade, A36, except for the tubular shapes, which are commonly fabricated in a different grade of steel.

In many cases, the simple, axial compression capacity of a column is all that is involved in its design selection. However, columns are also frequently subjected to bending and shear, and in some cases to torsion. In combined actions, however, the axial load capacity is usually included, so that its singular determination is still a factor—thus, the safe compression load under simple axial application conditions is included in some way in just about every column design situation.

Single Rolled Shapes as Columns

The single rolled shape most commonly used as a column is the squarish, H-shaped element, with a nominal depth of 8 in. or more. Responding to this, steel rolling mills produce a wide range of sizes in this category. Most are in the W shape series, although some are designated as M shapes.

The AISC Manual (Ref. 5) provides extensive tables with factored loads for W and M shapes. The tables also yield considerable other data for the listed shapes. Table 10.2 summarizes data from the series of AISC tables for shapes ranging from the W 8 × 24 to the W 14 × 211 for steel with yield of 36 ksi. Table 10.3 has similar values for steel with yield of 50 ksi. Table values are based on the r value for the y-axis with $K = 1.0$.

TABLE 10.2 Safe Factored Loads for Selected A36 W Shapes[a]

	Ratio	Effective Length (KL) in feet																		
F_y = 36 ksi	r_x/r_y	0	6	8	10	12	14	16	18	20	22	24	26	28	30	32	34	36	38	40
W 14 × 211	1.61	1897	1866	1842	1812	1776	1734	1687	1636	1580	1520	1458	1392	1325	1257	1187	1118	1048	980	912
W 14 × 176	1.60	1585	1559	1538	1512	1482	1446	1406	1362	1314	1263	1210	1154	1097	1039	980	922	863	805	748
W 14 × 145	1.59	1307	1284	1267	1246	1220	1190	1156	1119	1079	1036	992	945	898	849	800	751	703	655	608
W 14 × 120	1.67	1080	1059	1043	1023	999	971	940	906	870	831	791	749	706	663	620	577	535	494	454
W 14 × 82	2.44	734	703	679	649	615	577	536	493	449	404	361	319	279	243	214	189	169	151	137
W 14 × 68	2.44	612	585	565	540	511	479	444	408	371	334	297	262	229	199	175	155	138	124	112
W 14 × 53	3.07	477	443	418	389	355	319	282	245	210	176	148	126	109	95	83				
W 12 × 336	1.85	3023	2956	2904	2839	2761	2672	2573	2465	2350	2229	2104	1975	1845	1716	1587	1460	1337	1218	1102
W 12 × 279	1.82	2506	2447	2402	2345	2278	2201	2115	2021	1922	1818	1710	1600	1490	1379	1270	1164	1061	960	866
W 12 × 230	1.75	2072	2021	1982	1933	1875	1809	1735	1656	1571	1482	1391	1298	1204	1111	1020	931	845	761	687
W 12 × 190	1.79	1707	1664	1631	1589	1540	1483	1421	1353	1281	1206	1129	1051	973	895	819	745	674	605	546
W 12 × 152	1.77	1368	1332	1304	1270	1229	1182	1130	1074	1015	954	891	827	763	700	638	578	520	467	421
W 12 × 120	1.76	1080	1051	1028	1000	966	928	886	841	793	743	692	640	589	538	489	442	395	355	320
W 12 × 96	1.76	863	839	820	797	770	739	704	667	628	588	546	505	463	422	383	345	308	276	249
W 12 × 79	1.75	710	689	674	654	631	605	576	545	512	479	444	409	375	341	308	277	247	221	200
W 12 × 53	2.11	477	457	441	422	400	375	348	320	292	263	235	207	181	158	139	123	110	98	89
W 12 × 45	2.64	401	373	353	328	301	271	241	210	181	152	128	109	94	82	72				
W 10 × 112	1.74	1007	969	941	906	865	819	768	715	660	604	548	493	440	389	342	303	270	242	219
W 10 × 88	1.73	793	762	739	710	677	639	599	556	511	466	422	378	336	295	259	230	205	184	166
W 10 × 68	1.71	612	588	569	547	520	490	458	424	389	354	319	285	252	221	194	172	153	138	124
W 10 × 54	1.71	483	464	449	431	409	385	360	332	304	276	248	221	195	170	150	133	118	106	96
W 10 × 45	2.15	407	380	361	337	311	282	252	222	192	164	138	118	102	88	78				
W 10 × 33	2.16	297	276	261	243	222	200	177	155	133	112	94	80	69	60	53				
W 8 × 58	1.74	523	492	469	441	409	374	337	300	263	228	194	165	143	124	109	97			
W 8 × 40	1.73	358	335	319	298	275	251	225	198	173	148	125	107	92	80	70	62			
W 8 × 31	1.72	279	261	248	232	214	194	173	153	133	114	96	82	70	61	54				
W 8 × 24	2.12	217	195	180	162	142	122	102	84	68	56	47	40							
Bending Factor (m)			2.2	2.1	2.0	1.9	1.8	1.7	1.6	1.5	1.3	1.3	1.3	1.3	1.3	1.3	1.3	1.3	1.3	1.3

[a] Factored nominal strength of columns in kips.

Source: Developed from data in the *Manual of Steel Construction* with permission of the publishers, American Institute of Steel Construction.

TABLE 10.3 Safe Factored Loads for Selected 50 ksi Yield Stress W Shapes[a]

	Ratio	Effective Length (KL) in feet																		
$F_y = 50$ ksi	r_x/r_y	0	6	8	10	12	14	16	18	20	22	24	26	28	30	32	34	36	38	40
W 14 × 211	1.61	2635	2575	2530	2473	2405	2326	2239	2145	2043	1937	1827	1715	1601	1487	1374	1264	1156	1052	951
W 14 × 176	1.60	2202	2150	2112	2063	2004	1938	1863	1783	1696	1606	1513	1417	1321	1225	1130	1037	946	859	775
W 14 × 145	1.59	1815	1772	1739	1698	1649	1593	1531	1463	1391	1316	1237	1158	1078	998	919	842	767	694	626
W 14 × 120	1.67	1500	1460	1430	1391	1346	1294	1237	1176	1110	1042	972	902	832	762	694	628	565	507	457
W 14 × 82	2.44	1020	959	914	860	797	729	658	586	514	445	380	324	279	243	214	189	169	151	137
W 14 × 68	2.44	850	798	760	714	662	604	544	484	424	366	311	265	229	199	175	155	138	124	112
W 14 × 53	3.07	663	598	552	498	439	379	319	263	213	176	148	126	109	95	83				
W 12 × 336	1.85	4199	4069	3970	3847	3702	3538	3357	3163	2960	2750	2537	2325	2116	1911	1715	1525	1360	1221	1102
W 12 × 279	1.82	3481	3367	3281	3174	3048	2906	2749	2582	2408	2228	2047	1867	1690	1519	1354	1199	1070	960	866
W 12 × 230	1.75	2877	2779	2706	2614	2505	2383	2250	2107	1959	1807	1654	1503	1354	1212	1073	951	848	761	687
W 12 × 190	1.79	2372	2288	2225	2147	2054	1951	1837	1717	1592	1464	1336	1209	1085	967	853	755	674	605	546
W 12 × 152	1.77	1900	1830	1778	1713	1637	1551	1458	1359	1256	1151	1047	944	844	749	658	583	520	467	421
W 12 × 120	1.76	1500	1443	1401	1347	1285	1215	1139	1059	976	892	808	726	646	569	500	443	395	355	320
W 12 × 96	1.76	1199	1152	1117	1073	1023	966	904	838	771	703	635	569	505	443	390	345	308	276	249
W 12 × 79	1.75	986	947	917	880	838	790	738	683	627	570	514	459	406	355	312	277	247	221	200
W 12 × 53	2.11	663	623	594	559	518	474	428	381	334	290	247	210	181	158	139	123	110	98	89
W 12 × 45	2.64	557	504	466	422	374	324	274	227	185	152	128	109	94	82	72				
W 10 × 112	1.74	1398	1326	1273	1208	1132	1049	961	870	778	688	601	518	447	389	342	303	270	242	219
W 10 × 88	1.73	1101	1042	999	945	884	817	746	672	599	527	458	393	339	295	259	230	205	184	166
W 10 × 68	1.71	850	803	769	727	678	625	569	511	454	398	344	294	254	221	194	172	153	138	124
W 10 × 54	1.71	672	634	606	572	533	490	445	399	353	309	266	227	196	170	150	133	118	106	96
W 10 × 45	2.15	565	515	478	436	388	339	290	243	199	164	138	118	102	88	78				
W 10 × 33	2.16	413	373	345	312	276	238	202	167	135	112	94	80	69	60	53				
W 8 × 58	1.74	727	667	624	572	515	455	394	335	279	231	194	165	143	124	109	97			
W 8 × 40	1.73	497	454	423	386	345	303	260	219	180	149	125	107	92	80	70	62			
W 8 × 31	1.72	388	353	329	299	267	234	200	168	138	114	96	82	70	61	54				
W 8 × 24	2.12	301	260	232	200	168	136	106	84	68	56	47	40							
Bending Factor (m)			2.1	2.0	1.9	1.8	1.7	1.6	1.4	1.3	1.2	1.2	1.2	1.2	1.2	1.2	1.2	1.2	1.2	1.2

[a]Factored nominal strength of columns in kips.

Source: Developed from data in the *Manual of Steel Construction* with permission of the publishers, American Institute of Steel Construction.

Also included in this table are values for the bending factor *(m)*, which is used for approximate design for combined compression and bending, as discussed in Section 10.4.

For an illustration of one use of Table 10.2, refer to Example 1 of Section 10.3. For the safe load for the W 12 × 53 with unbraced height of 16 ft, the table yields a value of 348 kips, which agrees closely with the computed load found in the example.

The real value of Tables 10.2 and 10.3, however, is for quick design selections. The basic equation for design of steel columns in LRFD is

$$\phi_c \times P_n \geq P_u$$

where ϕ_c = 0.85
P_n = nominal column strength
P_u = maximum factored load

Example 4. Using Table 10.2, select a W shape A36 steel column section for an axial load of 100 kips [445 kN] dead load and 150 kips [667 kN] live load if the unbraced length is 24 ft [7.32 m] and the end conditions are pinned at top and bottom.

Solution: The load must be factored using the load combinations to determine the maximum factored load on the column.

$$P_u = 1.4 \times (\text{dead load})$$
$$= 1.4 \times (100 \text{ kips}) = 140 \text{ kips [623 kN]}$$

or

$$P_u = 1.2 \times (\text{dead load}) + 1.6 \times (\text{live load})$$
$$= 1.2 \times (100 \text{ kips}) + 1.6 \times (150 \text{ kips}) = 360 \text{ kips [1600 kN]}$$

From the basic equation for the maximum column load:

$$\phi_c \times P_n \geq P_u = 360 \text{ kips}$$

From Table 10.2 some possible choices are as shown in the following:

Section	Design Load ($\phi_c \times P_u$)
W 10 × 88	422 kips [1880 kN]
W 12 × 79	444 kips [1970 kN]
W 14 × 82	361 kips [1610 kN]

With no additional parameters required by the problem, we will select the W 12 × 79 because it is the most economical section based upon material weight.

Tables 10.2 and 10.3 are set up to work when the y-axis is the most slender—that is, when $K_yL_y/r_y > K_xL_x/r_x$. This is always the case when the end conditions are the same with respect to both axes and when the unbraced lengths are also the same or when the unbraced length is greater about the y-axis than it is about the x-axis. It is a bit trickier if the unbraced length about the x-axis is greater than the unbraced length about the y-axis. For this we need to use another piece of information provided in Tables 10.2 and 10.3, namely, the ratio between r_x/r_y. This ratio varies with each section but is usually within the range of 1.6 to 3.1, with an average value equal to 1.75. If $L_x/L_y \leq r_x/r_y$, then the slenderness ratio about the y-axis still controls and the unbraced length about the y-axis is used. If $L_x/L_y > r_x/r_y$, then the slenderness about the x-axis controls. When this happens, a new equivalent length (KL_y') must be used to find the most appropriate steel section for a column. The new equivalent length can be found by using the following equation:

$$KL_y' = \frac{KL_x}{r_x/r_y}$$

Example 5. Using Table 10.2, select a A36 steel column section for an axial load of 100 kips [445 kN] dead load and 150 kips [667 kN] live load if the unbraced length is 24 ft [7.32 m] about the x-axis, the unbraced length is 8 ft [2.44 m] about the y-axis, and the end conditions are pinned at top and bottom.

Solution: The load is the same as it is in Example 4; therefore, we already know the maximum factored load is 360 kips. We then determine the ratio of the unbraced lengths about the two axes.

$$\frac{L_x}{L_y} = \frac{24 \text{ ft}}{8 \text{ ft}} = 3.0$$

By looking at Table 10.2, we can note that only one section has a ratio of r_x/r_y greater than 3.0; therefore, we will assume that the x-axis will control. Next, we determine a new effective length for the column and use it to determine the most appropriate columns. Because we don't know the exact value for the ratio of r_x/r_y, we will assume it to be 1.75 because that is an average value from Table 10.2.

$$KL_y' = \frac{KL_x}{r_x/r_y} = \frac{1 \times 24 \text{ ft}}{1.75} = 13.7, \text{ say } 14$$

Options from Table 10.2 are as follows:

Section	Design Load ($\phi_c \times P_n$)	r_x/r_y	Actual New Equivalent Length
W 8 × 58	374 kips [1660 kn]	1.74	13.8 ft [4.21 m]
W 10 × 54	385 kips [1710 kN]	1.71	14.0 ft [4.27 m]
W 12 × 53	375 kips [1670 kN]	2.11	11.4 ft [3.48 m]
W 14 × 68	479 kips [2130 kN]	2.44	9.84 ft [3.0 m]

The actual new equivalent length for each section chosen is less than or equal to the 14 ft [4.27 m] used in the calculation, and each is greater than the unbraced length (KL_y) about the y-axis. Thus, all options are acceptable. Barring other parameters on the design, the most economical section would be the W 12 × 53.

Problem 10.4.A. Using Table 10.2, select a column section for an axial dead load of 60 kips [267 kN] and an axial live load of 88 kips [391 kN] if the unbraced height about both the x and the y axes is 12 ft [3.66 m]. A36 steel is to be used, and K is assumed as 1.0.

Problem 10.4.B. Same data as in Problem 10.4.A, except the dead load is 103 kips [468 kN] and live load is 155 kips [689 kN]. The unbraced height about the x-axis is 16 ft [4.88 m], and the unbraced height about the y-axis is 12 ft [3.66 m].

Problem 10.4.C. Same data as in Problem 10.4.A, except the dead load is 142 kips [632 kN] and the live load is 213 kips [947 kN]. The unbraced height about the x-axis is 20 ft [6.10 m] and about the y-axis is 10 ft [3.05 m].

Problem 10.4.D. Using Table 10.3, select a column section for an axial dead load of 400 kips [1779 kN] and a live load of 600 kips [2669 kN]. The unbraced height about the x-axis is 16 ft [4.88 m] and about the y-axis is 4 ft [1.22 m]. The steel is to have a yield stress of 50 ksi, and K is assumed as 1.0.

Steel Pipe Columns

Round steel pipe columns most frequently occur as single-story columns, supporting either wood or steel beams that sit on top of the columns. Pipe is available in three weight categories: *standard* (Std), *extra strong* (XS), and *double-extra strong* (XXS). Pipe is designated with a nominal diameter, which is slightly less than the outside diameter. The outside diameter is the same for all three weight categories, with variation occurring in terms of the pipe wall thickness and inside diameter. See Table A.7 for properties of standard weight pipe. Table 10.4 gives safe loads for pipe columns of steel with a yield stress of 35 ksi.

Example 6. Using Table 10.4, select a standard weight steel pipe to carry a dead load of 15 kips [67 kN] and a live load of 26 kips [116 kN] if the unbraced height is 12 ft [3.66 m].

Solution: The load must be factored using the load combinations to determine the maximum factored load on the column.

$$P_u = 1.4 \times (\text{dead load})$$
$$= 1.4 \times (15\text{ kips}) = 21\text{ kips [93 kN]}$$

or

$$P_u = 1.2 \times (\text{dead load}) + 1.6 \times (\text{live load})$$
$$= 1.2 \times (15\text{ kips}) + 1.6 \times (26\text{ kips}) = 59.6\text{ kips [265 kN]}$$

For the height of 12 ft, the table yields a value of 95 kips as the design load for a 5-in. pipe. A 4-in. pipe is close, but its design strength is 59 kips [262 kN], just short of that required.

Problems 10.4.E, F, G, H. Select the minimum size standard weight steel pipe for an axial dead load of 20 kips, a live load of 30 kips, and the following unbraced heights: (e) 8 ft; (f) 12 ft; (g) 18 ft; (h) 25 ft.

Structural Tubing Columns

Structural tubing is used for building columns and for members of trusses. Members are available in a range of designated nominal sizes that indicate the actual outer dimensions of the rectangular tube shapes. Within these sizes, various wall thicknesses (the thickness of the steel plates used to make the tubes) are available. For building columns, sizes used range upward from the 3-in.-square tube to the largest sizes fabricated (48-in.-square at present). Tubing may be specified in various grades of steel. That used for the AISC table, as noted, is 46 ksi.

TABLE 10.4 Safe Factored Loads for Selected 35 ksi Yield Stress Pipe Columns[a]

F_y = 35 ksi			Area	Effective Length (*KL*) in feet																		
			(sq. in.)	0	6	8	10	12	14	16	18	20	22	24	26	28	30	32	34	36	38	40
Pipe	12	XS	19.2	571	563	557	549	540	529	517	503	488	472	455	438	420	401	382	363	343	324	305
Pipe	12	Std.	14.6	434	428	424	418	411	403	394	384	372	361	348	335	321	307	293	279	264	249	235
Pipe	10	XS	16.1	479	469	462	453	442	429	415	400	383	365	347	328	309	290	270	251	232	214	196
Pipe	10	Std.	11.9	354	347	342	335	327	318	308	297	284	272	258	245	231	216	202	188	174	161	148
Pipe	8	XXS	21.3	634	612	596	575	551	524	495	463	430	397	363	329	297	265	235	208	186	166	150
Pipe	8	XS	12.8	381	369	360	348	335	320	303	286	267	248	228	209	190	171	153	136	121	109	98
Pipe	8	Std.	8.4	250	242	237	229	221	211	201	190	178	165	153	140	128	116	104	93	83	75	67
Pipe	6	XXS	15.6	464	436	415	390	361	330	298	264	232	200	170	145	125	109	96	85			
Pipe	6	XS	8.4	250	236	226	214	200	185	169	152	135	119	103	88	76	66	58	52	46		
Pipe	6	Std.	5.6	166	158	151	144	135	125	114	104	93	82	72	62	53	47	41	36	32		
Pipe	5	XXS	11.3	336	307	287	262	235	206	178	150	124	102	86	73	63						
Pipe	5	XS	6.1	182	168	158	146	133	119	104	90	76	63	53	45	39	34					
Pipe	5	Std.	4.3	128	119	112	104	95	85	75	65	56	47	39	33	29	25					
Pipe	4	XXS	8.1	241	209	187	163	137	112	88	70	56	47									
Pipe	4	XS	4.4	131	116	106	94	81	68	55	44	36	30	25								
Pipe	4	Std.	3.2	94	84	77	68	59	50	41	33	27	22	19								
Pipe	3.5	XS	3.7	109	94	83	71	59	47	37	29	23										
Pipe	3.5	Std.	2.7	80	69	61	53	44	36	28	22	18	15									
Pipe	3	XXS	5.5	163	128	106	83	62	46	35												
Pipe	3	XS	3.0	90	73	62	51	40	30	23	18											
Pipe	3	Std.	2.2	66	54	47	38	30	23	17	14											

[a] Factored nominal strength of columns in kips.

Developed from data in the *Manual of Steel Construction* with permission of the publishers, American Institute of Steel Construction.

Table 10.5 yields design strengths for square tubes: 3 in. to 12 in. Use of the tables is similar to that for the other design strength tables.

Problem 10.4.I. A structural tubing column, designated as HHS 4 × 4 × 3⁄8, of steel with F_y = 46 ksi [317 MPa], is used with an effective unbraced length of 12 ft [3.66 m]. Find the maximum factored axial load.

Problem 10.4.J. A structural tubing column, designated as HSS 3 × 3 × 5⁄16, of steel with F_y = 46 ksi [317 MPa], is used with an effective unbraced length of 15 ft [4.57 m]. Find the maximum factored axial load.

Problem 10.4.K. Using Table 10.5, select the lightest structural tubing column to carry an axial dead load of 30 kips [133 kN] and a live load of 34 kips [151 kN] if the effective unbraced length is 10 ft [3.05 m].

Problem 10.4.L. Using Table 10.5, select the lightest structural tubing column to carry an axial dead load of 90 kips [400 kN] and a live load of 60 kips [267 kN] if the effective unbraced length is 12 ft [3.66 m].

Double-Angle Compression Members

Matched pairs of angles are frequently used for trusses or for braces in frames. The common form consists of the two angles placed back to back but separated a short distance to achieve joints by use of gusset plates or by sandwiching the angles around the web of a structural tee. Compression members that are not columns are frequently called *struts*.

The AISC Manual contains safe load tables for double angles with an assumed average separation distance of 3⁄8 in. [9.5 mm]. For angles with unequal legs, two back-to-back arrangements are possible, described either as *long legs back to back* or as *short legs back to back*. Table 10.6 has been adapted from data in the AISC tables for selected pairs of double angles with long legs back to back. Note that separate data is provided for the variable situation of either axis being used for the determination of the effective unbraced length. If conditions relating to the unbraced length are the same for both axes, then the lower value for the safe load from Table 10.6 must be used. Properties for selected double angles with long legs back to back are given in Table A.6.

TABLE 10.5 Safe Factored Loads for Selected 46 ksi Yield Stress Square Tube Steel Columns[a]

	Area	Effective Length (KL) in feet																		
F_y = 46 ksi	(sq. in.)	0	6	8	10	12	14	16	18	20	22	24	26	28	30	32	34	36	38	40
HHS 12 × 12 × 5/8	25.7	1005	989	976	960	941	919	895	867	838	807	774	739	704	668	631	595	558	522	486
HHS 12 × 12 × 3/8	16	626	616	609	599	588	575	560	544	526	507	488	467	446	424	402	379	357	335	313
HHS 12 × 12 × 1/4	10.8	422	416	411	405	397	389	379	368	357	344	331	317	303	289	274	259	244	230	215
HHS 10 × 10 × 1/2	17.2	673	657	645	630	612	592	569	545	519	491	462	433	404	375	346	317	290	263	237
HHS 10 × 10 × 5/16	11.1	434	424	417	408	397	384	370	355	338	321	303	285	266	248	229	211	193	176	160
HHS 10 × 10 × 3/16	6.76	264	259	254	249	242	235	226	217	207	197	186	176	164	153	142	131	121	110	100
HHS 8 × 8 × 1/2	13.5	528	508	494	475	454	430	404	376	347	318	289	260	232	205	181	160	143	128	116
HHS 8 × 8 × 5/16	8.76	343	331	322	310	297	282	266	249	231	212	194	176	158	141	124	110	98	88	79
HHS 8 × 8 × 3/16	5.37	210	203	197	191	183	174	164	154	143	132	121	110	99	89	79	70	62	56	50
HHS 7 × 7 × 5/8	14	547	519	499	473	444	412	377	342	306	271	237	204	176	153	135	119	107	96	86
HHS 7 × 7 × 3/8	8.97	351	334	322	307	289	270	249	227	205	183	162	142	123	107	94	83	74	67	60
HHS 7 × 7 × 1/4	6.17	241	230	222	212	201	188	174	159	145	130	115	101	88	77	68	60	53	48	43
HHS 6 × 6 × 1/2	9.74	381	355	336	313	288	260	231	203	175	148	125	106	92	80	70	62	55		
HHS 6 × 6 × 5/16	6.43	251	236	224	210	194	176	158	140	122	104	88	75	65	56	50	44	39	35	
HHS 6 × 6 × 3/16	3.98	156	146	139	131	121	111	100	89	78	68	58	49	42	37	32	29	26	23	
HHS 5 × 5 × 3/8	6.18	242	219	202	183	162	140	119	98	80	66	56	47	41	36					
HHS 5 × 5 × 1/4	4.3	168	153	142	130	116	101	86	72	59	49	41	35	30	26	23				
HHS 5 × 5 × 1/8	2.23	87	80	75	68	61	54	47	39	33	27	23	19	17	15	13				
HHS 4 × 4 × 3/8	4.78	187	159	140	119	98	78	60	47	38	32	27								
HHS 4 × 4 × 1/4	3.37	132	113	101	87	72	58	45	36	29	24	20								
HHS 4 × 4 × 1/8	1.77	69	60	54	47	40	32	26	20	16	14	11	10							
HHS 3 × 3 × 5/16	3.52	138	116	101	85	69	54	41	32	26	22									
HHS 3 × 3 × 3/16	2.24	88	75	66	57	47	37	29	23	18	15	13								

[a] Factored nominal strength of columns in kips.

Source: Developed from data in the *Manual of Steel Construction* with permission of the publishers, American Institute of Steel Construction.

TABLE 10.6 Safe Factored Loads for Double Angle Compression Members of A36 Steel[a]

Size (in.)			4 × 3			3.5 × 2.5			3 × 2		2.5 × 2	
Thickness (In.)			3/8	5/16		5/16	1/4		5/16	1/4	5/16	1/4
Weight (Lb/ft)			16.9	14.2		12.2	9.88		10.1	8.18	8.97	7.30
Area (in.^{2})			4.98	4.19		3.58	2.90		2.96	2.40	2.64	2.14
r_x			1.26	1.27		1.11	1.12		0.945	0.953	0.774	0.782
r_y			1.30	1.29		1.09	1.08		0.897	0.883	0.943	0.930
Effective Buckling Length in Feet (L/r) with Respect to Indicated Axis	X-X Axis	0	152	128	0	110	86	0	90.6	73.4	80.8	65.5
		4	141	119	2	107	84	2	87.6	71.0	76.8	62.3
		6	128	108	4	99	78	3	83.9	68.1	72.1	58.6
		8	112	95	6	88	69	4	79.1	64.3	66.0	53.7
		10	94	80	8	74	59	5	73.3	59.6	58.9	48.0
		12	77	65	10	59	48	6	66.7	54.4	51.2	41.9
		14	60	51	12	45	37	8	52.6	43.0	35.9	29.6
		16	46	39	14	33	27	10	38.8	31.9	23.4	19.4
		18	36	31	16	25	21	12	27.2	22.4	16.3	13.5
		20	29	25	18	20	17	14	20.0	16.5	-	-
	Y-Y Axis	0	152	128	0	110	86	0	90.6	73.4	80.8	65.5
		4	130	104	2	96	71	2	80.7	62.4	74.4	58.6
		6	118	95	4	89	66	3	76.8	59.5	71.0	55.9
		8	104	84	6	78	59	4	71.6	55.6	66.5	52.4
		10	88	72	8	65	50	5	65.4	50.9	61.1	48.2
		12	71	58	10	51	40	6	58.5	45.7	55.1	43.5
		14	55	46	12	38	30	8	43.9	34.4	42.2	33.4
		16	43	35	14	29	22	10	31.9	24.9	31.5	24.8
		18	34	28	16	23	17	12	22.3	17.5	22.0	17.4
		20	28	23	18	18	14	14	16.5	12.9	16.2	12.8

Size (in.)			8 × 6			6 × 4			5 × 3.5		5 × 3	
Thickness (In.)			3/4	1/2		1/2	3/8		1/2	3/8	3/8	5/16
Weight (Lb/ft)			68.0	46.3		32.3	24.6		27.2	20.8	19.5	16.4
Area (in.^{2})			20.0	13.6		9.50	7.22		8.01	6.10	5.73	4.81
r_x			2.52	2.55		1.91	1.93		1.58	1.59	1.60	1.61
r_y			2.47	2.43		1.64	1.61		1.48	1.46	1.22	1.21
Effective Buckling Length in Feet (L/r) with Respect to Indicated Axis	X-X Axis	0	612	379	0	291	201	0	245	183	172	134
		10	543	341	10	236	167	6	220	165	155	122
		12	515	325	12	216	154	8	202	152	143	113
		14	484	308	14	193	140	10	181	137	129	103
		16	451	289	16	171	125	12	158	120	113	91
		20	380	248	18	148	110	14	135	103	97	79
		24	308	206	20	127	96	16	113	86	82	68
		28	240	165	22	106	82	18	91	70	67	57
		32	184	128	26	76	59	22	61	47	45	38
		36	145	101	30	57	44	26	44	34	32	27
	Y-Y Axis	0	612	379	0	291	201	0	245	183	172	134
		10	493	276	6	231	148	4	213	149	135	100
		12	467	264	10	194	128	6	199	140	123	92
		14	438	251	12	172	115	8	180	127	108	82
		16	406	236	14	148	101	10	158	113	90	70
		20	338	203	16	125	87	12	135	97	72	57
		24	269	167	18	102	73	14	111	81	55	45
		28	205	131	20	83	60	16	89	65	43	35
		32	158	102	22	69	50	18	71	52	34	28
		36	126	82	26	50	36	22	48	35		

[a]Factored nominal axial compression strength for members in kips.

Source: Developed from data in the *Manual of Steel Construction* with permission of the publishers, American Institute of Steel Construction.

Like other members that lack biaxial symmetry, such as the structural tee, there may be some reduction applicable as a result of the slenderness of the thin elements of the cross section. This reduction is incorporated in the values provided in the AISC safe load tables.

Problem 10.4.M. A double-angle compression member 8 ft [2.44 m] long is composed of two A36 steel angles 4 × 3 × 3/8 in. with the long legs back to back. Determine the maximum factored axial compression load for the angles.

Problem 10.4.N. A double-angle compression member 12 ft [3.66 m] long is composed of two A36 steel angles 6 × 4 × 1/2 in. with the long legs back to back. Determine the maximum factored axial compression load for the angles.

Problem 10.4.O. Using Table 10.6, select a double-angle compression member for an axial compression dead load of 25 kips [111 kN] and a live load of 25 kips [111 kN] if the effective unbraced length is 10 ft [3.05 m].

Problem 10.4.P. Using Table 10.6, select a double-angle compression member for an axial compression dead load of 75 kips [334 kN] and a live load of 100 kips [445 kN] if the effective unbraced length is 16 ft [4.88 m].

10.5 COLUMNS WITH BENDING

Steel columns must frequently sustain bending in addition to the usual axial compression. Figures 10.6*a* through *c* show three of the most common situations that result in this combined effect. When loads are supported on a bracket at the column face, the eccentricity of the compression adds a bending effect (Figure 10.6*a*). When moment-resistive connections are used to produce a rigid frame, any load on the beams will induce a twisting (bending) effect on the columns (Figure 10.6*b*). Columns built into exterior walls (a common occurrence) may become involved in the spanning effect of the wall in resisting wind forces (Figure 10.6*c*).

Adding bending to a direct compression effect results in a combined stress, or net stress, condition, with something other than an even distribution of stress across the column cross section. The two effects may be investigated separately and the stresses added to determine the net effect. The procedures for this investigation are described in Section 3.11.

However, the two *actions*—compression and bending—are essentially different, so that a combination of the separate actions—not just the

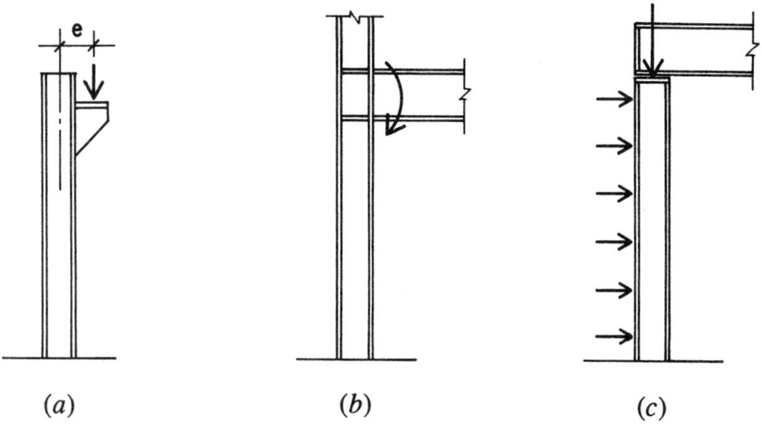

Figure 10.6 Considerations for development of bending in steel columns. (a) Bending induced by eccentric load. (b) Bending transferred to column in a rigid frame. (c) Combined loading condition, separately producing axial compression and bending

stresses—is more significant. This combination is accomplished with the so-called *interaction* analysis that takes the form of

$$\frac{P_u}{\phi_c P_n} + \frac{M_{ux}}{\phi_b M_{nx}} + \frac{M_{uy}}{\phi_b M_{ny}} \leq 1$$

The procedure for this analysis is also described in Section 3.11.

On a graph, the interaction formula describes a straight line, as shown in Figure 3.52*a*, which is the classical form of the relationship in elastic theory. However, variations from the straight-line form occur because of special conditions having to do with the nature of the materials, the usual form of columns, and some recognition of usual fabrication and construction practices.

For steel columns, major issues include slenderness of column flanges and webs (for W shapes), ductility of the steel, and overall column slenderness that affects potential buckling in both axial compression and bending. Understandably, the AISC formulas are considerably more complex than the preceding simple interaction formula, reflecting these as well as other concerns. Following are the AISC interaction formulas for compression and bending:

If $P_u/\phi_c P_n \geq 0.2$,

$$\frac{P_u}{\phi_c P_n} + \frac{8}{9}\left[\frac{M_{ux}}{\phi_b M_{nx}} + \frac{M_{uy}}{\phi_b M_{ny}}\right] \leq 1.0$$

If $P_u/\phi_c P_n < 0.2$,

$$\frac{P_u}{2 \times \phi_c P_n} + \left[\frac{M_{ux}}{\phi_b M_{nx}} + \frac{M_{uy}}{\phi_b M_{ny}}\right] \leq 1.0$$

Another potential problem with combined compression and bending is that of the *P*-delta effect, as discussed in Section 3.11. This occurs when a relatively slender column that is subjected to compression plus bending is curved significantly by the bending effect. The deflection due to this curvature (called delta, Δ) results in an eccentricity of the compression force and thus in an added bending moment equal to the product of *P* times delta. Any bending of the column can produce this effect; an especially critical case is a freestanding, towerlike column with no top restraint and a load on its top. Obviously, a very stiff column with little deflection will not suffer much from the *P*-delta effect, whereas a very slender column may be quite vulnerable.

For use in preliminary design work, or to quickly obtain a first trial section for use in a more extensive design investigation, a procedure developed by Vang, Wattar, and Leet (Ref. 13) may be used that involves the determination of an equivalent axial load that incorporates the bending effect. This is accomplished by use of a bending factor (*m*), which is listed here for the W shapes in Tables 10.2 and 10.3 at the bottom of the tables. Using this factor, we obtain the equivalent axial load P' as

$$P_u' = P_u + m \times M_{ux} + (2 \times m) \times M_{uy}$$

If $P_u/P_u' < 0.2$ recalculate P_u' using the following equation:

$$P_u' = \frac{P_u}{2} + \frac{9}{8} \times [m \times M_{ux} + (2 \times m) \times M_{uy}]$$

where P_u' = the equivalent factored axial compression load for design in kips

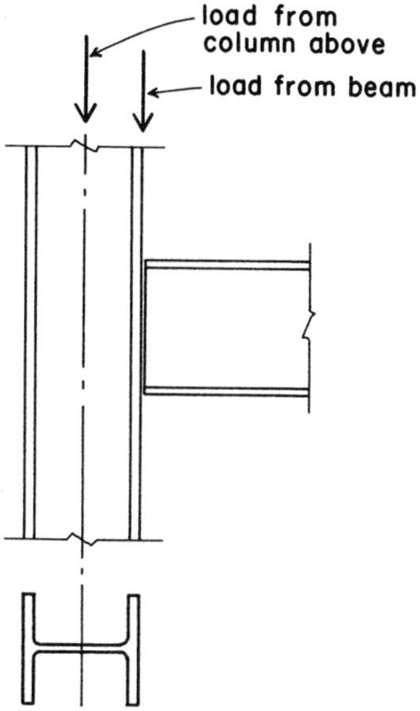

Figure 10.7 Development of an eccentric loading condition with steel framing.

P_u = the actual factored compression load in kips

m = the bending factor

M_x = the bending moment about the column x-axis in kip-ft

M_y = the bending moment about the column y-axis in kip-ft

The following examples illustrate the use of this approximation method.

Example 7. It is desired to use a 10-in. W shape for a column in a situation such as that shown in Figure 10.7. The factored axial load from above on the column is 175 kips [778 kN], and the factored beam load at the column face is 35 kips [156 kN]. The column has an unbraced height of 16 ft [4.88 m] and a K factor of 1.0. Select a trial section for the column.

Solution: Because the loading has already been factored, in order not to confuse matters more with this equation, we do not need to factor it as we have in earlier examples. Determining the axial load capacity and bending capacities will also not be illustrated. They were determined using methods shown earlier in this text.

First, we determine an equivalent axial load. The bending factor (m) is equal to 1.7 and is taken from Table 10.2.

$$
\begin{aligned}
P_u' &= P_u + m \times M_{ux} \\
&= (175 \text{ kips} + 35 \text{ kips}) + (1.7 \times 35 \text{ kips} \times 5 \text{ in.} \times \text{ft}/12 \text{ in.}) \\
&= 210 \text{ kips} + 24.8 \text{ kips} = 235 \text{ kips } [1050 \text{ kN}]
\end{aligned}
$$

Check to see if the correct equation for equivalent axial load was used.

$$
\frac{P_u}{P_u'} = \frac{210 \text{ kips}}{235 \text{ kips}} = 0.9 > 0.2
$$

Therefore, correct equation was used.

Using $P_u' = 235$ kips, we use Table 10.2 to determine a trial member size of W 10 × 45.

Next, we will check the section for compliance with the AISC interaction formula for axial compression and bending

$$
\phi_c P_n = 252 \text{ kips (Table 10.2)}
$$

$$
\phi_b M_{nx} = 175 \text{ kip-ft } (L_p < [L = 16 \text{ ft}] < L_r)
$$

$$
\frac{P_u}{\phi_c P_n} = \frac{210 \text{ kips}}{252 \text{ kips}} = 0.833 > 0.2
$$

$$
\begin{aligned}
&\frac{P_u}{\phi_c P_n} + \frac{8}{9}\left[\frac{M_{ux}}{\phi_b M_{nx}} + \frac{M_{uy}}{\phi_b M_{ny}}\right] \\
&= \frac{210 \text{ kips}}{252 \text{ kips}} + \frac{8}{9}\left[\frac{14.6 \text{ kip-ft}}{175 \text{ kip-ft}} + 0\right] \\
&= 0.907 < 1
\end{aligned}
$$

A W 10 × 45 is an acceptable option for this loading. Although this process may seem laborious, be assured that direct use of the AISC formulas is considerably more laborious.

When bending occurs about both axes, as it does in full three-dimensional rigid frames, all three parts of the approximation formula must be used. The following example demonstrates the process for this.

Example 8. It is desired to use a 12-in. [305 mm] W shape for a column that sustains the following factored loading: axial load of 85 kips [378 kN], M_{ux} = 100 kip-ft [136 kN-m], M_{uy} = 75 kip-ft [102 kN-m]. Select a column made from 50-ksi steel for an unbraced height of 12 ft [3.66 m].

Solution: In Table 10.3, observe that bending factor for 12 ft length is 1.8. Thus:

$$\begin{aligned} P_u' &= P_u + m \times M_{uy} + (2 \times m) \times M_{uy} \\ &= (85 \text{ kips}) + (1.8 \times 100 \text{ kip-ft}) + (2 \times 1.8 \times 75 \text{ kip-ft}) \\ &= 85 \text{ kips} + 180 \text{ kips} + 270 \text{ kips} = 535 \text{ kips } [2380 \text{ kN}] \end{aligned}$$

Check to see if the correct equation for equivalent axial load was used.

$$\frac{P_u}{P_u'} = \frac{85 \text{ kips}}{535 \text{ kips}} = 0.159 < 0.2$$

Therefore, incorrect equation was used.

Recalculate P_u':

$$\begin{aligned} P_u' &= \frac{P_u}{2} + \frac{9}{8} \times [m \times M_{ux} + (2 \times m) \times M_{uy}] \\ &= \frac{85 \text{ kips}}{2} + \frac{9}{8} \times [(1.8 \times 100 \text{ kip-ft}) + (2 \times 1.8 \times 75 \text{ kip-ft})] \\ &= 42.5 \text{ kips} + \frac{9}{8} \times (180 \text{ kips} + 270 \text{ kips}) = 549 \text{ kips } [2440 \text{ kN}] \end{aligned}$$

Using P_u' = 549 kips, we use Table 10.3 to determine a trial member size of W 12 × 79.

Next, we check the section for compliance with the AISC interaction formula for axial compression and bending.

$$\phi_c P_n = 838 \text{ kips (Table 10.3)}$$

$$\phi_b M_{nx} = 706 \text{ kip-ft } (L_p < [L = 12 \text{ ft}] < L_r)$$

$$\phi_b M_{ny} = 201 \text{ kip-ft } (\Phi_b M_{py})$$

$$\frac{P_u}{\phi_c P_n} = \frac{85 \text{ kips}}{838 \text{ kips}} = 0.101 < 0.2$$

$$\frac{P_u}{2\phi_b P_n} + \left[\frac{M_{ux}}{\phi_b M_{nx}} + \frac{M_{uy}}{\phi_b M_{ny}}\right]$$

$$= \frac{85 \text{ kips}}{2 \times 838 \text{ kips}} + \left[\frac{100 \text{ kip-ft}}{706 \text{ kip-ft}} + \frac{75 \text{ kip-ft}}{201 \text{ kip-ft}}\right]$$

$$= 0.0507 + 0.142 = 0.373$$

$$= 0.565 < 1$$

A W 12 × 79 is an acceptable option for this loading, though this loading is using only 56.5 percent of its capacity. Try a W 12 × 53 and check it against the AISC interaction formulas.

$$\phi_c P_n = 518 \text{ kips (Table 10.3)}$$

$$\phi_b M_{nx} = 287 \text{ kip-ft } (L_p < [L = 12 \text{ ft}] < L_r)$$

$$\phi_b M_{ny} = 108 \text{ kip-ft } (\Phi_b M_{py})$$

$$\frac{P_u}{\phi_c P_n} = \frac{85 \text{ kips}}{518 \text{ kips}} = 0.164 < 0.2$$

$$\frac{P_u}{2\phi_b P_n} + \left[\frac{M_{ux}}{\phi_b M_{nx}} + \frac{M_{uy}}{\phi_b M_{ny}}\right]$$

$$= \frac{85 \text{ kips}}{2 \times 518 \text{ kips}} + \left[\frac{100 \text{ kip-ft}}{287 \text{ kip-ft}} + \frac{75 \text{ kip-ft}}{108 \text{ kip-ft}}\right]$$

$$= 0.0820 + 0.348 = 0.694$$

$$= 1.124 > 1$$

The W 12 × 53 does not work for this given loading. Therefore, the W 12 × 79 is the most appropriate section. (*Note:* We have limited choices for the example to those listed in Tables 10.2 and 10.3. However, there are many more shapes listed in the reference document, AISC Manual, Ref. 5).

A major occurrence of the condition of columns with bending is that of columns in rigid frame bents. For steel columns, this typically means the use of a whole steel frame, with beams in two directions and with columns having beams framing into both sides—or indeed, into *all* sides for interior columns.

Problem 10.5.A. It is desired to use a 12-in. [305 mm] W shape for a column to support a beam as shown in Figure 10.7. Select a trial size for the column for the following data: column factored axial load from above = 200 kips [890 kN], factored beam reaction = 30 kips [133 kN], unbraced column height is 14 ft [4.27 m].

Problem 10.5.B. Check the section found in Problem 10.5.A to see if it complies with the AISC interaction formulas for axial compression and bending.

Problem 10.5.C. Same as Problem 10.5.A, except factored axial load is 485 kips [2157 kN], factored beam reaction is 100 kips [445 kN], unbraced height is 18 ft [5.49 m].

Problem 10.5.D. Check the section found in Problem 10.5.C to see if it complies with the AISC interaction formulas for axial compression and bending.

Problem 10.5.E. A 14-in. [356 mm] W shape is to be used for a column that sustains bending in on both axes. Select a trial section for the column for the following factored data: total axial load = 80 kips [356 kN], M_x = 85 kip-ft [115 kN-m], M_y = 64 kip-ft [87 kN-m], unbraced column height is 16 ft [4.88 m].

Problem 10.5.F. Same as Problem 10.5.E, except axial load is 200 kips [890 kN], M_x is 45 kip-ft [61 kN-m], M_y is 30 kip-ft [41 kN-m], unbraced height is 35 ft [10.7 m].

11

BOLTED CONNECTIONS FOR STEEL STRUCTURES

Making a steel structure for a building typically involves the connecting of many parts. The technology available for achieving connections is subject to considerable variety, depending on the form and size of the connected parts, the structural forces transmitted between parts, and the nature of the connecting materials. At the scale of building structures, the primary connecting methods utilized presently are those using electric arc welding and high-strength steel bolts.

11.1 BOLTED CONNECTIONS

Elements of steel are often connected by mating flat parts with common holes and inserting a pin-type device to hold them together. In times past, the device was a rivet; today it is usually a bolt. Many types and sizes of bolt are available, as are many connections in which they are used.

Structural Actions of Bolted Connections

Figures 11.1*a* and *b* show the plan and section of a simple connection between two steel bars that functions to transfer a tension force from one bar to the other. Although this is a tension-transfer connection, it is also referred to as a shear connection because of the manner in which the connecting device (the bolt) works in the connection (see Figure 11.1*c*). For structural connections, this type of joint is now achieved mostly with so-called *high-strength bolts*, which are special bolts that are tightened in a controlled manner that induces development of yield stress in the bolt shaft. For a connection using such bolts, many possible forms of failure must be considered, including the following:

Bolt Shear. In the connection shown in Figure 11.1*a* and *b*, the failure of the bolt involves a slicing (shear) failure that is developed as a shear stress on the bolt cross section. The resistance factor (ϕ_v) is taken as 0.75. The design shear strength ($\phi_v R_n$) of the bolt can be expressed as a nominal shear stress (F_v) times the nominal cross-sectional area of the bolt, or

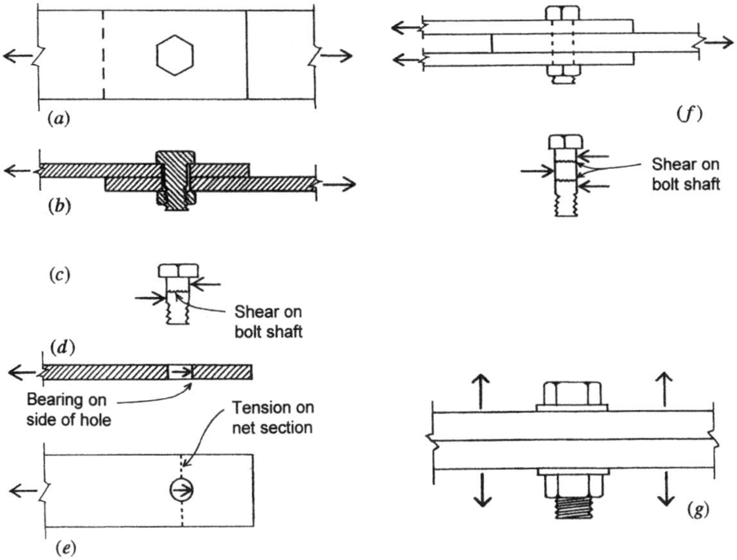

Figure 11.1 Actions of bolted joints.

$$\phi_v R_n = F_v A_b$$

With the size of the bolt and the grade of steel known, it is a simple matter to establish this limit. In some types of connections, it may be necessary to slice the same bolt more than once to separate the connected parts. This is the case in the connection shown in Figure 11.1*f*, in which it may be observed that the bolt must be sliced twice to make the joint fail. When the bolt develops shear on only one section (Figure 11.1*c*), it is said to be in *single shear*; when it develops shear on two sections (Figure 11.1*f*), it is said to be in *double shear.*

Bearing. If the bolt tension (due to tightening of the nut) is relatively low, the bolt serves primarily as a pin in the matched holes, bearing against the sides of the holes, as shown in Figure 11.1*d*. When the bolt diameter is larger or the bolt is made of very strong steel, the connected parts must be sufficiently thick if they are to develop the full capacity of the bolts. The design bearing strength ($\phi_v R_n$) permitted for this situation by the AISC Specification is

$$\phi_v R_n = L_c \times t \times F_u \leq 3.0 \times d \times t \times F_u$$

where ϕ_v = 0.75
L_c = distance between edge of hole and edge of next hole in material (in.)
t = thickness of connected material (in.)
F_u = ultimate tensile strength of connected material (ksi)
d = diameter of bolt (in.)

Tension on Net Section of Connected Parts. For the connected bars in Figure 11.1*b*, the tension stress in the bars will be a maximum at a section across the bar at the location of the hole. This reduced section is called the *net section* for tension resistance. Although this is indeed a location of critical stress, it is possible to achieve yield here without serious deformation of the connected parts. For this reason, design strength ($\phi_t P_n$) at the net section is based on the ultimate—rather than the yield—strength of the connected parts. The value used for the design tensile strength is

$$\phi_v P_n = \phi_t \times F_u \times A_e$$

where ϕ_t = 0.75

F_u = ultimate tensile strength of the connected material (ksi)

A_e = reduced (or net) cross-sectional area (in.^{2})

Bolt Tension. Even though the shear (slip-resisting) connection shown in Figure 11.1*a* and *b* is common, some joints employ bolts for their resistance in tension, as shown in Figure 11.1*g*. For the threaded bolt, the maximum tension stress is developed at the net section through the cut threads. However, it is also possible for the bolt to have extensive elongation if yield stress develops in the bolt shaft (at an unreduced section). However stress is computed, bolt tension resistance is established on the basis of data from destructive tests.

Bending in the Connection. Whenever possible, bolted connections are designed to have a bolt layout that is symmetrical with regard to the directly applied forces. This is not always possible because, in addition to the direct force actions, the connection may be subjected to twisting due to a bending moment or torsion induced by the loads. Figure 11.2 shows some examples of this situation.

In Figure 11.2*a*, two bars are connected by bolts, but the bars are not aligned in a way to transmit tension directly between the bars. This may induce a rotational effect on the bolts, with a torsional twist equal to the product of the tension force and the eccentricity caused by misalignment of the bars. Shearing forces on individual bolts will be increased by this twisting action. And, of course, the ends of the bars will also be twisted.

Figure 11.2*b* shows the single-shear joint, as shown in Figure 11.1*a* and *b*. When viewed from the top, such a joint may appear to have the bars aligned; however, the side view shows that the basic nature of the single-shear joint is such that a twisting action is inherent in the joint. This twisting increases with thicker bars. It is usually not highly critical for steel structures, where connected elements are usually relatively thin; for connecting of wood elements, however, it is not a favored form of joint.

Figure 11.2*c* shows a side view of a beam end with a typical form of connection that employs a pair of angles. As shown, the angles grasp the beam web between their legs and turn the other legs out to fit flat against a column or the web of another beam. Vertical load from the beam, vested in the shear in the beam web, is transferred to the angles by the connection of the angles to the beam web—with bolts as shown here. This load is then transferred from the angles at their outward-turned face, resulting

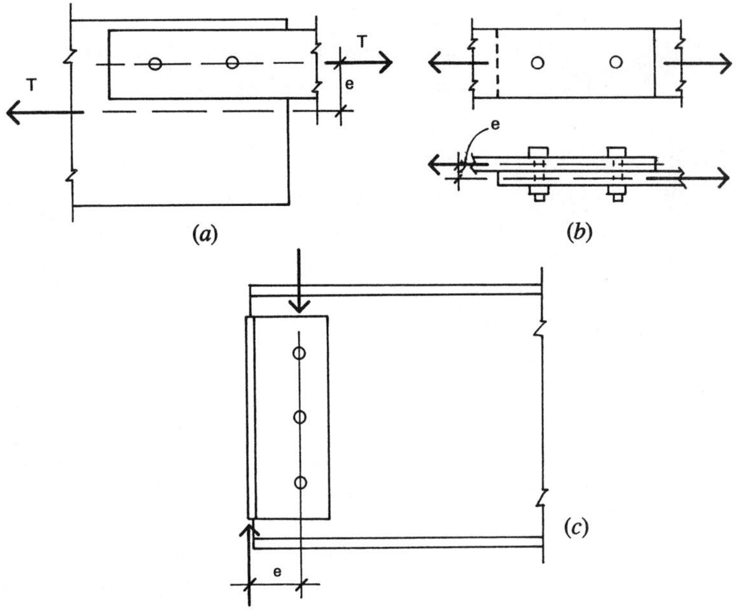

Figure 11.2 Development of bending in bolted joints.

in a separated set of forces caused by the eccentricity shown. This action must be considered with others in design of these connections.

Slipping of Connected Parts. Highly tensioned, high-strength bolts develop a very strong clamping action on the mated flat parts being connected, analogous to the situation shown in Figure 11.3*a*. As a result, there is a strong development of friction at the slip face, which is the initial form of resistance in the shear-type joint. Development of bolt shear, bearing, and even tension on the net section will not occur until this slipping is allowed. For service level loads, therefore, this is the *usual* form of resistance, and the bolted joint with high-strength bolts is considered to be a very rigid form of joint.

Block Shear. One possible form of failure in a bolted connection is that of tearing out the edge of one of the attached members. This is called a *block shear* failure. The diagrams in Figure 11.3*b* show this potentiality in a connection between two plates. The failure in this case involves

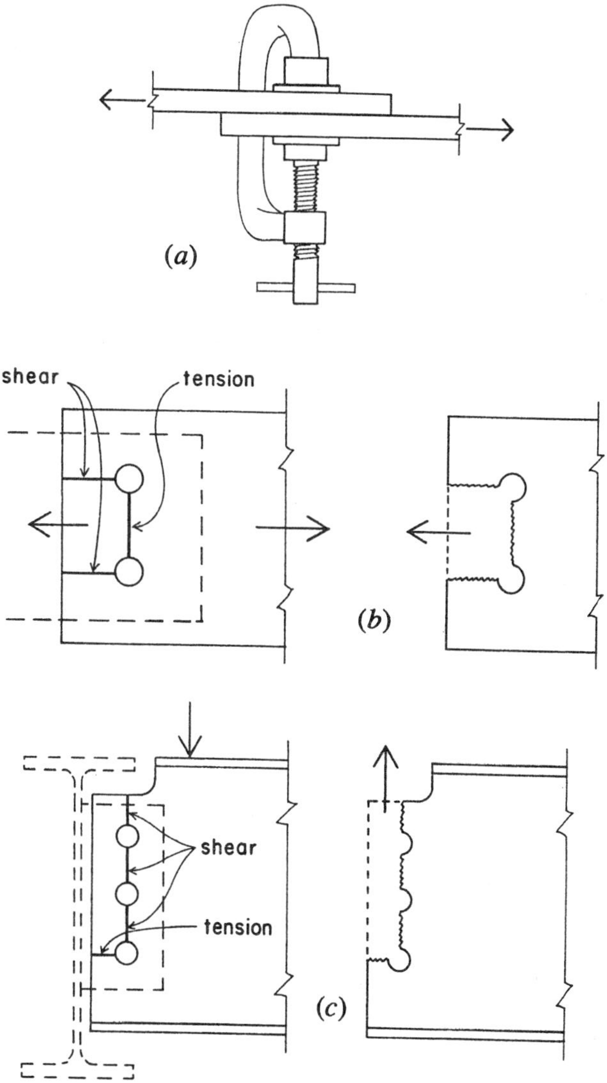

Figure 11.3 Special actions of bolted joints. (*a*) Nature of slip-resisting clamping action of the highly-tightened bolt. (*b*) Tearing failure in a bolted joint, consisting of a combination of shear and tension failures in the connected material. (*c*) Tearing failure in a bolted beam connection.

a combination of shear and tension to produce the torn-out form shown. The total tearing force is computed as the sum required to cause both forms of failure. The design strength ($\phi_t P_n$) of the net tension area is computed as described before for net cross sections. The design strength ($\phi_v R_n$) of the shear areas is specified as $0.75F_v A_c$, where A_c is the cross-sectional area experiencing shear stress.

With the edge distance, hole spacing, and diameter of the holes known, the net widths for tension and shear are determined and multiplied by the thickness of the part in which the tearing occurs. These areas are then multiplied by the appropriate stress to find the total tearing force that can be resisted. If this force is greater than the connection design load, the tearing problem is not critical.

Another case of potential tearing is shown in Figure 11.3*c*. This is the common situation for the end framing of a beam in which support is provided by another beam, whose top is aligned with that of the supported beam. The end portion of the top flange of the supported beam must be cut back to allow the beam web to extend to the side of the supporting beam. With the use of a bolted connection, the tearing condition shown is developed.

Types of Steel Bolts

Bolts used for the connection of structural steel members come in two basic types. Bolts designated A307 and called *unfinished* have the lowest load capacity of the structural bolts. The nuts for these bolts are tightened just enough to secure a snug fit of the attached parts; because of this low resistance to slipping, plus the oversizing of the holes to achieve practical assemblage, there is some movement in the development of full resistance. These bolts are generally not used for major connections, especially when joint movement or loosening under vibration or repeated loading may be a problem. They are, however, used extensively for temporary connections during erection of frames.

Bolts designated A325, F1852, or A490 are called *high-strength bolts.* The nuts of these bolts are tightened to produce a considerable tension force, which results in a high degree of friction resistance between the attached parts. Different specifications for installation of these bolts result in different classifications of their strength, relating generally to the critical mode of failure.

When loaded in shear-type connections, bolt capacities are based on the development of shearing action in the connection. The shear capacity of a single bolt is further designated as *S* for single shear (Figure 11.1*c*) or *D* for double shear (Figure 11.1*f*). In high-strength bolts, the shear capacity is affected by the bolt threads. If the threads are present in the shear plane being considered, the cross-sectional area is reduced and, therefore, the capacity of the bolt is also reduced. The capacities of structural bolts in both tension and shear are given in Table 11.1. These bolts range in size from ½ to 1½ in. in diameter, and properties for these sizes are given in the AISC Manual. The most commonly used sizes for light structural steel framing are ¾ and ⅞ in. However, for larger connections and large frameworks, sizes of 1 to 1¼ are also used. This is the size range for which data is given in Table 14.1: ¾ to 1¼.

Bolts are ordinarily installed with a washer under both head and nut. Some manufactured high-strength bolts have specially formed heads or nuts that in effect have self-forming washers, eliminating the need for a separate, loose washer. When a washer is used, it is sometimes the limiting dimensional factor in detailing for bolt placement in tight locations, such as close to the fillet (inside radius) of angles or other rolled shapes.

TABLE 11.1 Design Strength of Structural Bolts (kips)[a]

ASTM Designation	Loading Condition[b]	Thread Condition	Nominal Diameter of Bolts (in.)				
			¾	⅞	1	1⅛	1¼
A307	S		7.95	10.8	14.1	17.9	22.1
	D		15.9	21.6	28.3	35.8	44.2
	T		14.9	20.3	26.5	33.5	41.4
A325	S	Included	15.9	21.6	28.3	35.8	44.2
		Excluded	19.9	27.1	35.3	44.7	55.2
	D	Included	31.8	43.3	56.5	71.6	88.4
		Excluded	39.8	54.1	70.7	89.5	110
	T		29.8	40.6	53	67.1	82.8
A490	S	Included	19.9	27.1	35.3	44.7	55.2
		Excluded	24.9	33.8	44.2	55.9	69
	D	Included	39.8	54.1	70.7	89.5	110
		Excluded	49.7	67.6	88.4	112	138
	T		37.4	51	66.6	84.2	104

[a] Slip-critical connections; assuming there is no bending in the connection and that bearing on connected materials is not critical.

[b] *S* = single shear; *D* = double shear; *T* = tension.

Source: Compiled from data in the *Manual of Steel Construction,* with permission of the publishers, American Institute of Steel Construction.

For a given diameter of bolt, a minimum thickness is required for the bolted parts in order to develop the full shear capacity of the bolt. This thickness is based on the bearing stress between the bolt and the side of the hole. The stress limit for this situation may be established by either the bolt steel or the steel of the bolted parts.

Steel rods are sometimes threaded for use as anchor bolts or tie rods. When they are loaded in tension, their capacities are usually limited by the stress on the reduced section at the threads. Tie rods are sometimes made with *upset ends*, which consist of larger-diameter portions at the ends. When these enlarged ends are threaded, the net section at the thread is the same as the gross section in the remainder of the rods; the result is no loss of capacity for the rod.

Layout of Bolted Connections

Design of bolted connections generally involves a number of considerations in the dimensional layout of the bolt-hole patterns for the attached structural members. The material in this section presents some basic factors that often must be included in the design of bolted connections. In some situations, the ease or difficulty of achieving a connection may affect the choice for the form of the connected members.

Figure 11.4*a* shows the layout of a bolt pattern with bolts placed in two parallel rows. Two basic dimensions for this layout are limited by the size (nominal diameter) of the bolt. The first is the center-to-center spacing of the bolts, usually called the *pitch*. The AISC Specification limits this dimension to an absolute minimum of 2½ times the bolt diameter. The preferred minimum, however, which is used in this book, is three times the diameter.

The second critical layout dimension is the *edge distance*, which is the distance from the centerline of the bolt to the nearest edge of the member containing the bolt hole. There is also a specified limit for this as a function of bolt size and the nature of the edge; the latter refers to whether the edge is formed by rolling or cutting. Edge distance may also be limited by edge tearing in block shear, which is discussed later.

Table 11.2 gives the recommended limits for pitch and edge distance for the bolt sizes used in ordinary steel construction.

In some cases, bolts are staggered in parallel rows (Figure 11.4 *b*). In this case, the diagonal distance, labeled *m* in the illustration, must also be considered. For staggered bolts, the spacing in the direction of the rows is usually referred to as the *pitch*; the spacing of the rows is called the

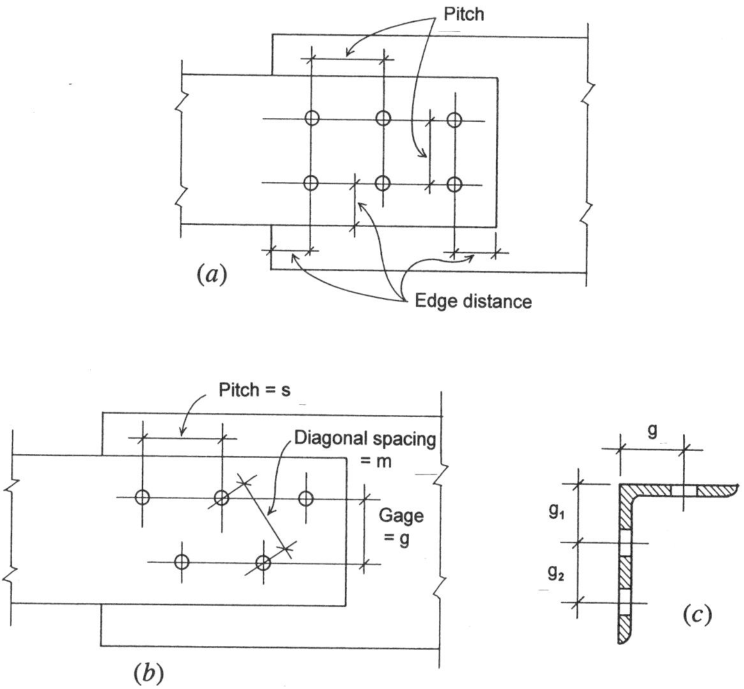

Figure 11.4 Layout considerations for bolted connections: (a) pitch and end distances; (b) bolt spacing; (c) gage distances for angle legs.

gage. The usual reason for staggering the bolts is that sometimes the rows must be spaced closer (gage spacing) than the minimum spacing required for the bolts selected. However, staggering the bolt holes also helps to create a slightly less critical net section for tension stress in the steel member with the holes.

Location of bolt lines is often related to the size and type of structural members being attached. This is especially true of bolts placed in the legs of angles or in the flanges of W-, M-, S-, C-, and structural tee shapes. Figure 11.4*c* shows the placement of bolts in the legs of angles. When a single row is placed in a leg, its recommended location is at the distance labeled g from the back of the angle. When two rows are used, the first row is placed at the distance g_1 and the second row is spaced a distance g_2 from the first. Table 11.3 gives the recommended values for these distances.

TABLE 11.2 Pitch and Edge Distances for Bolts

Rivet or Bolt Diameter d (in.)	Minimum Edge Distance for Punched, Reamed, or Drilled Holes (in.)		Minimum Recommended Pitch, Center-to-Center (in.)	
	At Sheared Edges	At Rolled Edges of Plates, Shapes,or Bars, or Gas-Cut Edges[a]	$2.667d$	$3d$
0.625	1.125	0.875	1.67	1.875
0.750	1.25	1.0	2.0	2.25
0.875	1.5[b]	1.125	2.33	2.625
1.000	1.75[b]	1.25	2.67	3.0

[a]May be reduced 0.125 in. when the hole is at a point where stress does not exceed 25% of the maximum allowed in the connected element.

[b]May be 1.25 in. at the ends of beam connection angles.

Source: Adapted from data in the *Manual of Steel Construction*, 8th ed., with permission of the publishers, American Institute of Steel Construction.

When placed at the recommended locations in rolled shapes, bolts will end up a certain distance from the edge of the part. Based on the recommended edge distance for rolled edges given in Table 11.2, it is thus possible to determine the maximum size of bolt that can be accommodated. For angles, the maximum fastener may be limited by the edge distance, especially when two rows are used; however, other factors may in some cases be more critical. The distance from the center of the bolts to the inside fillet of the angle may limit the use of a large washer where one is required. Another consideration may be the stress on the net section of the angle, especially if the member load is taken entirely by the attached leg.

TABLE 11.3 Usual Gage Dimensions for Angles (in.)

Gage Dimension	Width of Angle Leg								
	8	7	6	5	4	3.5	3	2.5	2
g	4.5	4.0	3.5	3.0	2.5	2.0	1.75	1.375	1.125
g_1	3.0	2.5	2.25	2.0					
g_2	3.0	3.0	2.5	1.75					

Source: Adapted from data in the *Manual of Steel Construction*, 8th edition, with permission of the publishers, American Institute of Steel Construction.

Tension Connections

When tension members have reduced cross sections, two stress investigations must be considered. This is the case for members with holes for bolts. For the member with a hole, the design tension strength ($\phi_t P_n$) at the reduced cross section through the hole is

$$\phi_t P_n = \phi_t \times F_u \times A_e$$

where A_e = reduced (or net) cross-sectional area
ϕ_t = 0.75

The resistance at the net section must be compared with the resistance at the unreduced section of the member for which the resistance factor (ϕ_t) is 0.90. For steel bolts, the design strength is specified as a value based on the type of bolt.

Angles used as tension members are usually connected by only one leg. In a conservative design, the effective net area is only that of the connected leg, less the reduction caused by bolt holes.

Rivet and bolt holes are punched larger in diameter than the nominal diameter of the fastener. The punching damages a small amount of the steel around the perimeter of the hole; consequently, the diameter of the hole to be deducted in determining the net section is ⅛ in. greater than the nominal diameter of the fastener.

When only one hole is involved, as in Figure 11.2, or in a similar connection with a single row of fasteners along the line of stress, the net area of the cross section of one of the plates is found by multiplying the plate thickness by its net width (width of member minus diameter of hole).

When holes are staggered in two rows along the line of stress (Figure 11.5), the net section is determined somewhat differently. The AISC Specification reads:

> In the case of a chain of holes extending across a part in any diagonal or zigzag line, the net width of the part shall be obtained by deducting from the gross width the sum of the diameters of all the holes in the chain and adding, for each gage space in the chain, the quantity $s^2/4g$, where
>
> s = longitudinal spacing (pitch) in inches or any two successive holes.
>
> g = transverse spacing (gage) in inches for the same two holes.

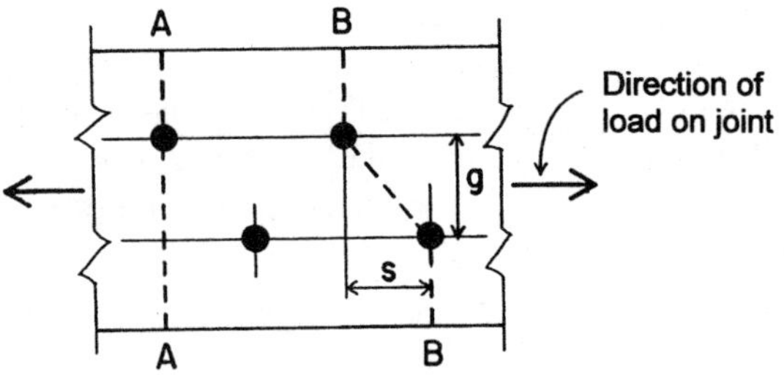

Figure 11.5 Determination of net cross-sectional area for connected members in a bolted connection.

The critical net section of the part is obtained from that chain which gives the least net width.

11.2 DESIGN OF A BOLTED CONNECTION

The issues raised in the preceding sections are illustrated in the following design example.

Example 1. The connection shown in Figure 11.6 consists of a pair of narrow plates that transfer a tension force that is produced by a dead load of 50 kips [222 kN] and a live load of 90 kips [495 kN]. All plates are of A36 steel with $F_y = 36$ ksi [250 MPa] and $F_u = 58$ ksi [400 MPa] and are attached with 3/4-in. A325 bolts placed in two rows with threads included in the planes of shear. Using data from Table 11.1, determine the number of bolts required, the width and thickness of the narrow plates, the thickness of the wide plate, and the layout for the connection.

Solution: The process begins with the determination of the ultimate load.

$$P_u = 1.2D + 1.6L = 1.2(50) + 1.6(90) = 204 \text{ kips [907 kN]}$$

From Table 11.1, the capacity of a single bolt in double shear is found as 31.8 kips [141 kN]. The required number of bolts for the connection is thus

$$n = \frac{204}{31.8} = 6.41, \text{ or } 7$$

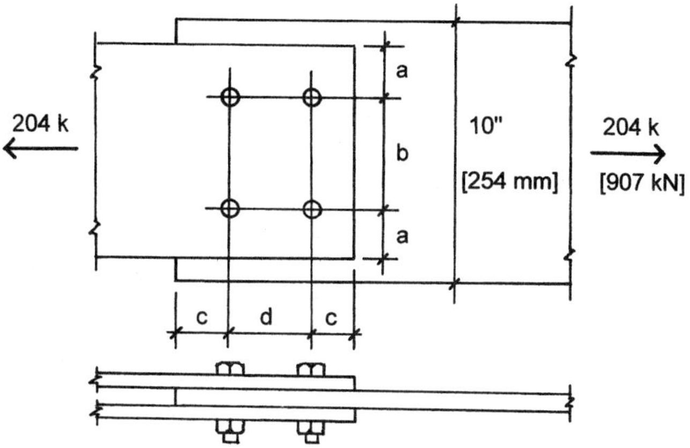

Figure 11.6 Reference Figure for the example problem.

Although placement of seven bolts in the connection is possible, most designers would choose to have a symmetrical arrangement with eight bolts, four to a row. The average bolt load is thus

$$V_u = \frac{204}{8} = 25.5 \text{ kips [113 kN]}$$

From Table 11.2, for the ¾-in. bolts, minimum edge distance for a cut edge is 1.25 in. and minimum recommended spacing is 2.25 in. The minimum required width for the plates is thus (see Figure 11.6)

$$w = b + 2(a) = 2.25 + 2(1.25) = 4.75 \text{ in. [121 mm]}$$

If space is tightly constrained, this actual width could be specified for the narrow plates. For this example, a width of 6 in. is used. Checking for the requirement of stress on the gross area of the plate cross section:

$$A_g = \frac{Pu}{\phi_t F_y} = \frac{204}{0.9(36)} = 6.30 \text{ in.}^2 \text{ [4070 mm}^2\text{]}$$

and, with the 6-in. width, the required thickness is

$$t = \frac{6.30}{2 \times 6} = 0.525 \text{ in. [13 mm]}$$

This permits the use of a minimum thickness of 9⁄16 in. [14 mm]. The next step is to check the stress on the net section. For the computations it is recommended to use a bolt-hole size at least 1⁄8-in. larger than the bolt diameter. This allows for the true oversize (usually 1⁄16-in.) and some loss due to the roughness of the hole edges. Thus, the hole is assumed to be 7⁄8-in. (0.875) in diameter, and the net width and area are

$$w = 6 - 2(0.875) = 4.25 \text{ in. [108 mm]}$$

$$A_e = w \times t = 4.25 \times \frac{9}{16} = 2.39 \text{ in.}^2$$

and the design strength at the net section is

$$\phi_t P_n = 0.75 \times F_u \times A_e = 0.75 \times 58 \times (2 \times 2.39) = 208 \text{ kips [935 kN]}$$

Because this is greater than the factored load (P_u), the narrow plates are adequate for tension stress.

The bolt capacities in Table 11.1 are based on a slip-critical condition, which assumes a design failure limit to be that of the friction resistance (slip resistance) of the bolts. However, the backup failure mode is the one in which the plates slip to permit development of the pin-action of the bolts against the sides of the holes; this then involves the shear capacity of the bolts and the bearing resistance of the plates. Bolt shear capacities are higher than the slip failures, so the only concern for this is the bearing on the plates.

Bearing design strength ($\phi_v R_n$) is computed for a single bolt as

$$\phi_v R_n = 1.5 \times L_c \times t \times F_u = 1.5 \times 1.5 \times \frac{9}{16} \times 58 = 73.3 \text{ kips [326 kN]}$$

which is clearly not a critical concern because the factored load on each bolt is 25.5 kips.

For the middle plate the procedure is essentially the same, except that the width is given and there is a single plate. As before, the stress on the unreduced cross section requires an area of 6.30 in.^{2}, so the required thickness of the 10-in.-wide plate is

$$t = \frac{6.30}{10} = 0.630 \text{ in. [16 mm]}$$

which indicates the use of a ⅝-in. thickness.

For the middle plate, the width and cross-sectional area at the net section are

$$w = 10 - (2 \times 0.875) = 8.25 \text{ in. [210 mm]}$$

and the design strength at the net section is

$$\phi_t P_n = 0.75 \times F_u \times A_e = 0.75 \times 58 \times 5.16 = 224 \text{ kips [9296 kN]}$$

which is greater than the factored load of 204 kips.

The computed bearing design strength on the sides of the holes in the middle plate is

$$\phi_v R_n = 1.5 \times L_c \times t \times F_u = 1.5 \times 1.5 \times \frac{5}{8} \times 58 = 81.5 \text{ kips [363 kN]}$$

which is greater than the factored load of 25.5 kips, as determined previously.

A final problem that must be considered is the possibility for tearing out of the two bolts at the end of a plate in a block shear failure (Figure 11.3*a*). Because the combined thicknesses of the outer plates is greater than that of the middle plate, the critical case for this connection is that of the middle plate. Figure 11.7 shows the condition for tearing, which involves a combination of tension on the section labeled "1" and shear on the two sections labeled "2." For the tension section

$$\text{Net } w = 3 - 0.875 = 2.125 \text{ in. [54 mm]}$$

$$A_e = 2.125 \times \frac{5}{8} = 1.328 \text{ in.}^2 \text{ [857 mm}^2\text{]}$$

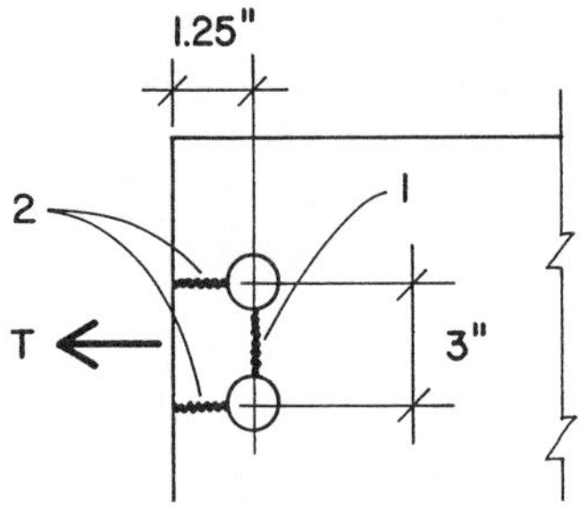

Figure 11.7 Tearing in the example problem.

and the design strength for tension is

$$\phi_t P_n = \phi_t \times F_u \times A_e = 0.75 \times 58 \times 1.328 = 57.8 \text{ kips [257 kN]}$$

For the two shear sections:

$$\text{Net } w = 2\left(1.25 - \frac{0.875}{2}\right) = 1.625 \text{ in. [41.3 mm]}$$

$$A_e = 1.625 \times \frac{5}{8} = 1.016 \text{ in.}^2 \text{ [655 mm}^2\text{]}$$

and the design strength for shear is

$$\phi_v R_n = \phi_v \times F_u \times A_c = 0.75 \times 58 \times 1.016 = 44.2 \text{ kips [197 kN]}$$

The total resistance to tearing is thus

$$T = 57.8 + 44.2 = 102 \text{ kips [454 kN]}$$

Because this is greater than the combined load on the two end bolts (51 kips), the plate is not critical for tearing in block shear.

The solution for the connection is displayed in the top and side views in Figure 11.8.

Connections that transfer compression between the joined parts are essentially the same with regard to the bolt stresses and bearing on the parts. Stress on the net section in the joined parts is not likely to be criti-

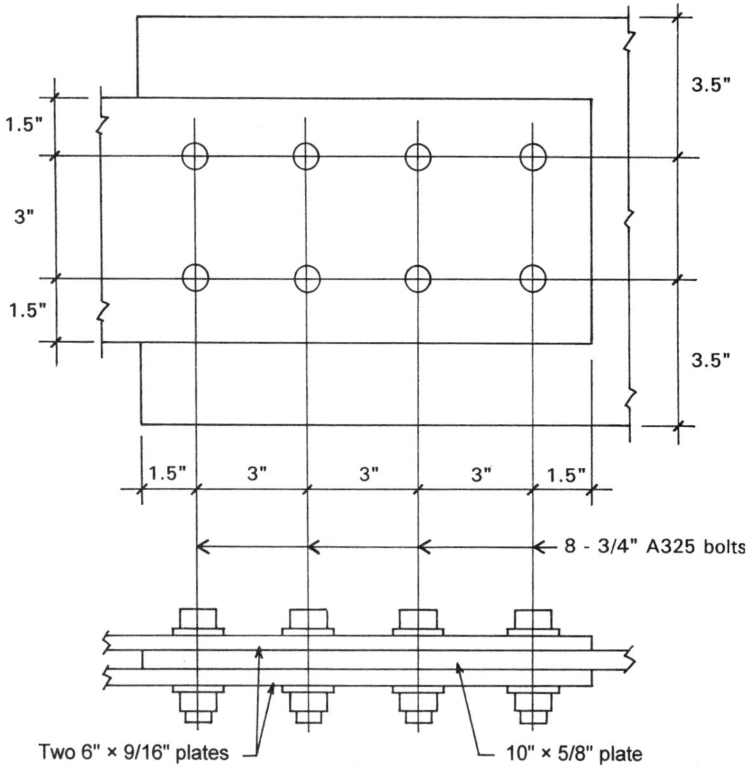

Figure 11.8 Solution for the example problem.

cal, since the compression members are likely to be designed for a relatively low stress due to column action.

Problem 11.2.A. A bolted connection of the general form shown in Figure 11.6 is to be used to transmit a tension force of 75 kips [334 kN] dead load and 100 kips live load by using ⅞-in. A325 bolts and plates of A36 steel. The outer plates are to be 8 in. wide [200 mm] and the center plate is to be 12 in. wide [300 mm]. Find the required thicknesses of the plates and the number of bolts needed if the bolts are placed in two rows. Sketch the final layout of the connection.

Problem 11.2.B. Design the connection for the data in Problem 11.2.A, except that the outer plates are 9 in. wide and the bolts are placed in three rows.

11.3 BOLTED FRAMING CONNECTIONS

The joining of structural steel members in a structural system generates a wide variety of situations, depending on the form of the connected parts, the type of connecting device used, and the nature and magnitude of the forces that must be transferred between the members. Framing connections quite commonly involve the use of welding and bolting in a single connection. In general, welding is favored for fabrication in the shop and bolting for erection in the field. If this practice is recognized, the connections must be developed with a view to the overall fabrication and erection process and some decision made regarding what is to be done where. With the best of designs, however, the contractor who is awarded the work may have some ideas about these procedures and may suggest alterations in the details.

Development of connection details is particularly critical for structures in which a great number of connections occur. The truss is one such structure.

Framed Beam Connections

The connection shown in Figure 11.9*a* is the type used most frequently in the development of framed structures that consist of I-shaped beams and H-shaped columns. This device is referred to as a *framed beam connection*, for which there are several design considerations:

1. *Type of fastening.* Fastening of the angles to the supported beam and to the support may be accomplished with welds or with any of several types of structural bolt. The most common practice is to weld the angles to the supported beam's web in the fabricating shop and to bolt the angles to the support (column face or supporting beam's web) in the field (the erection site).
2. *Number of fasteners.* If bolts are used, this refers to the number of bolts used on the supported beams web; twice this number of bolts is used in the outstanding legs of the angles. The capacities are matched, however, because the web bolts are in double shear and the others in single shear. For smaller beams, or for light loads in general, angle leg sizes are typically narrow, being just enough to accommodate a single row of bolts, as shown in Figure 11.9*b*. However, for very large beams and for greater loads, a wider leg may be used to accommodate two rows of bolts.

3. *Size of the angles.* Leg width and thickness of the angles depend on the size of fasteners and the magnitude of loads. Width of the outstanding legs may also depend on space available, especially if attachment is to the web of a column.
4. *Length of the angles.* Length must be that required to accommodate the number of bolts. Standard layout of the bolts is that shown in Figure 11.9, with bolts at 3-in. spacing and end distance of 1.25 in. This will accommodate up to 1-in.-diameter bolts. However, the angle length is also limited to the distance available on the beam web—that is, the total length of the flat portion of the beam web (see Figure 11.9*a*).

The AISC Manual (Ref. 5) provides considerable information to support the design of these frequently used connecting elements. Data is provided for both bolted and welded fastenings. Predesigned connections are tabulated and can be matched to magnitudes of loadings and to sizes (primarily depths) of the beams (mostly W shapes) that can accommodate them.

Although there is no specified limit for the minimum size of a framed connection to be used with a given beam, a general rule is to use one with the angle length at least one half of the beam depth. This rule is intended in the most part to ensure some minimum stability against rotational effects at the beam ends. For very shallow beams, the special connector

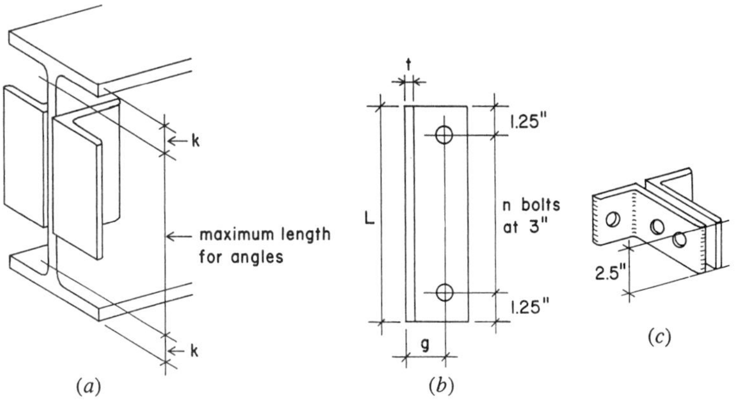

Figure 11.9 Framed beam connections for rolled shapes, using intermediate connecting angles.

shown in Figure 11.9*c* may be used. Regardless of loading, this requires the angle leg at the beam web to accommodate two rows of bolts (one bolt in each row) simply for the stability of the angles.

There are many structural effects to consider for these connections. Of special concern is the inevitable bending in the connection that occurs as shown in Figure 11.2*c*. The bending moment arm for this twisting action is the gage distance of the angle leg, dimension *g* as shown in Figure 11.9*b*. It is a reason for choosing a relatively narrow angle leg.

If the top flange of the supported beam is cut back—as it commonly is when connection is to another beam—either vertical shear in the net cross section or block shear failure (Figure 11.3*c*) may be critical. Both of these conditions will be aggravated when the supported beam has a very thin web, which is a frequent condition because the most efficient beam shapes are usually the lightest shapes in their nominal size categories.

Another concern for the thin beam web is the possibility of critical bearing stress in the bolted connection. Combine a choice for a large bolt with one for a beam with a thin web, and this is likely to be a problem.

11.4 BOLTED TRUSS CONNECTIONS

A major factor in the design of trusses is the development of the truss joints. Because a single truss typically has several joints, the joints must be relatively easy to produce and economical, especially if there is a large number of trusses of a single type in the building structural system. Considerations involved in the design of connections for the joints include the truss configuration, member shapes and sizes, and the fastening method—usually welding or high-strength bolts.

In most cases, the preferred method of fastening for connections made in the fabricating shop is welding. Trusses are usually shop-fabricated in the largest units possible, which means the whole truss for modest spans or the maximum-sized unit that can be transported for large trusses. Bolting is mostly used for connections made at the building site. For the small truss, bolting is usually done only for the connections to supports and to supported elements or bracing. For the large truss, bolting may also be done at splice points between shop-fabricated units. All of this is subject to many considerations relating to the nature of the rest of the building structure, the particular location of the site, and the practices of local fabricators and erectors.

Two common forms for light steel trusses are shown in Figure 11.10. In Figure 11.10*a*, the truss members consist of pairs of angles and the joints are achieved by using steel gusset plates to which the members are

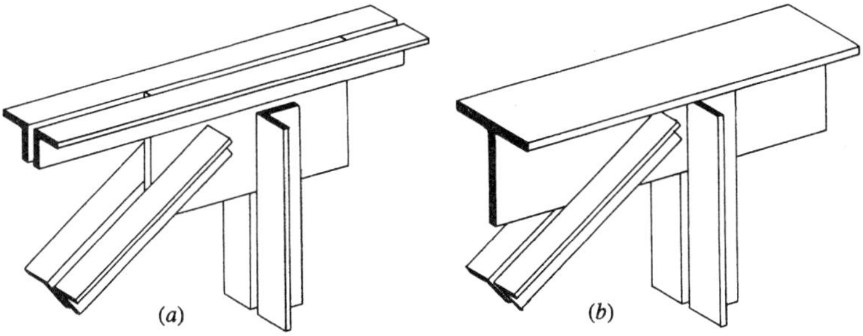

Figure 11.10 Common framing details for light steel trusses. (a) With chords consisting of double-angles and joints using gusset plates. (b) With chords consisting of structural tees.

attached. For top and bottom chords, the angles are often made continuous through the joint, reducing the number of connectors required and the number of separate cut pieces of the angles. For flat-profiled, parallel-chord trusses of modest size, the chords are sometimes made from tees, with interior members fastened directly to the tee web (Figure 11.10*b*).

Figure 11.11 shows a layout for several joints of a light roof truss, employing the system shown in Figure 11.10*a*. This is a form commonly used in the past for roofs with high slopes, with many short-span trusses fabricated in a single piece in the shop, usually with riveted joints. Trusses of this form are now mostly welded or use high-strength bolts as shown in Figure 11.11.

Development of the joint designs for the truss shown in Figure 11.11 would involve many considerations, including the following:

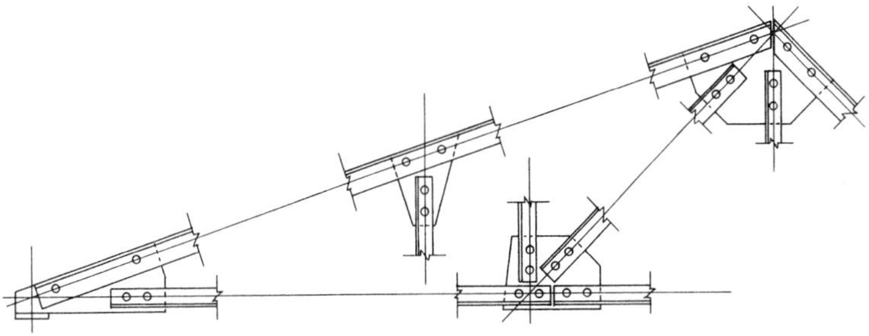

Figure 11.11 Typical form of a light steel truss with double-angle truss members and bolted joints with gusset plates.

1. *Truss member size and load magnitude.* This determines primarily the size and type of connector (bolt) required, based on individual connector capacity.
2. *Angle leg size.* This relates to the maximum diameter of bolt that can be used, based on angle gages and minimum edge distances. (See Table 11.3.)
3. *Thickness and profile size of gusset plates.* The preference is to have the lightest weight added to the structure (primarily for the cost per pound of the steel), which is achieved by reducing the plates to a minimum thickness and general minimum size.
4. *Layout of members at joints.* The aim is to have the action lines of the forces (vested in the rows of bolts) all meet at a single point, thus avoiding twisting in the joint.

Many of the points mentioned are determined by data. Minimum edge distances for bolts (Table 11.2) can be matched to usual gage dimensions for angles (Table 11.3). Forces in members can be related to bolt capacities in Table 11.1, the general intent being to keep the number of bolts to a minimum in order to make the required size of the gusset plate smaller.

Other issues involve some judgment or skill in the manipulation of the joint details. For really tight or complex joints, it is often necessary to study the form of the joint with carefully drawn large-scale layouts. Actual dimensions and form of the member ends and gusset plates may be derived from these drawings.

The truss shown in Figure 11.11 has some features that are quite common for small trusses. All member ends are connected by only two bolts, the minimum required by the specifications. This simply indicates that the minimum-sized bolt chosen has sufficient capacity to develop the forces in all members with only two bolts. At the top chord joint between the support and the peak, the top chord member is shown as being continuous (uncut) at the joint. This is quite common where the lengths of members available are greater than the joint-to-joint distances in the truss, a cost savings in member fabrication as well as connection.

If there are only one or a few of the trusses shown in Figure 11.11 to be used in a building, the fabrication may indeed be as shown in the illustration. However, if there are many such trusses, or the truss is actually a manufactured, standardized product, it is much more likely to be fabricated employing welding for shop work and bolting only for field connections.

12

LIGHT-GAGE FORMED STEEL STRUCTURES

Many structural elements are formed from sheet steel. Elements formed by the rolling process must be heat-softened, whereas those produced from sheet steel are ordinarily made without heating the steel. Thus, the common description for these elements is *cold-formed.* Because they are typically formed from thin sheet stock, they are also referred to as *light-gage* steel products.

12.1 LIGHT-GAGE STEEL PRODUCTS

Figure 12.1 illustrates the cross sections of some common products formed from sheet steel. Large corrugated or fluted panels are in wide use for wall paneling and for structural decks for roofs and floors (Figure 12.1*a*). These products are made by a number of manufacturers, and information regarding their structural properties may be obtained directly from the manufacturer. General information on structural decks may also be obtained from the Steel Deck Institute (see Ref. 7).

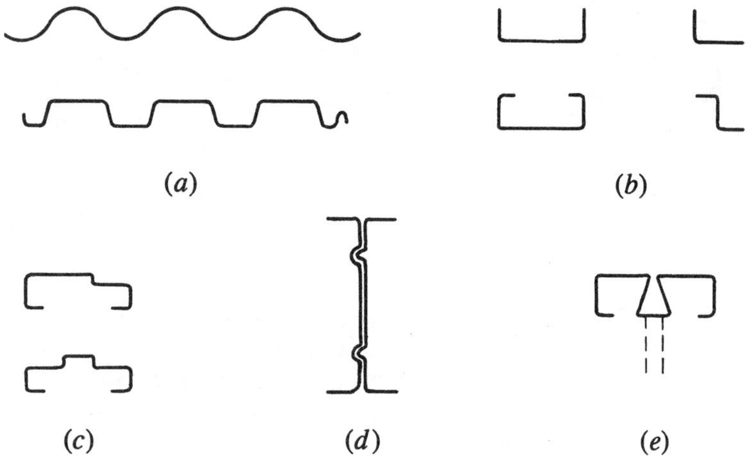

Figure 12.1 Cross-sectional shapes of common cold-formed sheet steel products: (a) panels for walls and decks; (b) stock for light steel framing; (c) door and window frames; (d) formed structural framing products; (e) chords for light steel trusses.

Cold-formed shapes range from the simple L, C, U, and so on (Figure 12.1*b*) to the special forms produced for various construction systems, such as door and window frames (Figure 12.1*c*). Structures for some buildings may be almost entirely composed of cold-formed products. Design of cold-formed elements is described in the *Cold-Formed Steel Design Manual,* published by the American Iron and Steel Institute.

Although some cold-formed and fabricated elements of sheet steel may be used for parts of structural systems, a major use of these products is for the formation of structural frames for partitions, curtain walls, suspended ceilings, and door and window framing. In large buildings, fire safety requirements usually prevent use of wood for these applications, so the noncombustible steel products are widely chosen.

12.2 LIGHT-GAGE STEEL DECKS

Steel decks consisting of formed sheet steel are produced in a variety of configurations, as shown in Figure 12.2. The simplest is the corrugated sheet, shown in Figure 12.2*a*. This may be used as the single, total surface for walls and roofs of utilitarian buildings (tin shacks). For more demanding usage, it is used mostly as the surfacing of a built-up panel or

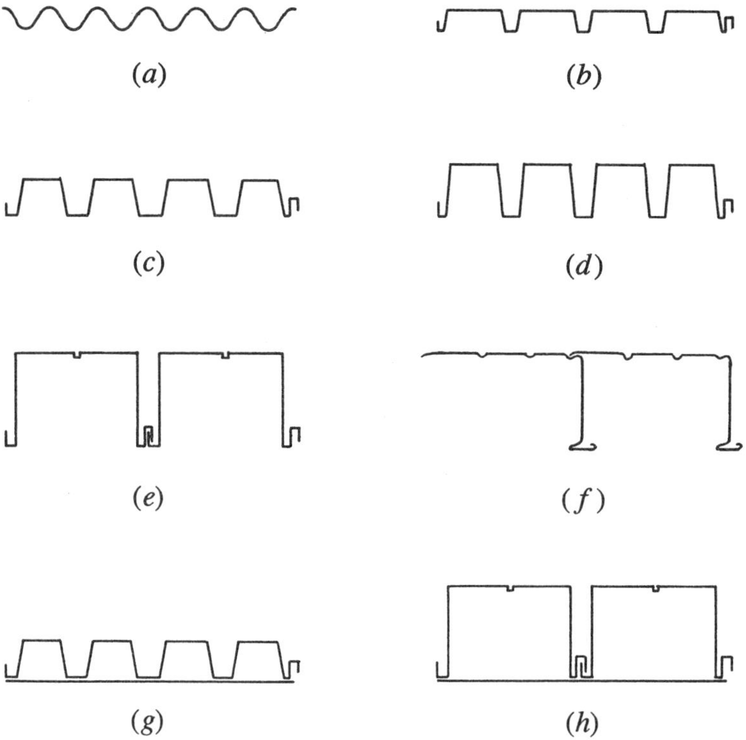

Figure 12.2 Cross-sectional shapes of formed sheet steel deck units.

general sandwich-type construction. As a structural deck, the simple corrugated sheet is used for very short spans, typically with a structural-grade concrete fill that effectively serves as the spanning deck—the steel sheet serving primarily as forming for the concrete.

A widely used product is that shown in three variations in Figure 12.2*b* through *d*. When used for roof deck, where loads are light, a flat top surface is formed with a very lightweight fill of foamed concrete or gypsum concrete or a rigid sheet material that also serves as insulation. For floors, especially those with heavier loads, a need for a relatively hard surface, and a concern for bouncing, a structural-grade concrete fill is used, and the deeper ribs of the units shown in Figure 12.2*c* and *d* may be selected to achieve greater spans with wide-spaced beams. Common overall deck heights are 1.5, 3, and 4.5 in.

There are also formed sheet steel decks produced with greater depth, such as those shown in Figure 12.2*e* and *f*. These can achieve considerable span, generally combining the functions of joist and deck in a single unit.

Although used somewhat less now, with the advent of different wiring products and techniques for facilitating the need for frequent and rapid change of building wiring, a possible use for steel deck units is as a conduit for power, signal, or communication wiring. This can be accomplished by closing the deck cells with a flat sheet of steel, as shown in Figures 12.2*g* and *h*. This usually provides for wiring in one direction in a grid; the perpendicular wiring is achieved in conduits buried in the concrete fill.

Decks vary in form (shape of the cross section) and in the thickness (gage) of the steel sheet used to form them. Design choices relate to the desire for a particular form and to the load and span conditions. Units are typically available in lengths of 30 ft or more, which usually permits design for a multiple-span condition; this reduces bending effects only slightly but has a considerable influence on the reduction of deflection and bouncing.

Fire protection for floor decks is provided partly by the concrete fill on top of the deck. Protection of the underside is achieved by sprayed-on materials (as also used on beams) or by the use of a permanent fire-rated ceiling construction. The latter is no longer favored, however, because many disastrous fires have occurred in the void space between ceilings and overhead floors or roofs.

Other structural uses of the deck must also be considered. The most common use is as a horizontal diaphragm for distribution of lateral forces from wind and earthquakes. Lateral bracing of beams and columns is also often assisted or completely achieved by structural decks.

When structural-grade concrete is used as a fill, there are three possibilities for its relationship to a forming steel deck:

1. The concrete serves strictly as a structurally inert fill, providing a flat surface, fire protection, added acoustic separation, and so on, but no significant structural contribution.
2. The steel deck functions essentially only as a forming system for the concrete fill; where the concrete is reinforced and designed as a spanning structural deck.
3. The concrete and sheet steel work together in what is described as *composite structural action*. In effect, the sheet steel on the bottom serves as the reinforcement for midspan bending stresses, leaving a need only for top reinforcement for negative bending moments over the deck supports.

Table 12.1 presents data relating to the use of the type of deck unit shown in Figure 12.2*b* for roof structures. This data is adapted from a publication distributed by an industry-wide organization referred to in the table footnotes and is adequate for preliminary design work. The reference publication also provides considerable information and standard specifications for usage of the deck. For any final design work for actual construction, structural data for any manufactured products should be obtained directly from the suppliers of the products.

The common usage for roof decks with units as shown in Table 12.1 is that described earlier as option 1: structural dependence strictly on the steel deck units. That is the basis for the data in the table given here.

Three different rib configurations are shown for the deck units in Table 12.1, described as *narrow*, *intermediate*, and *wide* rib deck. This has some effect on the properties of the deck cross section and thus produces three separate sections in the table. Even though structural performance may be a factor in choosing the rib width, other reasons usually predominate. If the deck is to be welded to its supports (usually required for good diaphragm action), this is done at the bottom of the ribs and the wide rib is required. If a relatively thin topping material is used, the narrow rib is favored.

Rusting is a critical problem for the very thin sheet steel deck. With its top usually protected by other construction, the main problem is the treatment of the underside of the deck. A common practice is to provide the appropriate surfacing of the deck units in the factory. The deck weights in Table 12.1 are based on simple painted surfaces, which is usually the least expensive surface. Surfaces consisting of bonded enamel or galvanizing are also available, adding somewhat to the deck weight.

As described previously, these products are typically available in lengths up to 30 ft or more. Depending on the spacing of supports, therefore, various conditions of continuity of the deck may occur. Recognizing this condition, the table provides three cases for continuity: simple span (one span), two span, and three or more spans.

Problem 12.2.A, B, C, D, E, F. Using data from Table 12.1, select the lightest steel deck for the following.

A. Simple span of 7 ft, total load of 45 psf

B. Simple span of 5 ft, total load of 50 psf

C. Two-span condition, span of 8.5 ft, total load of 45 psf

D. Two-span condition, span of 6 ft, total load of 50 psf

TABLE 12.1 Safe Service Load Capacity of Formed Steel Roof Deck

Deck Type[a]	Span Condition	Weight[b] (lb/sq ft)	Total Safe Service Load (Dead + Live)[c] for spans indicated in ft												
			4.0	4.5	5.0	5.5	6.0	6.5	7.0	7.5	8.0	8.5	9.0	9.5	10.0
NR22	Simple	1.6	73	58	47										
NR20		2.0	91	72	58	48	40								
NR18		2.7	125	99	80	66	55	47							
NR22	Two	1.6	80	63	51	42									
NR20		2.0	97	76	62	51	43								
NR18		2.7	128	101	82	68	57	48	42						
NR22	Three and +	1.6	100	79	64	53	44								
NR20		2.0	121	96	77	64	54	46							
NR18		2.7	160	126	102	85	71	61	52	45					
IR22	Simple	1.6	84	66	54	44									
IR20		2.0	104	82	67	55	46								
IR18		2.7	142	112	91	75	63	54	46	40					
IR22	Two	1.6	90	71	58	48	40								
IR20		2.0	110	87	70	58	49	41							
IR18		2.7	145	114	93	77	64	55	47	40					
IR22	Three and +	1.6	113	89	72	60	50	43							
IR20		2.0	137	108	88	72	61	52	45						
IR18		2.7	181	143	116	96	81	69	59	52	45	40			
WR22	Simple	1.6			90	70	56	46							
WR20		2.0			113	88	70	57	48	40					
WR18		2.7			159	122	96	77	64	54	46	40			
WR22	Two	1.6			96	79	67	57	49	43					
WR20		2.0			123	102	86	73	63	55	48	43			
WR18		2.7			164	136	114	98	84	73	64	57	51	46	41
WR22	Three and +	1.6			119	99	83	71	61	53	47	41	36		
WR20		2.0			153	127	107	91	79	68	58	50	43		
WR18		2.7			204	169	142	121	105	91	79	67	58	51	43

[a] Letters refer to rib type (see Fig. 12.3). Numbers indicate gage (thickness) of deck sheet steel.

[b] Approximate weight with paint finish; other finishes available.

[c] Total safe allowable service load in lb/sq ft. Loads in parentheses are governed by live load deflection not in excess of 1/240 of the span, assuming a dead load of 10 lb/sq ft.

Source: Adapted from the *Steel Deck Institute Design Manual for Composite Decks, Form Decks, and Roof Decks* (Ref. 9), with permission of the publishers, the Steel Deck Institute.

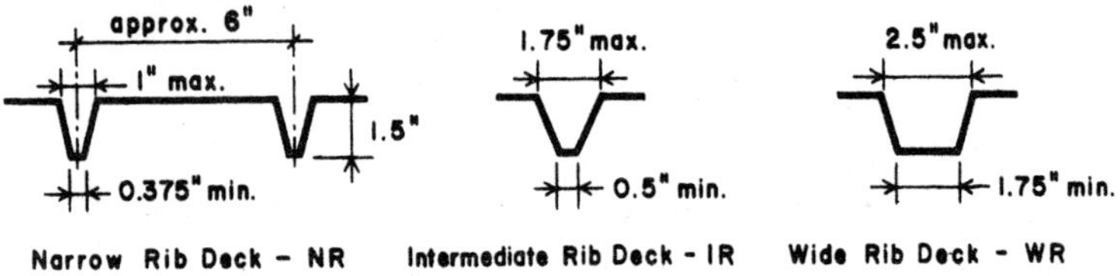

Figure 12.3 Reference for Table 12.1.

E. Three-span condition, span of 6 ft, total load of 50 psf

F. Three-span condition, span of 8 ft, total load of 50 psf

12.3 LIGHT-GAGE STEEL SYSTEMS

Proprietary steel structural systems are produced by many manufacturers for various applications. Even though some systems exist for developing structures for entire buildings, a larger market is that for the systems used for wall framing, ceiling structures, and supports for building service elements. With systems for large structures, rolled shapes or trusses may be used for larger elements, with light-gage elements creating the infilling structure, bracing, and various secondary framing.

The light-gage elements and systems widely developed for partition and ceiling framing for large buildings can be utilized to produce a stud/rafter/joist system that emulates the classic light wood frame with 2-in. nominal lumber elements.

IV

CONCRETE CONSTRUCTION

The term *concrete* covers a variety of products that share a common character: They consist of a mass of loose particles (called the *aggregate)* that is bound together by some cementing material. Included in this group are asphalt paving and precast shingle tiles, but the material in this part deals primarily with the more familiar material described by the term—that produced with Portland cement as the binder and sand and gravel as the inert mass of loose particles. When used for building structures, provisions are made with concrete construction to compensate for the low tensile strength of the material. Three different methods are currently in use: the addition of fibrous material to the concrete mix, prestressing to add a compressive stress to counteract tension stresses, and insertion of steel reinforcing rods to the cast concrete. When reinforced with steel rods, the construction is described as *reinforced concrete*, which is the form of construction treated in this part.

13

REINFORCED CONCRETE STRUCTURES

This part deals primarily with concrete formed with the common binding agent of *Portland cement*, and a loose mass consisting of sand and gravel. With minor variations, this is the material used mostly for structural concrete—to produce building structures, pavements, and foundations.

13.1 GENERAL CONSIDERATIONS

Concrete made from natural materials was used by ancient builders thousands of years ago. Modern concrete, made with industrially produced cement, was developed in the early part of the nineteenth century when the process for producing Portland cement was developed. Because of its lack of tensile strength, however, concrete was used principally for crude, massive structures—foundations, bridge piers, and heavy walls.

In the mid to late nineteenth century, several builders experimented with the technique of inserting iron or steel rods into relatively thin structures of concrete to enhance their ability to resist tensile forces. This was the beginning of what we now know as *reinforced concrete.*

From ancient times until now, there has been a steady accumulation of experience derived from experiments, research, and, most recently, intense development of commercial products. As a result, the designer now has available an immense variety of products under the general classification of concrete, although the range is somewhat smaller if structural usage is required.

Forms of Concrete Structures

For building structures, concrete is mostly used with one of three basic construction methods. The first is called *sitecast concrete,* in which the wet concrete mix is deposited in some forming at the location where it is to be used. This method is also described as *cast-in-place* or *in situ* construction.

A second method consists of casting portions of the structure at a location away from the desired location of the construction. These elements, described as *precast concrete*, are then moved into position, much as are blocks of stone or parts of steel frames.

Finally, concrete may be used for masonry construction—in one of two ways. Precast units of concrete, called concrete masonry units (CMUs), may be used in a manner similar to bricks or stones. Or, concrete fill may be used to produce solid masonry by being poured into cavities in the masonry construction produced with bricks, stone, or CMUs. The latter technique, combined with the insertion of steel reinforcement into the cavities, is widely used for masonry structures today. The use of concrete-filled masonry, however, is one of the oldest forms of concrete construction; it was used extensively by the Romans and the builders of early Christian churches.

Concrete is produced in great volume for various forms of construction. Other than for pavements, the widest general use of concrete for building construction is for foundations. Almost every building has a concrete foundation, whether the major aboveground construction is concrete, masonry, wood, steel, aluminum, or fabric. For small buildings with shallow footings and no basement, the total foundation system may be modest, but for large buildings and those with many below-ground levels, there may well be a gigantic underground concrete structure.

For aboveground building construction, concrete is generally used in situations that fully realize the various advantages of the basic material and the common systems that derive from it. For structural applications, this means using the major compressive resistance of the material and in some situations its relatively high stiffness and inertial resistance (major dead weight). However, in many applications, the nonrotting, vermin- and insect-resistive, and fire-resistive properties may be of major significance. And for many uses, its relatively low bulk volume cost is important.

Design Methods

Traditional structural design was developed primarily with a method now referred to as *stress design*. This method utilizes basic relationships derived from classic theories of elastic behavior of materials, and the adequacy or safety of designs are measured by comparison with two primary limits: an acceptable level for maximum stress and a tolerable limit for the extent of deformation (deflection, stretch, etc.). These limits are calculated as they occur in response to the *service loads*—that is, the loads caused by the normal usage conditions visualized for the structure. This method is also called the *working stress method*, the stress limits are called *allowable working stresses*, and the tolerable movements are called *allowable deflection, allowable elongation*, and the like.

The second method of analysis used with concrete design was first formulated in the 1930s. This method utilizes the basic relationship of elastic behavior of materials by comparisons of two primary limits: the failure or ultimate strength of the material that is factored based upon its use in the structure and the tolerable limit for the extent of deformation. These limits are calculated as they occur in response to *factored loads*—that is, loads that have been factored based upon the predictability of the load, in other words, dead or live loading. This method in concrete design is referred to as the *ultimate strength method* or just *strength design*.

The Stress Method

The stress method generally consists of the following:

1. The service (working) load conditions are visualized and quantified as intelligently as possible. Adjustments may be made here by determining various statistically likely load combinations (dead load plus live load plus wind load, etc.), by consideration of load duration, and so on.

2. Stress, stability, and deformation limits are set by standards for the various responses of the structure to the loads: in tension, bending, shear, buckling, deflection, and so on.
3. The structure is then evaluated (investigated) for its adequacy or is proposed (designed) for an adequate response.

An advantage obtained in working with the stress method is that the real usage condition (or at least an intelligent guess about it) is kept continuously in mind. The principal disadvantage comes from its detached nature regarding real failure conditions, because most structures develop much different forms of stress and strain as they approach their failure limits.

The Strength Method

In essence, the working stress method consists of designing a structure to *work* at some established appropriate percentage of its total capacity. The strength method consists of designing a structure to *fail*, but at a load condition well beyond what it should have to experience in use. A major reason for favoring of strength methods is that the failure of a structure is relatively easily demonstrated by physical testing. What is truly appropriate as a working condition, however, is pretty much a theoretical speculation. The strength method is now largely preferred in professional design work. Initially largely developed for design of concrete structures, it is now generally taking over all areas of structural design.

Nevertheless, it is considered necessary to study the classic theories of elastic behavior as a basis for visualization of the general ways that structures work. Ultimate responses are usually some form of variant from the classic responses (because of inelastic materials, secondary effects, multimode responses, etc.). In other words, the usual study procedure is to first consider a classic, elastic response and then to observe (or speculate about) what happens as failure limits are approached.

For the strength method, the process is as follows:

1. The service loads are quantified as in Step 1 in the previous section and then are multiplied by an adjustment factor to produce the *factored load*. These adjustment factors are different for different types of loading and are based upon how accurate the loading is to the actual loading the member will experience during its useful life. Dead loading is fairly accurate to predict, whereas the predic-

tion of live load is more difficult. Therefore, the adjustment factor for dead load is less than it is for live load.

2. The form of response of the structure is visualized, and its ultimate (maximum, failure) resistance is quantified in appropriate terms (resistance to compression, to buckling, to bending, etc.). Sometimes this quantified resistance is also subject to an adjustment factor called the *resistance factor*.
3. The usable resistance of the structure is then compared to the ultimate resistance required (an investigation procedure), or a structure with an appropriate resistance is proposed (a design procedure).

When the design process using the strength method employs both load and resistance factors, it is now sometimes called *load and resistance factor design* (abbreviated *LRFD*). This is the method that will be employed in this book.

Strength of Concrete

The quality of concrete of greatest significance for structural purposes is its resistance to compressive stress. As such, the common practice is to specify a desired limiting capacity of compressive stress, to design a concrete mix to achieve that limit, and to test samples of cast and hardened concrete to verify its true capacity for compression. This stress is given the symbol f'_c.

For design work, the capacity of concrete for all purposes is established as some percentage of f'_c. Attainment of a quality of concrete to achieve a particular level of compressive resistance generally also serves to certify various other properties, such as hardness, density, and durability. Choice for the desired strength is typically based on the form of construction. For most purposes, a strength of 3000 to 5000 psi [21 to 34 MPa] for f'_c is usually adequate. However, strengths of 20,000 psi [135 MPa] and higher have recently been achieved for lower columns in very tall structures. At the other end of this spectrum is the situation where quality control may be less reliable, and the designer may assume the attainment of a relatively low strength, basing conservative design computations on strength as low as 2000 psi [14 MPa], while doing all possible to achieve a better concrete.

Because it makes up the major bulk of the finished concrete, the aggregate is of primary importance for stress resistance. It must be hard and durable, and must be graded in size so that small particles fill the voids between larger ones, producing a dense mass before the cement and water are added. The weight of concrete is usually determined primarily by the density of the aggregate.

The other major factor for concrete strength is the amount of water used for mixing. The basic idea is to use as little water as possible, because an excess will water down the water-cement mixture and produce very weak, porous concrete. However, this must be balanced against the need for a wet mix that can be easily placed in forms and finished. A lot of skill and some science is involved in producing an ideal mix.

A final consideration for strength is the control of conditions during the early life of a cast mix. The mobile wet mix hardens relatively quickly but gains its highest potential strength over some period of time. It is important to control the water content and the temperature of the hardened concrete during this critical period, if the best-quality concrete is to be expected.

Stiffness of Concrete

As with other materials, the stiffness of concrete is measured by the *modulus of elasticity*, designated E. This modulus is established by tests and is the ratio of stress to strain. Because strain has no unit designation (measured as inch/inch, etc.), the unit for E thus becomes the unit for stress, usually lb/in.^{2} [MPa].

The modulus of elasticity for concrete, E_c, depends on the weight of the concrete and its strength. For values of unit weight between 90 and 155 lb/ft^3 or pcf [1440 and 2480 kg/m^3], the value of E_c is determined as

$$E_c = w^{1.5} \times 33\sqrt{f'_c}$$

The unit weight for ordinary stone-aggregate concrete is usually assumed to be an average of 145 pcf [2323 kg/m^3]. Substituting this value for w in the equation, an average concrete modulus of $E_c = 57{,}000\sqrt{f'_c}$. For metric units, with stress measured in MPa, the expression becomes $E_c = 4730\sqrt{f'_c}$.

Distribution of stresses and strains in reinforced concrete is dependent on the concrete modulus, where the steel modulus is a constant. In the design of reinforced concrete members, the term n is employed. This is the ratio of the modulus of elasticity of steel to that of concrete, or $n = E_s/E_c$. E_s is taken as 29,000 ksi [200,000 MPa], a constant.

Creep

When subjected to long-duration stress at a high level, concrete has a tendency to *creep*, a phenomenon in which strain increases over time under constant stress. This has effects on deflections and on the distributions of stresses between the concrete and reinforcing. Some of the implications of this for design are discussed in the chapters dealing with design of beams and columns.

Cement

The cement used most extensively in building construction is *Portland cement*. Of the five types of standard Portland cement generally available in the United States and for which the American Society for Testing and Materials has established specifications, two types account for most of the cement used in buildings. These are a general-purpose cement for use in concrete designed to reach its required strength in about 28 days, and a high-early-strength cement for use in concrete that attains its design strength in a period of a week or less.

All Portland cements set and harden by reacting with water, and this hydration process is accompanied by generation of heat. In massive concrete structures such as dams, the resulting temperature rise of the materials becomes a critical factor in both design and construction, but the problem is usually not significant in building construction. A low-heat cement is designed for use where the heat rise during hydration is a critical factor. It is, of course, essential that the cement actually used in construction corresponds to that employed in designing the mix, to produce the specified compressive strength of the concrete.

Air-entrained concrete is produced by using special cement or by introducing an additive during mixing of the concrete. In addition to improving workability (mobility of the wet mix), air entrainment permits lower water-cement ratios and significantly improves the durability of the concrete. Air-entraining agents produce billions of microscopic air cells throughout the concrete mass. These minute voids prevent accumulation of water in cracks and other large voids, which, on freezing, would permit the water to expand and result in spalling away of the exposed surface of the concrete.

Reinforcement

The steel used in reinforced concrete consists of round bars, mostly of the deformed type, with lugs or projections on their surfaces. The surface deformations help to develop a greater bond between the steel rods and the enclosing concrete mass.

Purpose of Reinforcement. The essential purpose of steel reinforcing is to reduce the failure of the concrete as a result of tensile stresses. Structural actions are investigated for the development of tension in the structural members, and steel reinforcement in the proper amount is placed within the concrete mass to resist the tension. In some situations, steel reinforcement may also be used to increase compressive resistance because the ratio of magnitudes of strength of the two materials is quite high. Thus, the steel displaces a much weaker material and the member gains significant strength.

Tension stress can also be induced by shrinkage of the concrete during its drying out from the initial wet mix. Temperature variations may also induce tension in many situations. To provide for these latter actions, a minimum amount of reinforcing is used in surface-type members such as walls and paving slabs, even when no structural action is visualized.

Stress-Strain Considerations. The most common grades of steel used for ordinary reinforcing bars are Grade 40 and Grade 60, having yield strengths of 40 ksi [276 MPa] and 60 ksi [414 MPa], respectively. The yield strength of the steel is of primary interest for two reasons. Plastic yielding of the steel generally represents the limit of its practical utilization for reinforcing of the concrete because the extensive deformation of the steel in its plastic range results in major cracking of the concrete. Thus, for service load conditions, it is desirable to keep the stress in the steel within its elastic range of behavior where deformation is minimal.

The second reason for the importance of the yield character of the reinforcing is its ability to impart a generally yielding nature (plastic deformation character) to the otherwise typically very brittle concrete structure. This is of particular importance for dynamic loading and is a major consideration in design for earthquake forces. Also of importance is the residual strength of the steel beyond its yield stress limit. The steel continues to resist stress in its plastic range and then gains a second, higher, strength before failure. Thus, the failure induced by yielding is only a first-stage response, and a second level of resistance is reserved.

Cover. Ample concrete protection, called *cover*, must be provided for the steel reinforcement. This is important to protect the steel from rusting and to be sure that it is well engaged by the mass of concrete. Cover is measured as the distance from the outside face of the concrete to the edge of the reinforcing bar.

Code minimum requirements for cover are ¾ in. [19 mm] for walls and slabs and 1.5 in. [38 mm] for beams and columns. Additional distance of cover is required for extra fire protection or for special conditions of exposure of the concrete surface to weather or by contact with the ground.

Spacing of Bars. Where multiple bars are used in concrete members (which is the common situation), there are both upper and lower limits for the spacing of the bars. Lower limits are intended to facilitate the flow of wet concrete during casting and to permit adequate development of the concrete-to-steel stress transfers for individual bars.

Maximum spacing is generally intended to ensure that some steel relates to a concrete mass of limited size; that is, no extensive mass of concrete should be without reinforcement. For relatively thin walls and slabs, there is also a concern of scale of spacing related to the thickness of the concrete.

Amount of Reinforcement. For structural members, the amount of reinforcement is determined from structural computations as that required for the tension force in the member. This amount (in total cross-sectional area of the steel) is provided by some combination of bars. In various situations, however, a minimum amount of reinforcement is desirable, which may on occasion exceed the amount determined by computation.

Minimum reinforcement may be specified as a minimum number of bars or as a minimum amount, the latter usually based on the amount of the cross-sectional area of the concrete member. These requirements are discussed in the sections that deal with the design of the various types of structural members.

Standard Reinforcing Bars. In early concrete work, reinforcing bars took various shapes. An early problem that emerged was the proper bonding of the steel bars within the concrete mass because the bars tended to slip or pull out of the concrete. This issue is still a critical one and is discussed in Section 13.6.

In order to anchor the bars in the concrete, various methods were used to produce something other than the usual smooth surfaces on bars. After much experimentation and testing, a single set of bars was developed with surface deformations consisting of ridges. These deformed bars were produced in graduated sizes with bars identified by a single number (see Table 13.1).

TABLE 13.1 Properties of Deformed Reinforcing Bars

Bar Size Designation	Nominal Weight		Nominal Dimensions			
			Diameter		Cross Sectional Area	
	lb/ft	kg/m	in.	mm	in.^{2}	mm^2
No. 3	0.376	0.560	0.375	9.5	0.11	71
No. 4	0.668	0.994	0.500	12.7	0.20	129
No. 5	1.043	1.552	0.625	15.9	0.31	200
No. 6	1.502	2.235	0.750	19.1	0.44	284
No. 7	2.044	3.042	0.875	22.2	0.60	387
No. 8	2.670	3.974	1.000	25.4	0.79	510
No. 9	3.400	5.060	1.128	28.7	1.00	645
No. 10	4.303	6.404	1.270	32.3	1.27	819
No. 11	5.313	7.907	1.410	35.8	1.56	1006
No. 14	7.650	11.390	1.693	43.0	2.25	1452
No. 18	13.600	20.240	2.257	57.3	4.00	2581

For bars numbered 2 through 8, the cross-sectional area is equivalent to a round bar having a diameter of as many eighths of an inch as the bar number. Thus, a No. 4 bar is equivalent to a round bar of 4/8 or 0.5 in. diameter. Bars numbered from 9 up lose this identity and are essentially identified by the tabulated properties in a reference document.

The bars in Table 13.1 are developed in U.S. units but can, of course, be used with their properties converted to metric units. However, a new set of bars has been developed that derive their properties more logically from metric units. The general range of sizes is similar for both sets of bars, and design work can readily be performed with either set. Metric-based bars are obviously more popular outside the United States, but for domestic (nongovernment) use in the United States, the old bars are still in wide use. This is part of a wider conflict over units, which is still going on.

The work in this book uses the old inch-based bars simply because the computational examples are done in U.S. units. In addition, many of the references still in wide use have data presented basically with U.S. units and the old bar sizes.

13.2 GENERAL APPLICATION OF STRENGTH METHODS

Application of the working stress method consists of designing members to work in an adequate manner (without exceeding established stress limits) under actual service load conditions. Strength design in effect consists of designing members to fail; thus, the ultimate strength of the member at failure (called its *design strength*) is the only type of resistance considered. The basic procedure of the strength method consists of determining a factored (increased) design load and comparing it to the factored (usually reduced) ultimate resistance of the structural member.

The ACI Code provides various combinations of loads that must be considered for design. Each type of load (live, dead, wind, earthquake, snow, etc.) is given an individual factor in these load equations.

$$U = 1.4D$$

$$U = 1.2D + 1.6L + 0.5(L_r \text{ or } S)$$

$$U = 1.2D + 1.0L + 1.6W + 0.5(L_r \text{ or } S)$$

$$U = 1.2D + 1.0L + 1.4E + 0.2S$$

where D = the effect of dead load
L = the effect of live load
L_r = the effect of roof live load
S = the effect of snow load
W = the effect of wind load
E = the effect of earthquake (seismic) load

The design strength of individual members (i.e., their usable ultimate strength) is determined by applying assumptions and requirements given in the code and is further modified by using a strength reduction factor ϕ as follows:

ϕ = 0.90 for flexure, axial tension, and combinations of flexure and tension
= 0.70 for columns with spirals
= 0.65 for columns with ties
= 0.75 for shear and torsion
= 0.65 for compressive bearing
= 0.55 for flexure in plain (not reinforced) concrete

13.3 BEAMS: ULTIMATE STRENGTH METHOD

The primary concerns for beams relate to their necessary resistance to bending and shear and some limitations on their deflection. For wood or steel beams, the usual concern is only for the singular maximum values of bending and shear in a given beam. For concrete beams, on the other hand, it is necessary to provide for the values of bending and shear as they vary along the entire length of a beam—even through multiple spans in the case of continuous beams, which are a common occurrence in concrete structures. For simplification of the work, designers must consider the actions of a beam at a specific location, but bear in mind that this action must be integrated with all the other effects on the beam throughout its length.

When a member is subjected to bending, such as the beam shown in Figure 13.1*a*, internal resistances of two basic kinds are generally required. Internal actions are "seen" by visualizing a cut section, such as that taken at *X-X* in Figure 13.1*a*. When the portion of the beam to the left of the cut section is removed, its free-body actions are as shown in Figure 13.1*b*. At the cut section, consideration of static equilibrium requires the development of the internal shear force (*V* in the figure) and the internal resisting moment (represented by the force couple: *C* and *T* in the figure).

If a beam consists of a simple rectangular concrete section with tension reinforcement only, as shown in Figure 13.1*c*, the force *C* is considered to be developed by compressive stresses in the concrete—indicated by the shaded area above the neutral axis. The tension force, however, is considered to be developed by the steel alone, ignoring the tensile resistance of the concrete. For low-stress conditions, the latter is not true, but

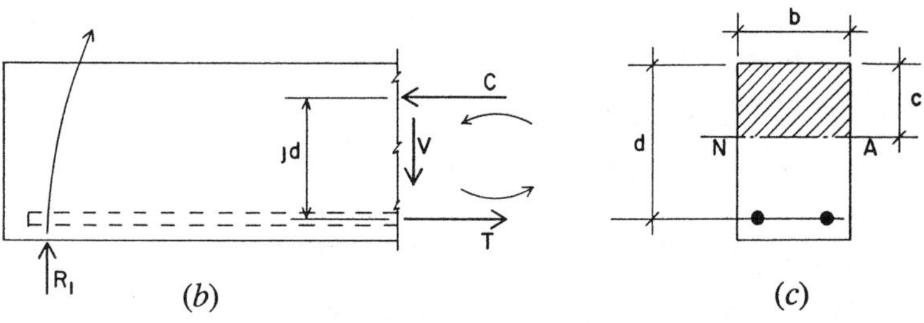

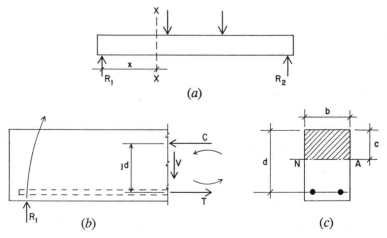

Figure 13.1 Bending action in a reinforced concrete beam.

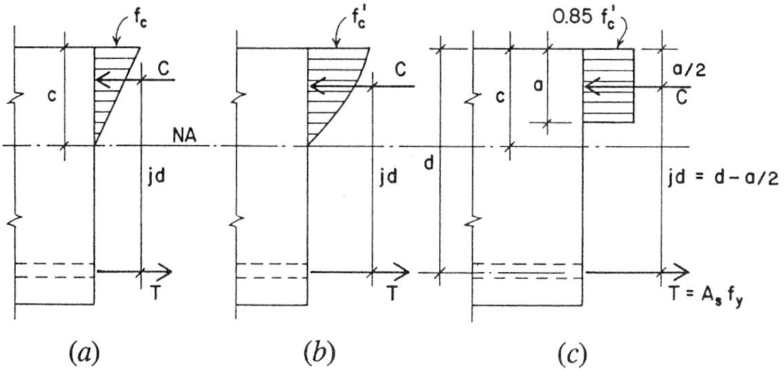

Figure 13.2 Development of bending stress actions in a reinforced concrete beam.

at a serious level of stress, the tension-weak concrete will indeed crack, virtually leaving the steel unassisted, as assumed.

At moderate levels of stress, the resisting moment is visualized as shown in Figure 13.2*a*, with a linear variation of compressive stress from zero at the neutral axis to a maximum value of f_c at the edge of the section. As stress levels increase, however, the nonlinear stress-strain character of the concrete becomes more significant, and we need to acknowledge a more realistic form for the compressive stress variation, such as that shown in Figure 13.2*b*. As stress levels approach the limit of the concrete, the compression becomes vested in an almost constant magnitude of unit stress, concentrated near the top of the section. For strength design, in which the moment capacity is expressed at the ultimate limit, it is common to assume the form of stress distribution shown in Figure 13.2*c*, with the limit for the concrete stress set at 0.85 times f'_c. Expressions for the moment capacity derived from this assumed distribution compare reasonably with the response of beams tested to failure in laboratory experiments.

Response of the steel reinforcement is more simply visualized and expressed. Because the steel area in tension is concentrated at a small location with respect to the size of the beam, the stress in the bars is considered to be a constant. Thus, at any level of stress, the total value of the internal tension force may be expressed as

$$T = A_s f_s$$

and for the practical limit of T as

$$T = A_s f_y$$

The following is a presentation of the formulas and procedures used in the strength method. The discussion is limited to a rectangular beam section with tension reinforcement only.

Referring to Figure 13.3, note that the following are defined:

b = width of the concrete compression zone

d = effective depth of the section for stress analysis; from the centroid of the steel to the edge of the compressive zone

h = overall depth (height) of the section

A_s = cross-sectional area of reinforcing bars

p = percentage of reinforcement, defined as

$$p = \frac{A_s}{bd}$$

Figure 13.2*c* shows the rectangular "stress block" that is used for analysis of the rectangular section with tension reinforcing only by the strength method. This is the basis for investigation and design as provided for in the ACI Code.

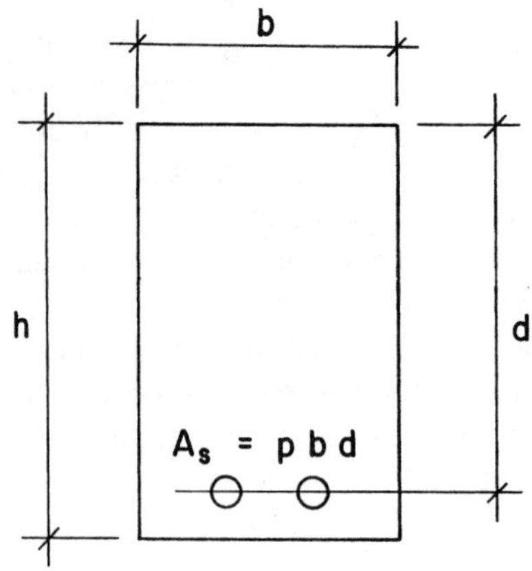

Figure 13.3 Reference notation for a reinforced concrete beam section.

The rectangular stress block is based on the assumption that a concrete stress of $0.85f'_c$ is uniformly distributed over the compression zone, which has dimensions equal to the beam width b and the distance a, which locates a line parallel to and above the neutral axis. The value of a is determined from the expression $a = \beta_1 \times c$, where β_1 (beta one) is a factor that varies with the compressive strength of the concrete, and c is the distance from the extreme fiber to the neutral axis. For concrete having f'_c equal to or less than 4000 psi [27.6 MPa], the ACI Code gives a maximum value for $a = 0.85c$.

With the rectangular stress block, the magnitude of the compressive force in the concrete is expressed as

$$C = (0.85f'_c)(b)(a)$$

and it acts at a distance of $a/2$ from the top of the beam. The arm of the resisting force couple then becomes $d - (a/2)$, and the developed resisting moment as governed by the concrete is

$$M_c = C\left(d - \frac{a}{2}\right) = 0.85f'_c ba\left(d - \frac{a}{2}\right) \tag{13.3.1}$$

With T expressed as $A_s \times f_y$, the developed moment as governed by the reinforcement is

$$M_t = T\left(d - \frac{a}{2}\right) = A_s f_y\left(d - \frac{a}{2}\right) \tag{13.3.2}$$

A formula for the dimension a of the stress block can be derived by equating the compression and tension forces; thus:

$$0.85f'_c ba = A_s f_y, \qquad a = \frac{A_s f_y}{0.85f'_c b} \tag{13.3.3}$$

By expressing the area of steel in terms of a percentage p, the formula for a may be modified as follows:

$$p = \frac{A_s}{bd}, \qquad A_s = pbd$$

$$a = \frac{(pbd)f_y}{0.85f'_c b} = \frac{pdf_y}{0.85f'_c} \quad \text{or} \quad \frac{a}{d} = \frac{pf_y}{0.85f'_c} \tag{13.3.4}$$

A useful reference is the so-called *balanced section,* which occurs when using the exact amount of reinforcement results in the simultaneous development of the limiting stresses in the concrete and steel. The balanced section for strength design is visualized in terms of strain rather than stress. The limit for a balanced section is expressed in the form of the percentage of steel required to produce balanced conditions. The formula for this percentage is

$$p_b = \frac{0.85f'_c}{f_y} \times \frac{87}{87 + f_y} \tag{13.3.5}$$

in which f'_c and f_y are in units of ksi.

Returning to the formula for the developed resisting moment, as expressed in terms of the steel, we can derive a useful formula as follows:

$$M_t = A_s f_y \left(d - \frac{a}{2}\right)$$

$$= (pbd)(f_y)\left(d - \frac{a}{2}\right)$$

$$= (pbd)(f_y)(d)\left(d - \frac{a}{2d}\right)$$

$$= (bd^2)\left[pf_y\left(1 - \frac{a}{2d}\right)\right]$$

Thus:

$$M_t = Rbd^2 \tag{13.3.6}$$

where:

$$R = pf_y\left(1 - \frac{a}{2d}\right) \tag{13.3.7}$$

With the reduction factor applied, the design moment for a section is limited to nine-tenths of the theoretical resisting moment.

Values for the balanced section factors (p, R, and a/d) are given in Table 13.2 for various combinations of f'_c and f_y. The balanced section,

TABLE 13.2 Balanced Section Properties for Rectangular Sections with Tension Reinforcement Only

f_y		f'_c				R	
ksi	MPa	ksi	MPa	p	a/d	ksi	kPa
40	276	2	13.8	0.0291	0.685	0.766	5280
		3	20.7	0.0437	0.685	1.149	7920
		4	27.6	0.0582	0.685	1.531	10600
		5	34.5	0.0728	0.685	1.914	13200
60	414	2	13.8	0.0168	0.592	0.708	4890
		3	20.7	0.0252	0.592	1.063	7330
		4	27.6	0.0335	0.592	1.417	9770
		5	34.5	0.0419	0.592	1.771	12200

as discussed in the preceding section, is not necessarily a practical one for design. In most cases, economy will be achieved by using less than the balanced reinforcing for a given concrete section. In special circumstances, it may also be possible, or even desirable, to use compressive reinforcing in addition to tension reinforcing. Nevertheless, the balanced section is often a useful reference when design is performed.

Beams with reinforcement less than that required for the balanced moment are called *under balanced sections* or *under reinforced sections.* If a beam must carry bending moment in excess of the balanced moment for the section, it is necessary to provide some compressive reinforcement, as discussed in Section 13.4. The balanced section is not necessarily a design ideal, but it is useful in establishing the limits for the section.

In the design of concrete beams, there are two situations that commonly occur. The first occurs when the beam is entirely undetermined; that is, the concrete dimensions and the reinforcement are unknown. The second occurs when the concrete dimensions are given and the required reinforcement for a specific bending moment must be determined. The following examples illustrate the use of the formulas just developed for each of these problems.

Example 1. The service load bending moments on a beam are 58 kip-ft [78.6 kN-m] for dead load and 38 kip-ft [51.5 kN-m] for live load. The beam is 10 in. [254 mm] wide, f'_c is 3000 psi [27.6 MPa], and f_y is 60 ksi [414 MPa]. Determine the depth of the beam and the tensile reinforcing required.

Solution: The first step is to determine the required moment, using the load factors. Thus:

$$U = 1.2(D) + 1.6(L)$$

$$M_u = 1.2(M_{DL}) + 1.6(M_{LL})$$

$$= 1.2(58 \text{ kip-ft}) + 1.6(38 \text{ kip-ft})$$

$$= 130.4 \text{ kip-ft } [177 \text{ kN-m}]$$

With the capacity reduction of 0.90 applied, the desired moment capacity of the section is determined as

$$M_t = \frac{M_u}{0.90} = \frac{130.4 \text{ kip-ft}}{0.90} = 145 \text{ kip-ft}$$

$$= 145 \text{ kip-ft} \times \left(\frac{12 \text{ in.}}{1 \text{ ft}}\right) = 1740 \text{ kip-in. } [197 \text{ kN-m}]$$

The reinforcement ratio as given in Table 13.2 is $p = 0.0252$. The required area of reinforcement for this section may thus be determined from the relationship:

$$A_s = pbd$$

While there is nothing especially desirable about a balanced section, it does represent the beam section with least depth if tension reinforcing only is used. Therefore proceed to find the required balanced section for this example.

To determine the required effective depth d, use Eq. (13.3.6); thus:

$$M_t = Rbd^2$$

With the value of $R = 1.063$ ksi from Table 13.2,

$$M_t = 1740 \text{ kip-in.} = 1.063 \text{ ksi } (10 \text{ in.})(d)^2$$

and

$$d = \sqrt{\frac{1740 \text{ kip-in.}}{1.063 \text{ ksi } (10 \text{ in.})}} = \sqrt{164 \text{ in.}^2} = 12.8 \text{ in. } [325 \text{ mm}]$$

If this value is used for *d*, the required steel area may be found as

$$A_s = pbd = 0.0252(10 \text{ in.})(12.8 \text{ in.}) = 3.23 \text{ in.}^2 \ [2080 \text{ mm}^2]$$

Although they are not given in this example, there are often some considerations other than flexural behavior alone that influence the choice of specific dimensions for a beam. These may include the following:

Design for shear
Coordination of the depths of a set of beams in a framing system
Coordination of the beam dimensions and placement of reinforcement in adjacent beam spans
Coordination of beam dimensions with supporting columns
Limiting beam depth to provide overhead clearance beneath the structure

If the beam is of the ordinary form shown in Figure. 13.4, the specified dimension is usually that given as *h*. Assuming the use of a No. 3 U-stirrup, a cover of 1.5 in. [38 mm], and an average-size reinforcing bar of 1-in. [25-mm] diameter (No. 8 bar), the design dimension *d* will be less than *h* by 2.375 in. [60 mm]. Lacking other considerations, the depth of the beam will be 16 in.

Next, select a set of reinforcing bars to obtain this area. For the purpose of the example, select bars all of a single size (see Table 13.1); the number required will be as follows:

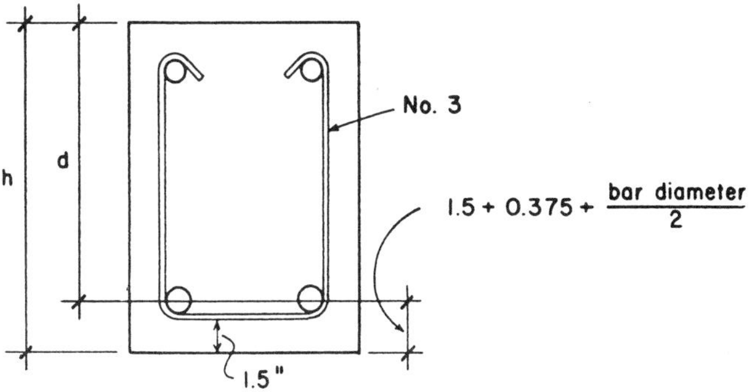

Figure 13.4 Common form of reinforced concrete beam.

No. 6 bars: 3.23/0.44 = 7.3, or 8 [2085/284 = 7.3]

No. 7 bars: 3.23/0.60 = 5.4, or 6 [2085/387 = 5.4]

No. 8 bars: 3.23/0.79 = 4.1, or = [2085/510 = 4.1]

No. 9 bars: 3.23/1.00 = 3.3, or 4 [2085/645 = 3.3]

No. 10 bars: 3.23/1.27 = 2.5, or 3 [2085/819 = 2.5]

No. 11 bars: 3.23/1.56 = 2.1, or 3 [2085/1006 = 2.1]

In real design situations, various additional considerations always influence the choice of the reinforcing bars. One general desire is that of having the bars in a single layer because this keeps the centroid of the steel as close as possible to the edge (bottom in this case) of the member, giving the greatest value for d with a given height (h) of a concrete section. With the section as shown in Figure 13.4, a beam width of 10 in. will yield a net width of 6.25 in. inside the No. 3 stirrups. (Outside width of 10 less 2×1.5 cover and 2×0.375 stirrup diameter.) Applying the code criteria for minimum spacing for this situation, we can determine the required width for the various bar combinations. Minimum space required between bars is one bar diameter or a lower limit of 1 in. (See discussion in Section 14.2.) Two examples for this are shown in Figure 13.5. It will be found that a greater beam width is required.

If there are reasons, as there often are, for not selecting the least deep section with the greatest amount of reinforcing, a slightly different procedure must be used, as illustrated in the following example.

Example 2. Using the same data as in Example 1, find the reinforcement required if the desired beam section has b = 10 in. [254 mm] and d = 18 in. [457 mm].

Solution: The first two steps in this situation would be the same as in Example 1—to determine M_u and M_t. The next step is to determine whether the given section is larger than, smaller than, or equal to a balanced section. Because this investigation has already been done in Example 1, observe that the 10-in. × 18-in. section is larger than a balanced section. Thus, the actual value of a/d will be less than the balanced section value of 0.592. The next step would then be as follows.

Estimate a value for a/d—something smaller than the balanced value. For example, try $a/d = 0.3$. Then

$$a = 0.3d = 0.3(18 \text{ in.}) = 5.4 \text{ in. } [137 \text{ mm}]$$

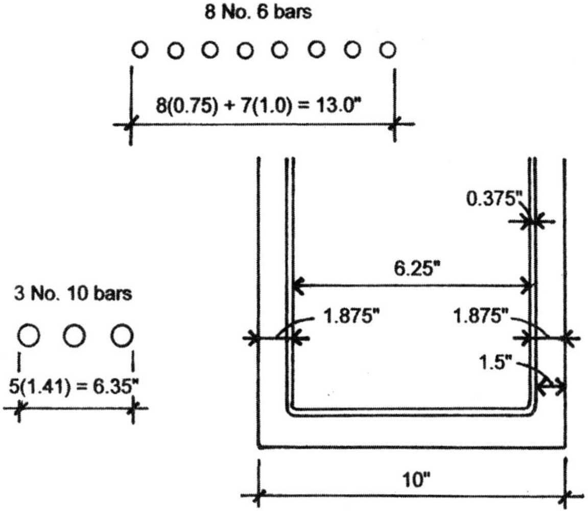

Figure 13.5 Consideration of beam width for proper spacing of a single layer of reinforcement.

With this value for a, use Eq. (13.2.2) to find a required value for A_s.

Referring to Figure. 13.2:

$$M_t = T(jd) = (A_s f_y)\left(d - \frac{a}{2}\right)$$

$$A_s = \frac{M_t}{f_y\left(d - \frac{a}{2}\right)} = \frac{1740 \text{ kip-in.}}{60 \text{ ksi } (15.3 \text{ in.})} = 1.90 \text{ in.}^2 \ [1226 \text{ mm}^2]$$

Next, test to see if the estimate for a/d was close by finding a/d using Eq. (13.2.4). Thus:

$$p = \frac{A_s}{bd} = \frac{1.90 \text{ in.}^2}{10 \text{ in.} \times (18 \text{ in.})} = 0.011$$

and from Eq. (13.2.4):

$$\frac{a}{d} = \frac{pf_y}{0.85f'_c} = \frac{0.011 \times (60 \text{ ksi})}{0.85 \times (3 \text{ ksi})} = 0.259$$

Thus:

$$a = 0.259 \times (18 \text{ in.}) = 4.66 \text{ in.}, \qquad d - \frac{a}{2} = 15.7 \text{ in. [399 mm]}$$

If this value for $d - a/2$ is used to replace that used earlier, the required value of A_s will be slightly reduced. In this example, the correction will be only a few percent. If the first guess of a/d had been way off, it may justify another run through the analysis to get closer to an exact answer.

For beams that are classified as underreinforced (section dimensions larger than thee limit for a balanced section), check for the minimum required reinforcement. For a rectangular section, the ACI Code specifies that a minimum area be

$$A_s = \frac{3\sqrt{f'_c}}{f_y}(bd)$$

but not less than

$$A_s = \frac{200}{f_y}(bd)$$

On the basis of these requirements, values for minimum reinforcement for rectangular sections with tension reinforcement only are given in Table 13.3 for two grades of steel and three concrete strengths. For the example, with a concrete strength of 3000 psi and f_y of 60 ksi, the minimum area of steel is thus

$$A_s = 0.00333(bd) = 0.00333(10 \times 18) = 0.60 \text{ in.}^2 \text{ [387 mm}^2\text{]}$$

which is clearly not critical in this case.

Problem 13.3.A. A rectangular concrete beam has $f'_c = 3000$ psi [20.7 MPa] and steel with $f_y = 40$ ksi [276 MPa]. Select the beam dimensions and reinforcement for a balanced section if the beam sustains a moment as a result of dead load of 60 kip-ft [81.3 kN-m] and a moment as a result of live load of 90 kip-ft [122 kN-m].

TABLE 13.3 Minimum Required Tension Reinforcement for Rectangular Sections[a]

f'_c (psi)	f_y = 40 ksi	f_y = 60 ksi
3000	0.0050	0.00333
4000	0.0050	0.00333
5000	0.0053	0.00354

[a]Required A_s equals table value times bd of the beam section.

Problem 13.3.B. Same as Problem 13.3.A, except f'_c = 4000 psi [27.6 MPa], f_y = 60 ksi[414 MPa], M_{DL} = 36 kip-ft [48.9 kN-m], and M_{LL} = 65 kip-ft 88.2 kN-m].

Problem 13.3.C. Find the area of steel reinforcement required and select the bars for the beam in Problem 13.3.A if the section dimensions are b = 16 in [406 mm]. and d = 32 in [813 mm].

Problem 13.3.D. Find the area of steel reinforcement required and select the bars for the beam in Problem 13.3.B if the section dimensions are b = 14 in. [356 mm] and d = 25 in. [635 mm].

13.4 SPECIAL BEAMS

Beams in Sitecast Systems

In sitecast construction, it is common to cast as much of the total structure as possible in a single, continuous pour. The length of the workday, the size of the available work crew, and other factors may affect this decision. Other considerations involve the nature, size, and form of the structure. For example, a convenient single-cast unit may consist of the whole floor structure for a multistory building, if it can be cast in a single workday.

Planning of the concrete construction is itself a major design task. The issue in consideration here is that such work typically results in achieving continuous beams and slabs versus the common condition of simple-spanning elements in wood and steel construction. The design of continuous-span elements involves more complex investigation for behaviors caused by the statically indeterminate nature of internal force resolution involving bending moments, shears, and deflections. For concrete structures, additional complexity results from the need to consider conditions all along the beam length, not just at locations of maximum responses.

In the upper part of Figure 13.6 is shown the condition of a simple-span beam subjected to a uniformly distributed loading. The typical bending moment diagram showing the variation of bending moment along the beam length takes the form of a parabola, as shown in Figure 13.6*b*. As with a beam of any material, the maximum effort in bending resistance must respond to the maximum value of the bending moment; here occurring at the midspan. For a concrete beam, the concrete section and its reinforcement must be designed for this moment value. However, for a long span and a large beam with a lot of reinforcement, it may be possible to reduce the amount of steel at points nearer to the beam ends. That is, some steel bars may be full length in the beam, whereas some others are only partial length and occur only in the midspan portion (see Figure 13.6*c*).

Figure 13.6*d* shows the typical situation for a continuous beam in a sitecast slab and beam framing system. For a single uniformly distributed loading, the moment diagram takes a form as shown in Figure 13.6*e*, with positive moments near the beam's midspan and negative moments at the supports. Locations for reinforcement that respond to the sign of these moments are shown on the beam elevation in Figure 13.6*f*.

For the continuous beam, it is obvious that separate requirements for the beam's moment resistance must be considered at each of the locations of peak values on the moment diagram. However, there are many additional concerns as well. Principal considerations include the following:

T-beam action. At points of positive bending moment (midspan), the beam and slab monolithic construction must be considered to function together, giving a T-shaped form for the portion of the beam section that resists compression.

Use of compression reinforcement. If the beam section is designed to resist the maximum bending moment, which occurs at only one point along the beam length, the section will be overstrong for all other locations. For this or other reasons, it may be advisable to use compressive reinforcement to reduce the beam size at the singular locations of maximum bending moment. This commonly occurs at support points and refers to the negative bending moment requiring steel bars in the top of the beam. At these points, an easy way to develop compressive reinforcement is to simply extend the bottom steel bars (used primarily for positive moments) through the supports.

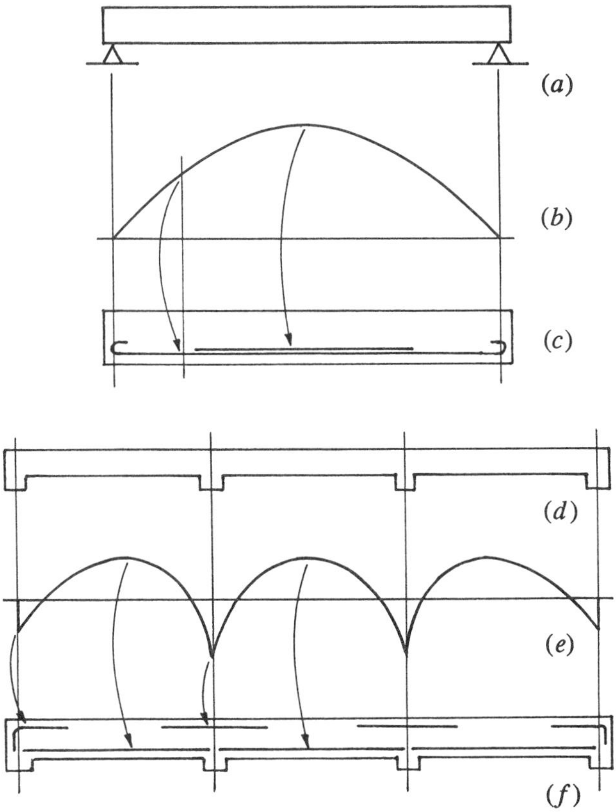

Figure 13.6 Utilization of reinforcement in concrete beams: (*a*) simple beam; (*b*) form of moment diagram for uniformly distributed loading on a simple beam; (*c*) use of reinforcing for a simple beam; (*d*) continuous beam, typical of concrete construction; (*e*) form of moment diagram for uniformly distributed loading on a continuous beam; (*f*) use of reinforcing for a continuous beam.

Spanning slabs. Design of sitecast beams must usually be done in conjunction with the design of the slabs that they support. Basic considerations for the slabs are discussed in Section 13.5. The whole case for the slab-beam sitecast system is discussed in Chapter 14.

Beam shear. While consideration for bending is a major issue, beams must also be designed for shear effects. This is discussed in Section 13.6. While special reinforcement is typically added for shear resistance, its interaction and coexistence in the beam with the flexural reinforcement must be considered.

Development of reinforcement. This refers generally to the proper anchorage of the steel bars in the concrete so that their resistance to tension can be developed. At issue primarily are the exact locations of the end cutoffs of the bars and the details such as that for the hooked bar at the discontinuous end at the far left support shown in Figure 13.6*f*. It also accounts for the hooking of the ends of some of the bars in the simple beam. The general problems of bar development are discussed in Section 13.7.

T-Beams

When a floor slab and its supporting beams are cast at the same time, the result is monolithic construction in which a portion of the slab on each side of the beam serves as the flange of a T-beam. The part of the section that projects below the slab is called the web or stem of the T-beam. This type of beam is shown in Figure 13.7*a*. For positive moment, the flange is in compression and there is ample concrete to resist compressive stresses, as shown in Figure 13.7*b* or *c*. However, in a continuous beam, there are negative bending moments over the supports, and the flange here is in the tension stress zone with compression in the web. For this situation, the beam is assumed to behave essentially as a rectangular section with dimensions b_w and d, as shown in Figure 13.7*d*. This section is also used in determining resistance to shear. The required dimensions of the beam are often determined by the behavior of this rectangular section. What remains for the beam is the determination of the reinforcement required at the midspan where the T-beam action is assumed.

The effective flange width (b_f) to be used in the design of symmetrical T-beams is limited to one-fourth the span length of the beam. In addition, the overhanging width of the flange on either side of the web is limited to eight times the thickness of the slab or half the clear distance to the next beam.

In monolithic construction with beams and one-way solid slabs, the effective flange area of the T-beams is usually quite capable of resisting the compressive stresses caused by positive bending moments. With a large flange area, as shown in Figure 13.7*a*, the neutral axis of the section usually occurs quite high in the beam web. If the compression developed in the web is ignored, the net compression force may be considered to be located at the centroid of the trapezoidal stress zone that

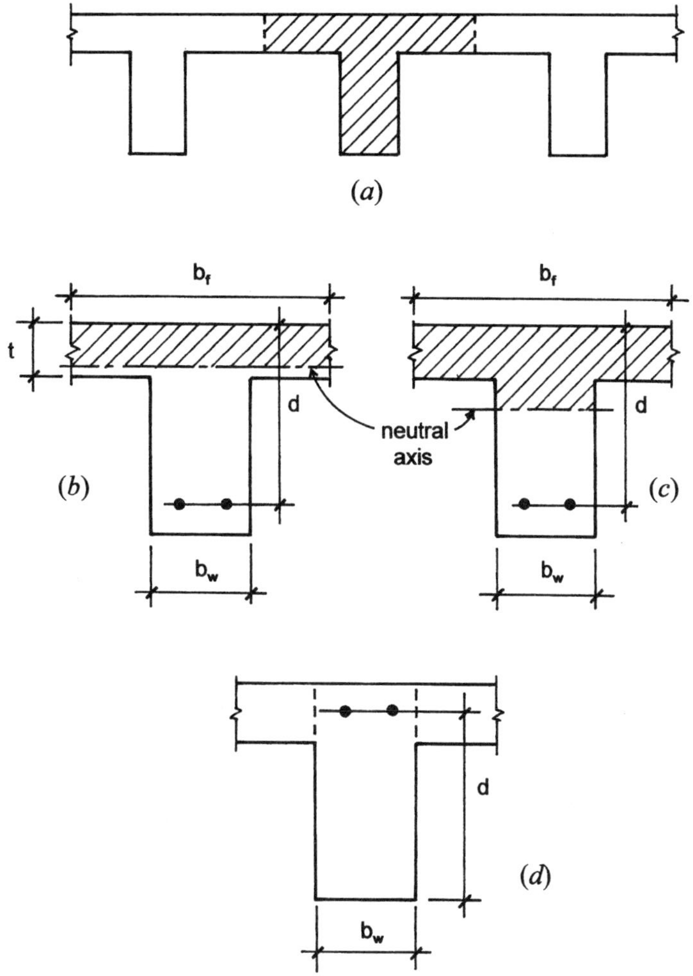

Figure 13.7 Considerations for T-beams.

represents the stress distribution in the flange and the compression force is located at something less than $t/2$ from the top of the beam.

An approximate analysis of the T-section by the strength method that avoids the need to find the location of the neutral axis and the centroid of the trapezoidal stress zone consists of the following steps:

1. Determine the effective flange width for the T, as previously described.
2. Ignore compression in the web and assume a constant value for compressive stress in the flange (see Figure 13.8). Thus:

$$jd = d - \frac{t}{2}$$

Then, find the required steel area as

$$M_r = \frac{M_u}{0.9} = T(jd) = A_s f_y \left(d - \frac{t}{2}\right)$$

$$A_s = \frac{M_t}{f_y \left(d - \frac{a}{2}\right)}$$

3. Check the compressive stress in the concrete as

$$f_c = \frac{C}{b_f t} \leq 0.85 f'_c$$

where

$$C = \frac{M_r}{jd} = \frac{M_r}{d - \frac{t}{2}}$$

The value of maximum compressive stress will not be critical if this computed value is significantly less than the limit of $0.85f'_c$.
4. T-beams ordinarily function for positive moments in continuous beams. Because these moments are typically less than those at the beam supports, and the required section is typically derived for the more critical bending at the supports, the T-beam is typically considerably under reinforced. This makes it necessary to consider the problem of minimum reinforcement, as discussed for the rectangular section. The ACI Code provides special requirements for this

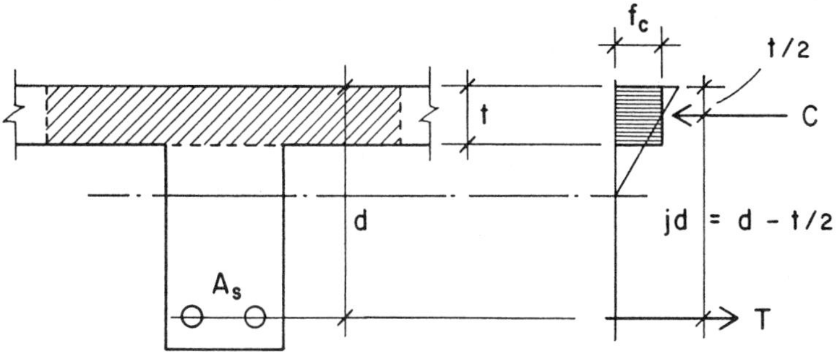

Figure 13.8 Basis for simplified analysis of a T-beam.

for the T-beam, for which the minimum area required is defined as the greater value of

$$A_s = \frac{6\sqrt{f'_c}}{F_y}(b_w d)$$

or

$$A_s = \frac{3\sqrt{f'_c}}{F_y}(b_f d)$$

where b_w = the width of the beam web
b_f = the effective width of the T flange

The following example illustrates the use of this procedure. It assumes a typical design situation in which the dimensions of the section (b_f, b_w, d and t; see Figure 13.7) are all predetermined by other design considerations and the design of the T-section is reduced to the requirement to determine the area of tension reinforcement.

Example 3. A T-section is to be used for a beam to resist positive moment. The following data are given: beam span is 18 ft [5.49 m], beams are 9 ft [2.74 m] center to center, slab thickness is 4 in. [0.102 m], beam stem dimensions are b_w = 15 in. [0.381 m] and d = 22 in. [0.559 m], f'_c = 4 ksi [27.6 MPa], f_y = 60 ksi [414 MPa]. Find the required area of steel and select the reinforcing bars for a dead load moment of 125 kip-ft [170 kN-m] plus a live load moment of 100 kip-ft [136 kN-m].

Solution: Determine the effective flange width (necessary only for a check on the concrete stress). The maximum value for the flange width is

$$b_f = \frac{\text{span}}{4} = \frac{18 \times 12}{4} = 54 \text{ in. [1.37 m]}$$

or

$$b_f = \text{center-to-center beam spacing} = 9 \times 12 = 108 \text{ in. [2.74 m]}$$

or

$$b_f = \text{beam stem width plus 16 times the slab thickness}$$
$$= 15 + (16 \times 4) = 79 \text{ in. [201 m]}$$

The limiting value is therefore 54 in. [1.37 m]. Next find the required steel area:

$$M_u = 1.2(125) + 1.6(100) = 310 \text{ kip-ft}$$

$$M_r = \frac{M_u}{0.9} = 344 \text{ kip-ft}$$

$$A_s = \frac{M_r}{f_y\left(d - \dfrac{t}{2}\right)} = \frac{344 \times 12}{60\left(22 - \dfrac{4}{2}\right)} = 3.44 \text{ in.}^2 \text{ [2219 mm}^2\text{]}$$

Select bars using Table 13.4, which incorporates consideration for the adequacy of the stem width. From the table, choose four No. 9 bars, actual $A_s = 4.00$ in.^{2}. From Table 14.1, the required width for four No. 9 bars is 12 in., less than the 15 in. provided.

Check the concrete stress:

$$C = \frac{M_r}{jd} = \frac{344 \times 12}{20} = 206.4 \text{ kips [918 kN]}$$

$$f_c = \frac{C}{b_f t} = \frac{206.4}{54 \times 4} = 0.956 \text{ ksi [6.69 MPa]}$$

TABLE 13.4 Options for the T-Beam Reinforcement

Bar Size	No. of Bars	Actual Area Provided (in.^{2})	Width Required[a] (in.)
7	6	3.60	14
8	5	3.95	13
9	4	4.00	12
10	3	3.81	11
11	3	4.68	11

[a] From Table 14.1.

Compare this to the limiting stress of

$$0.85f'_c = 0.85(4) = 3.4 \text{ ksi } [23.44 \text{ MPa}]$$

Thus, compressive stress in the flange is clearly not critical.

Using the beam stem width of 15 in. and the effective flange width of 54 in., the minimum area of reinforcement is determined as the greater of

$$A_s = \frac{6\sqrt{f'_c}}{F_y}(b_w d) = \frac{6\sqrt{4000}}{60{,}000}(15 \times 22) = 2.09 \text{ in.}^2 \ [1350 \text{ mm}^2]$$

or

$$A_s = \frac{3\sqrt{f'_c}}{F_y}(b_f \times d) = \frac{3\sqrt{4000}}{60{,}000}(54 \times 22) = 3.76 \text{ in.}^2 \ [2430 \text{ mm}^2]$$

The minimum area required is thus greater than the computed area of 3.44 in.^{2}. The choice of using 4 No. 9 bars is not affected by this development because its area is 4.00 in.^{2}. Some of the other data in Table 13.4 must be adjusted if other bar choices are considered. The examples in this section illustrate procedures that are reasonably adequate for beams that occur in ordinary beam and slab construction. When special T-sections occur with thin flanges (t less than $d/8$ or so), these methods may not be valid. In such cases, more accurate investigation should be performed, using the requirements of the ACI Code.

Problem 13.4.A. Find the area of steel reinforcement required for a concrete T-beam for the following data: f'_c = 3 ksi [20 MPa], f_y = 50 ksi [344 MPa], d = 28 in. [711 mm], t = 6 in. [152 mm]. b_w = 16 in. [406 mm], b_f = 60 in. [1524 mm], and the section sustains a factored bending moment of M_u = 360 kip-ft [488 kN-m].

Problem 13.4.B. Same as Problem 13.5.A, except f'_c = 4 ksi [28 MPa], f_y = 60 ksi [414 MPa], d = 32 in. [813 mm], t = 5 in. [127 mm], b_w = 18 in. [457 mm], b_f = 54 in. [1372 mm], M_u = 500 kip-ft [678 kN-m].

Beams with Compression Reinforcement

Steel reinforcement is used in many situations on both sides of the neutral axis in a beam. When this occurs, the steel on one side of the axis will be in tension and the steel on the other side will be in compression. Such a beam is referred to as a doubly reinforced beam or simply as a beam with compressive reinforcement (it being naturally assumed that there is also tensile reinforcement). Various situations involving such reinforcement have been discussed in the preceding sections. In summary, the most common occasions for such reinforcement include the following:

1. The desired resisting moment for the beam exceeds that for which the concrete alone is capable of developing the necessary compressive force.
2. Other functions of the section require the use of reinforcement on both sides of the beam. These include the need for bars to support U-stirrups and situations when torsion is a major concern.
3. It is desired to reduce deflections by increasing the stiffness of the compressive side of the beam. This is most significant for reduction of long-term creep deflections.
4. The combination of loading conditions on the structure result in reversal moments on the section at a single location; that is, the section must sometimes resist positive moment and other times resist negative moment.
5. Anchorage requirements (for development of reinforcement) require that the bottom bars in a beam be extended a significant distance into the supports.

The precise investigation and accurate design of doubly reinforced sections, whether performed by the working stress or by strength design methods, are quite complex and are beyond the scope of work in this book. The following discussion presents an approximation method that is adequate for preliminary design of a doubly reinforced section. For real design situations, this method may be used to establish a first trial design, which may then be more precisely investigated using more rigorous methods.

For the beam with double reinforcement, as shown in Figure 13.9*a*, consider the total resisting moment for the section to be the sum of the following two component moments.

M_1 (Figure 13.9*b*) is composed of a section with tension reinforcement only (A_{s1}). This section is subject to the usual procedures for design and investigation, as discussed in Section 13.3.

M_2 (Figure 13.9*c*) is composed of two opposed steel areas (A_{s2} and A'_s) that function in simple moment couple action, similar to the flanges of a steel beam or the top and bottom chords of a truss.

The limit for M_1 is the so-called "balanced moment," as described in Section 13.2. Given the values for steel yield stress and the specified strength of the concrete, the factors for definition of the properties of this balanced section can be obtained from Table 13.2. Given the dimensions for the concrete section, *b* and *d*, the limiting moment resistance for the section can be determined using the value of *R* from the table. If the capacity of the section as thus determined is less than the required factored moment (M_r), compressive reinforcing is required. If the balanced resistance is larger than the required factored moment, compressive reinforcement is not required. Although not actually required, the compressive reinforcement may be used for any of the reasons previously mentioned.

When M_1 is less than the factored required moment, it may be determined using the balanced value for *R* in Table 13.2. Then the required value for M_2 may be determined as $M_2 = M_r - M_1$. This is seldom the case in design work because the amount of reinforcement required to achieve the balanced moment capacity of the section is usually not practical to be fit into the section.

When the potential balanced value for M_1 is greater than the required factored moment, the properties given in Table 13.2 may be used for an approximation of the value for the required steel area defined as A_{s1}. For the true value of the required M_1 the table values may be adjusted as follows:

1. The actual value for *a*/*d* will be smaller than the table value.
2. The actual value for *p* will be less than the table value.

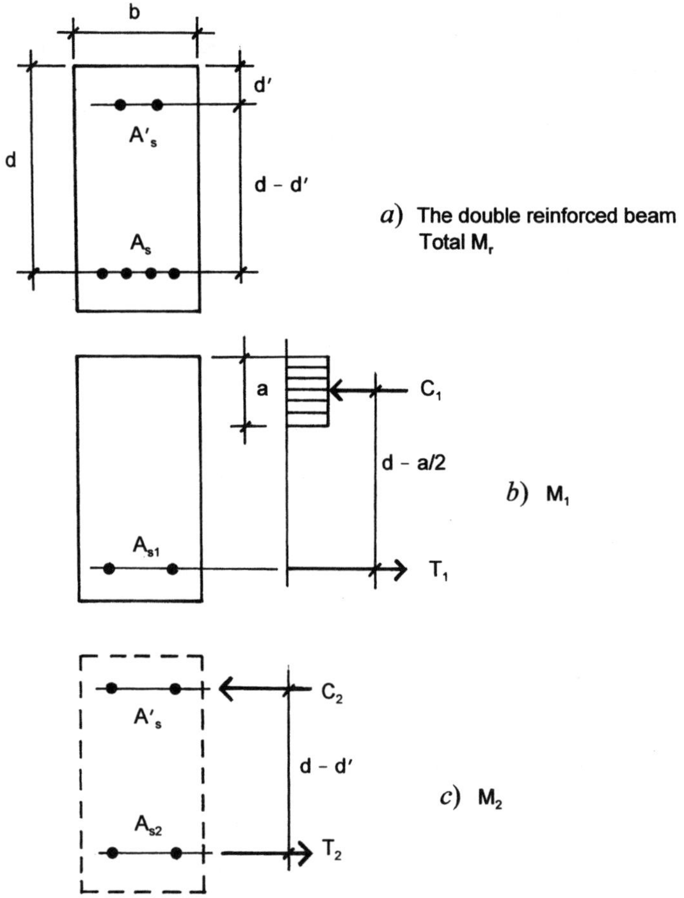

Figure 13.9 Basis for simplified analysis of a doubly-reinforced beam.

The approximation procedure that follows starts with the assumption that the concrete dimensions describe a section with the potential M_r greater than that required. The first step of the procedure is therefore to establish what is actually an arbitrary amount of compressive reinforcement. Using this reinforcement as A_{s2}, a value for M_2 is determined. The value for M_1 is then found by subtracting M_2 from the required moment for the section.

Ordinarily, we expect that $A_{s2} = A'_s$ because the same grade of steel is usually used for both. However, two special considerations must be made. The first involves the fact that A_{s2} is in tension, while A'_s is in compression. A'_s must therefore be dealt with in a manner similar to that for column reinforcement. This requires, among other things, that the compressive reinforcement be braced against buckling, using ties similar to those in a tied column.

The second consideration for A'_s involves the distribution of stress and strain on the section. Referring to Figure 13.2*a*, observe that, under normal circumstances (c less than $0.5d$), A'_s will be closer to the neutral axis than A_{s2}. Thus, the stress in A'_s will be lower than that in A_{s2} if pure elastic conditions are assumed. It is common practice to assume steel to be doubly stiff when sharing stress with concrete in compression as a result of shrinkage and creep effects. Thus, in translating from linear strain conditions to stress distribution, use the relation $f_c = f'_s/2n$ (where $n = E_s/E_c$).

For the approximate method, it is really not necessary to find separate values for A_{s1} and A_{s2}. This is because of an additional assumption that the value for a is $2 \times d'$. Thus, the moment arm for both A_{s1} and A_{s2} is the same and the value for the total tension reinforcement can be simply determined as

$$A_s = \frac{\text{Required } M_r}{f_y \times (d - d')}$$

This value for A_s can actually be determined as soon as the values for d and d' are established.

With the total tension reinforcement established, the next step involves the determination of A'_s and A_{s2}. Compression reinforcement in beams ordinarily ranges from 0.2 to 0.4 times the total tension reinforcement. For this approximation method, we will determine the area to be $0.3 \times A_s$. We will also assume the stress in the compression reinforcement to be one half of the yield stress. This permits a definition of the resisting moment M_2 as follows:

$$M_2 = A'_s\left(\frac{f_y}{2}\right)(d - d') = 0.3A_s\left(\frac{f_y}{2}\right)(d - d') = 0.15A_sf_y(d - d')$$

Using this moment, we find that the amount of the tension reinforcement that is required for the development of the steel force couple is twice A'_s, and the amount of the tension steel devoted to development of M_2 is found by subtracting this from the total tension reinforcement.

For a final step, the value of A_{s1} may be used to compute a value for the percentage of steel relating to the moment resistance of the section without the compression reinforcement. If this percentage is less than that listed in Table 13.2, the concrete stress will not be critical. However, this situation is predetermined if the total potential balanced resisting moment is determined first and compared to the required factored resisting moment.

For the section that is larger than the balanced section defined by Table 13.2, the procedure can be shortened. The following example illustrates this procedure.

Example 4. A concrete section with b = 18 in. [0.457 m] and d = 21.5 in. [0.546 m] is required to resist service load moments as follows: dead load moment = 175 kip-ft [237 kN-m], live load moment = 160 kip-ft [217 kN-m]. Using strength methods, find the required reinforcement. Use f'_c = 3 ksi [21 MPa], and f_y = 60 ksi [414 MPa].

Solution: Using Table 13.2, find a/d = 0.3773, p = 0.0214, and R = 1.041 in kip-in. units [2028 in kN-m units].

The factored moment for the section is

$$M_u = 1.2(175) + 1.6(160) = 466 \text{ kip-ft}$$

And the required factored resisting moment is

$$M_r = \frac{M_u}{0.9} = \frac{466}{0.9} = 518 \text{ kip-ft}$$

Using the R value for the balanced section, we find the maximum resisting moment of the section is

$$M_B = Rbd^2 = \frac{1.063}{12}(18)(21.5)^2 = 737 \text{ kip-ft [999 kN-m]}$$

Because this is considerable larger than the required resisting moment, the section is qualified as "underbalanced"—that is, it will be understressed as relates to the compression resistance of the section. It is reasonable therefore to use the simplified formula for the tension reinforcing, thus:

$$A_s = \frac{\text{Required } M_r}{f_y \times (d - d')} = \frac{518 \times 12}{60(21.5 - 2.5)} = 5.45 \text{ in.}^2$$

And a reasonable assumption for the compressive reinforcement is

$$A'_s = 0.3A_s = 0.3(5.45) = 1.63 \text{ in.}^2 \text{ [1052 mm}^2\text{]}$$

Bar combinations may next be found for these two steel areas.

If the situation exists in which it is desired to use a given concrete section for a resisting moment that exceeds the balanced limit described by the values in Table 13.2, a different procedure is required. For this case, the first two steps are the same as in the preceding example: Determine the required resisting moment M_r and the limiting balanced moment M_B. The tension reinforcement required for the balanced moment can be determined with the balanced percentage p from Table 13.2. M_B then becomes the moment M_1, as shown in Figure 13.9*b*. M_2, as shown in Figure 13.9c, is determined as the difference between M_r and M_B. The compression reinforcement and the additional tension reinforcement required for M_2 can then be determined. The following example illustrates this procedure.

Example 5. Find the reinforcement required for the beam in Example 4 if the required resisting moment M_r is 900 kip-ft [1052 kN-m].

Solution: The first step is to determine the limiting balanced moment M_B. For this section, this value was computed in Example 4 as 737 kip-ft. Because the required moment exceeds this value, compression reinforcement is required and the moment for this is determined as

$$M_2 = M_r - M_B = 900 - 737 = 163 \text{ kip-ft [221 kN-m]}$$

For the balanced moment the required tension reinforcement can be computed using the balanced p from Table 13.2. Thus:

$$A_{s1} = pbd = 0.0252 \times 18 \times 21.5 = 9.75 \text{ in.}^2 \text{ [6290 mm}^2\text{]}$$

To determine the tension reinforcement required for M_2, the procedure involves the use of the moment arm $d - d'$. Thus:

$$A_{s2} = \frac{M_2}{f_y(d - d')} = \frac{163 \times 12}{60(19)} = 1.72 \text{ in.}^2 \text{ [1110 mm}^2\text{]}$$

and the total required tension reinforcing is

$$A_s = 9.75 + 1.72 = 11.5 \text{ in.}^2 \text{ [7419 mm}^2\text{]}$$

TABLE 13.5 Options for the Tension Reinforcement—Example 5

Bar Size	Area of One Bar (in.^{2})	No. of Bars	Actual Area Provided (in.^{2})	Width Required[a] (in.)
8	0.79	15	11.85	33
9	1.00	12	12.00	30
10	1.27	10	12.70	28
11	1.56	8	12.48	25

[a] See Section 14.2.

For the compressive reinforcement, we assume a stress approximately equal to one half the yield stress. Thus, the area of steel required is twice the value of A_{s2}, 2(1.72) = 3.44 in.^{2} This requirement can be met with 3 No. 10 bars providing an area of 3.81 in.^{2}.

Options for the tension reinforcement are given in Table 13.5. As the table indicates, it is not possible to get the bars into the 18-in.-wide beam by placing them in a single layer. Options are to use two layers of bars or to increase the beam width. Frankly, this is not a good beam design and would most likely not be acceptable unless extreme circumstances force the use of the limited beam size. This is pretty much the typical situation for beams required to develop resisting moments larger than the balanced moment.

Problem 13.4.C. A concrete section with b = 16 in. [0.406 m] and d = 19.5 in. [0.495 m] is required to develop a factored resisting moment (M_r) of 400 kip-ft [542 kN-m]. Use of compressive reinforcement is desired. Find the required reinforcement. Use f'_c = 4 ksi [27.6 MPa] and f_y = 60 ksi [414 Mpa].

Problem 13.4.D. Same as Problem 13.4.C, except required M_r = 1000 kip-ft [1356 kN-m], b = 20 in. [508 mm], d = 27 in. [686 mm].

Problem 13.4.E. Same as Problem 13.4.C, except M_r = 640 kip-ft [868 kN-m].

Problem 13.4.F. Same as Problem 13.4.D, except M_r = 1400 kip-ft [1898 kN-m].

13.5 SPANNING SLABS

Concrete slabs are frequently used as spanning roof or floor decks, often occurring in monolithic, cast-in-place slab and beam framing systems. There are generally two basic types of slabs: one-way spanning and two-way spanning slabs. The spanning condition is not so much determined by the slab as by its support conditions. As part of a general framing system, the one-way spanning slab is discussed in Section 14.1. The following discussion relates to the design of one-way solid slabs using procedures developed for the design of rectangular beams.

Solid slabs are usually designed by considering the slab to consist of a series of 12-in.-wide planks. Thus, the procedure consists of simply designing a beam section with a predetermined width of 12 in. Once the depth of the slab is established, the required area of steel is determined, specified as the number of square inches of steel required per foot of slab width.

Reinforcing bars are selected from a limited range of sizes, appropriate to the slab thickness. For thin slabs (4 to 6 in. thick), bars may be of a size from No. 3 to No. 6 or so (nominal diameters from ⅜ to ¾ in.). The bar size selection is related to the bar spacing, the combination resulting in the amount of reinforcing in terms of so many square inches per foot of slab width. Spacing is limited by code regulation to a maximum of three times the slab thickness. There is no minimum spacing, other than that required for proper placing of the concrete; however, a very close spacing indicates a very large number of bars, making for laborious installation.

Every slab must be provided with two-way reinforcement, regardless of its structural functions. This is partly to satisfy requirements for shrinkage and temperature effects. The amount of this minimum reinforcement is specified as a percentage p of the gross cross-sectional area of the concrete as follows:

1. For slabs reinforced with grade 40 or grade 50 bars:

$$p = \frac{A_s}{bt} = 0.002 \quad (0.2\%)$$

2. For slabs reinforced with grade 60 bars:

$$p = \frac{A_s}{bt} = 0.0018 \quad (0.18\%)$$

Center-to-center spacing of this minimum reinforcement must not be greater than five times the slab thickness or 18 in.

Minimum cover for slab reinforcement is normally 0.75 in., although exposure conditions or need for a high fire rating may require additional cover. For a thin slab reinforced with large bars, there will be a considerable difference between the slab thickness and the effective depth—*t* and *d*, as shown in Figure 13.10. Thus, the practical efficiency of the slab in flexural resistance decreases rapidly as the slab thickness is decreased. For this and other reasons, very thin slabs (less than 4 in. thick) are often reinforced with wire fabric rather than sets of loose bars.

Shear reinforcement is seldom used in one-way slabs, and consequently, the maximum unit shear stress in the concrete must be kept within the limit for the concrete without reinforcement. This is usually not a concern because unit shear is usually low in one-way slabs, except for exceptionally high loadings.

Table 13.6 gives data that are useful in slab design, as demonstrated in the following example. Table values indicate the average amount of steel area per foot of slab width provided by various combinations of bar size and spacing. Table entries are determined as follows:

$$A_s/\text{ft} = (\text{single bar area})\frac{12}{\text{bar spacing}}$$

Thus, for No. 5 bars at 8-in. centers:

$$A_s/\text{ft} = (0.31)\left(\frac{12}{8}\right) = 0.465 \text{ in.}^2/\text{ft}$$

It may be observed that the table entry for this combination is rounded off to a value of 0.46 in.^{2}/ft.

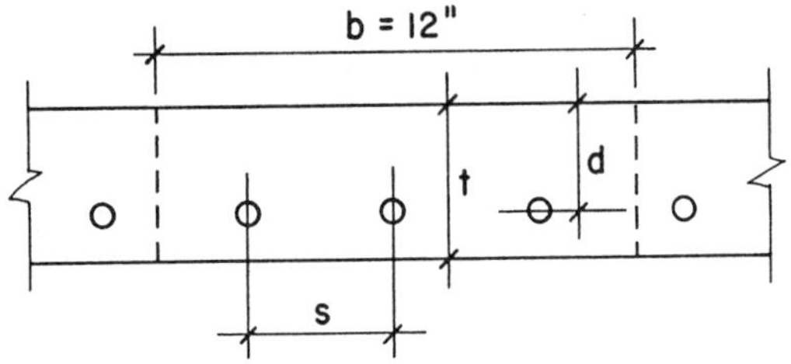

Figure 13.10 Reference for slab design.

TABLE 13.6 Areas Provided By Spaced Reinforcement

Bar Spacing (in.)	Area Provided (in.²/ft width)									
	No. 2	No. 3	No. 4	No. 5	No. 6	No. 7	No. 8	No. 9	No. 10	No. 11
3	0.20	0.44	0.80	1.24	1.76	2.40	3.16	4.00		
3.5	0.17	0.38	0.69	1.06	1.51	2.06	2.71	3.43	4.35	
4	0.15	0.33	0.60	0.93	1.32	1.80	2.37	3.00	3.81	4.68
4.5	0.13	0.29	0.53	0.83	1.17	1.60	2.11	2.67	3.39	4.16
5	0.12	0.26	0.48	0.74	1.06	1.44	1.89	2.40	3.05	3.74
5.5	0.11	0.24	0.44	0.68	0.96	1.31	1.72	2.18	2.77	3.40
6	0.10	0.22	0.40	0.62	0.88	1.20	1.58	2.00	2.54	3.12
7	0.08	0.19	0.34	0.53	0.75	1.03	1.35	1.71	2.18	2.67
8	0.07	0.16	0.30	0.46	0.66	0.90	1.18	1.50	1.90	2.34
9	0.07	0.15	0.27	0.41	0.59	0.80	1.05	1.33	1.69	2.08
10	0.06	0.13	0.24	0.37	0.53	0.72	0.95	1.20	1.52	1.87
11	0.05	0.12	0.22	0.34	0.48	0.65	0.86	1.09	1.38	1.70
12	0.05	0.11	0.20	0.31	0.44	0.60	0.79	1.00	1.27	1.56
13	0.05	0.10	0.18	0.29	0.40	0.55	0.73	0.92	1.17	1.44
14	0.04	0.09	0.17	0.27	0.38	0.51	0.68	0.86	1.09	1.34
15	0.04	0.09	0.16	0.25	0.35	0.48	0.63	0.80	1.01	1.25
16	0.04	0.08	0.15	0.23	0.33	0.45	0.59	0.75	0.95	1.17
18	0.03	0.07	0.13	0.21	0.29	0.40	0.53	0.67	0.85	1.04
24	0.02	0.05	0.10	0.15	0.22	0.30	0.39	0.50	0.63	0.78

Example 6. A one-way solid concrete slab is to be used for a simple span of 14 ft [4.27 m]. In addition to its own weight, the slab carries a superimposed dead load of 30 psf [1.44 kPa] plus a live load of 100 psf [4.79 kPa]. Using $f'_c = 3$ ksi [20.7 MPa] and $f_y = 40$ ksi [276 MPa], design the slab for minimum overall thickness.

Solution: Using the general procedure for design of a beam with rectangular section (Section 13.2), we first determine the required slab thickness. Thus, for deflection, from Table 13.10:

$$\text{Minimum } t = \frac{L}{25} = \frac{14 \times 12}{25} = 6.72 \text{ in. [171 mm]}$$

For flexure, first determine the maximum bending moment. The loading must include the weight of the slab, for which the thickness required for deflection may be used as a first estimate. Assuming a 7-in. [178 mm]-

thick slab, the slab weight is (7/12) (150 pcf) = 87.5 psf, say 88 psf, and the total dead load is 30 + 88 = 118 psf.

The factored load is thus

$$U = 1.2(\text{dead load}) + 1.6(\text{live load}) = 1.2(118) + 1.6(100)$$
$$= 302 \text{ psf } [14.46 \text{ kPa}]$$

The maximum bending moment on a 12-in.-wide strip of the slab thus becomes

$$M_u = \frac{wL^2}{8} = \frac{302(14)^2}{8} = 7399 \text{ ft-lb } [10.03 \text{ kN-m}]$$

And the required factored resisting moment is

$$M_r = \frac{7399}{0.9} = 8221 \text{ ft-lb } [11.15 \text{ kN-m}]$$

For a minimum slab thickness, we consider the use of a balanced section, for which Table 13.2 yields the following properties: $a/d = 0.685$ and $R = 1.149$ (in kip and inch units). Then the minimum value for bd^2 is

$$bd^2 = \frac{M_r}{R} = \frac{8.221 \times 12}{1.149} = 85.9 \text{ in.}^3$$

and because b is the 12-in. design strip width,

$$d = \sqrt{\frac{85.9}{12}} = \sqrt{7.16} = 2.68 \text{ in. } [68 \text{ mm}]$$

Assuming an average bar size of a No. 6 (¾ in. nominal diameter) and cover of ¾ in., we see that the minimum required slab thickness based on flexure becomes

$$t = d + \tfrac{1}{2}(\text{bar diameter}) + \text{cover}$$

$$t = 2.68 + \frac{0.75}{2} + 0.75 = 3.8 \text{ in. } [97 \text{ mm}]$$

The deflection limitation thus controls in this situation, and the minimum overall thickness is the 6.72-in. dimension. Staying with the 7-in overall thickness, the actual effective depth with a No. 6 bar will be

$$d = 7.0 - 1.125 = 5.875 \text{ in. } [149 \text{ mm}]$$

Because this d is larger than that required for a balanced section, the value for a/d will be slightly smaller than 0.685, as found from Table 13.2. Assume a value of 0.4 for a/d and determine the required area of reinforcement as

$$a = 0.4d = 0.4(5.875) = 2.35 \text{ in. } [59.7 \text{ mm}]$$

$$A_s = \frac{M}{f_y\left(d - \dfrac{a}{2}\right)} = \frac{8.221 \times 12}{40(5.875 - 1.175)} = 0.525 \text{ in.}^2 \ [339 \text{ mm}^2]$$

Using data from Table 13.6, the optional bar combinations shown in Table 13.7 will satisfy this requirement. Note that for bars larger then the assumed No. 6 bar (0.75 in. diameter) d will be slightly less and the required area of reinforcement slightly higher.

The ACI Code permits a maximum center-to-center bar spacing of three times the slab thickness (21 in. in this case) or 18 in., whichever is smaller. Minimum spacing is largely a matter of the designer's judgment. Many designers consider a minimum practical spacing to be one approximately equal to the slab thickness. Within these limits, any of the bar size and spacing combinations listed are adequate.

As described previously, the ACI Code requires a minimum reinforcement for shrinkage and temperature effects to be placed in the direction perpendicular to the flexural reinforcement. With the grade 40 bars in this

TABLE 13.7 Alternatives for the Slab Reinforcement

Bar Size	Spacing of Bars Center to Center (in.)	Average A_s in a 12-in. Width
5	7	0.53
6	10	0.53
7	13	0.55
8	18	0.53

example, the minimum percentage of this steel is 0.0020, and the steel area required for a 12-in. strip thus becomes

$$A_s = p(bt) = 0.0020(12 \times 7) = 0.168 \text{ in.}^2\text{/ft}$$

From Table 13.6, this requirement can be satisfied with No. 3 bars at 8-in. centers or No. 4 bars at 14-in. centers. Both of these spacings are well below the maximum of five times the slab thickness (35 in.) or 18 in.

Although simply supported single slabs are sometimes encountered, the majority of slabs used in building construction are continuous through multiple spans. An example of the design of such a slab is given in Chapter 14.

Problem 13.5.A. A one-way solid concrete slab is to be used for a simple span of 16 ft [4.88 m]. In addition to its own weight, the slab carries a superimposed dead load of 40 psf [1.92 kPa] and a live load of 100 psf [4.79 kPa]. Using the strength method with f'_c = 3 ksi [20.7 MPa], and f_y = 40 ksi [276 MPa], design the slab for minimum overall thickness.

Problem 13.5.B. Same as Problem 13.5.A, except span = 18 ft [5.49 m], superimposed dead load = 50 psf [2.39 kPa], live load = 75 psf [3.59 kPa], f'_c = 4 ksi [27.6 MPa], f_y = 60 ksi [414 MPa].

13.6 SHEAR IN BEAMS

From general consideration of shear effects, as developed in the science of mechanics of materials, we can make the following observations:

1. Shear is an ever-present phenomenon, produced directly by slicing actions, by lateral loading in beams, and on oblique sections in tension and compression members.
2. Shear forces produce shear stress in the plane of the force and equal unit shear stresses in planes that are perpendicular to the shear force.
3. Diagonal stresses of tension and compression, having magnitudes equal to that of the shear stress, are produced in directions of 45° from the plane of the shear force.
4. Direct slicing shear force produces a constant magnitude shear stress on affected sections, but beam shear action produces shear stress that varies on the affected sections, having a magnitude of

zero at the edges of the section and a maximum value at the centroidal neutral axis of the section.

In the discussions that follow, it is assumed that the reader has a general familiarity with these relationships.

Consider the case of a simple beam with uniformly distributed load and end supports that provide only vertical resistance (no moment restraint). The distribution of internal shear and bending moment are as shown in Figure 13.11*a*. For flexural resistance, it is necessary to provide longitudinal reinforcing bars near the bottom of the beam. These bars are oriented for primary effectiveness in resistance to tension stresses that develop on a vertical (90°) plane (which is the case at the center of the span, where the bending moment is maximum and the shear approaches zero).

Under the combined effects of shear and bending, the beam tends to develop tension cracks as shown in Figure 13.11*b*. Near the center of the span, where the bending is predominant and the shear approaches zero, these cracks approach 90°. Near the support, however, where the shear predominates and bending approaches zero, the critical tension stress plane approaches 45°, and the horizontal bars are only partly effective in resisting the cracking.

Shear Reinforcement for Beams

For beams, the most common form of added shear reinforcement consists of a series of U-shaped bent bars (Figure 13.11*d*), placed vertically and spaced along the beam span, as shown in Figure 13.11*c*. These bars, called *stirrups*, are intended to provide a vertical component of resistance, working in conjunction with the horizontal resistance provided by the flexural reinforcement. In order to develop flexural tension near the support face, the horizontal bars must be bonded to the concrete beyond the point where the stress is developed. Where the ends of simple beams extend only a short distance over the support (a common situation), it is often necessary to bend or hook the bars as shown in Figure 13.11*c*.

The simple span beam and the rectangular section shown in Figure 13.11*d* occur only infrequently in building structures. The most common case is that of the beam section shown in Figure 13.12*a*, which occurs when a beam is cast continuously with a supported concrete slab. In addition, these beams normally occur in continuous spans with negative moments at the supports. Thus, the stress in the beam near the support is as shown in Figure 13.12*a*, with the negative moment producing com-

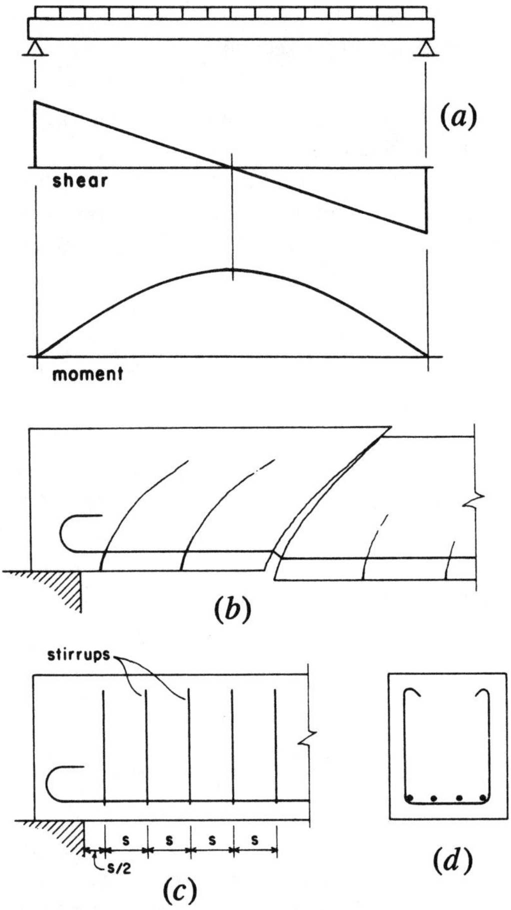

Figure 13.11 Considerations for shear in concrete beams.

pressive flexural stress in the bottom of the beam stem. This is substantially different from the case of the simple beam, where the moment approaches zero near the support.

For the purpose of shear resistance, the continuous, T-shaped beam is considered to consist of the section indicated in Figure 13.12*b*. The effect of the slab is ignored, and the section is considered to be a simple rectangular one. Thus, for shear design, there is little difference between the

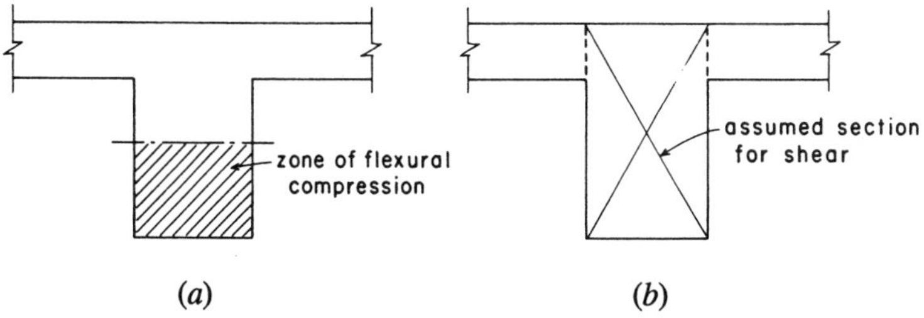

Figure 13.12 Development of negative bending and shear in concrete T-beams.

simple span beam and the continuous beam, except for the effect of the continuity on the distribution of shear along the beam span. It is important, however, to understand the relationships between shear and moment in the continuous beam.

Figure 13.13 illustrates the typical condition for an interior span of a continuous beam with uniformly distributed load. Referring to the portions of the beam span numbered 1, 2, and 3, note the following:

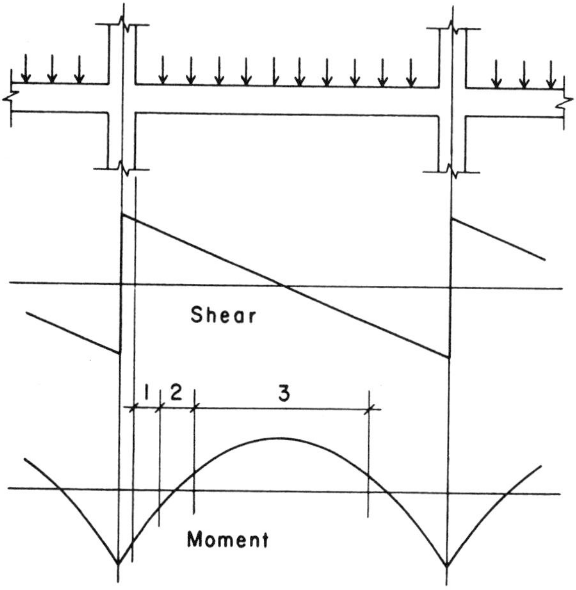

Figure 13.13 Shear and bending in continuous beams.

1. In this zone, the high negative moment requires major flexural reinforcing consisting of horizontal bars near the top of the beam.
2. In this zone, the moment reverses sign; moment magnitudes are low; and, if shear stress is high, the design for shear is a predominant concern.
3. In this zone, shear consideration is minor and the predominant concern is for positive moment requiring major flexural reinforcing in the bottom of the beam.

Vertical U-shaped stirrups, similar to those in Figure 13.14*a*, may be used in the T-shaped beam. An alternate detail for the U-shaped stirrup is shown in Figure 13.14*b*, in which the top hooks are turned outward; this makes it possible to spread the negative moment reinforcing bars to make placing of the concrete somewhat easier. Figures 13.14*c* and *d* show possibilities for stirrups in L-shaped beams that occur at the edges of large openings or at the outside edge of the structure. This form of stirrup is used to enhance the torsional resistance of the section and also assists in developing the negative moment resistance in the slab at the edge of the beam.

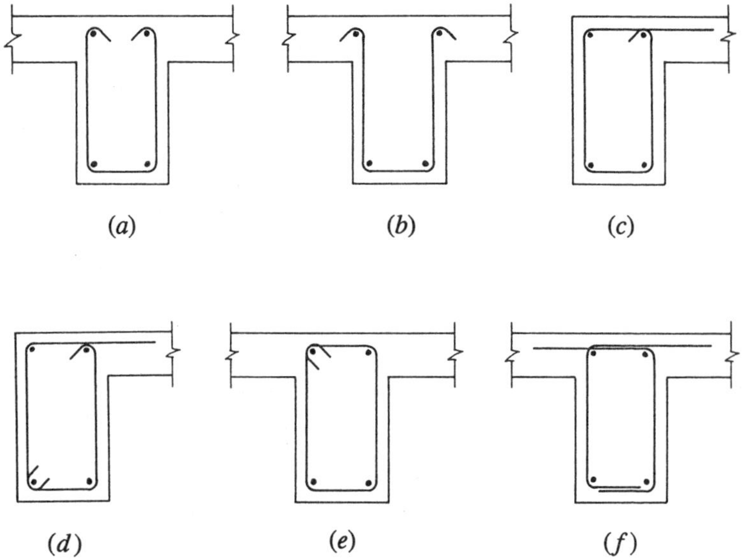

Figure 13.14 Forms for vertical stirrups.

So-called *closed stirrups*, similar to ties in columns, are sometimes used for T- and L-shaped beams, as shown in Figure 13.14*c* through *f*. These are generally used to improve the torsional resistance of the beam section.

Stirrup forms are often modified by designers or by the reinforcing fabricator's detailers to simplify the fabrication and/or the field installation. The stirrups shown in Figure 13.14*d* and *f* are two such modifications of the basic details in Figure 13.14*c* and *e*, respectively.

The following are considerations and code requirements that apply to design for beam shear.

Concrete Capacity. Whereas the tensile strength of the concrete is ignored in design for flexure, the concrete is assumed to take some portion of the shear in beams. If the capacity of the concrete is not exceeded—as is sometimes the case for lightly loaded beams—there may be no need for reinforcement. The typical case, however, is as shown in Figure 13.15, where the maximum shear V exceeds the capacity of the concrete alone (V_c) and the steel reinforcement is required to absorb the excess, indicated as the shaded portion in the shear diagram.

Minimum Shear Reinforcement. Even when the maximum computed shear stress falls below the capacity of the concrete, the present code requires the use of some minimum amount of shear reinforcement. Exceptions are made in some situations, such as for slabs and very shallow beams. The objective is essentially to toughen the structure with a small investment in additional reinforcement.

Type of Stirrup. The most common stirrups are the simple U-shape or closed forms shown in Figure 13.14, placed in a vertical position at intervals along the beam. It is also possible to place stirrups at an incline

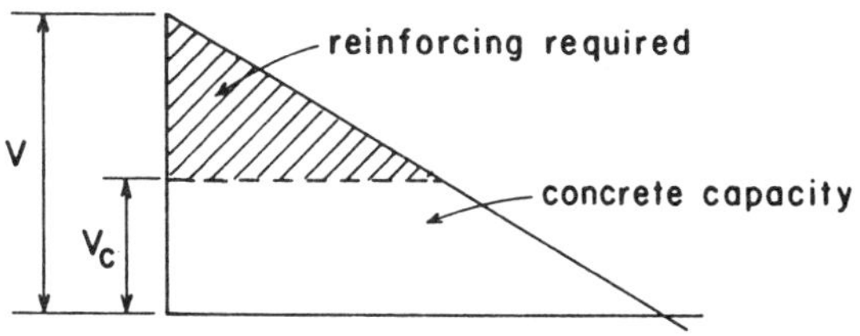

Figure 13.15 Sharing of shear resistance in reinforced concrete beams.

(usually 45°), which makes them somewhat more effective in direct resistance to the potential shear cracking near the beam ends (see Figure 13.11). In large beams with excessively high unit shear stress, both vertical and inclined stirrups are sometimes used at the location of the greatest shear.

Size of Stirrups. For beams of moderate size, the most common size for U-stirrups is a No. 3 bar. These bars can be bent relatively tightly at the corners (small radius of bend) in order to fit within the beam section. For larger beams, a No. 4 bar is sometimes used, its strength (as a function of its cross-sectional area) being almost twice that of a No. 3 bar.

Spacing of Stirrups. Stirrup spacings are computed (as discussed in the following sections) on the basis of the amount of reinforcing required for the unit shear stress at the location of the stirrups. A maximum spacing of $d/2$ (one half the effective beam depth d) is specified in order to ensure that at least one stirrup occurs at the location of any potential diagonal crack (see Figure 13.11). When shear stress is excessive, the maximum spacing is limited to $d/4$.

Critical Maximum Design Shear. Although the actual maximum shear value occurs at the end of the beam, the code permits the use of the shear stress at a distance of d (effective beam depth) from the beam end as the critical maximum for stirrup design. Thus, as shown in Figure 13.16, the shear requiring reinforcing is slightly different from that shown in Figure 13.15.

Total Length for Shear Reinforcement. On the basis of computed shear forces, reinforcement must be provided along the beam length for the distance defined by the shaded portion of the shear stress diagram shown in Figure 13.16. For the center portion of the span, the concrete is theoretically capable of the necessary shear resistance without the assistance of reinforcement. However, the code requires that some shear reinforcement be provided for a distance beyond this computed cutoff point. Earlier codes required that stirrups be provided for a distance equal to the effective depth of the beam beyond the computed cutoff point. Currently, codes require that minimum shear reinforcement be provided as long as the computed shear force exceeds one half of the capacity of the concrete ($\phi \times V_c/2$). However it is established, the total extended range over which reinforcement must be provided is indicated as R in Figure 13.16.

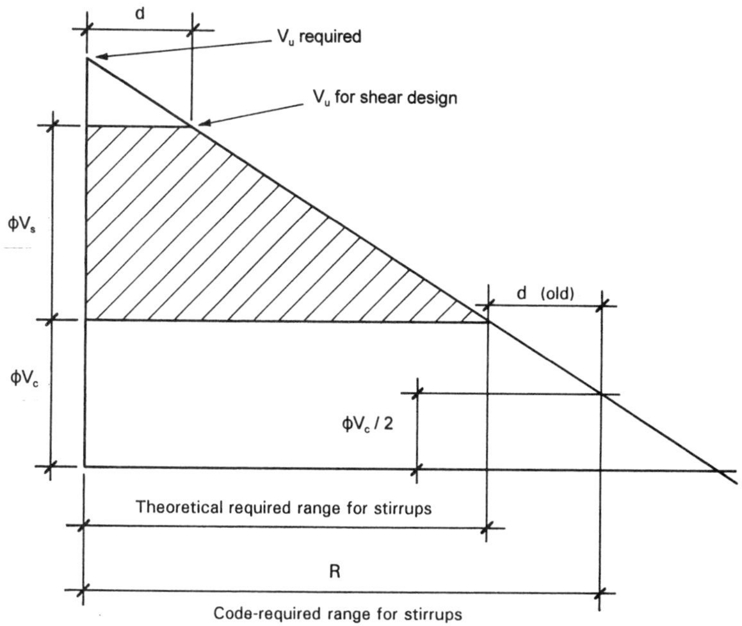

Figure 13.16 Layout for shear stress analysis: ACI Code requirements.

Design for Beam Shear

The following is a description of the procedure for design of shear reinforcement for beams that are experiencing flexural and shear stresses exclusively.

The ultimate shear force (V_u) at any cross section along a given beam as a result of factored loading must be less than the reduced shear capacity at the section. Mathematically this is represented as

$$V_u \leq \phi_v \times (V_c + V_s)$$

where V_u = ultimate shear force at the section as a result of factored loading

$\phi_v = 0.75$

V_c = shear capacity of the concrete

V_s = shear capacity of the steel reinforcing

For beams of normal weight concrete, subjected only to flexure and shear, shear stress in the concrete is limited to

$$V_c = 2\sqrt{f'_c}bd$$

where f'_c = specified strength of the concrete in psi
b = width of the cross section in in.
d = effective depth of the cross section in in.

When V_u exceeds the limit for $\phi_v V_c$, reinforcing must be provided, complying with the general requirements discussed previously. Thus:

$$V_s \geq \frac{V_u}{\phi_v} - V_c$$

Required spacing of shear reinforcement is determined as follows. Referring to Figure. 13.17, note that the capacity in tensile resistance of

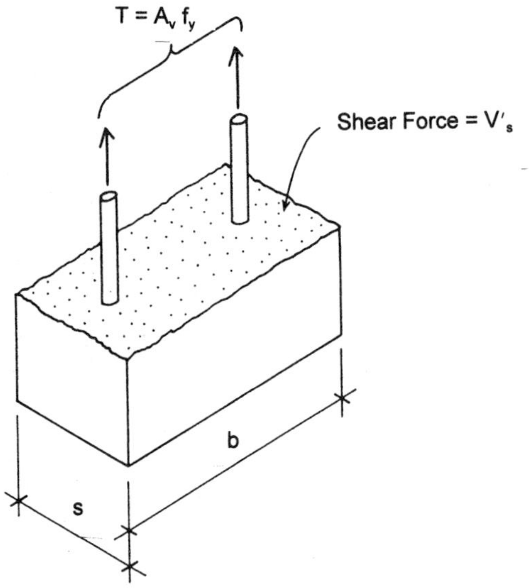

Figure 13.17 Consideration for spacing of a single stirrup.

a single, two-legged stirrup is equal to the product of the total steel cross-sectional area, A_v, times the yield steel stress. Thus:

$$T = A_v f_y$$

This resisting force opposes part of the steel shear force required, which we will refer to as V_s'. Equating the stirrup tension to this force, an equilibrium equation for one stirrup is obtained:

$$A_v f_y = V_s'$$

The total shear force capacity of the beam in excess of the concrete is determined by the number of stirrups encountered by the shear force acting at a 45-degree angle through the beam. The number of stirrups will be d/s. Thus, the equilibrium equation for the beam is

$$\left(\frac{d}{s}\right) A_v f_y = V_s'$$

From this equation, an expression for the required spacing can be derived; thus:

$$s \leq \frac{A_v f_y d}{V_s'}$$

The following example illustrates the design procedure for a simple beam.

Example 7. Design the required shear reinforcement for the simple beam shown in Figure 13.18. Use $f'_c = 3$ ksi [20.7 MPa] and $f_y = 40$ ksi [276 MPa] and single U-shaped stirrups.

Solution: First, the loading must be factored in order to determine the ultimate shear force.

$$w_u = 1.2 \times w_{DL} + 1.6 \times w_{LL} = 1.2 \times 2 \text{ klf} + 1.6 \times 3 \text{ klf} = 7.2 \text{ klf}$$

The maximum value for the shear is 57.6 kips [256 kN].

Now construct the ultimate shear force diagram (V_u) for one half of the beam, as shown in Figure13.18*c*. For the shear design, the critical shear

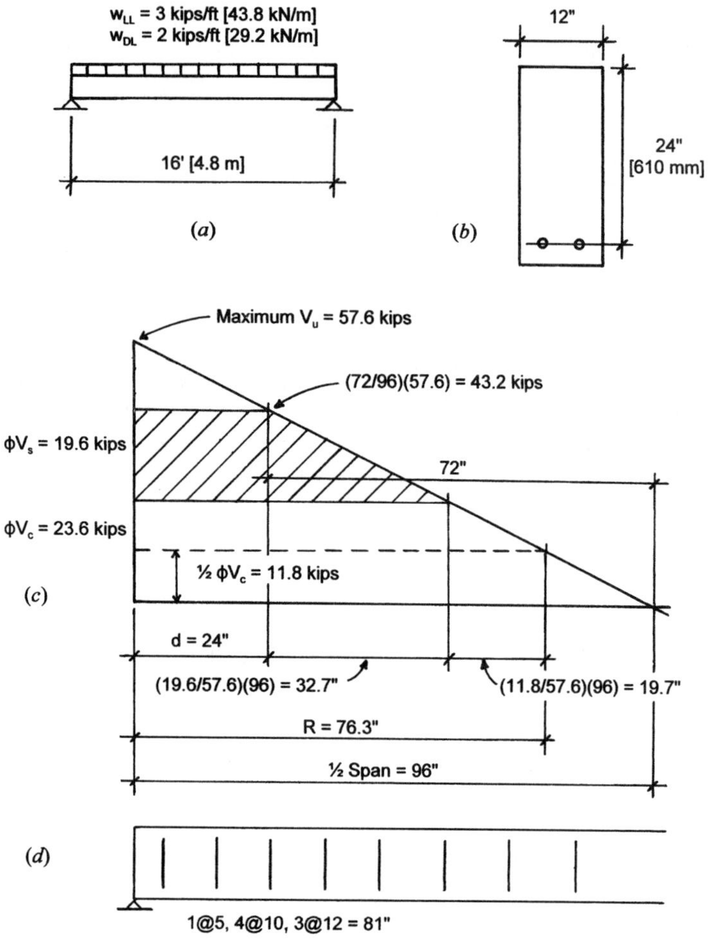

Figure 13.18 Stirrup design: Example 7.

force is at 24 in. (the effective depth of the beam) from the support. Using proportionate triangles, this value is

$$V_u = \left(\frac{72 \text{ in.}}{96 \text{ in.}}\right)(57.6 \text{ kips}) = 43.2 \text{ kips}$$

The shear capacity of the concrete without reinforcing is

$$\phi V_c = (0.75)2\sqrt{f'_c}(b)(d) = (0.75)2\sqrt{3000\text{ psi}}(12\text{ in.})(24\text{ in.}) = 23.6\text{ kips}$$

At the point of critical force, therefore, there is an excess shear force of 43.2 − 23.6 = 19.6 kips [87 kN] that must be carried by reinforcement. Next, complete the construction of the diagram in Figure 13.18*c* to define the shaded portion, which indicates the extent of the required reinforcement. Observe that the excess shear condition extends to 56.7 in. [1.44 m] from the support.

In order to satisfy the requirements of the ACI Code, shear reinforcement must be used wherever the shear force (V_u) exceeds one half of ϕV_c. As shown in Figure 13.18*c*, this is a distance of 76.3 in. from the support. The code further stipulates that the minimum cross-sectional area of this reinforcing be

$$A_v = 50\left[\frac{b \times s_{max}}{f_y}\right]$$

With f_y = 40 ksi [276 MPa] and the maximum allowable spacing of one half the effective depth, the required area is

$$A_v = 50\left[\frac{12\text{ in.} \times 12\text{ in.}}{40{,}000\text{ psi}}\right] = 0.18\text{ in.}^2$$

which is less than the area of 2 × 0.11 = 0.22 in.^{2} provided by the two legs of the No. 3 stirrup.

For the maximum V_s value of 19.6 kips, the maximum spacing permitted at the critical point 24 in. from the support is determined as

$$s \le \frac{A_v f_y d}{V_s} = \frac{(0.22\text{ in.}^2)(40\text{ ksi})(24\text{ in.})}{19.6\text{ kips}} = 10.8\text{ in.}$$

Because this is less than the maximum allowable of one half the depth or 12 in., it is best to calculate at least one more spacing at a short distance beyond the critical point. For example, at 36 in. from the support, the shear force is

$$V_u = \left(\frac{60\text{ in.}}{96\text{ in.}}\right)(57.6\text{ kips}) = 36.0\text{ kips}$$

and the value of V_s at this point is 36.0 − 23.6 kips = 12.4 kips. The spacing required at this point is thus

$$s \leq \frac{A_v f_y d}{V_s} = \frac{(0.22 \text{ in.}^2)(40 \text{ ksi})(24 \text{ in.})}{12.4 \text{ kips}} = 17 \text{ in.}$$

which indicates that the required spacing drops to the maximum allowed at less than 12 in. from the critical point. A possible choice for the stirrup spacings is shown in Figure 13.18*d*, with a total of eight stirrups that extend over a range of 81 in. from the support. There are thus a total of 16 stirrups in the beam, 8 at each end. Note that the first stirrup is placed at 5 in. from the support, which is one half the computed required spacing. This is a common practice with designers.

Example 8. Determine the required number and spacings for No. 3 U-stirrups for the beam shown in Figure 13.19. Use f'_c = 3 ksi [20.7 MPa] and f_y = 40 ksi [276 MPa].

Solution: As in Example 1, the shear values are determined, and the diagram in Figure 13.19*c*. is constructed. In this case, the maximum critical shear force of 28.5 kips and a shear capacity of concrete (ϕV_c) of 16.1 kips results in a maximum ϕV_s value to 12.4 kips, for which the required spacing is

$$s \leq \frac{A_v f_y d}{V_s} = \frac{(0.22 \text{ in.}^2)(40 \text{ ksi})(20 \text{ in.})}{12.4 \text{ kips}} = 14.2 \text{ in.}$$

Because this value exceeds the maximum limit of $d/2$ = 10 in., the stirrups may all be placed at the limited spacing, and a possible arrangement is as shown in Figure. 13.19*d*. As in Example 7, note that the first stirrup is placed at one half the required distance from the support.

Example 9. Determine the required number and spacings for No. 3 U-stirrups for the beam shown in Figure 13.20. Use f'_c = 3 ksi [20.7 MPa] and f_y = 40 ksi [276 MPa].

Solution: In this case, the maximum critical design shear force is found to be less than $V_{c,}$ which in theory indicates that reinforcement is not required. To comply with the code requirement for minimum reinforce-

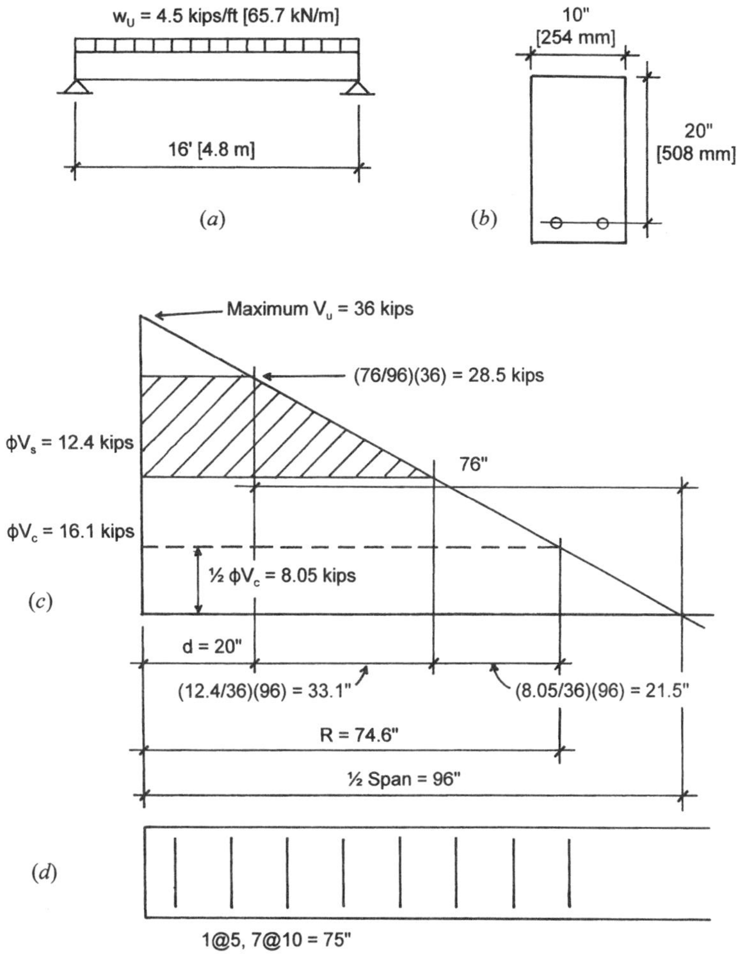

Figure 13.19 Stirrup design: Example 8.

ment, however, we must provide stirrups at the maximum permitted spacing out to the point where the shear stress drops to 8.05 kips (one half of ϕV_c). To verify that the No. 3 stirrup is adequate, compute

$$A_v = 50\left[\frac{10 \text{ in.} \times 10 \text{ in.}}{40{,}000 \text{ psi}}\right] = 0.125 \text{ in.}^2$$

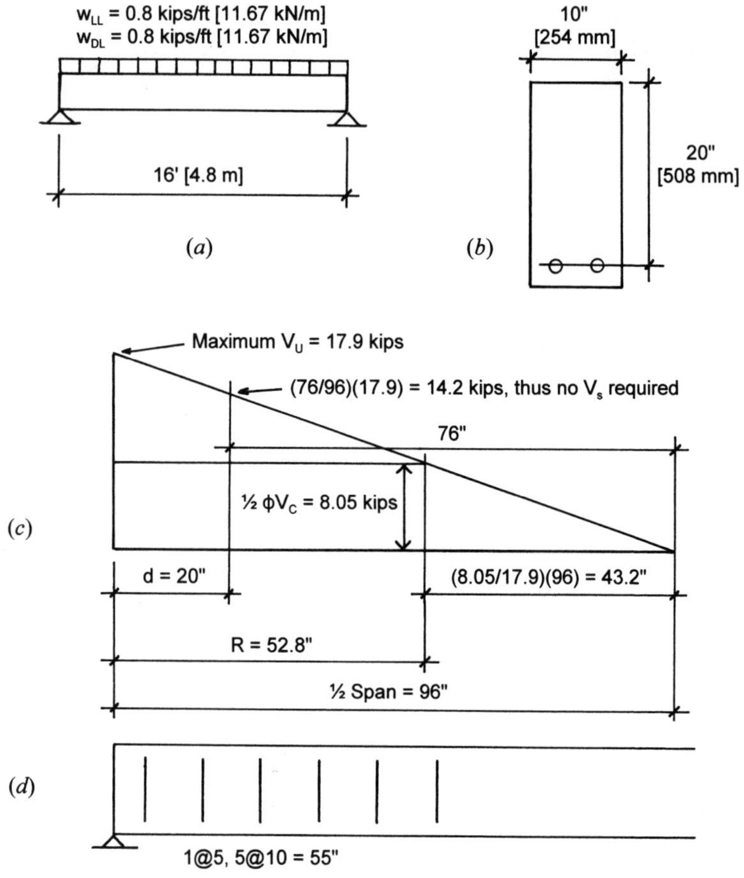

Figure 13.20 Stirrup design: Example 9.

which is less than the area of 0.22 in. provided, so the No. 3 stirrup at 10-in. is adequate.

Examples 7 through 9 have illustrated what is generally the simplest case for beam shear design—that of a beam with uniformly distributed load and with sections subjected only to flexure and shear. When concentrated loads or unsymmetrical loadings produce other forms for the shear diagram, these must be used for design of the shear reinforcing. In addition, where axial forces of tension or compression exist in the concrete frame, we must consider the combined effects when designing for shear.

When torsional moments exist (twisting moments at right angles to the beam), their effects must be combined with beam shear.

Problem 13.6.A. A concrete beam similar to that shown in Figure 13.18 sustains a uniform live load of 1.5 kips/ft and a uniform dead load of 1 kips/ft on a span of 24 ft [7.32 m]. Determine the layout for a set of No. 3 U-stirrups using the stress method with f_y = 40 ksi [276 MPa] and f'_c = 3000 psi [20.7 MPa]. The beam section dimensions are b = 12 in. [305 mm] and d = 26 in. [660 mm].

Problem 13.6.B. Same as Problem 13.6.A, except a span is 20 ft [6.1 m}, b = 10 in. [254 mm], d = 23 in. [584 mm].

Problem 13.6.C. Determine the layout for a set of No. 3 U-stirrups for a beam with the same data as Problem 13.6.A, except the uniform live load is 0.75 kips/ft and the uniform dead load is 0.5 kips/ft.

Problem 13.6.D. Determine the layout for a set of No. 3 U-stirrups for a beam with the same data as Problem 13.6.B, except the uniform live load is 1.875 kips/ft and the uniform dead load is 1.25 kips/ft.

13.7 DEVELOPMENT LENGTH FOR REINFORCEMENT

The ACI Code defines *development length* as the length of embedment required to develop the design strength of the reinforcement at a critical section. For beams, critical sections occur at points of maximum stress and at points within the span where some of the reinforcement terminates or is bent up or down. For a uniformly loaded simple span beam, one critical section is at midspan, where the bending moment is a maximum. The tensile reinforcement required for flexure at this point must extend on both sides a sufficient distance to develop the stress in the bars; however, except for very short spans with large bars, the bar lengths will ordinarily be more than sufficient.

In the simple beam, the bottom reinforcement required for the maximum moment at midspan is not entirely required as the moment decreases toward the end of the span. It is thus sometimes the practice to make only part of the midspan reinforcement continuous for the whole beam length. In this case, it may be necessary to ensure that the bars that are of partial length are extended sufficiently from the midspan point and that the bars remaining beyond the cutoff point can develop the stress required at that point.

When beams are continuous through the supports, top reinforcement is required for the negative moments at the supports. These top bars must be investigated for the development lengths in terms of the distance they extend from the supports.

For tension reinforcement consisting of bars of No. 11 size and smaller, the code specifies a minimum length for development (L_d) as follows:

For No. 6 bars and smaller:

$$L_d = \frac{f_y d_b}{25\sqrt{f'_c}}$$

(but not less than 12 in.)

For No. 7 bars and larger:

$$L_d = \frac{f_y d_b}{20\sqrt{f'_c}}$$

In these formulas d_b is the bar diameter.

Modification factors for L_d are given for various situations, as follows:

For top bars in horizontal members with at least 12 in. of concrete below the bars: increase by 1.3.

For flexural reinforcement that is provided in excess of that required by computations: decrease by ratio of (required A_s/provided A_s).

Additional modification factors are given for lightweight concrete, for bars coated with epoxy, for bars encased in spirals, and for bars with f_y in excess of 60 ksi. The maximum value to be used for $\sqrt{f'_c}$ is 100 psi.

Table 13.8 gives values for minimum development lengths for tensile reinforcement, based on the requirements of the ACI Code. The values listed under "Other Bars" are the unmodified length requirements; those listed under "Top Bars" are increased by the modification factor for this situation. Values are given for two concrete strengths and for the two most commonly used grades of tensile reinforcement.

The ACI Code makes no provision for a reduction factor for development lengths. As presented, the formulas for development length relate only to bar size, concrete strength, and steel yield strength. They are thus equally applicable for the stress method or the strength method with no further adjustment, except for the conditions previously described.

Example 10. The negative moment in the short cantilever shown in Figure 13.21 is resisted by the steel bars in the top of the beam. Determine whether the development of the reinforcement is adequate without hooked ends on the No. 6 bars, if $L_1 = 48$ in. and $L_2 = 36$ in. Use $f'_c = 3$ ksi [20.7 MPa] and $f_y = 60$ ksi [414 MPa].

Solution: At the face of the support, anchorage for development must be achieved on both sides: within the support and in the top of the beam. In the top of the beam, the condition is one of "Top Bars," as previously defined. Thus, from Table 13.8, a length of 43 in. is required for L_d, which is adequately provided, if cover is minimum on the outside end of the bars.

Within the support, the condition is one of "Other Bars" in the table reference. For this the required length for L_d is 33 in., which is also adequately provided.

Hooked ends are thus not required on either end of the bars, although most designers would probably hook the bars in the support just for the security of the additional anchorage.

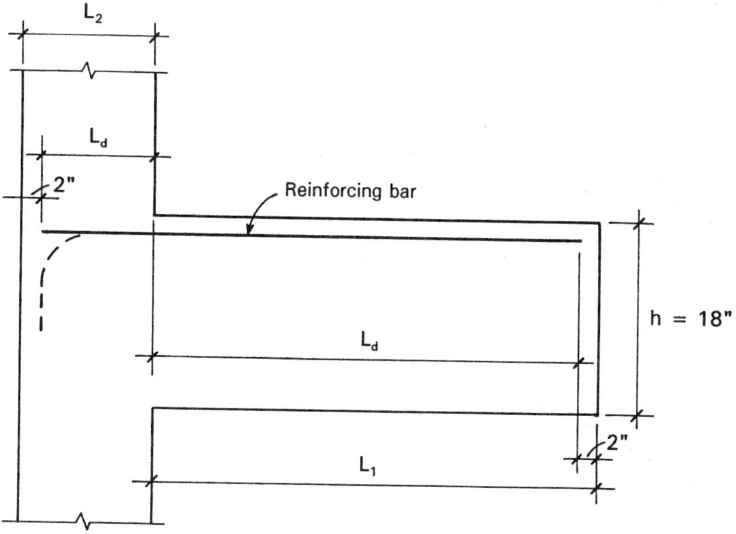

Figure 13.21 Reference for Example 10.

TABLE 13.8 Minimum Development Length for Tensile Reinforcement (in.)[a]

	f_y = 40 ksi [276 MPa]				f_y = 60 ksi [414 MPa]			
	f'_c = 3 ksi [20.7 MPa]		f'_c = 4 ksi [27.6 MPa]		f'_c = 3 ksi [20.7 MPa]		f'_c = 4 ksi [27.6 MPa]	
Bar Size	Top Bars[b]	Other Bars	Top Bars[b]	Other Bars	Top Bars[b]	Other Bars	Top Bars[b]	Other Bars
3	15	12	13	12	22	17	19	15
4	19	15	17	13	29	22	25	19
5	24	19	21	16	36	28	31	24
6	29	22	25	19	43	33	37	29
7	42	32	36	28	63	48	54	42
8	48	37	42	32	72	55	62	48
9	54	42	47	36	81	62	70	54
10	61	47	53	41	91	70	79	61
11	67	52	58	45	101	78	87	67

[a] Lengths are based on requirements of the ACI Code (Ref. 8).

[b] Horizontal bars with more than 12 in. of concrete cast below them in the member.

Problem 13.7.A. A short cantilever is developed as shown in Figure 13.21. Determine whether adequate development is achieved without hooked ends on the bars if L_1 is 36 inches, L_2 is 24 in., overall beam height is 16 in., bar size is No. 4, f'_c = 4 ksi, and f_y = 40 ksi.

Problem 13.7.B. Same as Problem 13.7.A, except L_1 = 40 in., L_2 = 30 in., No. 5 bar.

Hooks

When details of the construction restrict the ability to extend bars sufficiently to produce required development lengths, we can sometimes use a hooked end on the bar. So-called *standard hooks* may be evaluated in terms of a required development length, L_{dh}. Bar ends may be bent at 90°, 135°, or 180° to produce a hook. The 135° bend is used only for ties and stirrups, which normally consist of relatively small diameter bars.

Table 13.9 gives values for development length with standard hooks, using the same variables for f'_c and f_y that are used in Table 13.8. The table values given are in terms of the required development length as shown in Figure 13.22. Note that the table values are for 180° hooks and that values

TABLE 13.9 Required Development Length L_{dh} for Hooked Bars (in.)[a]

Bar Size	$f_y = 40$ ksi [276 MPa] $f'_c = 3$ ksi [20.7 MPa]	$f_y = 40$ ksi [276 MPa] $f'_c = 4$ ksi [27.6 MPa]	$f_y = 60$ ksi [414 MPa] $f'_c = 3$ ksi [20.7 MPa]	$f_y = 60$ ksi [414 MPa] $f'_c = 4$ ksi [27.6 MPa]
3	6	6	9	8
4	8	7	11	10
5	10	8	14	12
6	11	10	17	15
7	13	12	20	17
8	15	13	22	19
9	17	15	25	22
10	19	16	28	24
11	21	18	31	27

[a] See Fig. 13.22. Table values are for a 180° hook; values may be reduced by 30% for a 90° hook.

may be reduced by 30% for 90° hooks. The following example illustrates the use of the data from Table 13.9 for a simple situation.

Example 11. For the bars in Figure 13.21, determine the length L_d required for development of the bars with a 90° hooked end in the support. Use the same data as in Example 10.

Solution: From Table 13.9, the required length for the data given is 17 in. (No. 6 bar $f'_c = 3$ ksi, $f_y = 60$ ksi.) This may be reduced for the 90° hook to

$$L = 0.70(17) = 11.9 \text{ in.}$$

Problem 13.7.C. Find the development length required for the bars in Problem 13.7.A, if the bar ends in the support are provided with 90° hooks.

Problem 13.7.D. Find the development length required for the bars in Problem 13.7.B, if the bar ends in the support are provided with 90° hooks.

Bar Development in Continuous Beams

Development length is the length of embedded reinforcement required to develop the design strength of the reinforcement at a critical section. Critical sections occur at points of maximum stress and at points within the span at which adjacent reinforcement terminates or is bent up into the top of the beam. For a uniformly loaded simple beam, one critical section

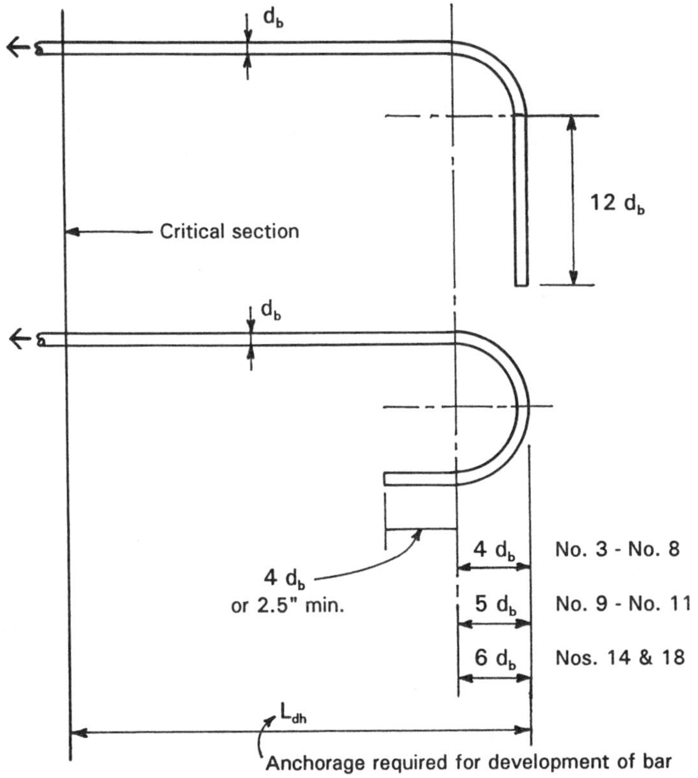

Figure 13.22 Detail requirements for standard hooks for use of values in Table 13.9.

is at midspan where the bending moment is maximum. This is a point of maximum tensile stress in the reinforcement (peak bar stress), and some length of bar is required over which the stress can be developed. Other critical sections occur between midspan and the reactions at points where some bars are cut off because they are no longer needed to resist the bending moment; such terminations create peak stress in the remaining bars that extend the full length of the beam.

When beams are continuous through their supports, the negative moments at the supports will require that bars be placed in the top of the beams. Within the span, bars will be required in the bottom of the beam for positive moments. Even though the positive moment will go to zero at some distance from the supports, the codes require that some of the positive moment reinforcement be extended for the full length of the span and a short distance into the support.

Figure 13.23 shows a possible layout for reinforcement in a beam with continuous spans and a cantilevered end at the first support. Referring to the notation in the illustration, note the following:

1. Bars *a* and *b* are provided for the maximum moment of positive sign that occurs somewhere near the beam midspan. If all these bars are made full length (as shown for bars *a*), the length L_1 must be sufficient for development (this situation is seldom critical.). If bars *b* are partial length as shown in the illustration, then length L_2 must be sufficient to develop bars *b* and length L_3 must be sufficient to develop bars *a*. As discussed for the simple beam, the partial length bars must actually extend beyond the theoretical cutoff point (*B* in the illustration) and the true length must include the dashed portions indicated for bars *b*.
2. For the bars at the cantilevered end, the distances L_4 and L_5 must be sufficient for development of bars *c*. L_4 is required to extend beyond the actual cutoff point of the negative moment by the extra length described for the partial length bottom bars. If L_5 is not adequate, the bar ends may be bent into the 90° hook as shown or the 180° hook shown by the dashed line.
3. If the combination of bars shown in the illustration is used at the interior support, L_6 must be adequate for the development of bars d and L_7 must be adequate for the development of bars *e*.

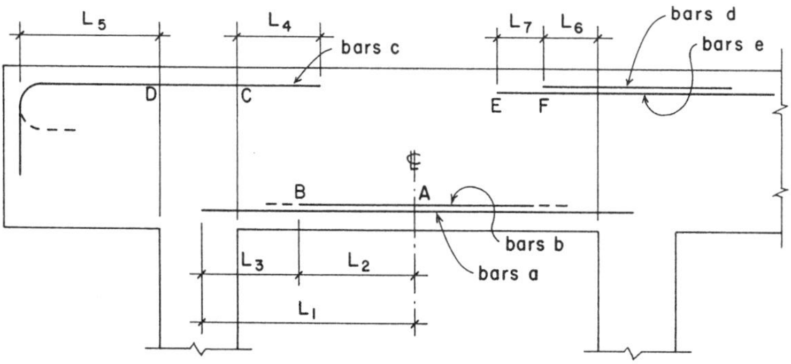

Figure 13.23 Development lengths in continuous beams.

For a single loading condition on a continuous beam, we can determine specific values of moment and their location along the span, including the locations of points of zero moment. In practice, however, most continuous beams are designed for more than a single loading condition, which further complicates the problems of determining development lengths required.

Splices in Reinforcement

In various situations in reinforced concrete structures, it becomes necessary to transfer stress between steel bars in the same direction. Continuity of force in the bars is achieved by splicing, which may be affected by welding, by mechanical means, or by the lapped splice. Figure 13.24 illustrates the concept of the lapped splice, which consists essentially of the development of both bars within the concrete. Because a lapped splice is usually made with the two bars in contact, the lapped length must usually be somewhat greater than the simple development length required in Table 13.8.

For a simple tension lap splice, the full development of the bars usually requires a lap length of 1.3 times that required for simple development of the bars. Lap splices are generally limited to bars of No. 11 size or smaller.

For pure tension members, lapped splicing is not permitted, and splicing must be achieved by welding the bars or by some other mechanical connection. End-to-end butt welding of bars is usually limited to compression splicing of large diameter bars with high f_y for which lapping is not feasible.

When members have several reinforcement bars that must be spliced, the splicing must be staggered. Splicing is generally not desirable and is to be avoided when possible, but because bars are obtainable only in lim-

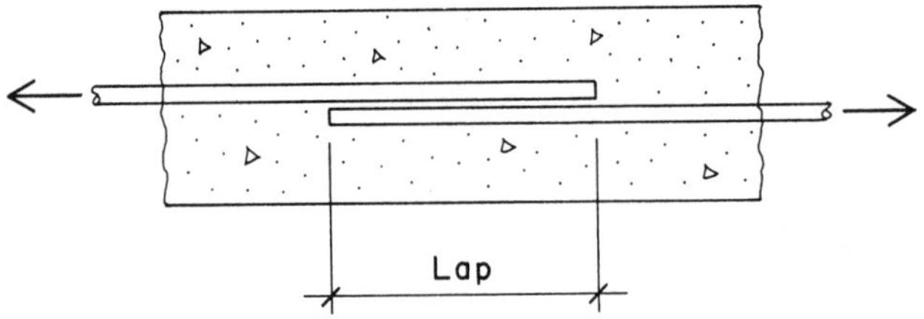

Figure 13.24 The lapped splice for steel reinforcing bars.

ited lengths, some situations unavoidably involve splicing. Horizontal reinforcement in walls is one such case. For members with computed stress, splicing should not be located at points of maximum stress—for example, at points of maximum bending.

Splicing of compression reinforcement for columns is discussed in the next section.

Development of Compressive Reinforcement

The discussion of development length so far has dealt with tension bars only. Development length in compression is, of course, a factor in column design and in the design of beams reinforced for compression.

The absence of flexural tension cracks in the portions of beams where compression reinforcement is employed, plus the beneficial effect of the end bearing of the bars on the concrete, permit shorter developmental lengths in compression than in tension. Development length for bars in compression is computed by the formula

$$L_d = \frac{0.02 f_y d_b}{\sqrt{f'_c}}$$

but not less than $0.0003 f_y d_b$ or 8 in., whichever is greater

In reinforced columns, both the concrete and the steel bars share the compression force. Ordinary construction practices require the consideration of various situations for development of the stress in the reinforcing bars. The various problems of bar development in columns are discussed in Chapter 15

13.8 DEFLECTION CONTROL

Deflection of spanning slabs and beams of cast-in-place concrete is controlled primarily by using recommended minimum thickness (overall height) expressed as a percentage of the span. Table 13.10 is adapted from a similar table given in the ACI Code and yields minimum thicknesses as a fraction of the span. Table values apply only for concrete of normal weight (made with ordinary sand and gravel) and for reinforcement with f_y of 40 ksi [276 MPa] and 60 ksi [414 MPa]. The ACI Code supplies correction factors for other concrete weights and reinforcing grades. The ACI Code further stipulates that these recommendations apply only where beam deflections are not critical for other elements of

TABLE 13.10 Minimum Thickness of Slabs or Beams Unless Deflections Are Computed[a]

Type of Member	End Conditions of Span	Minimum Thickness of Slab or Height of Beam	
		f_y = 40 ksi [276 MPa]	f_y = 60 ksi [414 MPa]
Solid one-way Slabs	Simple support	L/25	L/20
	One end continuous	L/30	L/24
	Both ends continuous	L/35	L/28
	Cantilever	L/12.5	L/10
Beams or Joists	Simple aupport	L/20	L/16
	One end continuous	L/23	L/18.5
	Both ends continuous	L/26	L/21
	Cantilever	L/10	L/8

[a]Refers to overall vertical dimension of concrete section. For normal weight concrete (145 pcf) only; code provides adjustment for other weights. Valid only for members not supporting or attached rigidly to partitions or other construction likely to be damaged by large deflections.

Source: Adapted from material in *Building Code Requirements for Structural Concrete (ACI 318–02)*, with permission of the publishers, American Concrete Institute.

the building construction, such as supported partitions subject to cracking caused by beam deflections.

Deflection of concrete structures presents a number of special problems. For concrete with ordinary reinforcement (not prestressed), flexural action normally results in some tension cracking of the concrete at points of maximum bending. Thus, the presence of cracks in the bottom of a beam at midspan points and in the top over supports is to be expected. In general, the size (and visibility) of these cracks will be proportional to the amount of beam curvature produced by deflection. Crack size will also be greater for long spans and for deep beams. If visible cracking is considered objectionable, more conservative depth-to-span ratios should be used, especially for spans over 30 ft and beam depths over 30 in.

Creep of concrete results in additional deflections over time. This is caused by the sustained loads—essentially the dead load of the construction. Deflection controls reflect concern for this as well as for the instantaneous deflection under live load, the latter being the major concern in structures of wood and steel.

In beams, deflections, especially creep deflections, may be reduced by the use of some compressive reinforcement. Where deflections are of concern, or where depth-to-span ratios are pushed to their limits, it is advisable to use some compressive reinforcement, consisting of continuous top bars.

When, for whatever reasons, deflections are deemed to be critical, computations of actual values of deflection may be necessary. The ACI Code provides directions for such computations; they are quite complex in most cases and beyond the scope of this work. In actual design work, however, they are required very infrequently.

14

FLAT-SPANNING CONCRETE SYSTEMS

Many different systems can be used to achieve flat spans. These are used most often for floor structures, which typically require a dead flat form. However, in buildings with an all-concrete structure, they may also be used for roofs. Sitecast systems generally consist of one of the following basic types:

1. One-way solid slab and beam
2. Two-way solid slab and beam
3. One-way joist construction
4. Two-way flat slab or flat plate without beams
5. Two-way joist construction, called waffle construction

Each system has its own distinct advantages and limits and some range of logical use, depending on required spans, general layout of supports, magnitude of loads, required fire ratings, and cost limits for design and construction.

The floor plan of a building and its intended usage determine loading conditions and the layout of supports. Also of concern are requirements for openings for stairs, elevators, large ducts, skylights, and so on, because they result in discontinuities in the otherwise commonly continuous systems. Whenever possible, columns and bearing walls should be aligned in rows and spaced at regular intervals in order to simplify design and construction and lower costs. However, the fluid concrete can be molded in forms not possible for wood or steel, and many very innovative, sculptural systems have been developed as takeoffs on these basic ones.

14.1 SLAB AND BEAM SYSTEMS

The most widely used and most adaptable cast-in-place concrete floor system is that which utilizes one-way solid slabs supported by one-way spanning beams. This system may be used for single spans but occurs more frequently with multiple-span slabs and beams in a system such as that shown in Figure 14.1. In the example shown, the continuous slabs are supported by a series of beams that are spaced at 10 ft. center to center. The beams, in turn, are supported by a girder and column system with columns at 30-ft centers, every third beam being supported directly by the columns and the remaining beams being supported by the girders.

Because of the regularity and symmetry of the system shown in Figure 14.1, there are relatively few different elements in the basic system, each being repeated several times. Although special members must be designed for conditions that occur at the outside edge of the system and at the location of any openings for stairs, elevators, and so on, the general interior portions of the structure may be determined by designing only six basic elements: S1, S2, B1, B2, G1, and G2, as shown in the framing plan.

In computations for reinforced concrete, the span length of freely supported beams (simple beams) is generally taken as the distance between centers of supports or bearing areas; it should not exceed the clear span plus the depth of beam or slab. The span length for continuous or restrained beams is taken as the clear distance between faces of supports.

In continuous beams, negative bending moments are developed at the supports and positive moments at or near midspan. This may be readily observed from the exaggerated deformation curve of Figure 14.2*a*. The exact values of the bending moments depend on several factors, but in the case of approximately equal spans supporting uniform loads, when the live load does not exceed three times the dead load, the bending moment values given in Figure 14.2 may be used for design.

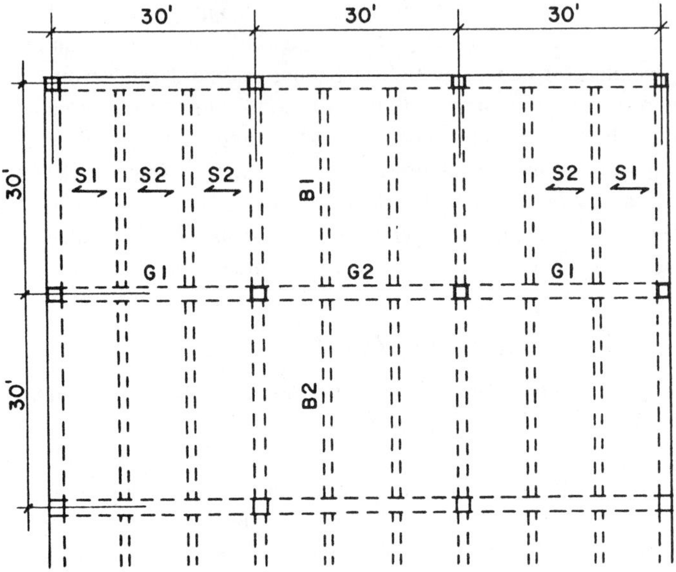

Figure 14.1 Framing layout for a typical slab and beam system.

The values in Figure 14.2 are in general agreement with the those given in Chapter 8 of the ACI Code. These values have been adjusted to account for partial live loading of multiple-span beams. Note that these values apply only to uniformly loaded beams. The ACI Code also gives some factors for end-support conditions other than the simple supports shown in Figure 14.2

Design moments for continuous-span slabs are given in Figure 14.3. With large beams and short slab spans, the torsional stiffness of the beam tends to minimize the continuity effect in adjacent slab spans. Thus, most slab spans in the slab-and-beam systems tend to function much like individual spans with fixed ends.

Design of a One-Way Continuous Slab

The general design procedure for a one-way solid slab was illustrated in Section 13.4. The example given there is for a simple span slab. The following example illustrates the procedure for the design of a continuous solid one-way slab.

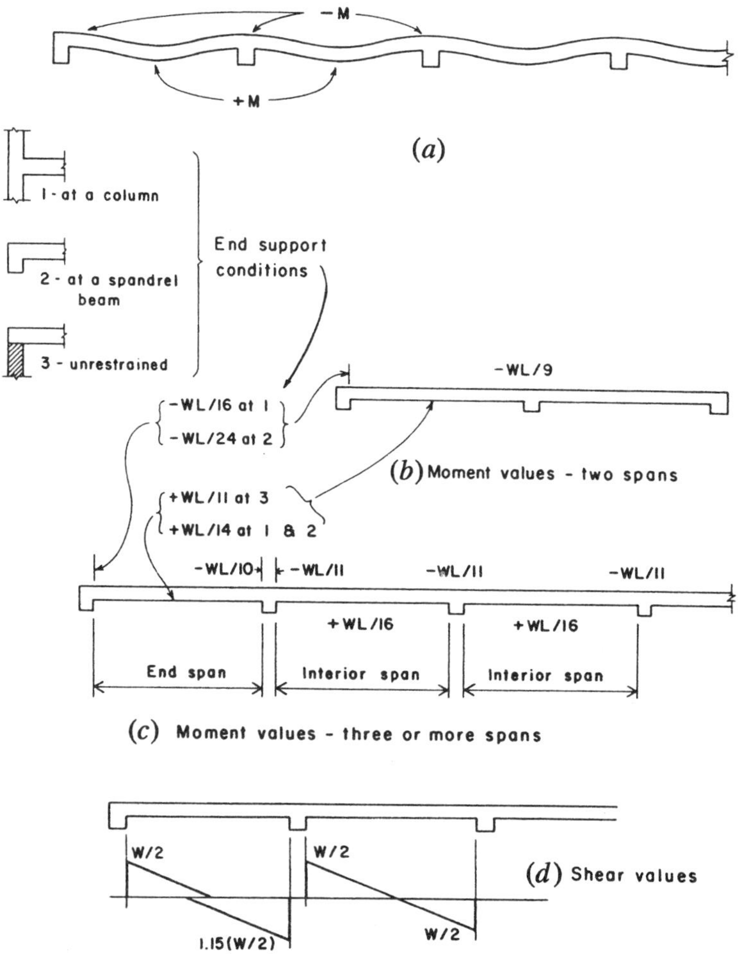

Figure 14.2 Approximate design factors for concrete beams.

Example 1. A solid one-way slab is to be used for a framing system similar to that shown in Figure 14.1. Column spacing is 30 ft. with evenly spaced beams occurring at 10 ft. center to center. Superimposed loads on the structure (floor live load plus other construction dead load) are a dead load of 38 psf [1.82 kPa] and a live load of 100 psf [4.79 kPa]. Use $f'_c = 3$ ksi [20.7 MPa] and $f_y = 40$ ksi [275 MPa]. Determine the thickness for the slab and select its reinforcement.

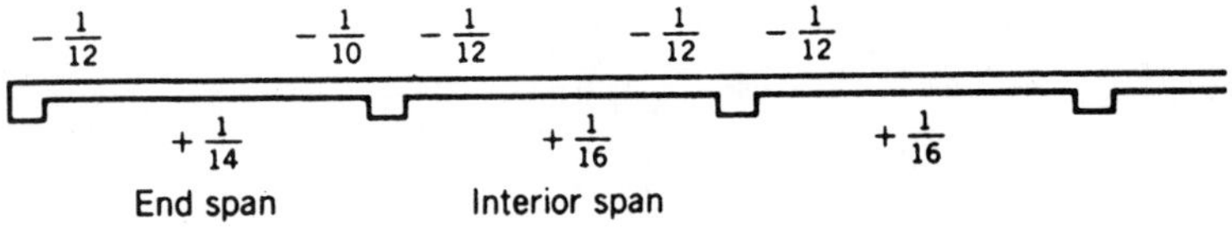

Figure 14.3 Approximate design factors for continuous slabs with spans of 10 ft or less.

Solution: To find the slab thickness, consider three factors: the minimum thickness for deflection, the minimum effective depth for the maximum moment, and the minimum effective depth for the maximum shear. For design purposes, the span of the slab is taken as the clear span, which is the dimension from face to face of the supporting beams. With the beams at 10-ft centers, this dimension is 10 ft, less the width of one beam. Because the beams are not given, a dimension must be assumed for them. For this example, assume a beam width of 12 in., yielding a clear span of 9 ft.

Consider first the minimum thickness required for deflection. If the slabs in all spans have the same thickness (which is the most common practice), the critical slab is the end span because there is no continuity of the slab beyond the end beam. Although the beam will offer some restraint, it is best to consider this as a simple support; thus, the appropriate factor is $L/30$ from Table 13.10, and

$$\text{Minimum } t = \frac{L}{30} = \frac{9 \times 12}{30} = 3.6 \text{ in.}$$

Assume here that fire-resistive requirements make it desirable to have a relatively heavy slab of 5-in. overall thickness, for which the dead weight of the slab is

$$w = \frac{5}{12} \times 150 = 62 \text{ psf}$$

The total dead load is thus $62 + 38 = 100$ psf and the factored total load is

$$U = 1.2(100) + 1.6(100) = 280 \text{ psf } [13.4 \text{ kPa}]$$

Next consider the maximum bending moment. Inspection of the moment values given in Figure 14.3 shows the maximum moment to be

$$M = \frac{1}{10} wL^2$$

With the clear span and the loading as determined, the maximum moment is thus

$$M = \frac{wL^2}{10} = \frac{280 \times (9)^2}{10} = 2268 \text{ ft-lb}$$

and the required resisting moment for the slab is

$$M_R = \frac{2268}{0.9} = 2520 \text{ ft-lb [3.417 kN-m]}$$

Now compare this moment value to the balanced capacity of the design section, using the relationships discussed for rectangular beams in Section 13.3. For this computation, an effective depth for the design section must be assumed. This dimension will be the slab thickness minus the concrete cover and one half the bar diameter. With the bars not yet determined, assume an approximate effective depth to be the slab thickness minus 1.0 in.; this will be exactly true with the usual minimum cover of 3/4 in. and a No. 4 bar. Then using the balanced moment R factor from Table 13.2, determine that the maximum resisting moment for the 12-in.-wide design section is

$$M_R = Rbd^2 = (1.149)(12)(4)^2 = 221 \text{ kip-in.}$$

or

$$M_R = 221 \times \frac{1000}{12} = 18{,}400 \text{ ft-lb}$$

Because this value is in excess of the required resisting moment, the slab is adequate for concrete flexural stress.

It is not practical to use shear reinforcement in one-way slabs; consequently, the maximum unit shear stress must be kept within the limit for the concrete alone. The usual procedure is to check the shear stress with the effective depth determined for bending before proceeding to find A_s.

Except for very short span slabs with excessively heavy loadings, shear stress is seldom critical.

For interior spans, the maximum shear will be $wL/2$, but for the end span, it is the usual practice to consider some unbalanced condition for the shear due to the discontinuous end. Use a maximum shear of $1.15(wL/2)$, or an increase of 15% over the simple beam shear value. Thus:

$$\text{Maximum shear} = V_u = 1.15 \times \frac{wL}{2} = 1.15 \times \frac{280 \times 9}{2} = 1449 \text{ lb}$$

and

$$\text{Required } V_r = \frac{1449}{0.75} = 1932 \text{ lb}$$

For the slab section with $b = 12$ in. and $d = 4$ in.:

$$V_c = 2\sqrt{f'_c}(b \times d) = 2\sqrt{3000}(12 \times 4) = 5258 \text{ lb}$$

This is considerably greater than the required shear resistance, so the assumed slab thickness is not critical for shear stress.

Having thus verified the choice for the slab thickness, we may now proceed with the design of the reinforcement. For a balanced section, Table 13.2 yields a value of 0.685 for the a/d factor. However, because all sections will be classified as underreinforced (actual moment less than the balanced limit), use an approximate value of 0.4 for a/d. Once the reinforcement for a section is determined, the true value of a/d can be verified using the procedures developed in Section 13.3.

For the slab in this example the following is computed:

$$\frac{a}{d} = 0.4, \quad \text{and} \quad a = 0.4(d) = 0.4(4) = 1.6 \text{ in.}$$

For the computation of required reinforcement, use

$$d - \frac{a}{2} = 4 - \frac{1.6}{2} = 3.2 \text{ in.}$$

Referring to Figure 14.3, note that there are five critical locations for which a moment must be determined and the required steel area computed. Reinforcement required in the top of the slab must be computed for the negative moments at the end support, at the first interior beam, and at the typical interior beam. Reinforcement required in the bottom of the slab must be computed for the positive moments at midspan locations in the first span and in the typical interior spans. The design for these conditions is summarized in Figure 14.4. For the data displayed in the figure, note the following:

Maximum spacing of reinforcement:

$$s = 3t = 3(5) = 15 \text{ in.}$$

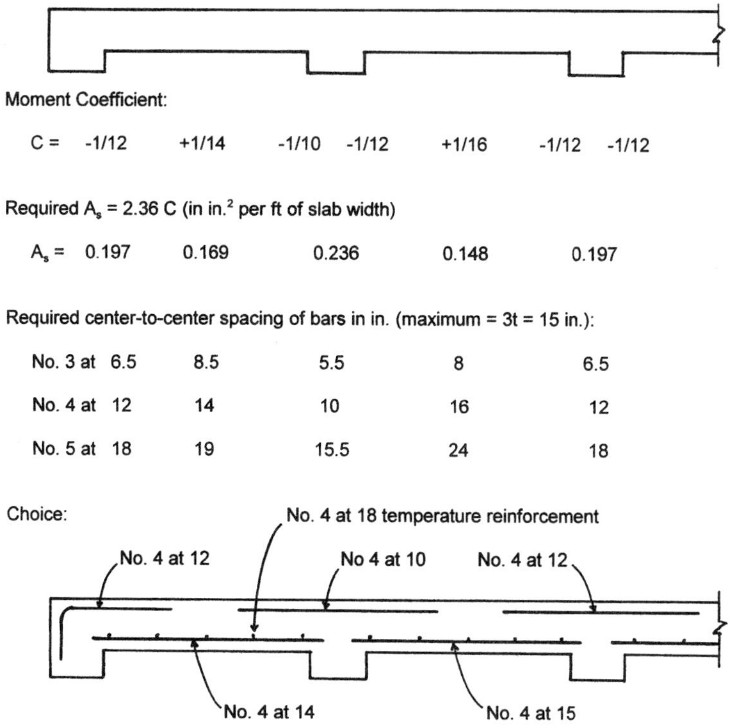

Figure 14.4 Summary of design for the continuous slab.

Maximum required bending moment:

$$M_R = (\text{Moment factor } C)\left(\frac{wL^2}{0.9}\right) = C\left(\frac{280 \times (9)^2}{0.9}\right) \times 12 = 302{,}400C$$

Note that the use of the factor 12 makes this value for the moment in in.-lb units.

Required area of reinforcement:

$$A_s = \frac{M}{f_y\left(d - \dfrac{a}{2}\right)} = \frac{302{,}400 \times C}{40{,}000 \times 3.2} = 2.36C$$

Using data from Table 13.6, Figure 14.4 shows required spacing for Nos. 3, 4, and 5 bars. A possible choice for the slab reinforcement, using straight bars, is shown at the bottom of Figure 14.4.

For required temperature reinforcement:

$$A_s = 0.002bt = 0.002(12 \times 5) = 0.12 \text{ in.}^2\text{/ft of slab width}$$

Using data from Table 13.6, we find that possible choices are #3 at 11 in. or #4 at 18 in.

Problem 14.1.A. A solid one-way slab is to be used for a framing system similar to that shown in Figure 14.1. Column spacing is 36 ft [11 m], with regularly spaced beams occurring at 12 ft [3.66 m] center to center. Superimposed dead load on the structure is 40 psf [1.92 kPa], and live load is 80 psf [3.83 kPa]. Use f'_c = 4 ksi [27.6 MPa] and f_y = 60 ksi [414 MPa]. Determine the thickness for the slab and select the size and spacing for the bars.

Problem 14.1.B. Same as Problem 14.1.A, except column spacing is 33 ft [10 m], beams are at 11 ft [3.33 m] centers, superimposed dead load is 50 psf [2.39 kPa] and live load is 75 psf [3.59 kPa].

14.2 GENERAL CONSIDERATIONS FOR BEAMS

The design of a single beam involves a large number of pieces of data, most of which are established for the system as a whole rather than individually for each beam. System-wide decisions usually include those for the type of concrete and its design strength (f'_c), the type of reinforcing steel (f_y), the cover required for the necessary fire rating, and various generally used details for forming the concrete and placing the reinforcement. Most beams occur in conjunction with solid slabs that are cast monolithically with the beams. Slab thickness is established by the structural requirements of the spanning action between beams and by various concerns, such as those for fire rating, acoustic separation, type of reinforcement, and so on. Design of a single beam is usually limited to determination of the following:

1. Choice of shape and dimensions of the beam cross section
2. Selection of the type, size, and spacing of shear reinforcement
3. Selection of the flexural reinforcement to satisfy requirements based on the variation of moment along the several beam spans

The following are some factors that must be considered in effecting these decisions.

Beam Shape

Figure 14.5 shows the most common shapes used for beams in sitecast construction. The single, simple rectangular section is actually uncommon, but it does occur in some situations. Design of the concrete section consists of selecting the two dimensions: the width b and the overall height or depth h.

As mentioned previously, beams occur most often in conjunction with monolithic slabs, resulting in the typical T-shape shown in Figure 14.5*b* or the L-shape shown in Figure 14.5*c*. The full T-shape occurs at the interior portions of the system, whereas the L shape occurs at the outside of the system or at the side of large openings. As shown in the illustration, there are four basic dimensions for the T and L that must be established in order to fully define the beam section:

t = the slab thickness; it is ordinarily established on its own, rather than as a part of the single beam design

h = the overall beam stem depth, corresponding to the same dimension for the rectangular section

b_w = the beam stem width, which is critical for consideration of shear and for problems of fitting reinforcing into the section

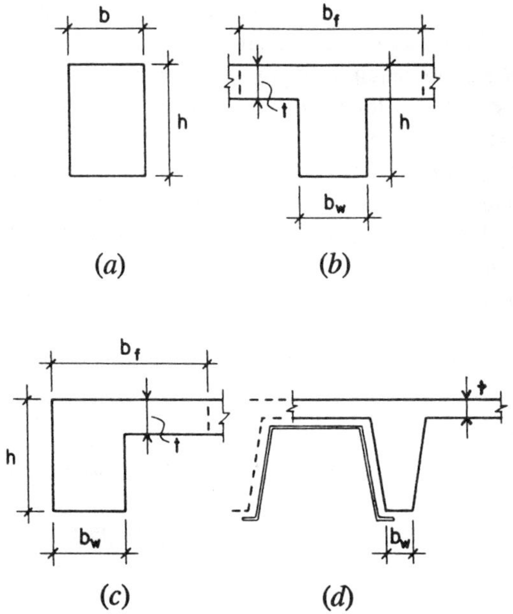

Figure 14.5 Common shapes for beams.

b_f = the so-called *effective width* of the flange, which is the portion of the slab assumed to work with the beam

A special beam shape is that shown in Figure 14.5*d*. This occurs in concrete joist and waffle construction when "pans" of steel or reinforced plastic are used to form the concrete, the taper of the beam stem being required for easy removal of the forms. The smallest width dimension of the beam stem is ordinarily used for the beam design in this situation.

Beam Width

The width of a beam will affect its resistance to bending. The flexure formulas given in Section 13.3 show that the width dimension affects the bending resistance in a linear relationship (double the width and you double the resisting moment, etc.). On the other hand, the resisting moment is affected by the *square* of the effective beam depth. Thus, efficiency, in terms of beam weight or concrete volume, will be obtained by striving for deep, narrow beams instead of shallow, wide ones. (Just as a 2×8 joist is more efficient than a 4×4 joist in wood).

Beam width also relates to various other factors, however, and these are often critical in establishing the minimum width for a given beam. The formula for shear capacity indicates that the beam width is equally as effective as the depth in shear resistance. Placement of reinforcing bars is sometimes a problem in narrow beams. Table 14.1 gives minimum beam widths required for various bar combinations, based on considerations of bar spacing, minimum concrete cover of 1.5 in., placement of the bars in a single layer, and use of a No. 3 stirrup. Situations requiring additional concrete cover, use of larger stirrups, or the intersection of beams with columns may necessitate widths greater than those given in Table 14.1.

Beam Depth

Although selection of beam depth is partly a matter of satisfying structural requirements, it is typically constrained by other considerations in the building design. Figure 14.6 shows a section through a typical building floor/ceiling with a concrete slab-and-beam structure. In this situation, the critical depth from a general building design point of view is the overall thickness of the construction, shown as H in the illustration. In addition to the concrete structure, this includes allowances for the floor finish, the ceiling construction, and the passage of an insulated air duct. The net usable portion of H for the structure is shown as the dimension

TABLE 14.1 Minimum Beam Widths[a]

Number of Bars	Bar Size								
	3	4	5	6	7	8	9	10	11
2	10	10	10	10	10	10	10	10	10
3	10	10	10	10	10	10	10	10	11
4	10	10	10	10	11	11	12	13	14
5	10	11	11	12	12	13	14	15	17
6	11	12	13	14	14	15	17	18	19
7	13	14	15	15	16	17	19	20	22
8	14	15	16	17	18	19	21	23	25
9	16	17	18	19	20	21	23	25	28
10	17	18	19	21	22	23	26	28	30

[a] Minimum width in inches for beams with 1.5-in. cover, No. 3 U-stirrups, clear spacing between bars of one bar diameter or minimum of 1 in. General minimum practical width for any beam with No. 3 U-stirrups is 10 in.

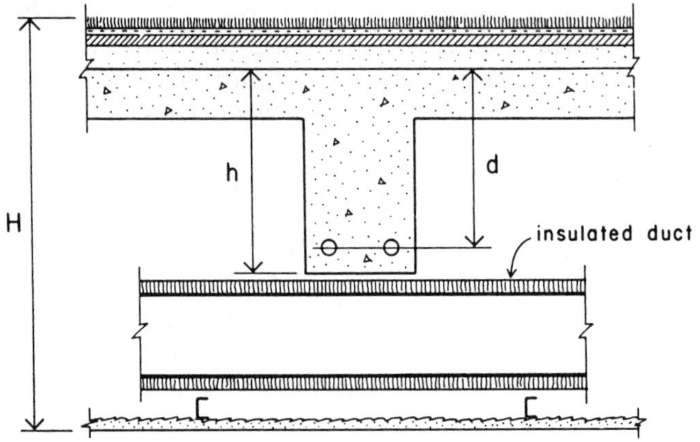

Figure 14.6 Concrete beam in typical multistory construction. Dimension *H* is critical for architectural planning; dimension *d* is critical for structural design.

h, with the effective structural depth *d* being something less than *h*. Because the space defined by *H* is not highly usable for the building occupancy, there is a tendency to constrain it, which works to limit any extravagant use of *d*.

Most concrete beams tend to fall within a limited range in terms of the ratio of width to depth. The typical range is for a width-to-depth ratio between 1:1.5 and 1: 2.5, with an average of 1:2. This is not a code requirement or a magic rule; it is merely the result of satisfying typical requirements for flexure, shear, bar spacing, economy of use of steel, and deflection.

Deflection Control

Deflection of spanning slabs and beams must be controlled for a variety of reasons. This topic is discussed in Section 13.7. Typically, the most critical decision factor relating to deflection is the overall vertical thickness or height of the spanning member. The ratio of this height dimension to the span length is the most direct indication of the degree of concern for deflection.

Design of Continuous Beams

Continuous beams are typically indeterminate and must be investigated for the bending moments and shears that are critical for the various loading conditions. When the beams are not involved in rigid-frame actions (as they are when they occur on column lines in multistory buildings), it may be possible to use approximate analysis methods, as described in the ACI Code and here in Section 14.1. An illustration of such a procedure is shown in the design of a concrete floor structure in Chapter 20.

In contrast to beams of wood and steel, those of concrete must be designed for the changing internal force conditions along their length. The single, maximum values for bending moment and shear may be critical in establishing the required beam size, but requirements for reinforcement must be investigated at all supports and midspan locations.

15

CONCRETE COLUMNS AND FRAMES

In view of the ability of concrete to resist compressive stress and its weakness in tension, it would seem to be apparent that its most logical use is for structural members whose primary task is the resistance of compression. This observation ignores the use of reinforcement to a degree, but it is nevertheless not without some note. And, indeed, major use is made of concrete for columns, piers, pedestals, posts, and bearing walls—all basically compression members. This chapter presents discussions of the use of reinforced concrete for such structural purposes, with emphasis on the development of columns for building structures. Concrete columns often exist in combination with concrete beam systems, forming rigid frames with vertical planar bents. This subject is addressed as part of the discussion of the concrete example structure in Chapter 20.

15.1 EFFECTS OF COMPRESSION FORCE

When concrete is subjected to a direct compressive force, the most obvious stress response in the material is one of compressive stress, as shown in Figure 15.1*a*. This response may be the essential one of concern, as it would be in a wall composed of flat, precast concrete bricks, stacked on top of each other. Direct compressive stress in the individual bricks and in the mortar joints between bricks would be a primary situation for investigation.

However, if the concrete member being compressed has some dimension in the direction of the compressive force, as in the case of a column or pier, there are other internal stress conditions that may well be the source of actual structural failure under the compressive force. Direct compressive force produces a three-dimensional deformation that includes a pushing out of the material at right angles to the force, actually producing tension stress in that direction, as shown in Figure 15.1*b*. In a tension-weak material, this tension action may produce a lateral bursting effect.

Because concrete as a material is also weak in shear, another possibility for failure is along the internal planes where maximum shear stress is developed. This occurs at a 45° angle with respect to the direction of the applied force, as shown in Figure 15.1*c*.

In concrete compression members, other than flat bricks, it is generally necessary to provide for all three stress responses shown in Figure 15.1. In fact, additional conditions can occur if the structural member is also subjected to bending or torsional twisting. Each case must be investigated individually for all the individual actions and the combinations in which they can occur. Design for the concrete member and its reinforcement will typically respond to several considerations of behavior, and the same member and reinforcement must function for all responses. The following discussions focus on the primary function of resistance to compression, but other concerns will also be mentioned. The basic consideration for combined compression and bending is discussed here because present codes require that all columns be designed for this condition.

Reinforcement of Columns

Column reinforcement takes various forms and serves various purposes; the essential consideration is to enhance the structural performance of the column. Considering the three basic forms of column stress failure shown in Figure 15.1, we can visualize basic forms of reinforcement for each condition. This is done in the illustrations in Figures 15.2*a*, *b*, and *c*.

To assist the basic compression function, steel bars are added with their linear orientation in the direction of the compression force. This is the fundamental purpose of the vertical reinforcing bars in a column.

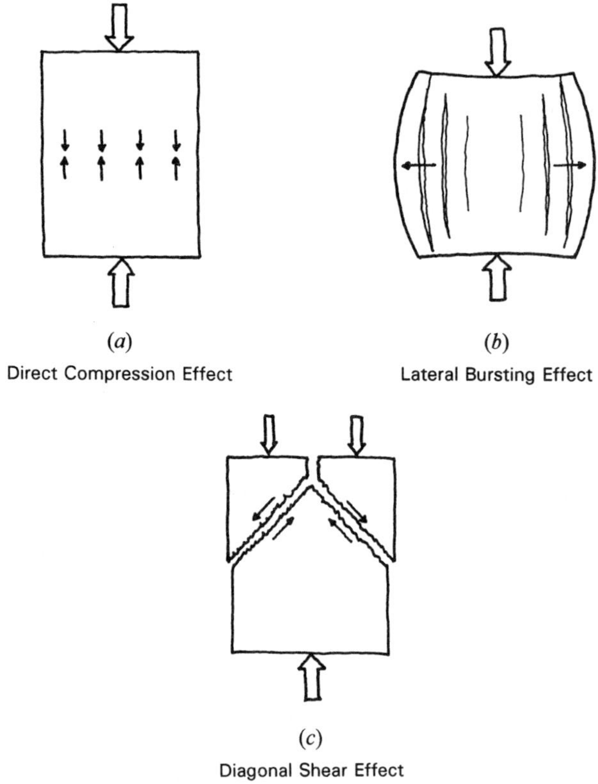

Figure 15.1 Fundamental failure modes of the tension-weak concrete.

Even though the steel bars displace some concrete, their superior strength and stiffness make them a significant improvement.

To assist in resistance to lateral bursting (Figure 15.2*b*), a critical function is to hold the concrete from moving out laterally, which may be achieved by so-called *containment* of the concrete mass, similar to the action of a piston chamber containing air or hydraulic fluid. If compression resistance can be obtained from air that is contained, surely it can be more significantly obtained from contained concrete. This is a basic reason for the traditional extra strength of the spiral column and one reason for now favoring very closely spaced ties in tied columns. In retrofitting columns for improved seismic resistance, a technique sometimes used is

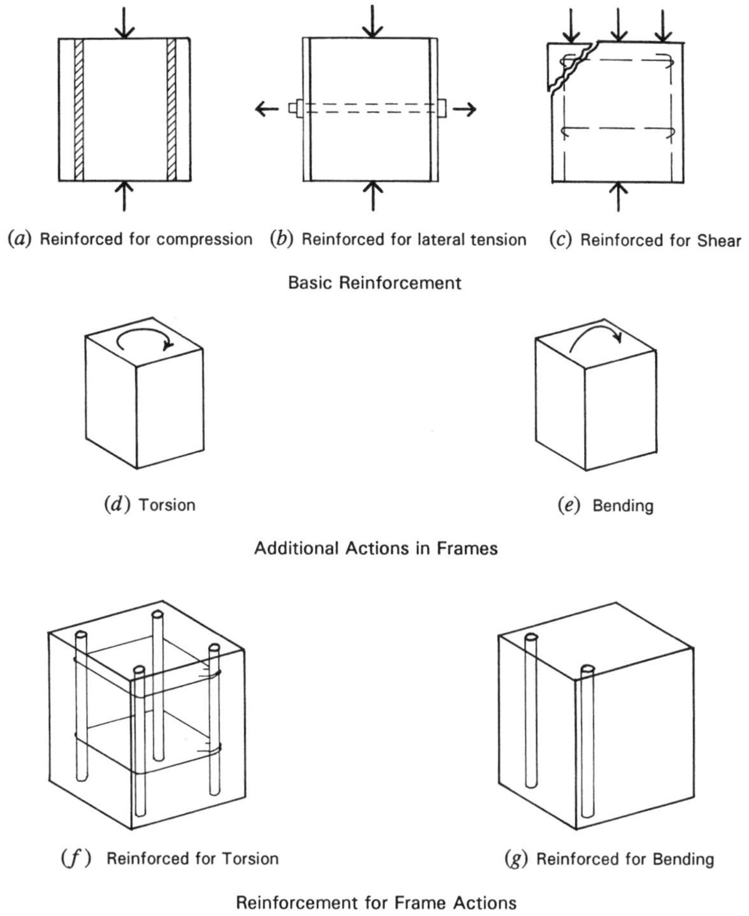

Figure 15.2 Forms and functions of column reinforcement.

to actually provide a confining, exterior jacket of steel or fiber strand, essentially functioning as illustrated in Figure 15.2*b*.

Natural shear resistance is obtained from the combination of the vertical bars and the lateral ties or spiral, as shown in Figure 15.2*c*. If this is a critical concern, improvements can be obtained by using closer-spaced ties and a larger number of vertical bars that spread out around the column perimeter.

When used as parts of concrete frameworks, columns are also typically subjected to torsion and bending, as shown in Figures 15.2*d* and *e*. Torsional twisting tends to produce a combination of longitudinal tension and lateral shear; thus, the combination of full perimeter ties or spirals and the perimeter vertical bars provide for this in most cases.

Bending, if viewed independently, requires tension reinforcement, just as in an ordinary beam. In the column, the ordinary section is actually a doubly reinforced one, with both tension and compression reinforcement for beam action. This function, combined with the basic axial compression, is discussed more fully in later sections of this chapter. An added complexity in many situations is the existence of bending in more than a single direction.

All of these actions can occur in various combinations due to different conditions of loading. Column design is thus a quite complex process if all possible structural functions are considered. A fundamental design principle becomes the need to make multiple usage of the simplest combination of reinforcing elements.

15.2 GENERAL CONSIDERATIONS FOR CONCRETE COLUMNS

Types of Columns

Concrete columns occur most often as the vertical support elements in a structure generally built of cast-in-place concrete (commonly called *sitecast*). This is the situation discussed in this chapter. Very short columns, called *pedestals*, are sometimes used in the support system for columns or other structures. The ordinary pedestal is discussed as a foundation transitional device in Chapter 16. Walls that serve as vertical compression supports are called *bearing walls.*

The sitecast concrete column usually falls into one of the following categories:

1. Square columns with tied reinforcement
2. Oblong columns with tied reinforcement
3. Round columns with tied reinforcement
4. Round columns with spiral-bound reinforcement
5. Square columns with spiral-bound reinforcement
6. Columns of other geometries (L-shaped, T-shaped, octagonal, etc.) with either tied or spiral-bound reinforcement

Obviously, the choice of column cross-sectional shape is an architectural, as well as a structural, decision. However, forming methods and costs, arrangement and installation of reinforcement, and relations of the column form and dimensions to other parts of the structural system must also be dealt with.

In tied columns, the longitudinal reinforcement is held in place by loop ties made of small-diameter reinforcement bars, commonly No. 3 or No. 4. Such a column is represented by the square section shown in Figure 15.3*a*. This type of reinforcement can quite readily accommodate other geometries as well as the square.

Spiral columns are those in which the longitudinal reinforcing is placed in a circle, with the whole group of bars enclosed by a continuous

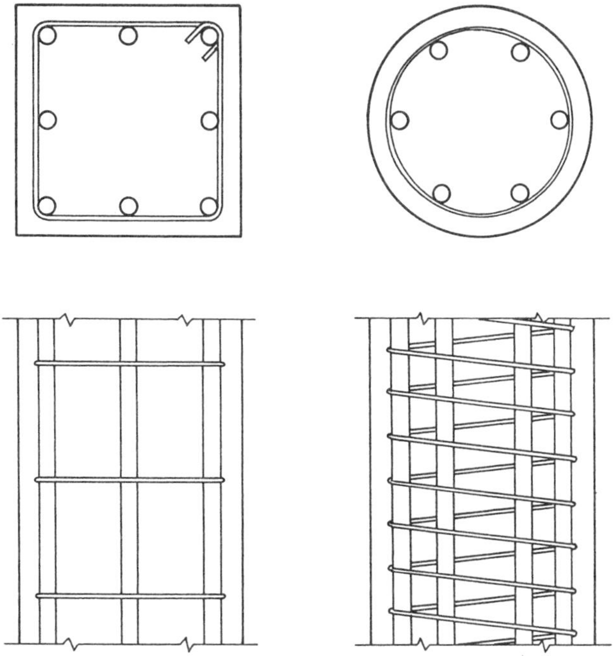

(*a*) (*b*)

Figure 15.3 Primary forms of column reinforcement: (*a*) rectangular layout of vertical bars with lateral ties; (*b*) circular layout of vertical bars with continuous helix (spiral) wrap.

cylindrical spiral made from steel rod or large-diameter steel wire. Although this reinforcing system obviously works best with a round column section, it can be used also with other geometries. A round column of this type is shown in Figure 15.3*b*.

Experience has shown the spiral column to be slightly stronger than an equivalent tied column with the same amount of concrete and reinforcement. For this reason, code provisions have traditionally allowed slightly more load on spiral columns. Spiral reinforcement tends to be expensive, however, and the round bar pattern does not always mesh well with other construction details in buildings. Thus, tied columns are often favored where restrictions on the outer dimensions of the sections are not severe.

A recent development is the use of tied columns with very closely spaced ties. A basic purpose for this is the emulation of a spiral column for achieving additional strength, although many forms of gain are actually obtained simultaneously, as discussed with respect to Figure 15.2.

General Requirements for Columns

Code provisions and practical construction considerations place a number of restrictions on column dimensions and choice of reinforcement.

Column Size. The current code does not contain limits for column dimensions. For practical reasons, the following limits are recommended. Rectangular tied columns should be limited to a minimum area of 100 in.^{2} and a minimum side dimension of 10 in. if square and 8 in. if oblong. Spiral columns should be limited to a minimum size of 12 in. if either round or square.

Reinforcement. Minimum bar size is No. 5. The minimum number of bars is four for tied columns, five for spiral columns. The minimum amount of area of steel is 1% of the gross column area. A maximum area of steel of 8% of the gross area is permitted, but bar spacing limitations makes this difficult to achieve; 4% is a more practical limit. The ACI Code stipulates that for a compression member with a larger cross section than required by considerations of loading, a reduced effective area not less than one half the total area may be used to determine minimum reinforcement and design strength.

Ties. Ties should be at least No. 3 for bars No. 10 and smaller. No. 4 ties should be used for bars that are No. 11 and larger. Vertical spacing of ties should be not more than 16 times the bar diameter, 48 times the tie diameter, or the least dimension of the column. Ties should be arranged so that every corner and alternate longitudinal bar is held by the corner of a tie with an included angle of not greater than 135°, and no bar should be farther than 6 in. clear from such a supported bar. Complete circular ties may be used for bars placed in a circular pattern.

Concrete Cover. A minimum of 1.5-in. cover is needed when the column surface is not exposed to weather and is not in contact with the ground. Cover of 2 in. should be used for formed surfaces exposed to the weather or in contact with ground. Cover of 3 in. should be used if the concrete is cast directly against earth without constructed forming, such as occurs on the bottoms of footings.

Spacing of Bars. Clear distance between bars should not be less than 1.5 times the bar diameter, 1.33 times the maximum specified size for the coarse aggregate, or 1.5 in.

Combined Compression and Bending

Because of the nature of most concrete structures, design practices generally do not consider the possibility of a concrete column with axial compression alone. This is to say that the existence of some bending moment is always considered together with the axial force. The general case of columns with bending is discussed in Section 3.11.

Figure 15.4 illustrates the nature of the so-called *interaction response* for a concrete column, with a range of combinations of axial load plus bending moment. In general, there are three basic ranges of this behavior, as follows (see the dashed lines in Figure 15.4):

1. *Large axial force, minor moment.* For this case, the moment has little effect, and the resistance to pure axial force is only negligibly reduced.
2. *Significant values for both axial force and moment.* For this case, the analysis for design must include the full combined force effects, that is, the interaction of the axial force and the bending moment.
3. *Large bending moment, minor axial force.* For this case, the column behaves essentially as a doubly reinforced (tension and compression reinforced) member, with its capacity for moment resistance affected only slightly by the axial force.

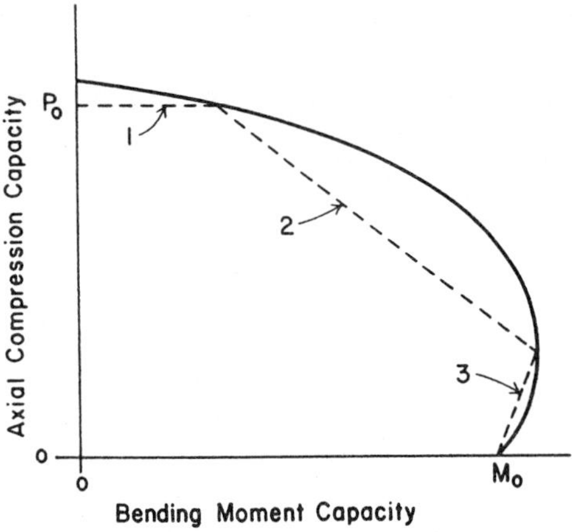

Figure 15.4 Interaction of axial compression (*P*) and bending moment (*M*) in a reinforced concrete column. Solid line indicates general form of response; dashed line indicates three separate zones of response: (1) dominant compression with minor bending; (2) significant compression plus bending interaction; (3) dominant bending in the cracked section.

In Figure 15.4, the solid line on the graph represents the true response of the column—a form of behavior verified by many laboratory tests. The dashed line represents the generalization of the three types of response just described.

The terminal points of the interaction response—pure axial compression or pure bending moment—may be reasonably easily determined (P_o and M_o in Figure 15.4). The interaction responses between these two limits require complex analyses beyond the scope of this book.

A special type of bending is generated when a relatively slender compression member develops some significant curvature due to the bending induced at its ends. In this case the center portion of the member's length literally moves away (deflects sideways, called *delta* for the deflected dimension) from a straight line. In this center portion, therefore, a bending is induced, consisting of the product of the compression force (*P*) and the deflected dimension (*delta*). Because this effect produces additional bending, and thus additional deflection, it may become a progressive failure condition, as it indeed is for very slender members. This is called the *P-delta effect*, and it is a critical consideration for relatively slender columns. Because concrete columns are not usually very slender, this effect is usually of less concern than it is for columns of wood and steel.

Considerations for Column Shape

Usually, a number of possible combinations of reinforcing bars may be assembled to satisfy the steel area requirement for a given column. Aside from providing for the required cross-sectional area, the number of bars must also work reasonably in the layout of the column. Figure 15.5 shows a number of columns with various numbers of bars. When a square tied column is small, the preferred choice is usually that of the simple four-bar layout, with one bar in each corner and a single perimeter tie. As the column gets larger, the distance between the corner bars gets larger, and it is best to use more bars so that the reinforcement is spread out around the column periphery. For a symmetrical layout and the simplest of tie layouts, the best choice is for numbers that are multiples of four, as shown in Figure 15.5*a*. The number of additional ties required for these layouts depends on the size of the column and the considerations discussed in Section 15.1.

An unsymmetrical bar arrangement (Figure 15.5*b*) is not necessarily bad, even though the column and its construction details are otherwise not oriented differently on the two axes. In situations where moments may be greater on one axis, the unsymmetrical layout is actually preferred; in fact, the column shape will also be more effective if it is unsymmetrical, as shown for the oblong shapes in Figure 15.5*c*.

Figures 15.5*d* through *g* show a number of special column shapes developed as tied columns. Although spirals could be used in some cases for such shapes, the use of ties allows much greater flexibility and simplicity of construction. One reason for using ties may be the column dimensions; there is a practical lower limit of about 12 in. in width for a spiral-bound column.

Round columns are frequently formed as shown in Figure 15.5*h*, if built as tied columns. This allows for a minimum reinforcement with four bars. If a round pattern is used (as it must be for a spiral-bound column), the usual minimum number recommended is six bars, as shown in Figure 15.5*i*. Spacing of bars is much more critical in spiral-bound circular arrangements, making it very difficult to use high percentages of steel in the column section. For very large-diameter columns, it is possible to use sets of concentric spirals, as shown in Figure 15.5*j*.

For cast-in-place columns, a concern that must be dealt with is that for vertical splicing of the steel bars. Two places where this commonly occurs are at the top of the foundation and at floors where a multistory column continues upward. At these points there are three ways to achieve

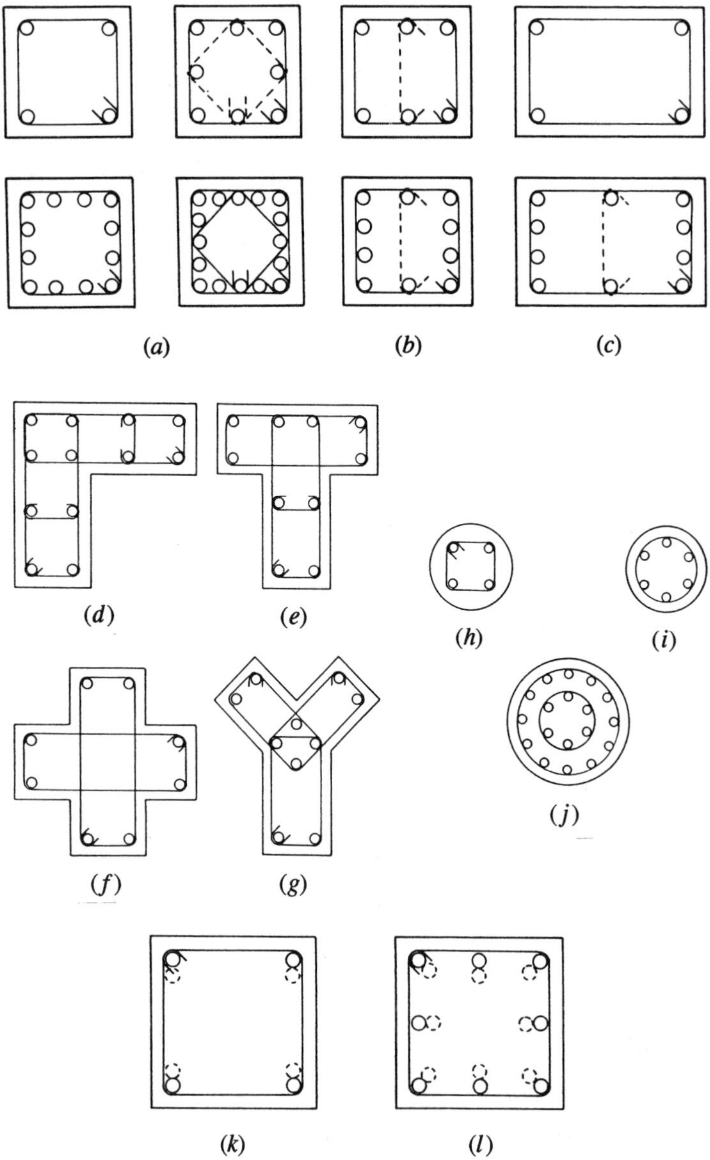

Figure 15.5 Considerations for bar layouts and tie patterns in tied concrete columns. (*a*) Square columns with symmetrical reinforcement. (*b*) Square columns with unsymmetrical reinforcement. (*c*) Oblong columns. (*d*)—(*g*) Oddly-shaped columns. (*h*)—(*j*) Round columns with tied or spiral reinforcement. (*k*), (*l*) Bar placement at location of lapped spice of vertical bars.

the vertical continuity (splicing) of the steel bars, any of which may be appropriate for a given situation.

1. Bars may be lapped the required distance for development of the compression splice. For bars of smaller dimension and lower yield strengths, this is usually the desired method.
2. Bars may have milled square-cut ends butted together with a grasping device to prevent separation in a horizontal direction.
3. Bars may be welded with full-penetration butt welds or by welding of the grasping device described for method 2.

The choice of splicing methods is basically a matter of cost comparison but is also affected by the size of the bars, the degree of concern for bar spacing in the column arrangement, and possibly a need for some development of tension through the splice if uplift or high magnitudes of moments exist. If lapped splicing is used, a problem that must be considered is the bar layout at the location of the splice, at which point there will be twice the usual number of bars. The lapped bars may be adjacent to each other, but the usual considerations for space between bars must be made. If spacing is not critical, the arrangement shown in Figure 15.5*k* is usually chosen, with the spliced sets of bars next to each other at the tie perimeter. If spacing limits prevent the arrangement in Figure 15.5*k*, that shown in Figure 15.5*l* may be used, with the lapped sets in concentric patterns. The latter arrangement is used for spiral columns, where spacing is often critical.

Bending of steel bars involves the development of yield stress to achieve plastic deformation (the residual bend). As bars get larger in diameter, they are more difficult—and less feasible—to bend. Also, as the yield stress increases, the bending effort gets larger. It is questionable to try to bend bars as large as No. 14 or No. 18 in any grade, and it is also not advised to bend any bars with yield stress greater than 75 ksi. Bar fabricators should be consulted for real limits of this nature.

15.3 DESIGN METHODS AND AIDS FOR CONCRETE COLUMNS

At present, design of concrete columns is mostly achieved by using either tabulations from handbooks or computer-aided procedures. Using the code formulas and requirements to design by "hand operation" with both axial compression and bending present at all times is prohibitively laborious. The number of variables present (column shape and size, f'_c,

f_y, number and size of bars, arrangement of bars, etc.) adds to the usual problems of column design to make for a situation much more complex than those for wood or steel columns.

The large number of variables also works against the efficiency of handbook tables. Even if a single concrete strength (f'_c) and a single steel yield strength (f_y) are used, tables would be very extensive if all sizes, shapes, and types (tied and spiral) of columns were included. Even with a very limited range of variables, handbook tables are much larger than those for wood or steel columns. They are, nevertheless, often quite useful for preliminary design estimation of column sizes.

The obvious preference when relationships are complex, requirements are tedious and extensive, and there are a large number of variables is for a computer-aided system. It is hard to imagine a professional design office that is turning out designs of concrete structures on a regular basis at the present without computer-aided methods. The reader should be aware that the software required for this work is readily available.

As in other situations, the common practices at any given time tend to narrow down to a limited usage of any type of construction, even though the potential for variation is extensive. It is thus possible to use some very limited but easy-to-use design aids to make early selections for design. These approximations may be adequate for preliminary building planning, cost estimates, and some preliminary structural analyses.

Approximate Design of Tied Columns

Tied columns are much preferred because of the relative simplicity and usually lower cost of their construction, plus their adaptability to various column shapes (square, round, oblong, T-shape, L-shape, etc.) Round columns—most naturally formed with spiral-bound reinforcing—are often made with ties instead, when the structural demands are modest.

The column with moment is often designed using the equivalent eccentric load method. The method consists of translating a compression plus bending situation into an equivalent one with an eccentric load, the moment becoming the product of the load and the eccentricity (see discussion in Section 3.11). This method is often used in presentation of tabular data for column capacities.

Figures 15.6 and 15.7 yield safe ultimate factored capacities for a selected number of sizes of square tied columns with varying percentages of reinforcement. Allowable axial compression loads are given for various degrees of eccentricity, which is a means for handling axial load and bend-

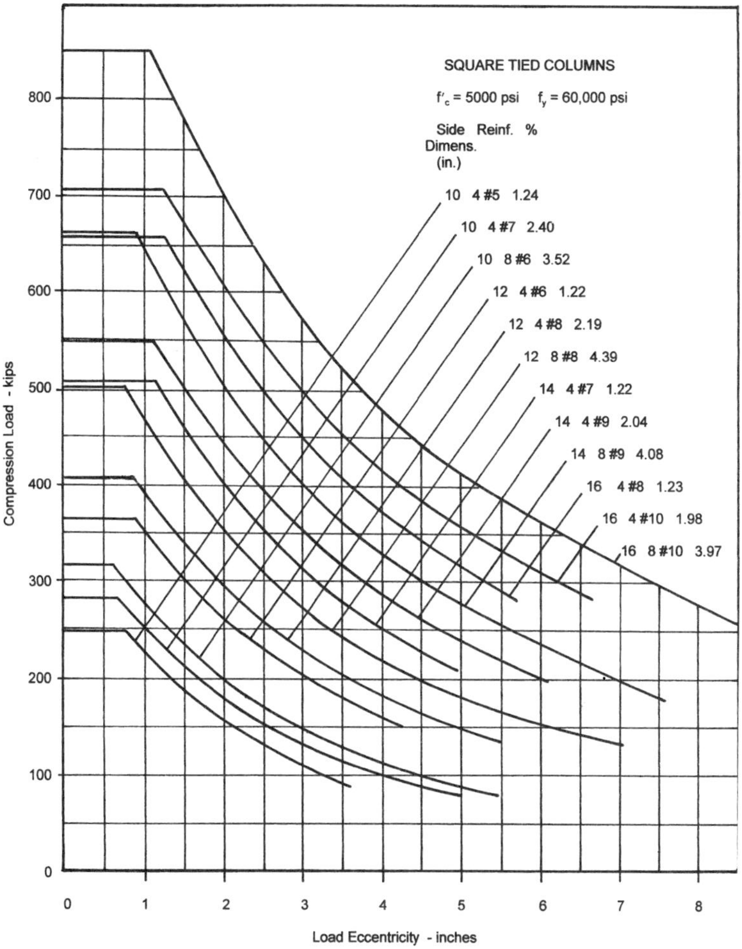

Figure 15.6 Maximum factored axial compression capacity for selected square tied columns.

ing moment combinations. The computed moment on the column is translated into an equivalent eccentric loading. Data for the curves were computed by strength design methods, as currently required by the ACI Code.

When bending moments are relatively high in comparison to axial loads, round or square column shapes are not the most efficient, just as they are not for spanning beams. Figure 15.8 gives safe ultimate factored

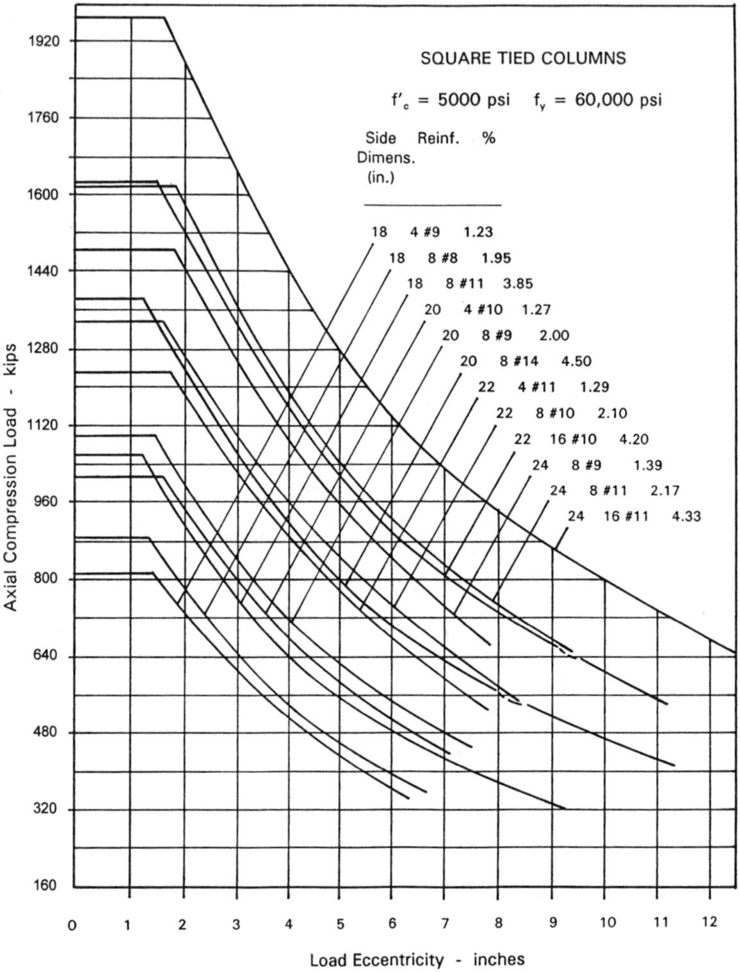

Figure 15.7 Maximum factored axial compression capacity for selected square tied columns.

capacities for columns with rectangular cross sections. To further emphasize the importance of major bending resistance, all the reinforcement is assumed to be placed on the narrow sides, thus utilizing it for its maximum bending resistance effect.

The following examples illustrate the use of Figures 15.6 through 15.8 for the design of square and rectangular tied columns.

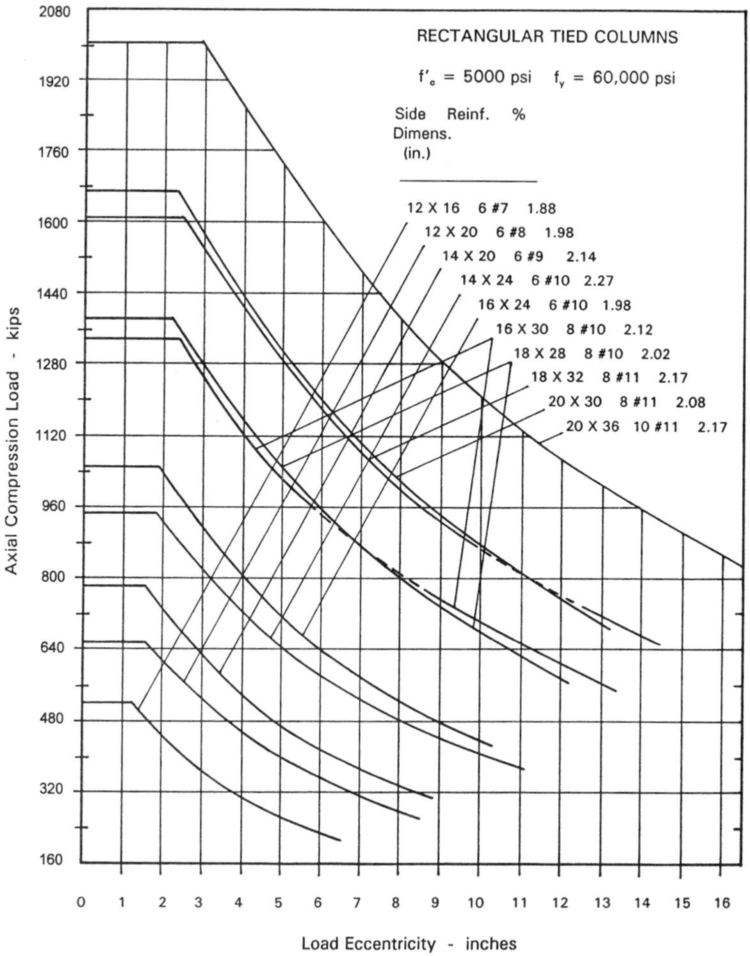

Figure 15.8 Maximum factored axial compression capacity for selected rectangular tied columns. Bending moment capacity determined for the major axis with reinforcement equally divided on the short sides of the section.

Example 1. A square tied column with $f'_c = 5$ ksi and steel with $f_y = 60$ ksi sustains an axial compression load of 150 kips dead load and 250 kips live load with no computed bending moment. Find the minimum practical column size if reinforcing is a maximum of 4% and the maximum size if reinforcing is a minimum of 1%.

Solution: As in all problems, this one begins with the determination of the factored ultimate axial load P_u.

$$P_u = 1.2P_{DL} + 1.6P_{LL} = (1.2 \times 150) + (1.6 \times 250) = 580 \text{ kips}$$

With no real consideration for bending, the maximum axial load capacity may be determined from the graphs by simply reading up the left edge of the figure. The curved lines actually end some distance from this edge because the code requires a minimum bending for all columns.

Using Figure 15.6, note that the minimum size is a 14-in. square column with eight No. 9 bars, for which the graph yields a maximum capacity of approximately 660 kips. Note that this column has a steel percentage of 4.08%.

What constitutes the maximum size is subject to some judement. Any column with a curve above that of the chosen minimum column will work. It becomes a matter of increasing redundancy of capacity. However, there are often other design considerations involved in developing a whole structural system, so these examples are quite academic. See the discussion for the example building in Chapter 20. For this example, it may be observed that the minimum choice (14-in. square) requires considerable reinforcement. Thus, going up to a 15-in. or 16-in. size will reduce the reinforcement. We may thus note from the limited choices in Figure 15.6: maximum size is 16 in. square with 4 No. 8 bars, capacity is 650 kips, $p_g = 1.23\%$. Because this is close to the usual recommended minimum reinforcement percentage (1%), columns of larger size will be increasingly redundant in strength (structurally oversized, in designer's lingo).

Example 2. A square tied column with $f'_c = 5$ ksi and steel with $f_y = 60$ ksi sustains an axial load of 150 kips dead load and 250 kips live load and a bending moment of 75 kip-ft dead load and 125 kip-ft live load. Determine the minimum size column and its reinforcement.

Solution: First determine the ultimate axial load and ultimate bending moment. From Example 1, $P_u = 580$ kips.

$$M_u = 1.2 \times M_{DL} + 1.6 \times M_{LL} = (1.2 \times 75) + (1.6 \times 125) = 290 \text{ kip-ft}$$

Next determine the equivalent eccentricity. Thus:

$$e = \frac{M_u}{P_u} = \frac{290 \times 12}{580} = 6 \text{ in.}$$

Then, from Figure 15.7, note that minimum size is 20-in. square with eight No. 14 bars; capacity at 6-in. eccentricity is approximately 700 kips. Also note that the steel percentage is 4.5%. If this is considered to be too high, use a 22 × 22 in. column with 4 No. 11 bars, which has a capacity of approximately 680 kips.

Example 3. Select the minimum size rectangular column from Figure 15.8 for the same data as used in Example 2.

Solution: With the factored axial load of 580 kips and the eccentricity of 6 in., Figure 15.8 yields the following: 16 × 24 in. column, six No. 10 bars, which has a capacity of approximately 640 kips.

It may be noted that there is a substantial savings in reinforcement with this selection, as compared to that from Example 2. The percentage of reduction may be determined as follows:

With eight No. 14 bars, $A_s = 8(2.25) = 18.0 \text{ in.}^2$

With six No. 10 bars, $A_s = 6(1.27) = 7.62 \text{ in.}^2$

Reduction in $A_s = 18.0 - 7.62 = 10.48 \text{ in.}^2$

% reduction = 100(10.48/18.0) = 58%

Round Columns

Round columns, as discussed previously, may be designed and built as spiral columns, or they may be developed as tied columns with bars in a rectangular layout or with the bars placed in a circle and held by a series of round circumferential ties. Because of the cost of spirals, it is usually more economical to use the tied column, so it is often used unless the additional strength or other behavioral characteristics of the spiral column are required.

Figure 15.9 gives safe loads for round columns that are designed as tied columns. As for the square and rectangular columns in Figures 15.6

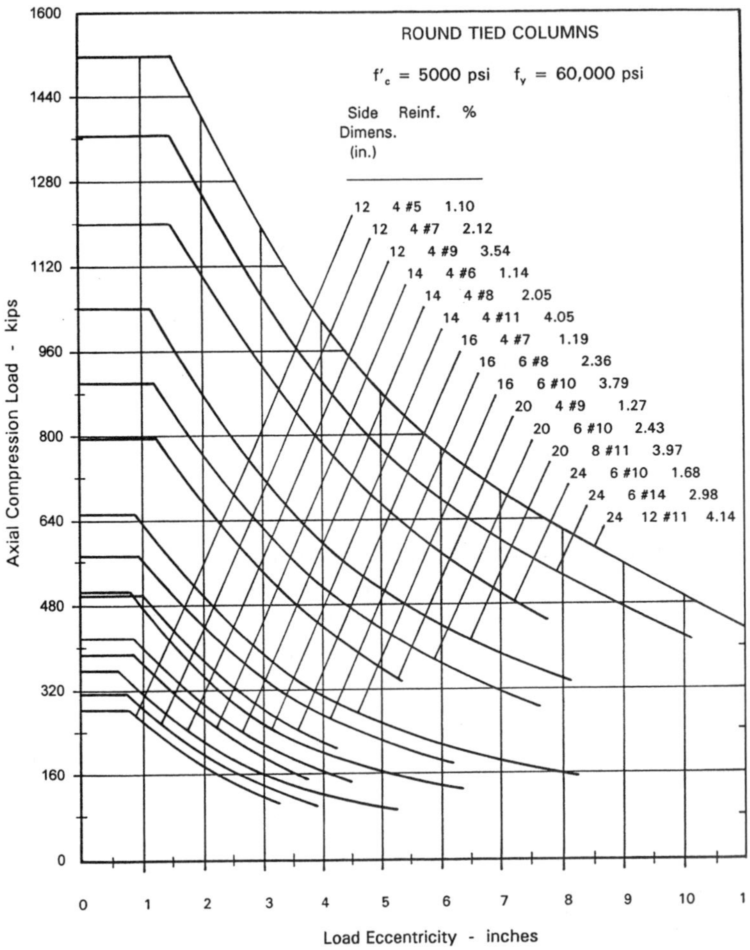

Figure 15.9 Maximum factored axial compression capacity for selected round tied columns.

through 15.8, load values have been adapted from values determined by strength design methods, and use is similar to that demonstrated in the preceding examples.

Problem 15.3.A, B, C. Using Figures 15.6 and 15.7, select the minimum size square tied column and its reinforcement for the following data.

Concrete Strength (psi)	Axial Load (kips)		Bending Moment (kip-ft)	
	Live	Dead	Live	Dead
A 5000	40	60	10	15
B 5000	40	60	15	25
C 5000	150	200	80	100

Problem 15.3.D, E, F. From Figure 15.8, determine minimum sizes for rectangular columns for the same data as in Problems 15.8A, B, and C. Also determine the percentage of savings in reinforcement as compared to the square columns, if any.

Problem 15.3.G, H ,I. Using Figure 15.9, pick the minimum size round column and its reinforcing for the load and moment combinations in Problems 15.2.A, B, and C.

15.4 SPECIAL PROBLEMS WITH CONCRETE COLUMNS

Slenderness

Cast-in-place concrete columns tend to be quite stout in profile so that slenderness related to buckling failure is much less often a critical concern than with columns of wood or steel. Earlier editions of the ACI Code provided for consideration of slenderness but permitted the issue to be neglected when the *L*/*r* of the column fell below a controlled value. For rectangular columns, this usually meant that the effect was ignored when the ratio of unsupported height to side dimension was less than about 12. This is roughly analogous to the case for the wood column with *L*/*d* less than 11.

Slenderness effects must also be related to the conditions of bending for the column. Because bending is usually induced at the column ends, the two typical cases are those shown in Figure 15.10. If a single end moment exists, or two equal end moments exist, as shown in Figure 15.10*a*, the buckling effect is magnified and the *P*-delta effect is maximum. The condition in Figure 15.10*a* is not the common case, however. More typical in framed structures is the condition shown in Figure 15.10*b*, for which the code treats the problem as one of moment magnification

When slenderness must be considered, the ACI Code provides procedures for a reduction of column axial load capacity. One should be aware, however, that reduction for slenderness is not considered in design aids such as tables or graphs.

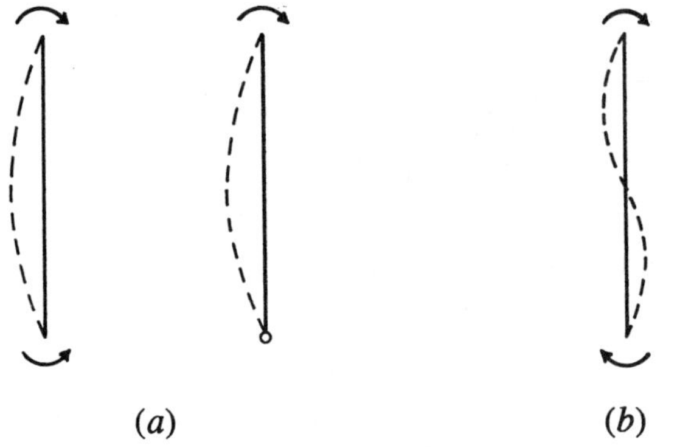

Figure 15.10 Assumed conditions of bending moments at column ends for consideration of column slenderness.

Development of Compressive Reinforcement

Development length in compression is a factor in column design and in the design of beams reinforced for compression.

The absence of flexural tension cracks in the portions of beams where compression reinforcement is employed, plus the beneficial effect of the end bearing of the bars on the concrete, permit shorter developmental lengths in compression than in tension. The ACI Code prescribes that l_d for bars in compression shall be computed by the formula

$$L_d = \frac{0.02 f_y d_b}{\sqrt{f'_c}}$$

but shall not be less than $0.0003 f_y d_b$ or 8 in., whichever is greater. Table 15.1 lists compression bar development lengths for a few combinations of specification data.

In reinforced columns, both the concrete and the steel bars share the compression force. Ordinary construction practices require the consideration of various situations for development of the stress in the reinforcing bars. Figure 15.11 shows a multistory concrete column with its base supported on a concrete footing. Referring to the illustration, note the following:

TABLE 15.1 Minimum Development Length for Compressive Reinforcement (in.)

	f_y = 40 ksi [276 MPa]		f_y = 60 ksi [414 MPa]		
Bar Size	f'_c = 3 ksi [20.7 MPa]	f'_c = 4 ksi [27.6 MPa]	f'_c = 3 ksi [20.7 MPa]	f'_c = 4 ksi [27.6 MPa]	f'_c = 5 ksi [34.5 MPa]
3	8	8	8	8	7
4	8	8	11	10	9
5	10	8	14	12	11
6	11	10	17	15	13
7	13	12	20	17	15
8	15	13	22	19	17
9	17	15	25	22	20
10	19	17	28	25	22
11	21	18	31	27	24
14			38	33	29
18			50	43	39

1. The concrete construction is ordinarily produced in multiple, separate pours, with construction joints between the separate pours occurring as shown in the illustration.
2. In the lower column, the load from the concrete is transferred to the footing in direct compressive bearing at the joint between the column and footing. The load from the reinforcing must be developed by extension of the reinforcing into the footing—distance L_1 in the illustration. Although it may be possible to place the column bars in position during casting of the footing to achieve this, the common practice is to use dowels, as shown in the illustration. These dowels must be developed on both sides of the joint: L_1 in the footing and L_2 in the column. If the f'_c value for both the footing and the column are the same, these two required lengths will be the same.
3. The lower column will ordinarily be cast together with the supported concrete framing above it, with a construction joint occurring at the top level of the framing (bottom of the upper column), as shown in the illustration. The distance L_3 is that required to develop the reinforcing in the lower column—bars *a* in the illustration. As for the condition at the top of the footing, the distance L_4 is required to develop the reinforcing in bars *b* in the upper column. L_4 is more likely to be the critical consideration for the determination of the extension required for bars *a*.

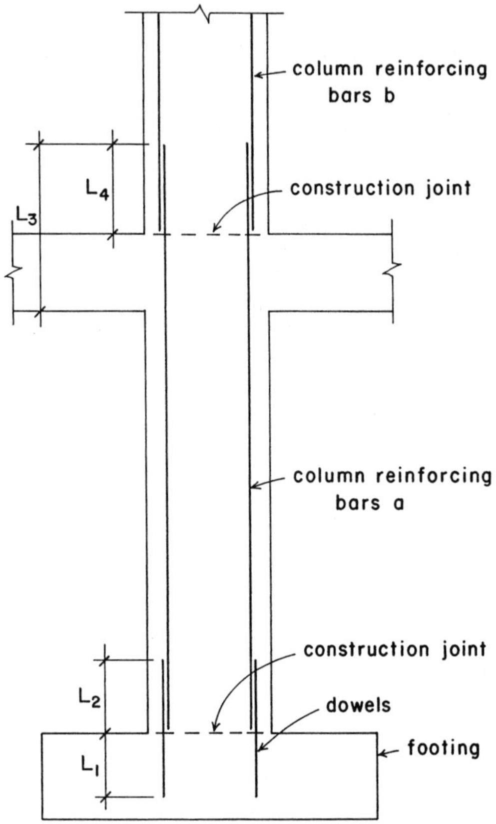

Figure 15.11 Considerations for bar development in concrete columns.

Developed Anchorage for Frame Continuity

In concrete rigid frame structures, the engagement of the frame members at joints between columns and beams requires special attention in the detailing of reinforcement. A particular concern is the potential for the beams to pull loose from the columns, an action typically resisted by the extended reinforcing bars from the beam ends. In addition to the usual concerns for bar development, some special detailing to enhance the anchoring of the bars may be indicated. This is a matter of particular note in resistance to seismic effects.

The concrete column as developed within a multistory rigid frame structure is presented in Chapter 20.

Vertical Concrete Compression Elements

Several types of construction elements are used to resist vertical compression for building structures. Dimensions of elements are used to differentiate between the defined elements. Figure 15.12 shows four such elements, described as follows:

Wall. Walls of one or more story height are often used as bearing walls, especially in concrete and masonry construction. Walls may be quite extensive in length but are also sometimes built in relatively short segments.

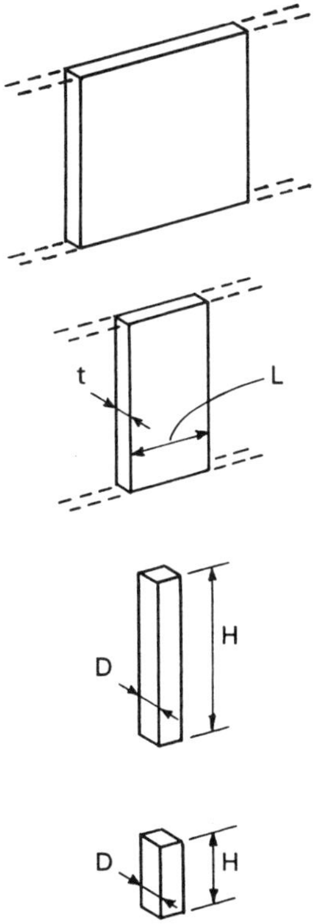

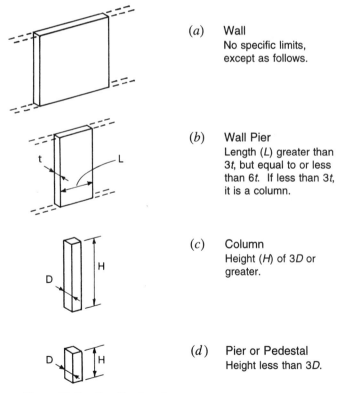

(*a*) Wall
No specific limits, except as follows.

(*b*) Wall Pier
Length (*L*) greater than 3*t*, but equal to or less than 6*t*. If less than 3*t*, it is a column.

(*c*) Column
Height (*H*) of 3*D* or greater.

(*d*) Pier or Pedestal
Height less than 3*D*.

Figure 15.12 Classification of concrete compression members.

Pier. When a segment of wall has a length that is less than six times the wall thickness, it is called a pier, or sometimes a wall pier.

Column. Columns come in many shapes but generally have some extent of height in relation to dimensions of the cross section. The usual limit for consideration as a column is a minimum height of three times the column diameter (side dimension, etc.). A wall pier may serve as a column, so the name distinction gets somewhat ambiguous.

Pedestal. A pedestal is really a short column (that is, a column with height not greater than three times its thickness). This element is also frequently called a pier, adding to the confusion of names.

To add more confusion, most large, relatively stout and massive concrete support elements are typically also called piers. These may be used to support bridges, longspan roof structures, or any other extremely heavy load. Identity in this case is more a matter of overall size than any specific proportions of dimensions. Bridge supports and supports for arch-type structures are also sometimes called *abutments*.

We also use the word *pier* for description of a type of foundation element that is also sometimes called a *caisson*. This consists essentially of a concrete column cast in a vertical shaft that is dug in the ground.

Walls, piers, columns, and pedestals may also be formed from concrete masonry units (CMUs). The pedestal, as used with foundation systems, is discussed in Chapter 16.

16

FOOTINGS

Almost every building has a foundation built into the ground as its base. In times past, stone or masonry construction was the usual form of this structure. Today, however, sitecast concrete is the basic choice. This is one of the most common and extensive uses for concrete in building construction.

16.1 SHALLOW BEARING FOUNDATIONS

The most common foundation consists of pads of concrete placed beneath the building. Because most buildings make a relatively shallow penetration into the ground, these pads, called footings, are generally classified as *shallow bearing foundations*. For simple economic reasons, shallow foundations are generally preferred. However, when adequate soil does not exist at a shallow location, driven piles or excavated piers (caissons), which extend some distance below the building, must be used; these are called *deep foundations*.

The two common footings are the wall footing and the column footing. Wall footings occur in strip form, usually placed symmetrically beneath the supported wall. Column footings are most often simple square pads supporting a single column. When columns are very close together or at the very edge of the building site, special footings that carry more than a single column may be used.

Two other basic construction elements that occur frequently with foundation systems are foundation walls and pedestals. Foundation walls may be used as basement walls or merely to provide a transition between more deeply placed footings and the aboveground building construction. Foundation walls are common with aboveground construction of wood or steel because these constructions must be kept from contact with the ground.

Pedestals are actually short columns used as transitions between the building columns and their bearing footings. These may also be used to keep wood or steel columns above ground, or they may serve a structural purpose to facilitate the transfer of a highly concentrated force from a column to a widely spread footing.

The remainder of this chapter presents design considerations for wall footings, column footings, and pedestals.

16.2 WALL FOOTINGS

Wall footings consist of concrete strips placed under walls. The most common type is that shown in Figure 16.1, consisting of a strip with a rectangular cross section placed in a symmetrical position with respect to the wall and projecting an equal distance as a cantilever from both faces of the wall. For soil pressure, the critical dimension of the footing is its width as measured perpendicular to the wall.

Footings ordinarily serve as construction platforms for the walls they support. Thus, a minimum width is established by the wall thickness plus a few inches on each side. The extra width is necessary because of the crude form of foundation construction but also may be required for support of forms for concrete walls. A minimum projection of 2 in. is recommended for masonry walls and 3 in. for concrete walls.

With relatively light vertical loads, the minimum construction width may be adequate for soil bearing. Walls ordinarily extend some distance below grade, and allowable bearing will usually be somewhat higher than for very shallow footings. With the minimum recommended footing width, the cantilever bending and shear will be negligible, so no trans-

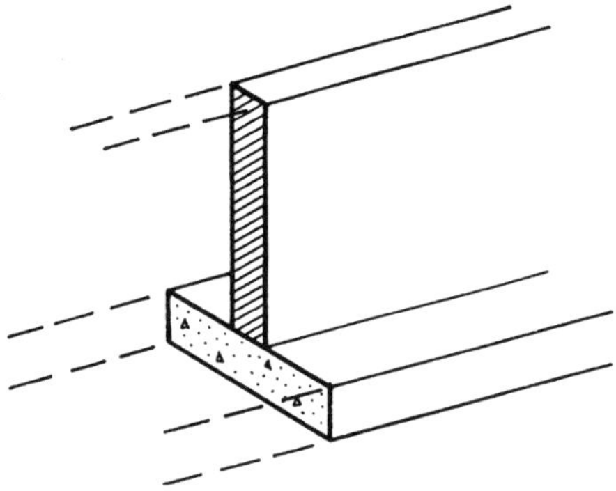

Figure 16.1 Typical form of a strip wall footing.

verse (perpendicular to the wall) reinforcement is used. However, some longitudinal reinforcement is recommended.

As the wall load increases and a wider footing is required, the transverse bending and shear require some reinforcement. At some point, the increased width also determines a required thickness. Otherwise, recommended minimum thickness is 8 in. for nonreinforced footings and 10 in. for reinforced footings.

Determination of the Footing Width

Footing width is determined by soil pressure, assuming that the minimum width required for construction is not adequate for bearing. Because footing weight is part of the total load on the soil, the required width cannot be precisely determined until the footing thickness is known. A common procedure is to assume a footing thickness, design for the total load, verify the structural adequacy of the thickness, and, if necessary, modify the width once the final thickness is determined. The current ACI Code (Ref. 8) calls for using the unfactored loading (service load) when determining footing width by soil pressure. Example 1 demonstrates this procedure.

Determination of the Footing Thickness

If the footing has no transverse reinforcing, the required thickness is determined by the tension stress limit of the concrete, in either flexural stress or diagonal stress due to shear. Transverse reinforcement is not required until the footing width exceeds the wall thickness by some significant amount, usually 2 ft or so. A good rule of thumb is to provide transverse reinforcement only if the cantilever edge distance for the footing (from the wall face to the footing edge) exceeds the footing thickness. For average conditions, this means for footings of about 3 ft width or greater.

If transverse reinforcement is used, the critical concerns become for shear in the concrete and tension stress in the reinforcing. Thicknesses determined by shear will usually ensure a low bending stress in the concrete, so the cantilever beam action will involve a very low percentage of steel. This is in keeping with the general rule for economy in foundation construction, which is to reduce the amount of reinforcement to a minimum.

Minimum footing thicknesses are a matter of design judgment, unless limited by building codes. The ACI Code recommends limits of 8 in. for unreinforced footings and 10 in. for footings with transverse reinforcement. Another possible consideration for the minimum footing thickness is the necessity for placing dowels for wall reinforcement.

Selection of Reinforcement

Transverse reinforcement is determined on the basis of flexural tension and development length resulting from the cantilever action. Longitudinal reinforcement is usually selected on the basis of providing minimum shrinkage reinforcement. A reasonable value for the latter is a minimum of 0.0015 times the gross concrete area (area of the cross section of the footing). Cover requirements are for 2 in. from formed edges and 3 in. from surfaces generated without forming (such as the footing bottom). For practical purposes, it may be desirable to coordinate the spacing of the footing transverse reinforcement with that of any dowels for wall reinforcement. Reinforcement for shear is required by code only when the ultimate shear capacity as a result of factored loading (V_u) is greater than the factored shear capacity of the concrete (ϕV_c).

The following example illustrates the design procedure for a reinforced wall footing. Data for predesigned footings are given in Table 16.1. Figure 16.2 provides an explanation of the table entries. Using unreinforced footings is not recommended for footings greater than 3 ft in width. *Note*: In using the strip method, a strip width of 12 in. is an obvious choice when using U.S. units, but not with metric units. To save space, the computations are performed with U.S. units only, but some metric equivalents are given for key data and answers.

Example 1. Design a wall footing with transverse reinforcement for the following data:

Footing design load = 3750 lb/ft [5.47 kN/m] dead load and 5000 lb/ft [7.30 kN/m] live load of wall length.

Wall thickness for design = 6 in. [150 mm].

Maximum soil pressure = 2000 psf [100 kPa].

Concrete design strength = 2000 psi [14 MPa].

Steel yield stress = 40,000 psi [280 MPa]

Solution: For the reinforced footing, shear is the only concrete stress of concern. Concrete flexural stress will be low because of the low percentage of reinforcement. As with the unreinforced footing, the usual design procedure consists of making a guess for the footing thickness, determining the required width for soil pressure, and then checking the footing stress.

Try h = 12 in. Then, footing weight = 150 psf, and the net usable soil pressure is 2000 − 150 = 1850 lb/ft^2

The footing design load is unfactored when we are determining footing width; therefore, the load is 8750 lb/ft [11.9 kN/m]. Required footing width is 8750/1850 = 4.73 ft, or 4.73(12) = 56.8 in.—say 57 in.—or 4 ft 9 in., or 4.75 ft. With this width, the design soil pressure for stress is 8750/4.75 = 1842 psf.

For the reinforced footing, we must determine the effective depth—that is, the distance from the top of the footing to the center of the steel bars. For a precise determination, this requires a second guess: the steel bar diameter (D). For the example a guess is made of a No. 6 bar with a diameter of 0.75 in. With the cover of 3 in., this produces an effective depth of $d = h - 3 - (D/2) = 12 - 3 - (0.75/2) = 8.625$ in.

Concern for precision is academic in footing design, however, considering the crude nature of the construction. The footing bottom is formed by a hand-dug soil surface, unavoidably roughed up during placing of the reinforcement and casting of the concrete. The value of d will therefore be taken as 8.6 in.

Next, we need to determine how much the soil is pushing back up on the footing using the factored loads. This load is

$$w_u = (1.2 \times 3750) + (1.6 \times 5000) = 12500 \text{ lb/ft}$$

TABLE 16.1 Allowable Loads on Wall Footings (see Figure 16.2)

Maximum Soil Pressure (lb/ft^2)	Minimum Wall Thickness, t (in.)		Allowable Load on Footing[a] (lb/ft)	Footing Dimensions (in.)		Reinforcement	
	Concrete	Masonry		h	w	Long Direction	Short Direction
1000	4	8	2,625	10	36	3 No. 4	No. 3 at 17
	4	8	3,062	10	42	2 No. 5	No. 3 at 12
	6	12	3,500	10	48	4 No. 4	No. 4 at 18
	6	12	3,938	10	54	3 No. 5	No. 4 at 13
	6	12	4,375	10	60	3 No. 5	No. 4 at 10
	6	12	4,812	10	66	5 No. 4	No. 5 at 13
	6	12	5,250	10	72	4 No. 5	No. 5 at 11
1500	4	8	4,125	10	36	3 No. 4	No. 3 at 11
	4	8	4,812	10	42	2 No. 5	No. 4 at 14
	6	12	5,500	10	48	4 No. 4	No. 4 at 11
	6	12	6,131	11	54	3 No. 5	No. 5 at 16
	6	12	6,812	11	60	5 No. 4	No. 5 at 12
	6	12	7,425	12	66	4 No. 5	No. 5 at 11
	8	16	8,100	12	72	5 No. 5	No. 5 at 10
2000	4	8	5,625	10	36	3 No. 4	No. 4 at 15
	6	12	6,562	10	42	2 No. 5	No. 4 at 12
	6	12	7,500	10	48	4 No. 4	No. 5 at 13
	6	12	8,381	11	54	3 No. 5	No. 5 at 12
	6	12	9,520	12	60	4 No. 5	No. 5 at 10
	8	16	10,106	13	66	4 No. 5	No. 5 at 10
	8	16	10,875	15	72	6 No. 5	No. 5 at 10
3000	6	12	8,625	10	36	3 No. 4	No. 4 at 11
	6	12	10,019	11	42	4 No. 4	No. 5 at 14
	6	12	11,400	12	48	3 No. 5	No. 5 at 11
	6	12	12,712	14	54	6 No. 4	No. 5 at 11
	8	16	14,062	15	60	5 No. 5	No. 5 at 10
	8	16	15,400	16	66	5 No. 5	No. 6 at 13
	8	16	16,725	17	72	6 No. 5	No. 6 at 11

[a] Allowable loads do not include the weight of the footing, which has been deducted from the total bearing capacity. Criteria: f'_c = 2000 psi, Grade 40 bars.

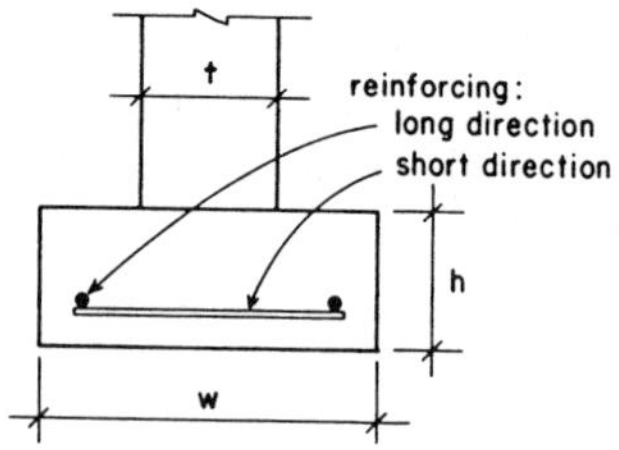

Figure 16.2 Reference Figure for Table 16.1.

With a width of 4.75 ft, the factored design soil pressure is

$$p_d = \frac{12{,}500}{4.75} = 2630 \text{ psf}$$

The critical section for shear stress is taken at a distance of *d* from the face of the wall. As shown in Figure 16.3a, this places the shear section at a distance of 16.9 in. from the footing edge. At this location, the shear force is determined as

$$V_u = (2630 \text{ plf}) \times (16.9 \text{ in.}) \times \left(\frac{1 \text{ ft}}{12 \text{ in.}}\right) = 3700 \text{ lb}$$

and the shear capacity of the concrete is

$$\phi V_c = 0.75\,(2\sqrt{f'_c})(b \times d) = 0.75(2\sqrt{2000})(12 \times 8.6) = 6920 \text{ lb}$$

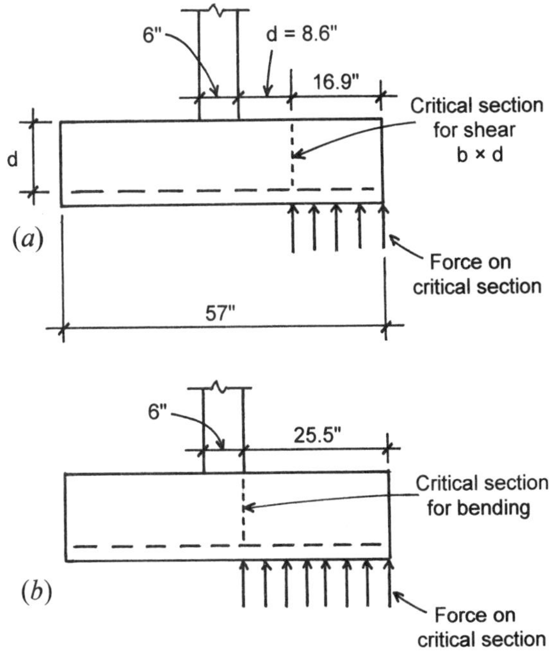

Figure 16.3 Shear considerations for the wall footing.

It is possible, therefore, to reduce the footing thickness. However, cost-effectiveness is usually achieved by reducing the steel reinforcement to a minimum. Low-grade concrete dumped into a hole in the ground is quite inexpensive, compared to the cost of steel bars. Selection of footing thickness therefore becomes a matter of design judgment, unless the footing width becomes as much as five times or so the wall thickness, at which point stress limits may become significant.

If a thickness of 11 in. is chosen for this example, the shear capacity will decrease only slightly and the required footing width will remain effectively the same. A new effective depth of 7.6 in. will be used, but the design soil pressure for stresses will remain the same because it relates only to the width of the footing.

The bending moment to be used for determination of the steel bars is computed as follows (see Figure 16.3*b*). The force on the cantilevered edge of the footing is

$$F = \frac{25.5}{12} \times 2630 = 5589 \text{ lb}$$

and the cantilever bending moment at the wall face is thus

$$M_u = 5589 \times \frac{25.5}{2} = 71{,}260 \text{ in.-lb}$$

$$M_t = \frac{M_u}{\phi} = \frac{71{,}260}{0.9} = 79{,}200 \text{ lb-in.}$$

and the required steel area per foot of wall length is

$$A_s = \frac{M_t}{f_y jd} = \frac{79{,}200}{40{,}000 \times 0.9 \times 7.6} = 0.290 \text{ in.}^2$$

The spacing required for a given bar size to satisfy this requirement can be derived as follows:

$$\text{Required spacing} = (\text{Area of bar}) \times \frac{12}{\text{Required area/ft}}$$

Thus, for a No. 3 bar:

$$s = 0.11 \times \frac{12}{0.290} = 4.6 \text{ in.}$$

Using this procedure, note that the required spacings for bar sizes 3 through 7 are shown in the fourth column of Table 16.2. Bar sizes and spacings can be most easily selected using handbook tables that yield the average steel areas for various combinations of bar size and spacing. One such table is Table 13.5, from which the spacings shown in the last column of Table 16.2 were selected, indicating a range of choices for the footing transverse reinforcement. Selection of the actual bar size and spacing is a matter of design judgment. Some considerations are the following:

1. Maximum recommended spacing is 18 in.
2. Minimum recommended spacing is 6 in., to reduce the number of bars and make placing of concrete easier.
3. Preference is for smaller bars as long as spacing is not too close.
4. A practical spacing may be that of the spacing of vertical reinforcement in the supported wall, for which footing dowels are required (or some full number multiple or division of the wall bar spacing).

With these considerations in mind, we may choose either the No. 5 bars at 12.5 in. spacing or the No. 4 bars at 8 in. spacing. The No. 6 bars at 24-in. spacing would not be a good choice because of the large distance between bars; thus, the practice is to use a maximum spacing of 18 in. for any spaced set of bars. Another consideration that must be made for the choice of reinforcement regards the required development length

TABLE 16.2 Selection of Reinforcement for Example 1

Bar Size	Area of Bar (in.^{2})	Area Required for Bending (in.^{2})	Bar Spacing Required (in.)	Bar Spacing Selected (in.)
3	0.11	0.290	4.6	4.5
4	0.20	0.290	8.3	8.0
5	0.31	0.290	12.8	12.5
6	0.60	0.290	24.8	18.0

for anchorage. With 2 in. of edge cover, the bars will extend 23.5 in. from the critical bending section at the wall face (see Figure 16.3*b*). Inspection of Table 13.7 will show that this is an adequate length for all the bar sizes used in Table 16.2. Note that the placement of the bars in the footing falls in the classification of "Other Bars" in Table 13.7.

For the longitudinal reinforcement the minimum steel area is

$$A_s = (0.0015)(11)(57) = 0.94 \text{ in.}^2$$

Using three No. 5 bars yields

$$A_s = (3)(0.31) = 0.93 \text{ in.}^2$$

Table 16.1 gives values for wall footings for four different soil pressures. Table data were derived using the procedures illustrated in the example. Figure 16.2 shows the dimensions referred to in the table.

Problem 16.2.A. Using concrete with a design strength of 2000 psi [13.8 MPa] and grade 40 bars with a yield strength of 40 ksi [276 MPa], design a wall footing for the following data: wall thickness = 8 in. [203 mm]; dead load on footing = 4000 lb/ft [58.4 kN/m] and live load = 6500 lb/ft [94.7 kN/m]; maximum soil pressure = 2000 psf [96 kN/m^2].

Problem 16.2.B. Same as Problem 16.2.A, except wall is 15 in. [380 mm] thick, dead load is 6000 lb/ft [87.5 kN/m] and live load is 8000 [117 kN/m], and maximum soil pressure is 3000 psf [144 kN/m^2].

16.3 COLUMN FOOTINGS

The great majority of independent or isolated column footings are square in plan, with reinforcement consisting of two equal sets of bars at right angles to each other. The column may be placed directly on the footing, or it may be supported by a pedestal, consisting of a short column that is wider than the supported column. The pedestal helps to reduce the so-called *punching shear* effect in the footing; it also slightly reduces the edge cantilever distance and thus the magnitude of bending in the footing. The pedestal thus allows for a thinner footing and slightly less footing reinforcement. However, another reason for using a pedestal may be to raise the bottom of the supported column above the ground, which is important for columns of wood and steel.

The design of a column footing is based on the following considerations:

Maximum soil pressure. The sum of the unfactored superimposed load on the footing and the unfactored weight of the footing must not exceed the limit for bearing pressure on the supporting soil material. The required total plan area of the footing is derived on this basis.

Design soil pressure. By itself, simply resting on the soil, the footing does not generate shear or bending stresses. These are developed only by the superimposed load. Thus, the soil pressure to be used for designing the footing is determined as the factored superimposed load divided by the actual chosen plan area of the footing.

Control of settlement. Where buildings rest on highly compressible soil, it may be necessary to select footing areas that ensure a uniform settlement of all the building foundation supports. For some soils, long-term settlement under dead load only may be more critical in this regard and must be considered as well as maximum soil pressure limits.

Size of the column. The larger the column, the less will be the shear and bending stresses in the footing, because these are developed by the cantilever effect of the footing projection beyond the edges of the column.

Shear capacity limit for the concrete. For square-plan footings, this is usually the only critical stress in the concrete. To achieve an economical design, the footing thickness is usually chosen to reduce the need for reinforcement. Although small in volume, the steel reinforcement is a major cost factor in reinforced concrete construction. This generally rules against any concerns for flexural compression stress in the concrete. As with wall footings, the factored load is used when determining footing thickness and any required reinforcement

Flexural tension stress and development length for the bars. These are the main concerns for the steel bars, on the basis of the cantilever bending action. It is also desired to control the spacing of the bars between some limits.

Footing thickness for development of column bars. When a footing supports a reinforced concrete or masonry column, the compressive force in the column bars must be transferred to the footing by development action (called *doweling*), as discussed in Chapter 13 and in Section 15.4. The thickness of the footing must be adequate for this purpose.

The following example illustrates the design process for a simple, square column footing.

Example 2. Design a square column footing for the following data:

Column load = 200 kips [890 kN] dead load and 300 kips [1334 kN] live load

Column size = 15 in. [380 mm] square

Maximum allowable soil pressure = 4000 psf [200 kPa]

Concrete design strength = 3000 psi [21 MPa]

Yield stress of steel reinforcement = 40 ksi [280 MPa]

Solution: A quick guess for the footing size is to divide the load by the maximum allowable soil pressure. Thus:

$$A = \frac{500}{4} = 125 \text{ ft}^2, \qquad w = \sqrt{125} = 11.2 \text{ ft}$$

This does not allow for the footing weight, so the actual size required will be slightly larger. However, it gets the guessing quickly into the approximate range.

For a footing this large, the first guess for the footing thickness is a real shot in the dark. However, any available references to other footings designed for this range of data will provide some reasonable first guess.

Try $h = 31$ in. Then, footing weight = (31/12)(150) = 388 psf. Net usable soil pressure = 4000 – 388 = 3612. The required plan area of the footing is thus:

$$A = \frac{500{,}000}{3612} = 138.4 \text{ ft}^2$$

and the required width for a square footing is

$$w = \sqrt{138.4} = 11.76 \text{ ft}$$

Try w = 11 ft 9 in., or 11.75 ft. Then, design soil pressure = 500,000/$(11.75)^2$ = 3622 psf.

For determining reinforcement and footing thickness, a factored soil pressure is needed. The weight of the footing needs to be added into the dead load.

$$P_u = 1.2 \times P_{DL} + 1.6 \times P_{LL} = 1.2(200 + 54) + 1.6(300) = 785 \text{ kips}$$

and

$$W_u = \frac{785}{(11.75)^2} = 5.69 \text{ ksf or } 5690 \text{ psf}$$

Determination of the bending force and moment are as follows (See Figure 16.4):

Bending force:

$$F = 5690 \times \frac{63}{12} \times 11.75 = 351{,}000 \text{ lb}$$

Bending moment:

$$M_u = 351{,}000 \times \frac{63}{12} \times \frac{1}{2} = 921{,}000 \text{ ft-lb}$$

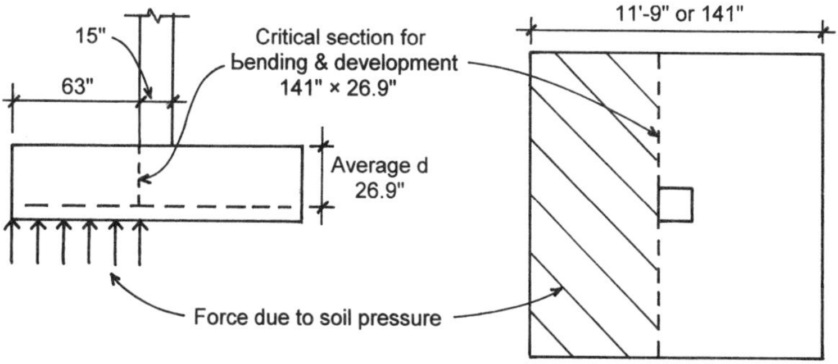

Figure 16.4 Considerations for bending and bar development in the column footing.

and for design:

$$M_t = \frac{M_u}{\phi} = \frac{921{,}000}{0.9} = 1{,}024{,}000 \text{ ft-lb}$$

We assume this bending moment operates in both directions on the footing and is provided for with similar reinforcement in each direction. However, it is necessary to place one set of bars on top of the perpendicular set, as shown in Figure. 16.5, and there are thus different effective depths in each direction. A practical procedure is to use the average of these two depths—that is, a depth equal to the footing thickness minus the 3-in. cover and one bar diameter. This will theoretically result in a minor overstress in one direction, which is compensated for by a minor understress in the other direction.

It is also necessary to assume a size for the reinforcing bar in order to determine the effective depth. As with the footing thickness, this must be a guess, unless some reference is used for approximation. Assuming a No. 9 bar for this footing, the effective depth thus becomes

$$d = h - 3 - (\text{bar } D) = 31 - 3 - 1.13 = 26.87 \text{ in., say } 26.9 \text{ in.}$$

The section resisting the bending moment is one that is 141 in. wide and has a depth of 26.9 in. Using a resistance factor for a balanced section from Table 13.2, the balanced moment capacity of this section is determined as follows:

$$M_R = Rbd^2 = \frac{1149 \times 141 \times (26.9)^2}{12} = 9{,}770{,}000 \text{ ft-lb}$$

which is almost ten times the required moment.

From this analysis, it may be seen that the compressive bending stress in the concrete is not critical. Furthermore, the section may be classified as considerably underreinforced, and a conservative value can be used for j in determining the required reinforcement.

The critical stress condition in the concrete is that of shear, either in beam-type action or in punching action. Referring to Figure 16.6, note that the investigation for these two conditions is as follows:

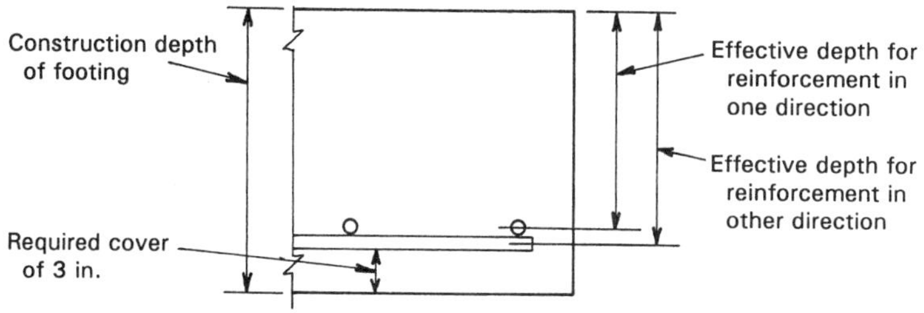

Figure 16.5 Consideration for effective depth of the column footing with two-way reinforcement.

For beam-type shear (Figure 16.6a):

$$V_u = 5690 \times 11.75 \times \frac{36.1}{12} = 201{,}000 \text{ lb}$$

For the shear capacity of the concrete:

$$V_c = 2\sqrt{f'_c}(b \times d) = 2\sqrt{3000}\,(141 \times 26.9) = 415{,}000 \text{ lb}$$

$$\phi V_c = 0.75(415{,}000) = 311{,}000 \text{ lb}$$

For punching shear (Figure 16.6b):

$$V_u = 5690\left[(11.75)^2 - \left(\frac{41.9}{12}\right)^2\right] = 716{,}000 \text{ lb}$$

Shear capacity of the concrete is

$$V_c = 4\sqrt{f'_c}(b \times d) = 4\sqrt{3000}\,(4 \times 41.9)(26.9) = 988{,}000 \text{ lb}$$

$$\phi V_c = 0.75 \times 988{,}000 = 741{,}000 \text{ lb}$$

Although the beam shear stress is low, the punching shear stress is just barely short of the limit, so the 31-in. thickness is indeed the minimum allowable dimension.

Using an assumed value of 0.9 for j, the area of steel required is determined as

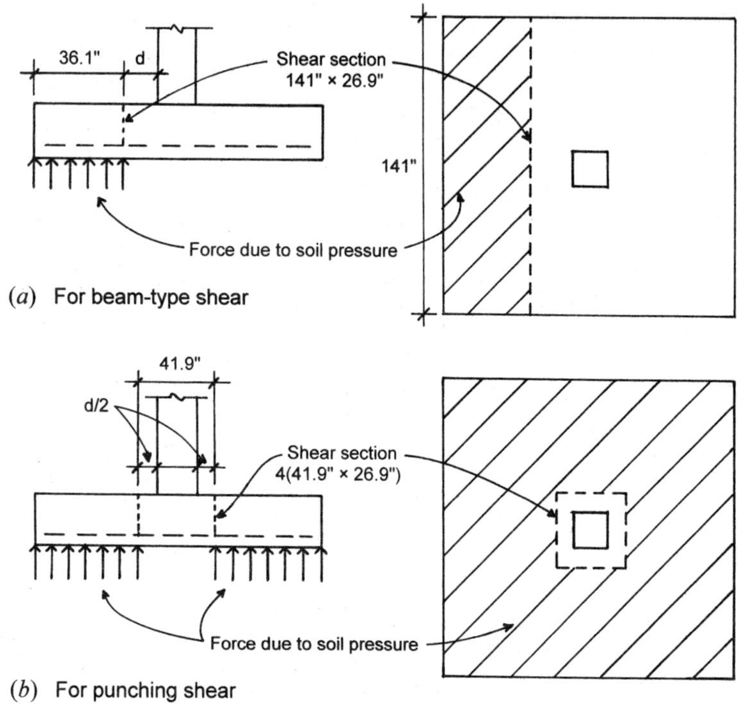

Figure 16.6 Considerations for the two forms of shear development in the column footing: (*a*) beam-type shear; and (*b*) punching shear.

$$A_s = \frac{M}{f_y jd} = \frac{1{,}024{,}000 \times 12}{40{,}000 \times 0.9 \times 26.9} = 12.7 \text{ in.}^2$$

We may select a number of combinations of bar size-and-number combinations to satisfy this area requirement. A range of possible choices is shown in Table 16.3. Also displayed in the table are data relating to two other considerations for the bar choice: the center-to-center spacing of the bars and the development lengths required. Spacings given in the table assume the first bar to be centered at 4 in. from the footing edge. Maximum spacing should be limited to 18 in. and minimum to about 6 in.

Required development lengths are taken from Table 13.7. The development length available is a maximum of the distance from the column face to the footing edge minus a 2-in. cover—in this case, a distance of 61 in.

TABLE 16.3 Reinforcement Alternatives for the Column Footing

Number and Size of Bars	Area of Steel Provided (Required = 12.7 in.^{2}) in.^{2}	mm^2	Required Development Length[a] in.	mm	Center-to-Center Spacing in.	mm
21 No. 7	12.6	8130	32	813	6.6	168
16 No. 8	12.6	8130	37	940	8.8	224
13 No. 9	13.0	8388	42	1067	11.0	279
10 No. 10	12.7	8194	47	1194	14.7	373
9 No. 11	14.0	9033	52	1321	16.5	419

[a]From Table 13.8, values for "other bars," $f_y = 40$ ksi, $f'_c = 3$ ksi.

Inspection of Table 16.3 reveals that all the combinations given are acceptable. In most cases, designers prefer to use the largest possible bar in the fewest number because handling of the bars is simplified with fewer bars, which is usually a savings of labor time and cost.

Although the computations have established that the 31-in. dimension is the least possible thickness, it may be more economical to use a thicker footing with less reinforcement, assuming the usual ratio of costs of concrete and steel. In fact, if construction cost is the major determinant, the ideal footing is the one with the lowest combined cost for excavation, forming, concrete, and steel.

One possible limitation for the footing reinforcement is the total percentage of steel. If this is excessively low, the section is hardly being reinforced. ACI Code stipulates that the minimum reinforcement be the same as that for temperature reinforcement in slabs, a percentage of $0.002A_g$ for Grade 40 bars and $0.0015A_g$ for Grade 60 bars. For this footing cross section of 141 in. by 31 in. with Grade 40 bars, this means an area of

$$A_s = 0.002(141 \times 31) = 8.74 \text{ in.}^2$$

A number of other considerations may affect the selection of footing dimensions, such as the following:

Restricted thickness. Footing thickness may be restricted by excavation problems, water conditions, or presence of undesirable soil materials at lower strata. Thickness may be reduced by use of pedestals, as discussed in Section 16.4.

Need for dowels. When the footing supports a reinforced concrete or masonry column, dowels must be provided for the vertical column reinforcement, with sufficient extension into the footing for development of the bars. This problem is discussed in Section 15.4.

Restricted footing width. Proximity of other construction or close spacing of columns sometimes makes it impossible to use the required square footing. For a single column, a possible solution is the use of an oblong (called a rectangular) footing. For multiple columns, a combined footing is sometimes used. A special footing is the cantilever footing, used when footings cannot extend beyond the building face. An extreme case occurs when the entire building footprint must be used in a single large footing, called a mat foundation.

Table 16.4 yields the allowable superimposed load for a range of predesigned footings and soil pressures. This material has been adapted from more extensive data in *Simplified Design of Building Foundations* (Ref. 10). Designs are given for footings using concrete strengths of 2000 psi and 3000 psi. Figure 16.7 indicates the symbols used for dimensions in Table 16.4. As discussed for the wall footings and elsewhere in this book, the low design strength of 2000 psi may sometimes be used to avoid the necessity for the usual code-required field testing of concrete. However, no structural concrete should be *specified* with a strength less than 3000 psi.

Problem 16.3.A. Design a square footing for a 14-in. [356 mm] square column and a superimposed dead load of 100 kips [445 kN] and a live load of 100 kips [445 kN]. The maximum permissible soil pressure is 3000 psf [144 kPa]. Use concrete with a design strength of 3 ksi [20.7 MPa] and grade 40 reinforcing bars with yield strength of 40 ksi [276 MPa].

Problem 16.3.B. Same as Problem 16.3.A, except column is 18-in. [457 mm], dead load is 200 kips [890 kN] and live load is 300 kips [1334 kN], and permissible soil pressure is 4000 psf [192 MPa].

16.4 PEDESTALS

A pedestal (also called a pier) is defined by the ACI Code as a short compression member whose height does not exceed three times its width. Pedestals are frequently used as transitional elements between columns and the bearing footings that support them. Figure 16.8 shows the use of pedestals with both steel and reinforced concrete columns. Following are the most common reasons for use of pedestals:

TABLE 16.4 Safe Loads for Square Column Footings[a] (see Figure 16.7)

Maximum Soil Pressure (psf)	Minimum Column Width, t (in.)	Service Load on Footing (kips)	Dimensions h (in.)	Dimensions w (ft)	Reinforcement Each Way
1000	8	7	10	3	3 No. 2
	8	10	10	3.5	3 No. 3
	8	14	10	4	4 No. 3
	8	17	10	4.5	4 No. 4
	8	21	10	5	4 No. 5
	8	31	10	6	4 No. 6
	8	42	11	7	6 No. 6
1500	8	12	10	3	3 No. 3
	8	16	10	3.5	3 No. 4
	8	22	10	4	4 No. 4
	8	27	10	4.5	4 No. 5
	8	34	10	5	5 No. 5
	8	49	12	6	5 No. 6
	8	65	13	7	5 No. 7
	8	84	15	8	7 No. 7
	8	105	17	9	8 No. 7
2000	8	16	10	3	3 No. 3
	8	23	10	3.5	3 No. 4
	8	30	10	4	5 No. 4
	8	38	10	4.5	5 No. 5
	8	46	11	5	4 No. 6
	8	66	13	6	6 No. 6
	8	89	15	7	6 No. 7
	8	114	17	8	8 No. 7
	8	143	19	9	7 No. 8
	10	175	20	10	9 No. 8
3000	8	25	10	3	3 No. 4
	8	35	10	3.5	3 No. 5
	8	45	11	4	4 No. 5
	8	57	12	4.5	4 No. 6
	8	71	13	5	5 No. 6
	8	101	15	6	7 No. 6
	10	136	17	7	7 No. 7
	10	177	20	8	7 No. 8
	12	222	21	9	9 No. 8
	12	272	24	10	9 No. 9
	12	324	26	11	10 No. 9
	14	383	28	12	10 No. 10

(*continued*)

TABLE 16.4 (*Continued*)

Maximum Soil Pressure (psf)	Minimum Column Width, *t* (in.)	Service Load on Footing (kips)	Dimensions *h* (in.)	Dimensions *w* (ft)	Reinforcement Each Way
4000	8	34	10	3	4 No. 4
	8	47	11	3.5	4 No. 5
	8	61	12	4	5 No. 5
	8	77	13	4.5	5 No. 6
	8	95	15	5	5 No. 6
	8	136	18	6	6 No. 7
	10	184	20	7	8 No. 7
	10	238	23	8	8 No. 8
	12	300	25	9	8 No. 9
	12	367	27	10	10 No. 9
	14	441	29	11	10 No. 10
	14	522	32	12	11 No. 10
	16	608	34	13	13 No. 10
	16	698	37	14	13 No. 11
	18	796	39	15	14 No. 11

[a]Service loads do not include the weight of the footing, which has been deducted from the total bearing capacity. Service load is considered 40% dead load and 60% live load. Grade 40 reinforcement. f'_c = 3 ksi.

1. To spread the load on top of the footing. This may relieve the intensity of direct bearing pressure on the footing or may simply permit a thinner footing with less reinforcing due to the wider column.
2. To permit the column to terminate at a higher elevation where footings must be placed at depths considerably below the lowest parts of the building. This is generally most significant for steel columns.

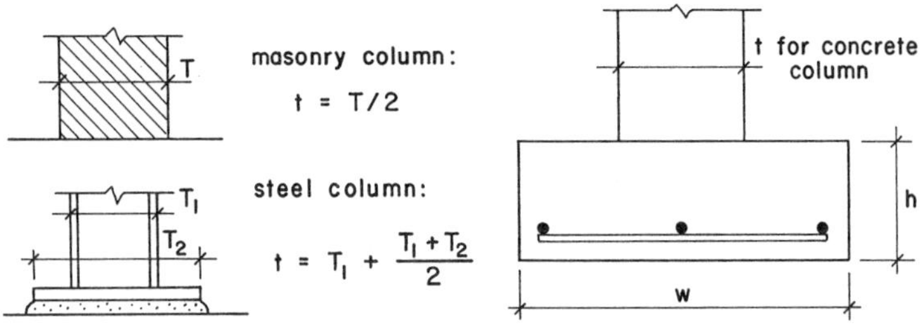

Figure 16.7 Reference Figure for Table 16.4.

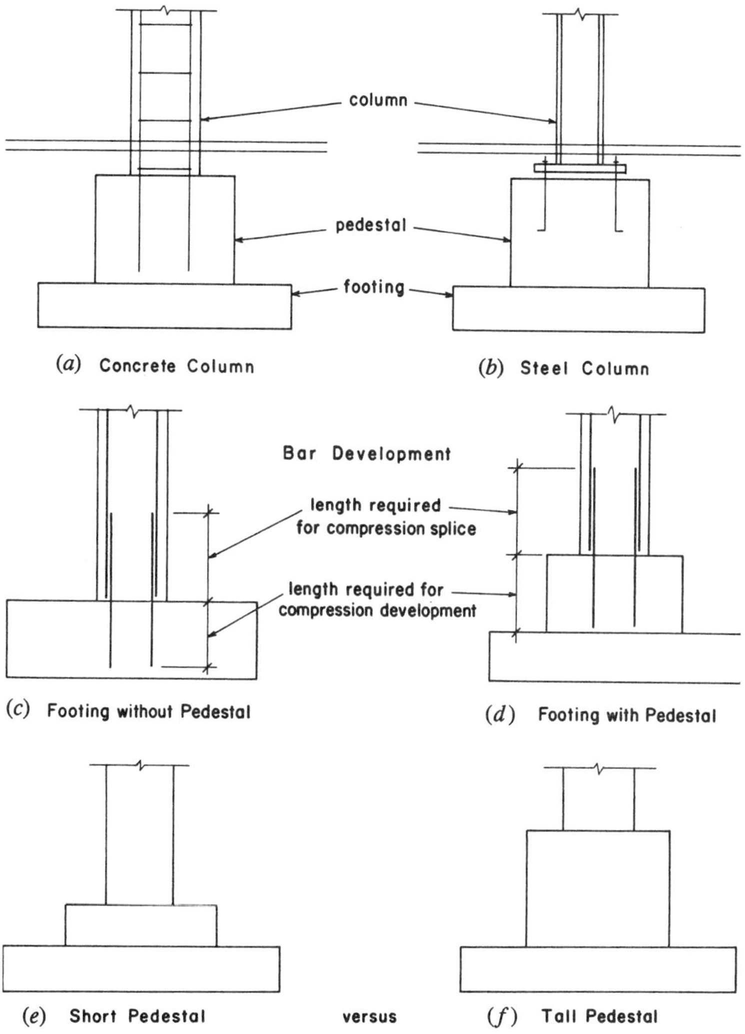

Figure 16.8 Usage considerations for column pedestals.

3. To provide for the required development length of reinforcing in reinforced concrete columns, where footing thickness is not adequate for development within the footing.

Figure 16.8*d* illustrates the third situation described. Referring to Table 15.1, observe that a considerable development length is required for large-diameter bars made from high grades of steel. If the minimum required footing does not have a thickness that permits this development, a pedestal may offer a reasonable solution. However, there are many other considerations to be made in the decision, and the column reinforcing problem is not the only factor in this situation.

If a pedestal is quite short with respect to its width (see Figure 16.8*e*), it may function essentially the same as a column footing, with significant values for shear and bending stresses. This condition is likely to occur if the pedestal width exceeds twice the column width and the pedestal height is less than one half of the pedestal width. In such cases, the pedestal must be designed by the same procedures used for an ordinary column footing.

V

STRUCTURAL SYSTEMS FOR BUILDINGS

This part contains examples of the design of structural systems for buildings. The buildings selected for design are not intended as examples of good architectural design, but rather have been selected to create a range of common situations to demonstrate the use of various structural components. Design of individual elements of the structural systems is based on the materials presented in earlier chapters. The purpose here is to show a broader context of design work by dealing with the whole structure and with the building in general.

17

GENERAL CONSIDERATIONS FOR STRUCTURES

This chapter contains discussions of some general issues relating to design of building structures. These concerns have mostly not been addressed in the presentations in earlier chapters but require consideration when you are dealing with whole building design situations. Application of these materials is illustrated in the design examples in Chapters 18 through 20.

17.1 CHOICE OF BUILDING CONSTRUCTION

Materials, methods, and details of building construction vary considerably on a regional basis. Many factors affect this situation, including the effects of response to climate and regional availability of construction materials. Even in a single region, differences occur among individual buildings, based on styles of architectural design and techniques of builders. Nevertheless, at any given time, there are usually a few predominant, popular methods of construction that are employed for most buildings of a given type and size. The construction methods and details shown here are reasonable, but in no way are they intended to illustrate a singular, superior style of building.

It is not possible to choose the materials and forms for a building structure without considering its integration with the general building construction. In some cases, it may also be necessary to consider the elements required for various building services, such as those for piping, electrical service, lighting, communication, roof drainage, and the HVAC (heating, ventilating, and air conditioning) systems.

For multistory buildings, it is necessary to accommodate the placement of stairs, elevators, and the vertical elements for various building services—particularly for air ducts between building levels. A major consideration for multistory buildings is the planning of the various levels so that they work when superimposed on top of each other. Bearing walls and columns must be supported from below.

Choice of the general structural system as well as the various individual elements of the system is typically highly dependent on the general architectural design of the building. Hopefully, the two issues—structural planning and architectural planning—are dealt with simultaneously from preliminary design to final construction drawings.

17.2 STRUCTURAL DESIGN STANDARDS

Use of methods, procedures, and reference data for structural design are subject to the judgment of the designer. Many guides exist, but some individual selection is often required. Strong influences on choices include the following:

- Building code requirements from the enforceable statutes relating to the location of the building
- Acceptable design standards as published by professional groups, such as the reference from the American Society of Civil Engineers (ASCE) referred to frequently in this book (Ref.1)
- Recommended design standards from industry organizations, such as the AISC and ACI
- The body of work from current texts and references produced by respected authors

Some reference is made to these sources in this book. However, much of the work is also simply presented in a manner familiar to the authors, based on their own experiences. If you pursue this subject, you are sure to encounter styles and opinions that differ from those presented here. Making one's own choices in face of those conflicts is part of the progress of professional growth.

17.3 LOADS FOR STRUCTURAL DESIGN

The issue of determining design loads is addressed in Chapter 4. In the main part, these must derive from enforceable building codes. However, the primary concern of codes is public health and safety. Performance of the structure for other concerns may not be adequately represented in the minimum requirements of the building code. Issues sometimes not included in code requirements are as follows:

Effects of deflection of spanning structures on nonstructural elements of the construction

Sensations of bounciness of floors by building occupants

Protection of structural elements from damage due to weather or normal usage

It is quite common for professional structural designers to have situations where they use their own judgment in assigning design loads. This ordinarily means using increased loads, because the minimum loads required by codes must always be recognized.

18

BUILDING ONE

The building in this chapter consists of a simple, single-story, box-shaped building. Lateral bracing is developed in response to wind load. Several alternatives are considered for the building structure.

18.1 GENERAL CONSIDERATIONS

Figure 18.1 shows the general form, the construction of the basic building shell, and the form of the wind-bracing shear walls for Building One. The drawings show a building profile with a generally flat roof (with minimal slope for drainage) and a short parapet at the roof edge. This structure is generally described as a *light wood frame* and is the first alternative to be considered for Building One. The following data is used for design:

Roof live load = 20 psf (reducible)
Wind load as determined from the AISC standard (Ref. 1)
Wood framing lumber of Douglas fir-larch

18.2 DESIGN OF THE WOOD STRUCTURE FOR GRAVITY LOADS

With the construction as shown in Figure 18.1*f*, the roof dead load is determined as follows:

Three-ply felt and gravel roofing	5.5 psf
Glass fiber insulation batts	0.5
½-in.-thick plywood roof deck	1.5
Wood rafters and blocking (estimate)	2.0
Ceiling framing	1.0
½-in.-thick drywall ceiling	2.5
Ducts, lights, etc.	3.0
Total roof dead load for design	16.0 psf

Assuming a partitioning of the interior as shown in Figure 18.2*a*, various possibilities exist for the development of the spanning roof and ceiling framing systems and their supports. Interior walls may be used for supports, but a more desirable situation in commercial uses is sometimes obtained by using interior columns that allow for rearrangement of interior spaces. The roof framing system shown in Figure 18.2*b* is developed with two rows of interior columns placed at the location of the corridor walls. If the partitioning shown in Figure 18.2*a* is used, these columns may be totally out of view (and not intrusive in the building plan) if they are incorporated in the wall construction. Figure 18.2*c* shows a second possibility for the roof framing using the same column layout as in Figure 18.2*b*. There may be various reasons for favoring one of these framing schemes over the other. Problems of installation of ducts, lighting, wiring, roof drains, and fire sprinklers may influence this structural design decision. For this example, the scheme shown in Figure 18.2*b* is arbitrarily selected for illustration of the design of the elements of the structure.

Installation of a membrane-type roofing ordinarily requires at least a ½-in.-thick roof deck. Such a deck is capable of up to 32-in. spans in a direction parallel to the face ply grain (the long direction of ordinary 4 × 8 ft panels). If rafters are not over 24 in. on center—as they are likely to be for the schemes shown in Figure 18.2*b* and *c*—the panels may be placed so that the plywood span is across the face grain. An advantage in the latter arrangement is the reduction in the amount of blocking between

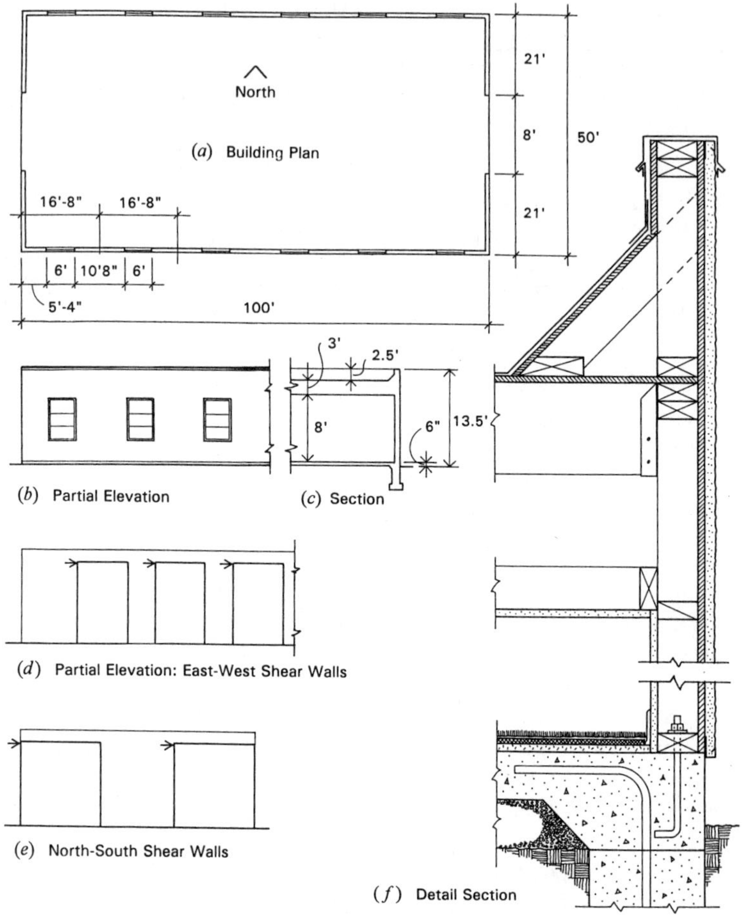

Figure 18.1 Building One: general form.

rafters that is required at panel edges not falling on a rafter. The reader is referred to the discussion of plywood decks in Section 5.8.

A common choice of grade for rafters is No. 2, for which Table 5.1 yields an allowable bending stress of 1035 psi (repetitive use) and a modulus of elasticity of 1,600,000 psi. Because the data for this case falls approximately within the criteria for Table 5.6, possible choices from the table are for either 2 × 10s at 12-in. centers or 2 × 12s at 16-in. centers.

A ceiling may be developed by direct attachment to the underside of the rafters. However, the construction as shown here indicates a ceiling

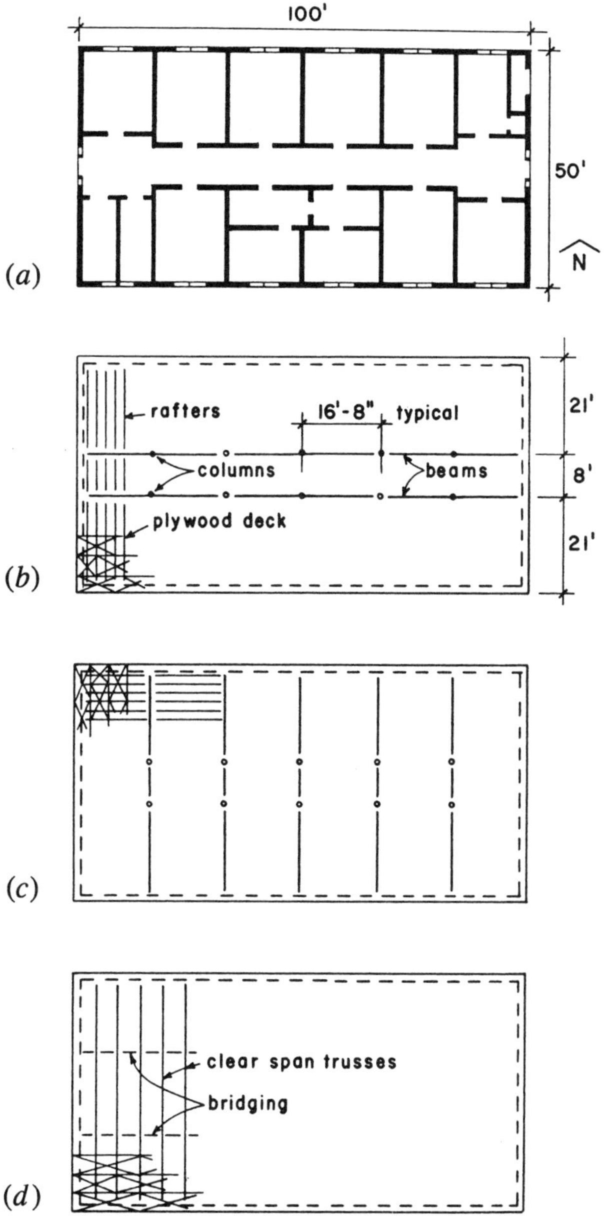

Figure 18.2 Developed plan for interior partitioning and alternatives for the roof framing.

at some distance below the rafters, allowing for various service elements to be incorporated above the ceiling. Such a ceiling might be framed independently for short spans (such as at a corridor) but is more often developed as a *suspended ceiling*, with hanger elements from the overhead structure used to shorten the span of ceiling framing.

The wood beams as shown in Figure.18.2*b* are continuous through two spans, with a total length of 33 ft 4 in. and beam spans of 16 ft 8 in. For the two-span beam, the maximum bending moment is the same as for a simple span, the principal advantage being a reduction in deflection. The total load area for one span is

$$A = \left(\frac{21 + 8}{2}\right) \times 16.67 = 242 \text{ ft}^2$$

The Uniform Building Code (Ref. 2) permits the use of a live load of 16 psf. Thus, the uniformly distributed load on the beam is found as

$$w = (16 \text{ psf LL} + 16 \text{ psf DL}) \times \frac{21 + 8}{2} = 464 \text{ lb/ft}$$

Adding a bit for the beam weight, a design for 480 lb/ft is reasonable, for which the maximum bending moment is

$$M = \frac{wL^2}{8} = \frac{480 \times (16.67)^2}{8} = 16{,}673 \text{ ft-lb}$$

A common minimum grade for beams is No. 1. The allowable bending stress depends on the beam size and the load duration. Assuming a 15% increase for load duration, Table 5.1 yields the following:

For a 4 × member: $F_b = 1.15(1000) = 1150$ psi

For a 5 × or larger: $F_b = 1.15(1350) = 1552$ psi

Then, for a 4 ×:

$$S = \frac{M}{F_b} = \frac{16{,}673 \times 12}{1150} = 174 \text{ in.}^3$$

From Table 4.10, the largest 4-in.-thick member is a 4 × 16 with $S = 135.7$ in.^{3}, which is not adequate. (*Note:* Deeper 4 × members are available but are quite laterally unstable and thus not recommended.) For a thicker member, the required S may be determined as

$$S = \frac{1150}{1552} \times 174 = 129 \text{ in.}^3$$

for which possibilities include a 6 × 14 with $S = 167$ in.^{3} or an 8 × 12 with $S = 165$ in.^{3}

Although the 6 × 14 has the least cross-sectional area and ostensibly the lower cost, various considerations of the development of construction details may affect the beam selection. This beam could also be formed as a built-up member from a number of 2 × members. Where deflection or long-term sag are critical, a wise choice might be to use a glued laminated section or even a steel rolled section. This may also be a consideration if shear is critical, as is often the case with heavily loaded beams.

A minimum slope of the roof surface for drainage is usually 2%, or approximately ¼ in. per ft. If drainage is achieved as shown in Figure 18.3*a*, this requires a total slope of ¼ × 25 = 6.25 in. from the center to the edge of the roof. There are various ways of achieving this sloped surface, including the simple tilting of the rafters.

Figure 18.3*b* shows some possibilities for the details of the construction at the center of the building. As shown here, the rafters are kept flat and the roof profile is achieved by attaching cut 2 × members to the tops of the long rafters and using a short profiled rafter at the corridor. Ceiling joists for the corridor are supported directly by the corridor walls. Other ceiling joists are supported at their ends by the walls and at intermediate points by suspension from the rafters.

The typical column at the corridor supports a load approximately equal to the entire spanning load for one beam, or

$$P = 480 \times 16.67 = 8000 \text{ lb}$$

This is a light load, but the column height requires something larger than a 4 × size. (See Table 6.1.) If a 6 × 6 is not objectionable, it is adequate in the lower stress grades. However, it is common to use a steel pipe or tubular section, either of which can probably be accommodated in a stud partition wall.

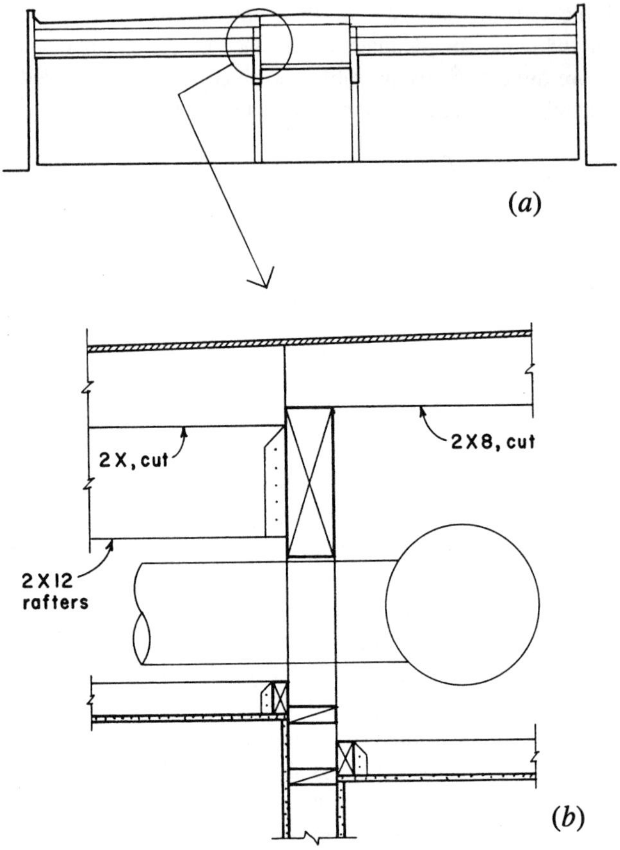

Figure 18.3 Construction details.

18.3 DESIGN FOR LATERAL LOADS

Design of the building structure for wind includes consideration for the following:

1. Inward and outward pressure on exterior building surfaces, causing bending of the wall studs and an addition to the gravity loads on roofs
2. Total lateral (horizontal) force on the building, requiring bracing by the roof diaphragm and the shear walls

3. Uplift on the roof, requiring anchorage of the roof structure to its supports
4. Total effect of uplift and lateral forces, possibly resulting in overturn (toppling) of the entire building

Uplift on the roof depends on the roof shape and the height above ground. For this low, flat-roofed building, the ASCE standard (Ref. 1) requires an uplift pressure of 10.7 psf. In this case, the uplift pressure does not exceed the roof dead weight of 16 psf, so anchorage of the roof construction is not required. However, common use of metal framing devices for light wood frame construction provides an anchorage with considerable resistance.

Overturning of the building is not likely critical for a building with this squat profile (50 ft wide × only 13.5 ft high). Even if the overturning moment caused by wind exceeds the restoring moment because of the building dead weight, the sill anchor bolts will undoubtedly hold the building down in this case. Overturn of the whole building is usually more critical for towerlike building forms or for extremely light construction. Of separate concern is the overturn of individual bracing elements—in this case the individual shear walls, which will be investigated later.

Wind Force on the Bracing System

The building's bracing system must be investigated for horizontal force in the two principal orientations: east-west and north-south. And, if the building is not symmetrical, in each direction on each building axis: east, west, north, and south.

The horizontal wind force on the north and south walls of the building is shown in Figure 18.4. This force is generated by a combination of positive (direct, inward) pressure on the windward side and negative (suction, outward) on the lee side of the building. The pressures shown as Case 1 in Figure 18.4 are obtained from data in the ASCE 2003 (Ref. 1) chapter on wind loads. (See discussion of wind loads in Chapter 4.) The single pressures shown in the figure are intended to account for the combination of positive and suction pressures. The ASCE standard provides for two zones of pressure: a general one and a small special increased area of pressure at one end. The values shown in Figure 18.4 for these pressures are derived by considering a critical wind velocity of 90 mph and an exposure condition B, as described in the standard.

The range for the increased pressure in Case 1 is defined by the dimension *a* and the height of the windward wall. The value of *a* is estab-

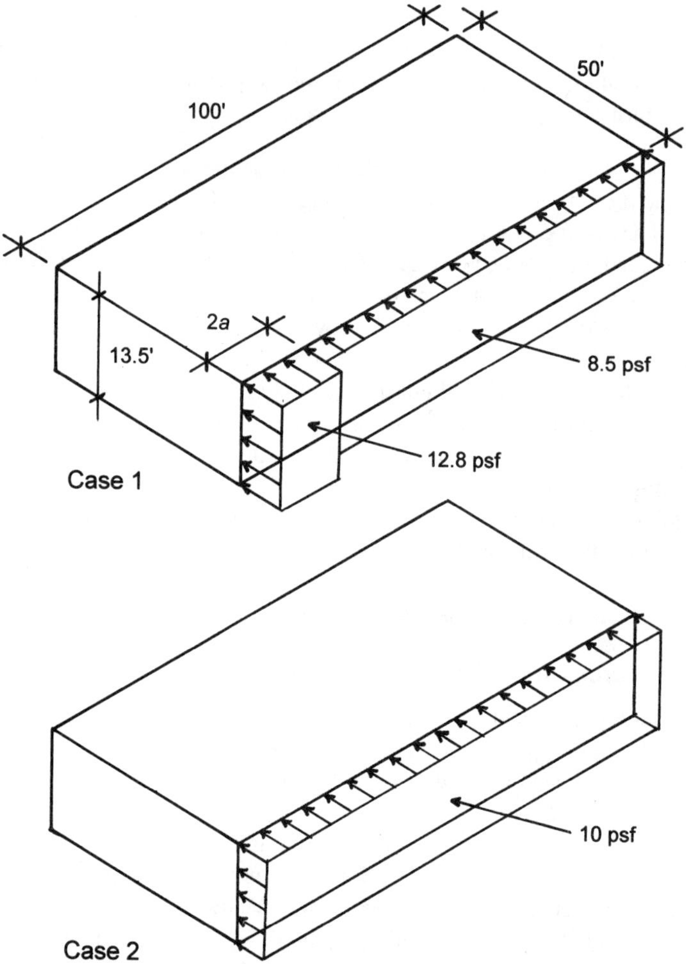

Figure 18.4 Wind pressure on the south wall, ASCE 2002 (Ref. 1).

lished as 10% of the least plan dimension of the building or 40% of the wall height, whichever is smaller, but not less than 3 ft. For this example, *a* is determined as 10% of 50 ft, or 5 ft. The distance for the pressure of 12.8 psf in Case 1 is thus $2(a) = 10$ ft.

The design standard also requires that the bracing system be designed for a minimum pressure of 10 psf on the entire area of the wall. This sets up two cases (Case 1 and Case 2 in Figure 18.4) that must be considered. Because the concern for the design is the generation of maximum effect on the roof diaphragm and the end shear walls, the critical conditions may be determined by considering the development of end reaction forces and maximum shear for an analogous beam subjected to the two loadings. This analysis is shown in Figure 18.5, from which it is apparent that the critical concern for the end shear walls and the maximum effect in the roof diaphragm is derived from Case 2 in Figure 18.4.

The actions of the horizontal wind force resisting system in this regard are illustrated in Figure 18.6. The initial force comes from wind pressure on the building's vertical sides. The wall studs span vertically to resist this uniformly distributed load, as shown in Figure 18.6*a*. Assuming the wall function to be as shown in Figure 18.6*a*, the north-south wind force delivered to the roof edge is determined as

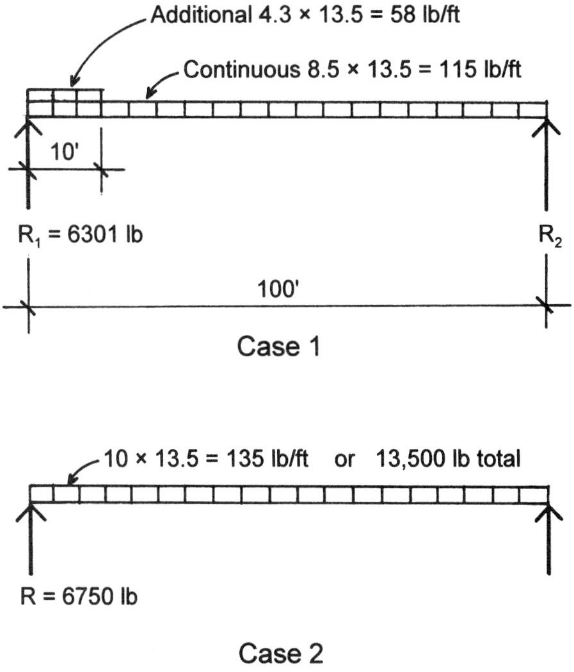

Figure 18.5 Resultant wind forces on the end shear walls.

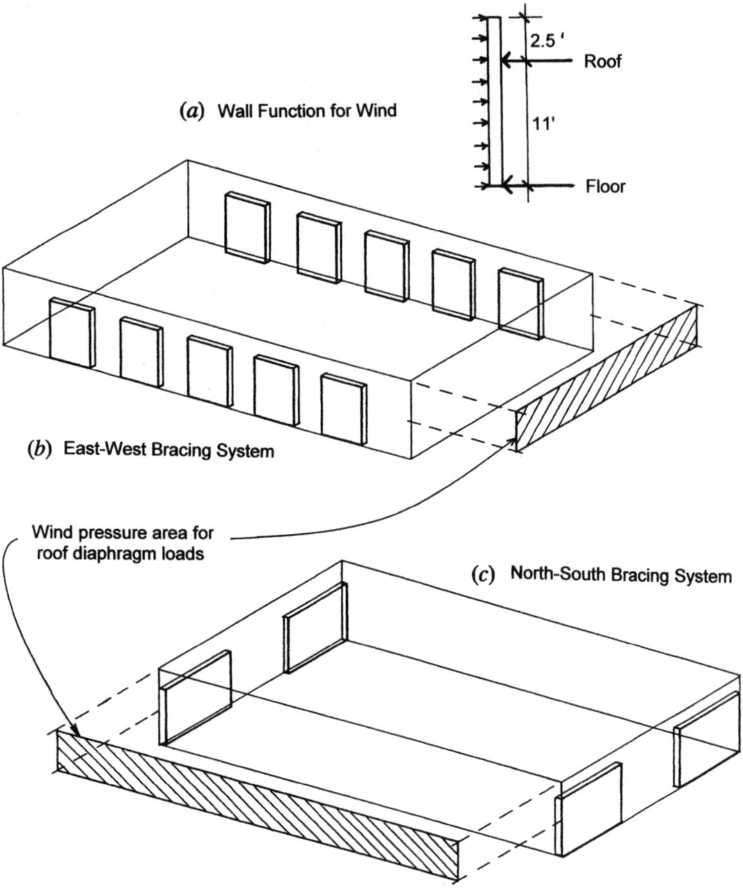

Figure 18.6 Wall functions and wind pressure development.

$$\text{Total } W = (10 \text{ psf})(100 \times 13.5) = 13{,}500 \text{ lb}$$

$$\text{Roof edge } W = 13{,}500 \times \frac{6.75}{11} = 8284 \text{ lb}$$

In resisting this load, the roof functions as a spanning member supported by the shear walls at the east and west ends of the building. The investigation of the diaphragm as a 100-ft simple span beam with uniformly distributed loading is shown in Figure 18.7. The end reaction and maximum diaphragm shear force is found as

$$R = V = \frac{8284}{2} = 4142 \text{ lb}$$

which produces a maximum unit shear in the 50-ft-wide diaphragm of

$$v = \frac{\text{Shear force}}{\text{Roof width}} = \frac{4142}{50} = 82.8 \text{ lb/ft}$$

From Table 18.1 (UBC Table 23-II-H) note that a variety of selections is possible. Variables include the class of the plywood, the panel thickness, the width of supporting rafters, the nail size and spacing, the use of blocking, and the layout pattern of the plywood panels. Assuming a minimum plywood thickness for the flat roof at ½-in. (given as 15/32 in the table), a possible choice is as follows:

C-D sheathing, $^{15}/_{32}$-in.-thick, 2 × rafters, 8d nails at 6 in. at all panel edges, blocked diaphragm

For these criteria, the table yields a capacity of 270 lb/ft.

In this example, if the need for the minimum thickness plywood is accepted, it turns out that the minimum construction is more than adequate for the required lateral force resistance. Had this not been the case, and the required capacity had resulted in considerable nailing beyond the minimum, it would be possible to graduate the nailing spacings from that required at the building ends to minimal nailing in the center portion of the roof. (See the form of the shear variation across the roof width.)

The moment diagram shown in Figure 18.7 indicates a maximum value of 104 kip-ft at the center of the span. This moment is used to determine the maximum force in the diaphragm chord at the roof edges. The force must be developed in both compression and tension as the wind direction reverses. With the construction as shown in Figure 18.1*f*, the top plate of the stud wall is the most likely element to be utilized for this function. In this case, the chord force of 2071 lb, as shown in Figure 18.7, is quite small, and the doubled 2 × member should be capable of resisting the force. However, the building length requires the use of several pieces to create this continuous plate, so the splices for the double member should be investigated.

TABLE 18.1 Allowable Shear in Pounds per Foot for Horizontal Plywood Diaphragms[1–3]

PANEL GRADE	COMMON NAIL SIZE	MINIMUM NAIL PENETRATION IN FRAMING (inches)	MINIMUM NOMINAL PANEL THICKNESS (inches)	MINIMUM NOMINAL WIDTH OF FRAMING MEMBER (inches)	BLOCKED DIAPHRAGMS				UNBLOCKED DIAPHRAGMS	
					Nail spacing (in.) at diaphragm boundaries (all cases), at continuous panel edges parallel to load (Cases 3 and 4) and at all panel edges (Cases 5 and 6) × 25.4 for mm				Nails spaced 6″ (152 mm) max. at supported edges	
					6	4	2 1/2[2]	2[2]	Case 1 (No unblocked edges or continuous joints parallel to load)	All other configurations (Cases 2, 3, 4, 5 and 6)
					Nail spacing (in.) at other panel edges × 25.4 for mm					
					6	6	4	3		
		× 25.4 for mm			× 0.0146 for N/mm					
Structural 1	6d	1 1/4	5/16	2	185	250	375	420	165	125
				3	210	280	420	475	185	140
	8d	1 1/2	3/8	2	270	360	530	600	240	180
				3	300	400	600	675	265	200
	10d[3]	1 5/8	15/32	2	320	425	640	730	285	215
				3	360	480	720	820	320	240
C-D, C-C, Sheathing, and other grades covered in UBC Standard 23-2 or 23-3	6d	1 1/4	5/16	2	170	225	335	380	150	110
				3	190	250	380	430	170	125
			3/8	2	185	250	375	420	165	125
				3	210	280	420	475	185	140
	8d	1 1/2	3/8	2	240	320	480	545	215	160
				3	270	360	540	610	240	180
			7/16	2	255	340	505	575	230	170
				3	285	380	570	645	255	190
			15/32	2	270	360	530	600	240	180
				3	300	400	600	675	265	200
	10d[3]	1 5/8	15/32	2	290	385	575	655	255	190
				3	325	430	650	735	290	215
			19/32	2	320	425	640	730	285	215
				3	360	480	720	820	320	240

[1]These values are for short-time loads due to wind or earthquake and must be reduced 25 percent for normal loading. Space nails 12 inches (305 mm) on center along intermediate framing members.

Allowable shear values for nails in framing members of other species set forth in Division III, Part III, shall be calculated for all other grades by multiplying the shear capacities for nails in Structural I by the following factors: 0.82 for species with specific gravity greater than or equal to 0.42 but less than 0.49, and 0.65 for species with a specific gravity less than 0.42.

[2]Framing at adjoining panel edges shall be 3-inch (76 mm) nominal or wider and nails shall be staggered where nails are spaced 2 inches (51 mm) or $2^1/_2$ inches (64 mm) on center.

[3]Framing at adjoining panel edges shall be 3-inch (76 mm) nominal or wider and nails shall be staggered where 10d nails having penetration into framing of more than $1^5/_8$ inches (41 mm) are spaced 3 inches (76 mm) or less on center.

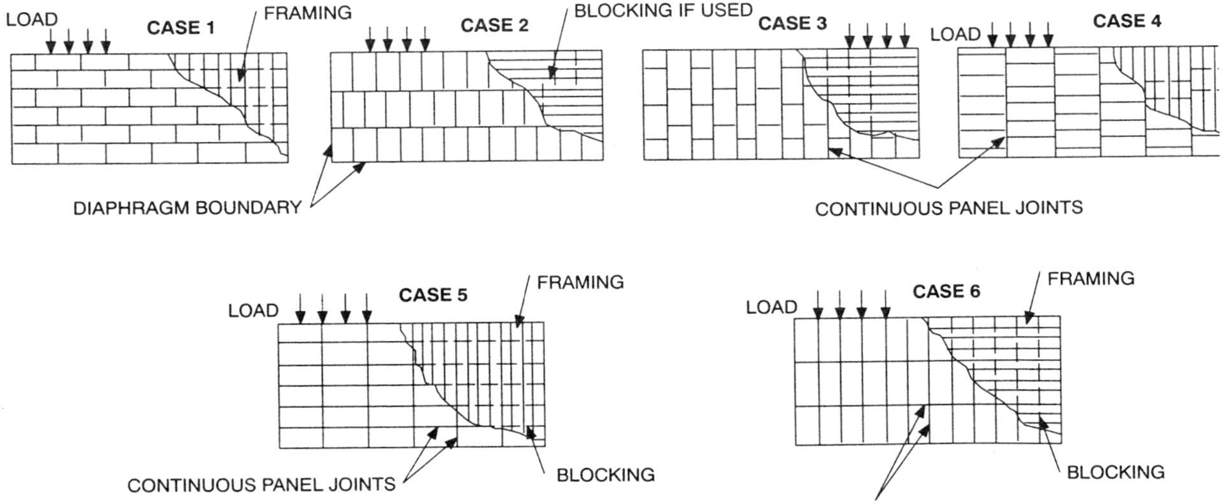

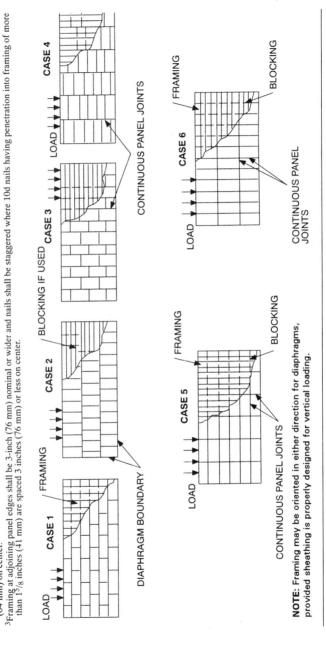

NOTE: **Framing may be oriented in either direction for diaphragms, provided sheathing is properly designed for vertical loading.**

Source: Table 23-II-H from the *Uniform Building Code*, 1997 ed. (Ref. 2), reproduced with permission of the publishers, International Conference of Building Officials.

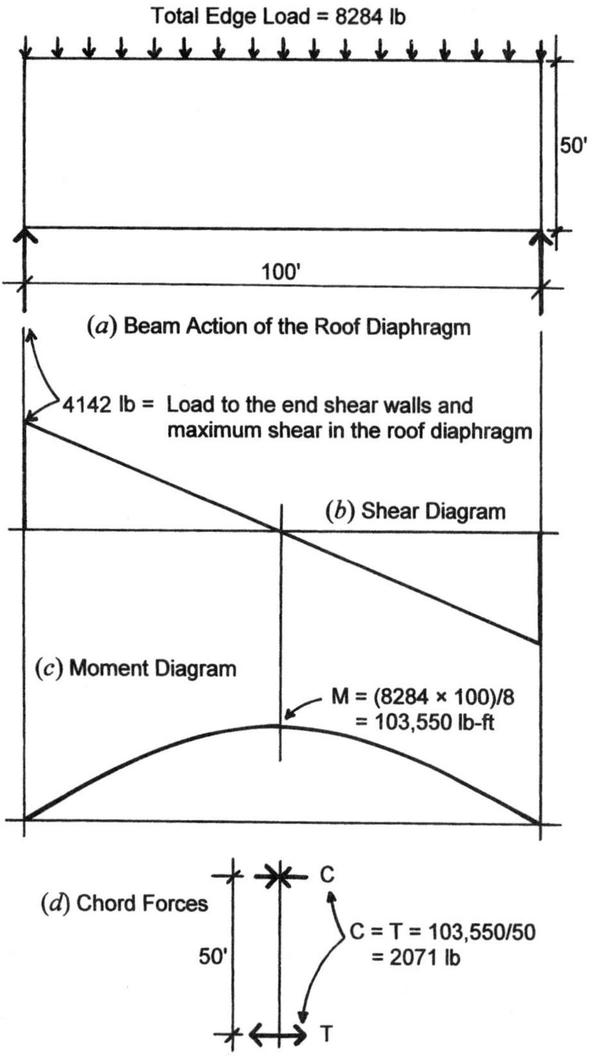

Figure 18.7 Spanning functions of the roof diaphragm.

The end reaction force for the roof diaphragm, as shown in Figure 18.7, must be developed by the end shear walls. As shown in Figure 18.1, there are two walls at each end, both 21 ft long in plan. Thus, the total shear force is resisted by a total of 42 ft of shear wall and the unit shear in the wall is

$$v = \frac{4142}{42} = 98.6 \text{ lb/ft}$$

As with the roof, there are various considerations for the selection of the wall construction. Various materials may be used on both the exterior and interior surfaces of the stud wall. The common form of construction shown in Figure 18.1*f* indicates gypsum drywall on the inside and a combination of plywood and stucco (cement plaster) on the outside of this wall. All three of these materials have rated resistances to shear wall stresses. However, with a combination of materials, it is common practice to consider only the strongest of the materials to be the resisting element. In this case that means the plywood sheathing on the exterior of the wall. From Table 18.2 (UBC Table 23-II-I-1), a possible choice is C-D sheathing, ⅜ in. thick, with 6d nails at 6-in. spacing at all panel edges.

For this criteria, the table allows a unit shear of 200 lb/ft. Again, this is minimal construction. For higher loadings, a greater resistance can be obtained by using better plywood, thicker panels, larger nails, closer nail spacing, and, sometimes, wider studs. Unfortunately, the nail spacing cannot be graduated—as it may be for the roof—because the unit shear is a constant value throughout the height of the wall.

Figure 18.8*a* shows the loading condition for investigation of the overturn effect on the end shear wall. Overturn is resisted by the so-called restoring moment caused by the dead load on the wall—in this case, a combination of the wall weight and the portion of roof dead load supported by the wall. Safety is considered adequate if the restoring moment is at least 1.5 times the overturning moment. A comparison is therefore made between the value of 1.5 times the overturning moment and the restoring moment as follows:

$$\text{Overturning moment} = (2.071)(11)(1.5) = 34.2 \text{ kip-ft}$$

$$\text{Restoring moment} = (3 + 6)(21/2) = 94.5 \text{ kip-ft}$$

This indicates that no tie down force is required at the wall ends (force T as shown in Figure 18.6). Details of the construction and other functions

TABLE 18.2 Allowable Shear in Pounds per Foot for Plywood Shear Walls[1–5]

PANEL GRADE	MINIMUM NOMINAL PANEL THICKNESS (inches)	MINIMUM NAIL PENETRATION IN FRAMING (inches)	PANELS APPLIED DIRECTLY TO FRAMING: Nail Size (Common or Galvanized Box)[5]	Nail Spacing at Panel Edges (in.) 6	4	3	2	PANELS APPLIED OVER 1/2-INCH (13 mm) OR 5/8-INCH (16 mm) GYPSUM SHEATHING: Nail Size (Common or Galvanized Box)[5]	Nail Spacing at Panel Edges (in.) 6	4	3	2
	× 25.4 for mm			× 25.4 for mm / × 0.0146 for N/mm					× 25.4 for mm / × 0.0146 for N/mm			
Structural I	5/16	1 1/4	6d	200	300	390	510	8d	200	300	390	510
	3/8			230[4]	360[4]	460[4]	610[4]					
	7/16	1 1/2	8d	255[4]	395[4]	505[4]	670[4]	10d	280	430	550	730
	15/32			280	430	550	730					
	15/32	1 5/8	10d	340	510	665	870	—	—	—	—	—
C-D, C-C Sheathing, plywood panel siding and other grades covered in UBC Standard 23-2 or 23-3	5/16	1 1/4	6d	180	270	350	450	8d	180	270	350	450
	3/8			200	300	390	510		200	300	390	510
	3/8			220[4]	320[4]	410[4]	530[4]					
	7/16	1 1/2	8d	240[4]	350[4]	450[4]	585[4]	10d	260	380	490	640
	15/32			260	380	490	640					
	15/32	1 5/8	10d	310	460	600	770	—	—	—	—	—
	19/32			340	510	665	870					
			Nail Size (Galvanized Casing)					Nail Size (Galvanized Casing)				
Plywood panel siding in grades covered in UBC Standard 23-2	5/16	1 1/4	6d	140	210	275	360	8d	140	210	275	360
	3/8	1 1/2	8d	160	240	310	410	10d	160	240	310	410

[1]All panel edges backed with 2-inch (51 mm) nominal or wider framing. Panels installed either horizontally or vertically. Space nails at 6 inches (152 mm) on center along intermediate framing members for 3/8-inch (9.5 mm) and 7/16-inch (11 mm) panels installed on studs spaced 24 inches (610 mm) on center and 12 inches (305 mm) on center for other conditions and panel thicknesses. These values are for short-time loads due to wind or earthquake and must be reduced 25 percent for normal loading.

Allowable shear values for nails in framing members of other species set forth in Division III, Part III, shall be calculated for all other grades by multiplying the shear capacities for nails in Structural I by the following factors: 0.82 for species with specific gravity greater than or equal to 0.42 but less than 0.49, and 0.65 for species with a specific gravity less than 0.42.

[2]Where panels are applied on both faces of a wall and nail spacing is less than 6 inches (152 mm) on center on either side, panel joints shall be offset to fall on different framing members or framing shall be 3-inch (76 mm) nominal or thicker and nails on each side shall be staggered.

[3]Where allowable shear values exceed 350 pounds per foot (5.11 N/mm), foundation sill plates and all framing members receiving edge nailing from abutting panels shall not be less than a single 3-inch (76 mm) nominal member. Nails shall be staggered.

[4]The values for 3/8-inch (9.5 mm) and 7/16-inch (11 mm) panels applied direct to framing may be increased to values shown for 15/32-inch (12 mm) panels, provided studs are spaced a maximum of 16 inches (406 mm) on center or panels are applied with long dimension across studs.

[5]Galvanized nails shall be hot-dipped or tumbled.

Source: Table 23-II-I-1 from the *Uniform Building Code*, 1997 ed. (Ref. 2), reproduced with permission of the publishers, International Conference of Building Officials.

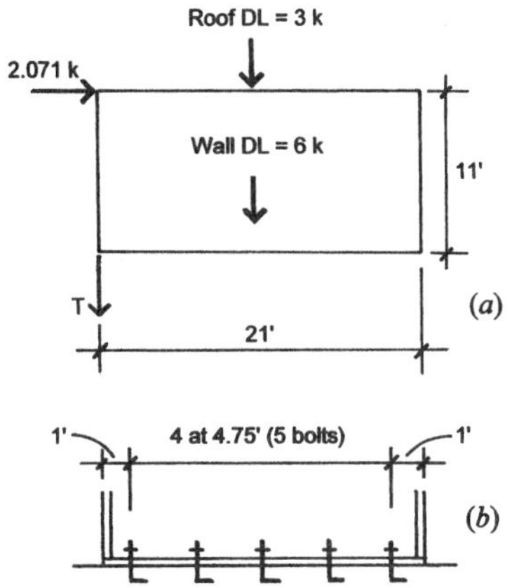

Figure 18.8 Functions of the end shear wall.

of the wall may provide additional resistances to overturn. However, some designers prefer to use end anchorage devices (called tie-down anchors) at the ends of all shear walls, regardless of loading magnitudes.

Finally, the walls will be bolted to the foundation with code-required sill bolts, which provide some resistance to uplift and overturn effects. At present, most codes do not permit sill bolts to be used for computed resistances to these effects because of the cross-grain bending that is developed in the wood sill members.

The sill bolts are used, however, for resistance to the sliding of the wall. The usual minimum bolting is with ½-in. bolts, spaced at a maximum of 6-ft centers, with a bolt not less than 12 in. from the wall ends. This results in a bolting for this wall as shown in Figure 18.8*b*. The five bolts shown should be capable of resisting the lateral force, using the values given by the codes.

For buildings with relatively shallow foundations, the effects of shear wall anchorage forces on the foundation elements should also be investigated. For example, the overturning moment is also exerted on the foundations and may cause undesirable soil stresses or require some structural resistance by foundation elements.

Another area of concern has to do with the transfer of forces from element to element in the whole lateral force resisting structural system. A critical point of transfer in this example is at the roof-to-wall joint. The force delivered to the shear walls by the roof diaphragm must actually be passed through this joint, by the attachments of the construction elements. The precise nature of this construction must be determined and must be investigated for these force actions.

18.4 ALTERNATIVE STEEL AND MASONRY STRUCTURE

Alternative construction for Building One is shown in Figure 18.9. In this case, the walls are made of concrete masonry units (CMU construction), and the roof structure consists of a formed sheet steel deck supported by open-web steel joists (light, prefabricated steel trusses). The following data is assumed for design:

Roof dead load = 15 psf, not including the weight of the structure
Roof live load = 20 psf, reducible for large supported areas

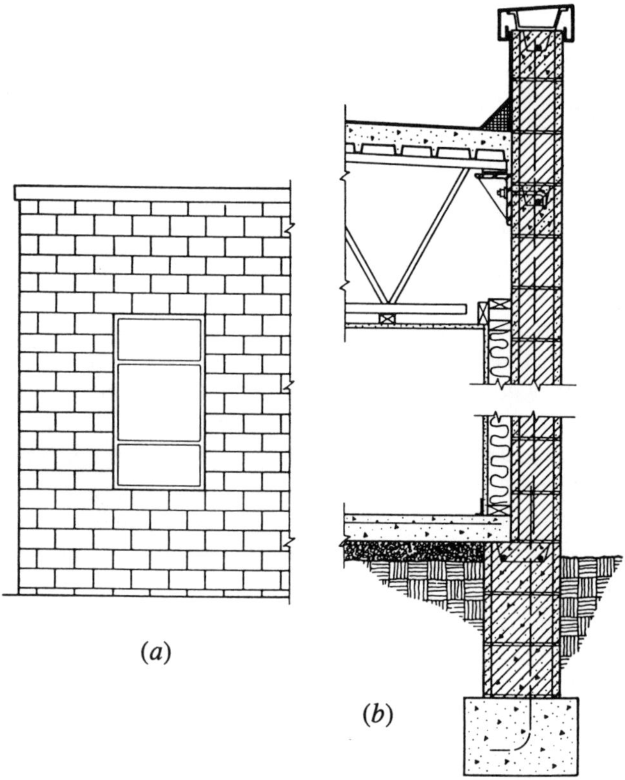

Figure 18.9 Building One: alternative steel and masonry structure.

Construction consists of:

K-Series open-web steel joists. (See Section 9.10.)
Reinforced hollow concrete masonry construction
Formed sheet steel deck (see Table 12.1)
Deck surfaced with lightweight insulating concrete fill
Multiple-ply, hot-mopped felt and gravel roofing
Suspended ceiling with gypsum drywall.

The section in Figure 18.9*b* indicates that the wall continues above the top of the roof to create a parapet, and that the steel trusses are supported at the wall face. The span of the joists is thus established as approximately 48 ft, which is used for their design.

As the construction section shows, the roof deck is placed directly on top of the trusses and the ceiling is supported by attachment to the bottom of the trusses. For reasonable drainage of the roof surface, a slope of at least ¼ inch per foot (2%) must be provided. The roof slope may be provided by tilting the trusses or by using a variable depth truss with the top chord sloped and the bottom chord horizontally flat. The following work assumes a constant depth of the trusses for design purposes.

Design of the Roof Structure

Spacing of the open web joists must be coordinated with the selection of the roof deck and the details for construction of the ceiling. For a trial design, a spacing of 4 ft is assumed. From Table 12.1, with deck units typically achieving three spans or more, the lightest deck in the table (22 gage) may be used. Choice of the deck configuration (rib width) depends on the type of materials placed on top of the deck and the means used to attach the deck to the supports.

Adding the weight of the deck to the other roof dead load produces a total dead load of 17 psf for the superimposed load on the joists. As illustrated in Section 9.10, the design for a K-series joist is as follows:

Joist dead load = 4(17) = 68 lb/ft (not including joist)
Joist live load = 4(20) = 80 lb/ft
Total factored load = 1.2(68) + 1.6(80) = 82 + 128 = 210 lb/ft + the joist weight

For the 48-ft span, the following alternative choices are obtained from Table 9.3:

24K9 at 12.0 plf, total load 1.2(12 + 68) + 128 = 96 + 128 = 224 lb/ft (less than the table value of 313 lb/ft)

26K5 at 10.6 plf, total load = 1.2(68 + 10.6) + 128 = 94 + 128 = 222 lb/ft (less than the table value of 233 lb/ft)

Live load capacity for L/360 deflection exceeds the requirement for both of these choices.

Although the 26K5 is the lightest permissible choice, there may be compelling reasons for using a deeper joist. For example, if the ceiling is directly attached to the bottoms of the joists, a deeper joist will provide more space for passage of building service elements. Deflection will also be reduced if a deeper joist is used. Pushing the live load deflection to the limit means a deflection of $(1/360)(48 \times 12) = 1.9$ in. Even though this may not be critical for the roof surface, it can present problems for the underside of the structure, involving sag of ceilings or difficulties with nonstructural walls built up to the ceiling. Choosing a 30K7 at 12.3 plf results in considerably less deflection at a small premium in additional weight.

Note that Table 9.3 is abridged from a larger table in the reference, and there are therefore many more choices for joist sizes. The example here is meant only to indicate the process for use of such references.

Specifications for open web joists give requirements for end support details and lateral bracing (see Ref. 6). If the 30K7 is used for the 48-ft span, for example, four rows of bridging are required.

Although the masonry walls are not designed for this example, note that the support indicated for the joists in Figure 18.9*b* results in an eccentric load on the wall. This induces bending in the wall, which may be objectionable. An alternative detail for the roof-to-wall joint is shown in Figure 18.10, in which the joists sit directly on the wall with the joist top chord extending to form a short cantilever. This is a common detail, and the reference supplies data and suggested details for this construction.

Alternative Roof Structure with Interior Columns

If a clear spanning roof structure is not required for this building, it may be possible to use some interior columns and a framing system for the roof with quite modest spans. Figure 18.11*a* shows a framing plan for a system that uses columns at 16 ft 8 in. on center in each direction. Although short span joists may be used with this system, it would also be possible to use a longer span deck, as indicated on the plan. This span ex-

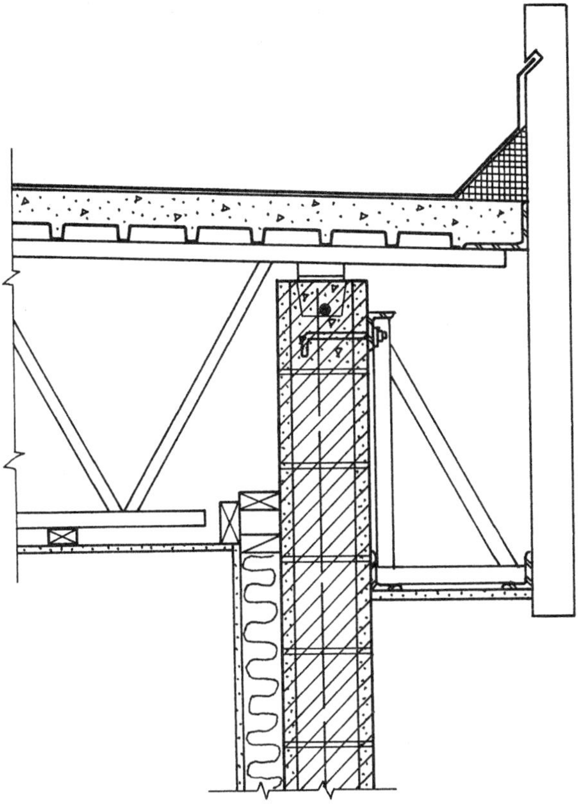

Figure 18.10 Building One: variation of the roof-to-wall joint.

ceeds the capability of the deck with 1.5-in. ribs, but decks with deeper ribs are available.

A second possible framing arrangement is shown in Figure 18.11*b*, in which the deck spans the other direction and only two rows of columns are used. This arrangement allows for wider column spacing. Although this arrangement increases the beam spans, a major cost savings is represented by the elimination of 60% of the interior columns and their footings.

Beams in continuous rows can sometimes be made to simulate a continuous beam action without the need for moment-resistive connections. Use of beam splice joints off the columns, as shown in Figure 18.11*c*, allows for relatively simple connections but some advantages of the continuous beam. A principal gain thus achieved is a reduction in deflections.

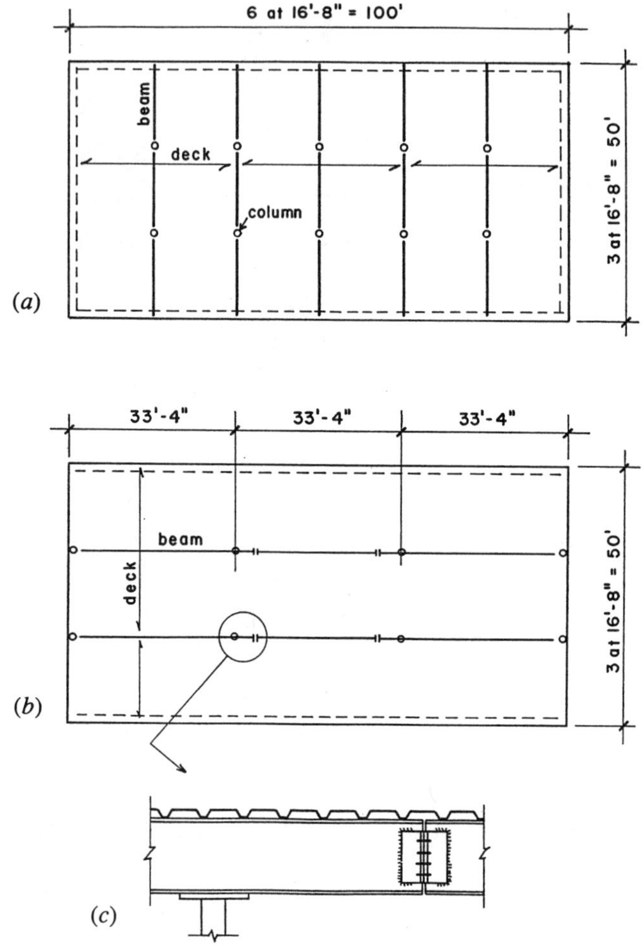

Figure 18.11 Building One: options for roof framing with interior columns.

For the beam in Figure 18.11*b*, assuming a slightly heavier deck, an approximate dead load of 20 psf will result in a beam factored load of

$$w = 16.67[1.2(20) + 1.6(16)] = 827 \text{ plf} + \text{the beam, say } 900 \text{ plf}$$

Note: The beam periphery of $33.3 \times 16.67 = 555 \text{ ft}^2$ qualifies the beam for a roof live load reduction, indicating the use of 16 psf as discussed in Section 4.1.)

The simple beam bending moment for the 33.3 ft span with the factored load is

$$M = \frac{wL^2}{8} = \frac{0.900 \times (33.3)^2}{8} = 125 \text{ kip-ft}$$

The required moment resistance of the beam is therefore

$$M = 125/0.9 = 139 \text{ kip-ft}$$

From Table 9.1, the lightest W-shape beam permitted is a W 16 31. From Figure 9.5note that the total load deflection will be approximately L/240, which is usually not critical for roof structures. Furthermore, the live load deflection will be less than one half this amount, which is quite a modest value.

It is assumed that the continuous connection of the deck to the beam top flange is adequate to consider the beam to have continuous lateral support, permitting the use of a resisting moment of $M_{p.}$ If the three-span beam is constructed with three simple-spanning segments, the detail at the top of the column will be as shown in Figure 18.12.

It is also possible to consider the use of the beam framing indicated in Figure 18.9b with a continuous beam having pinned connections off the columns. This will reduce both the maximum bending moment and the deflection for the beam and most likely permit a slightly lighter beam. Investigation of such a beam is illustrated in Section 3.10.

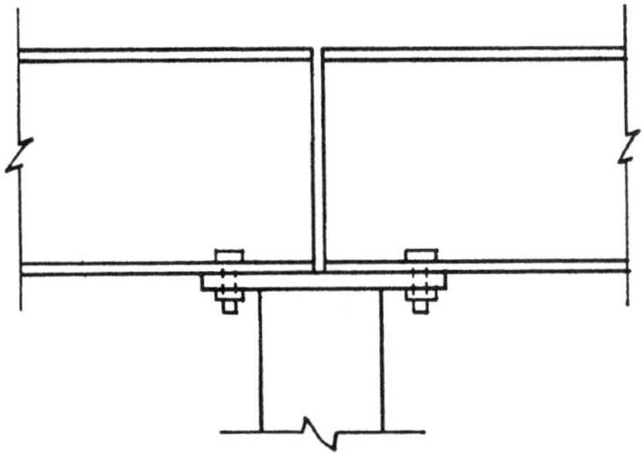

Figure 18.12 Framing detail at the top of the steel column with simple beam action.

The total load on the beam is the approximate load on the column. Thus, the column factored load is $0.900 \times 33.3 = 30$ kips. The required design strength of the column is therefore $30(1/0.85) = 35.3$ kips. Assuming an unbraced column height of 10 ft, the following choices may be found for the column:

From Table 10.4, a 3-in. pipe (nominal size, standard weight)

From Table 10.5, an HHS 3-in. square tube, with $\frac{3}{16}$-in. thick wall

18.5 ALTERNATIVE TRUSS ROOF

If a gabled (double sloped) roof form is desirable for Building One, a possible roof structure is shown in Figure 18.13. The building profile shown in Figure 18.13*a* is developed with a series of trusses, spaced at plan intervals, as shown for the beam-and-column rows in Figure 18.11*a*. The truss form is shown in Figure 18.13*b*. The complete results of an algebraic analysis for a unit loading on this truss are displayed in Figure 1.19. The true unit loading for the truss is derived from the form of construction and is approximately 10 times the unit load used in the example. This accounts for the values of the internal forces in the members as displayed in Figure 18.13*e*.

The detail in Figure 18.13*d* shows the use of double angle members with joints developed with gusset plates. The top chord is extended to form the cantilevered edge of the roof. For clarity of the structure, the detail shows only the major structural elements. Additional construction would be required to develop the roofing, ceiling, and soffit.

In trusses of this size, it is common to extend the chords without joints for as long as possible. Available lengths depend on the sizes of members and the usual lengths in stock by local fabricators. Figure 18.13*c* shows a possible layout that creates a two-piece top chord and a two-piece bottom chord. The longer top chord piece is thus 36 ft plus the overhang, which may be difficult to obtain if the angles are small.

The roof construction illustrated in Figure 18.13*d* shows the use of a longspan steel deck that bears directly on top of the top chord of the trusses. This option simplifies the framing by eliminating the need for intermediate framing between the trusses. For the truss spacing of 16 ft 8 in. as shown in Figure 18.11*a*, the deck will be quite light, and this is a feasible system. However, the direct bearing of the deck adds a spanning function to the top chord, and the chords must be considerably heavier to work for this added task.

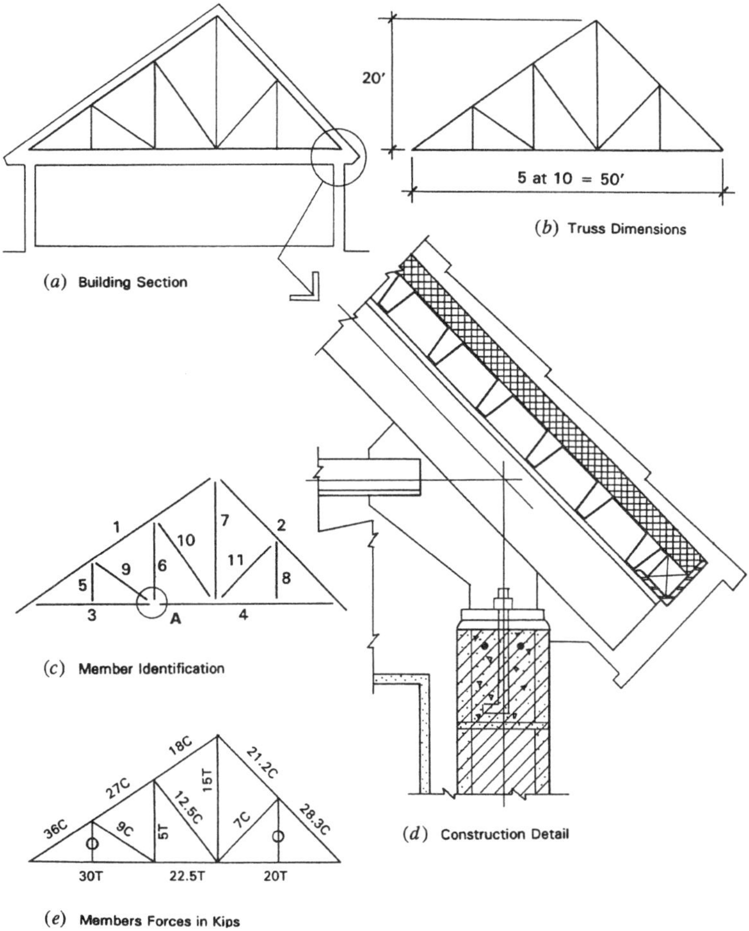

Figure 18.13 Building One: alternative truss roof structure.

The loading condition for the truss as shown in Section 1.6 indicates concentrated forces of 1000 lb each at the top chord joints. (*Note:* It is typical procedure to assume this form of loading, even though the actual load is distributed along the top chord, or roof load, and the bottom chord, or ceiling load.) If the total of the live load, roof dead load, ceiling dead load, and truss weight is approximately 60 psf, the single joint load is

$$P = (60)(10)(16.67) = 10{,}000 \text{ lb}$$

This is 10 times the load in the truss in Section 1.6, so the internal forces for the gravity loading will be 10 times those shown in Figure 1.19. These values are shown here in Figure 18.13*e*.

Various forms may be used for the members and the joints of this truss. The loading and span is quite modest here, so the truss members will be quite small and joints will have minimum forces. A common form for this truss would be one using tee shapes for the top and bottom chords and double angles for interior members with the angles welded directly to the tee stems (see Figure 11.10*b*). Bolted connections are possible but probably not practical for this size truss.

18.6 FOUNDATIONS

Foundations for Building One would be quite minimal. For the exterior bearing walls, the construction provided will depend on concerns for frost and the location below ground of suitable bearing material. Options for the wood structure are shown in Figure 18.14.

Where frost is not a problem and suitable bearing can be achieved at a short distance below the finished grade, a common solution is to use the construction shown in Figure 18.14*a*, called a *grade beam*. This is es-

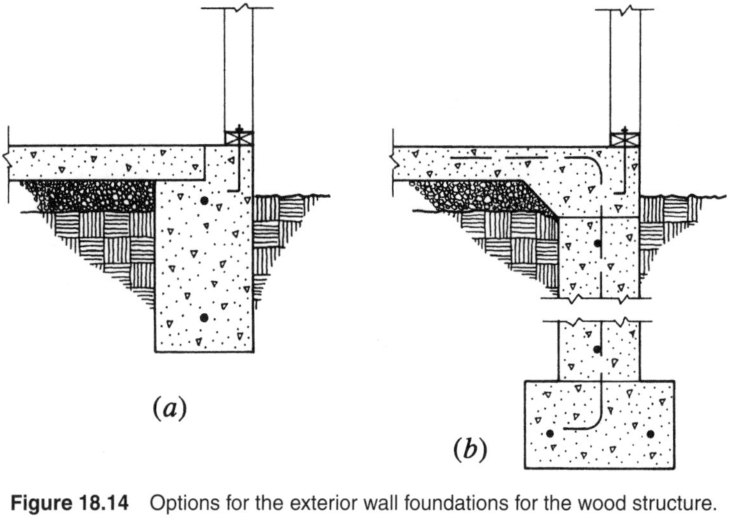

Figure 18.14 Options for the exterior wall foundations for the wood structure.

sentially a combined footing and short foundation wall in one. It is typically reinforced with steel bars in the top and bottom to give it some capacity as a continuous beam, capable of spanning over isolated weak spots in the supporting soil.

Where frost *is* a problem, local codes will specify a minimum distance from the finished grade to the bottom of the foundation. To reach this distance, it may be more practical to build a separate footing and foundation wall, as shown in Figure 18.14*b*. This short, continuous wall may also be designed for some minimal beamlike action, similar to that for the grade beam.

For either type of foundation, the light loading of the roof and the wood stud wall will require a very minimal width of foundation, if the bearing soil material is at all adequate. If bearing is not adequate, then this type of foundation (shallow bearing footings) must be replaced with some form of deep foundation (piles or caissons), which presents a major structural design problem.

Figure 18.15 shows foundation details for the masonry wall structure, similar to those for the wood structure. Here the extra weight of the masonry wall may require some more width for the bearing elements, but the general form of construction will be quite similar. An alternative for

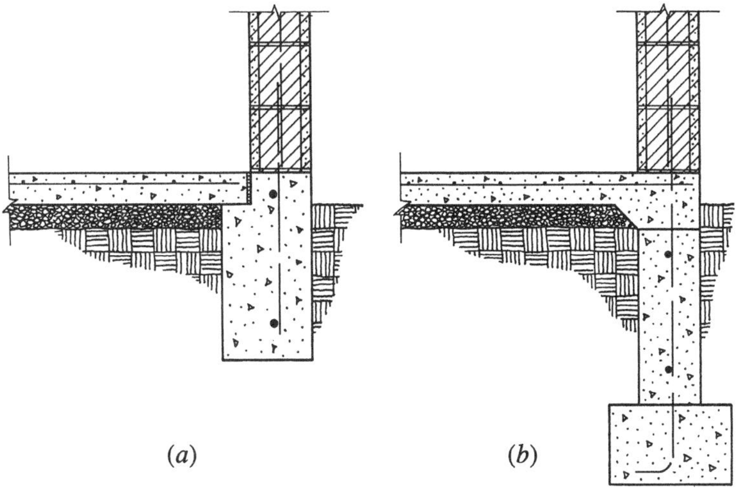

Figure 18.15 Options for the exterior wall foundation for the masonry wall structure.

the foundation wall in either Figure 18.14*b* or Figure 18.15*b* is to use grout-filled concrete blocks instead of the cast concrete wall.

Footings for any interior columns for Building One would also be minimal because of to the light loading from the roof structure and the low value of live load for the roof.

19

BUILDING TWO

Figure 19.1 shows a building that consists essentially of stacking the plan for Building One to produce a two-story building. The profile section of the building shows that the structure for the second story is developed the same as the roof structure and walls for Building One. Here, for both the roof and the second floor, the framing option chosen is that shown in Figure 18.2*b*.

Even though the framing layout is similar, the principal difference between the roof and floor structures has to do with the loadings. Both the dead load and live load are greater for the floor. In addition, the deflection of long-spanning floor members is a concern both for the dimension and for the bounciness of the structure.

The two-story building sustains a greater total wind load, although the shear walls for the second story will be basically the same as for Building One. The major effect in this building is the force generated in the first story shear walls. In addition, there is a second horizontal diaphragm: the second-floor deck.

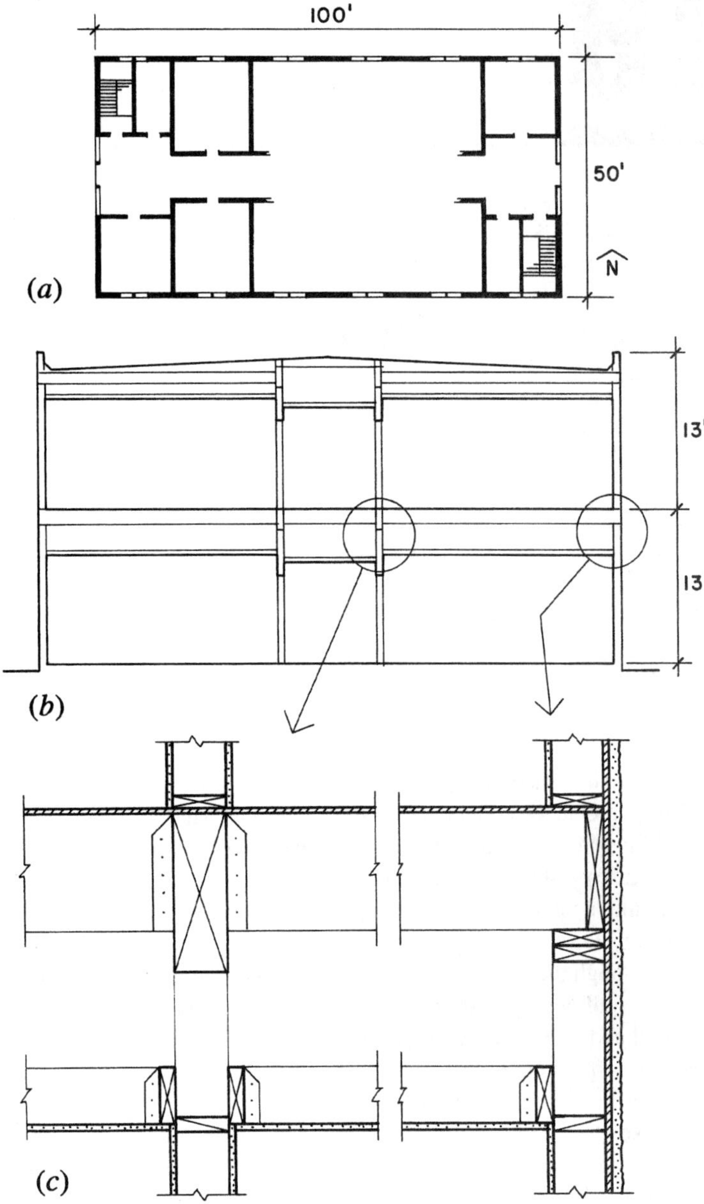

Figure 19.1 Building Two, general form and construction details.

Some details for the second-floor framing are shown in Figure 19.1*c*. Roof framing details are similar to those shown for Building One in Figure 18.3*b*. As with Building One, an option here is to use a clear-spanning roof structure—most likely with light trusses—which would eliminate the need for the corridor wall columns in the second floor.

19.1 DESIGN FOR GRAVITY LOADS

For design of the second-floor structure the following construction is assumed. The weight of the ceiling is omitted, assuming it to be supported by the first-story walls.

Carpet and pad	3.0 psf
Fiberboard underlay	3.0
Concrete fill, 1.5 in.	12.0
Plywood deck, ¾ in.	2.5
Ducts, lights, wiring	3.5
Total, without joists	24.0 psf

The minimum live load for office areas is 50 psf. However, the code requires the inclusion of an extra load to account for possible additional partitioning, usually 25 psf. Thus, the full design live load is 75 psf. At the corridor, the live load is 100 psf. Many designers would prefer to design the whole floor for a live load of 100 psf, thereby allowing for other arrangements or occupancies in the future. Because the added partition load is not required for this live load, it is only an increase of about 20% in the total load. With this consideration, the total design load for the joists is thus 124 psf.

With joists at 16-in. centers, the superimposed uniformly distributed load on a single joist is thus

$$\text{DL} = \frac{16}{12}(24) = 32 \text{ lb/ft} + \text{the joist, say } 40 \text{ lb/ft}$$

$$\text{LL} = \frac{16}{12}(100) = 133 \text{ lb/ft}$$

and the total load is 173 lb/ft. For the 21-ft-span joists the maximum bending moment is

$$M = \frac{wL^2}{8} = \frac{173 \times (21)^2}{8} = 9537 \text{ ft-lb}$$

For Douglas fir-larch joists of select structural grade and 2-in. nominal thickness, F_b from Table 5.1 is 1725 psi for repetitive member use. Thus the required section modulus is

$$S = \frac{M}{F_b} = \frac{9537 \times 12}{1725} = 66.3 \text{ in.}^3$$

Table A. 8 shows that there is no 2 × member with this value for section modulus. A possible choice is for a 3 × 14 with $S = 73.151$ in.^{3}. This is not a good design because the 3 × members in select structural grade are very expensive. Furthermore, as Figure 5.1 reveals, the deflection is considerable—not beyond code limits but sure to result in some sag and bounciness. A better choice for this span and load is probably for one of the proprietary fabricated joists (see discussion in Section 5.10).

The beams support both the 21-ft joists and the short 8-ft corridor joists. The total load carried by one beam is approximately 240 ft^2, for which a reduction of 7% is allowed for the live load (see discussion in Section 4.1). Using the same loading for both the corridor and offices, the beam load is determined as

DL = (30)(14.5)	= 435 lb/ft
+ beam weight	= 50
+ wall above	= 150
Total DL	= 635 lb/ft
LL = (0.93)(100)(14.5)	= 1349 lb/ft
Total load on beam	= 1984, say 2000 lb/ft

For the uniformly loaded simple beam with a span of 16.67 ft:

Total load = W = (2)(16.67) = 33.4 kips

End reaction = maximum beam shear = $W/2$ = 16.7 kips

Maximum bending moment is

$$M = \frac{WL}{8} = \frac{33.4 \times 16.67}{8} = 69.6 \text{ kip-ft}$$

For a Douglas fir-larch, dense No. 1 grade beam, Table 5.1 yields values of $F_b = 1550$ psi, $F_v = 85$ psi, and $E = 1{,}700{,}000$ psi. To satisfy the flexural requirement, the required section modulus is

$$S = \frac{M}{F_b} = \frac{69.6 \times 12}{1.550} = 539 \text{ in.}^3$$

From Table A.8, the least weight section is a 10 × 20 or a 12 × 18.

If the 20-in.-deep section is used, its effective bending resistance must be reduced (see discussion in Section 5.2). Thus, the actual moment capacity of the 10 × 20 is reduced by the size factor from Table 5.3 and is determined as

$$M = C_F \times F_b \times S = (0.947)(1.550)(602.1)(1/12) = 73.6 \text{ kip-ft}$$

Because this still exceeds the requirement, the selection is adequate. Similar investigations will show the other size options to also be acceptable.

If the actual beam depth is 19.5 in., the critical shear force may be reduced to that at a distance of the beam depth from the support. Thus, an amount of load equal to the beam depth times the unit load can be subtracted from the maximum shear. The critical shear force is thus

$$V = \text{(Actual end shear force)} - \text{(Beam depth in ft times unit load)}$$
$$= 16.67 - 2.0(19.5/12) = 16.67 - 3.25 = 13.42 \text{ kips}$$

For the 10 × 20 the maximum shear stress is thus

$$f_v = 1.5\frac{V}{A} = 1.5\frac{13{,}420}{185.25} = 108.7 \text{ psi}$$

This is less than the limiting stress of 170 psi as given in Table 5.1, so the beam is acceptable for shear resistance. However, this is still a really big piece of lumber, and questionably feasible, unless this building is in the heart of a major timber region. It is probably logical to modify the structure to reduce the beam span or to choose a steel beam or a glued laminated section in place of the solid-sawn timber.

Although deflection is often critical for long spans with light loads, it is seldom critical for the short-span, heavily loaded beam. The reader may verify this by investigating the deflection of this beam, but the computation is not shown here (see Section 5.5).

For the interior column at the first story, the design load is approximately equal to the total load on the second-floor beam plus the load from the roof structure. Because the roof loading is about one third of that for the floor, the design load is about 50 kips for the 10-ft-high column. Table 6.1 yields possibilities for an 8 × 10 or 10 × 10 section. For various reasons, it may be more practical to use a steel member here —a round pipe or a square tubular section—which may actually be accommodated within a relatively thin stud wall at the corridor.

Columns must also be provided at the ends of the beams in the east and west walls. Separate column members may be provided at these locations, but it is also common to simply build up a column from a number of studs.

19.2 DESIGN FOR LATERAL LOADS

Lateral resistance for the second story of Building Two is essentially the same as for Building One. Design consideration here will be limited to the diaphragm action of the second-floor deck and the two-story end shear walls.

The wind loading condition for the two-story building is shown in Figure 19.2*a*. For the same design conditions assumed for wind in Chapter 18, the pressure used for horizontal force on the building bracing system is 10 psf for the entire height of the exterior wall. At the second-floor level, the wind load delivered to the edge of the diaphragm is 120 lb/ft, resulting in the spanning action of the diaphragm as shown in Figure 19.2*b*. Referring to the building plan in Figure19.1*a*, observe that the opening required for the stairs creates a void in the floor deck at the ends of the diaphragm. The net width of the diaphragm is thus reduced to approximately 35 ft at this point, and the unit stress for maximum shear is

$$v = \frac{6000}{35} = 171 \text{ lb/ft}$$

From Table 18.1 we can determine that this requires only minimum nailing for a $^{19}/_{32}$-in.-thick plywood deck, which is the usual minimum thick-

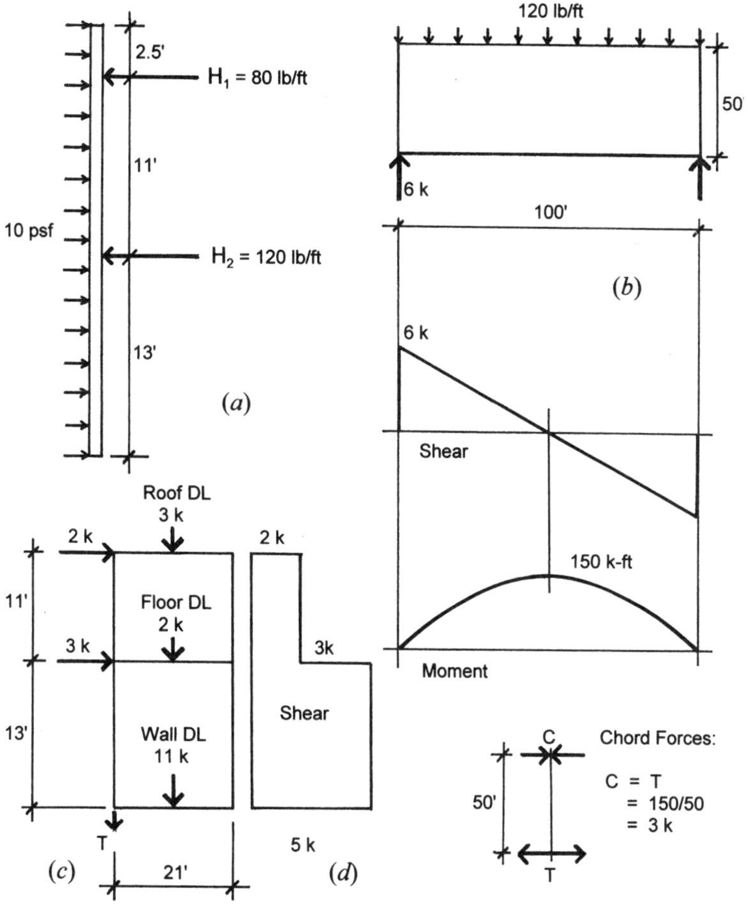

Figure 19.2 Building Two, development of lateral force due to wind: (a) functions of the exterior wall under wind pressure and bracing from the roof and floor diaphragms; (b) spanning functions of the second floor diaphragm; (c) loading of the two-story end shear wall; (d) shear diagram for the shear wall.

ness used for floor decks. As discussed for the roof diaphragm in Chapter 18, the chord at the edge of the floor diaphragm must be developed by framing members to sustain the computed tension/compression force of 3 kips. Ordinary framing members may be capable of this action, if attention is paid to splicing for full continuity of the 100-ft-long edge member.

The construction details for the roof, floor, and exterior walls must be carefully studied to ensure that the necessary transfers of force are achieved. These transfers include the following:

1. Transfer of the force from the roof plywood deck (the horizontal diaphragm) to the wall plywood sheathing (the shear walls)
2. Transfer from the second story shear wall to the first story shear wall that supports it
3. Transfer from the second-floor deck (horizontal diaphragm) to the first story wall plywood sheathing (the shear walls)
4. Transfer from the first story shear wall to the building foundations

In the first-story end shear walls, the total lateral load is 5000 lb, as shown in Figure 19.2*d*. For the 21-ft-wide wall the unit shear is

$$v = \frac{5000}{21} = 238 \text{ lb/ft}$$

From Table 18.2, note that this resistance can be achieved with C-D grade plywood of 3/8-in. thickness, although nail spacing closer than the minimum of 6 in. is required at the panel edges.

At the first-floor level, the investigation for overturn of the end shear wall is as follows (see Figure 19.2*c*):

Overturning moment = (2)(24)(1.5) + (3)(13)(1.5) = 72 kip-ft + 58.5 = 130.5 kip-ft

Restoring moment = (3 + 2 + 11)(21/2) = 168 kip-ft

Net overturning effect = 130.5 – 168 = –37.5 kip-ft

Because the restoring moment provides a safety factor greater than 1.5, there is no requirement for the anchorage force *T*.

In fact, there are other resisting forces on this wall. At the building corner, the end walls are reasonably well attached through the corner framing to the north and south walls, which would need to be lifted to permit overturning. At the sides of the building entrance, with the second-floor framing as described, there is a post in the end of the wall that supports the end of the floor beams. All in all, there is probably no computational basis for requiring an anchor at the ends of the shear walls. Nevertheless, many designers routinely supply such anchors.

19.3 ALTERNATIVE STEEL AND MASONRY STRUCTURE

As with Building One, an alternative construction for this building is one with masonry walls and interior steel framing, with ground floor and roof construction essentially the same as that shown for Building One in Section 18.7, The second floor may be achieved as shown in Figure 19.3. Because of heavier loads, the floor structure here consists of a steel framing system with rolled steel beams supported by steel columns on the interior and by pilasters in the exterior masonry walls. The plan detail in Figure 19.3*b* shows the typical pilaster as formed in the CMU construction.

The floor deck consists of formed sheet steel units with a structural-grade concrete fill. This deck spans between steel beams, which are in turn supported by larger beams that are supported directly by the columns. All

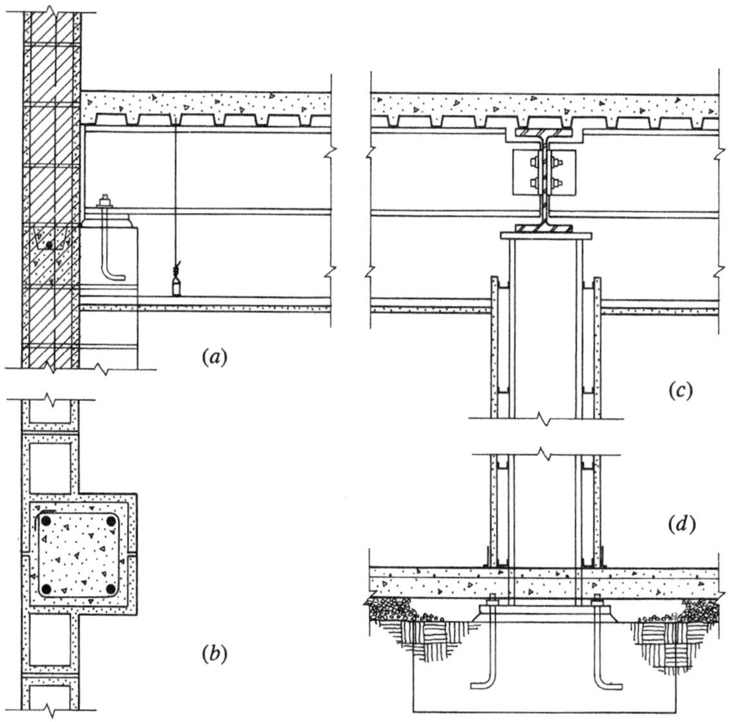

Figure 19.3 Building Two: details for the alternative steel and masonry structure.

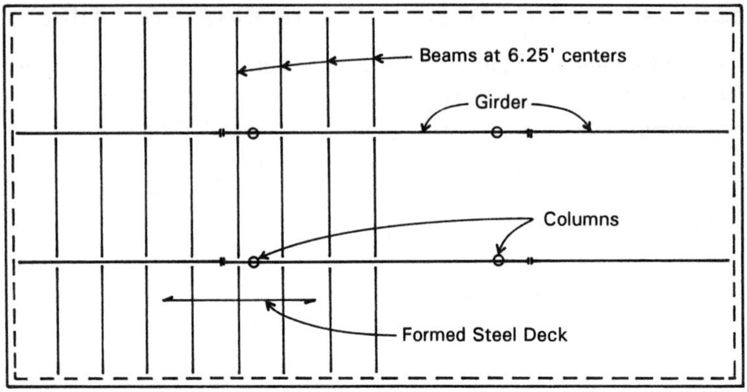

Figure 19.4 Second floor framing plan with steel beams.

of the elements of this steel structure can be designed by procedures described in Part III in this book.

Figure 19.4 shows a possible framing plan for the second floor of Building Two, consisting of a variation of the roof framing plan in Figure 18.9*b*. Here, instead of the longspan roof deck, a shorter-span floor deck is used with a series of beams at 6.25-ft spacing supported by the exterior walls and the two interior girders.

Fireproofing details for the steel structure would depend on the local fire zone and the building code requirements. Encasement of steel members in fire-resistive construction may suffice for this small building.

Although the taller masonry walls, carrying both roof and floor loads, have greater stress development than in Building One, it is still possible that minimal code-required construction may be adequate. Provisions for lateral load depend on load magnitudes and code requirements. Another possibility for Building Two, depending on fire-resistance requirements, is to use wood construction for the building roof and floor systems, together with the masonry walls. Many buildings were built through the nineteenth and early twentieth centuries with masonry walls and interior structures of timber, a form of construction referred to as mill construction. A similar system in present-day construction uses a combination of wood and steel structural elements with exterior masonry walls, as illustrated for Building Three in Section 20.7.

20

BUILDING THREE

This is a modest-size office building, generally qualified as being low rise (see Figure 20.1). In this category, there is a considerable range of choice for the construction, although in a particular place, at a particular time, a few popular forms of construction tend to dominate the field.

20.1 GENERAL CONSIDERATIONS

Some modular planning is usually required for this type of building, involving the coordination of dimensions for spacing of columns, window mullions, and interior partitions in the building plan. This modular coordination may also be extended to development of ceiling construction, lighting, ceiling HVAC elements, and the systems for access to electric power, phones, and other signal wiring systems. There is no single magic number for this modular system; all dimensions between 3 and 5 ft have been used and strongly advocated by various designers. Selection of a par-

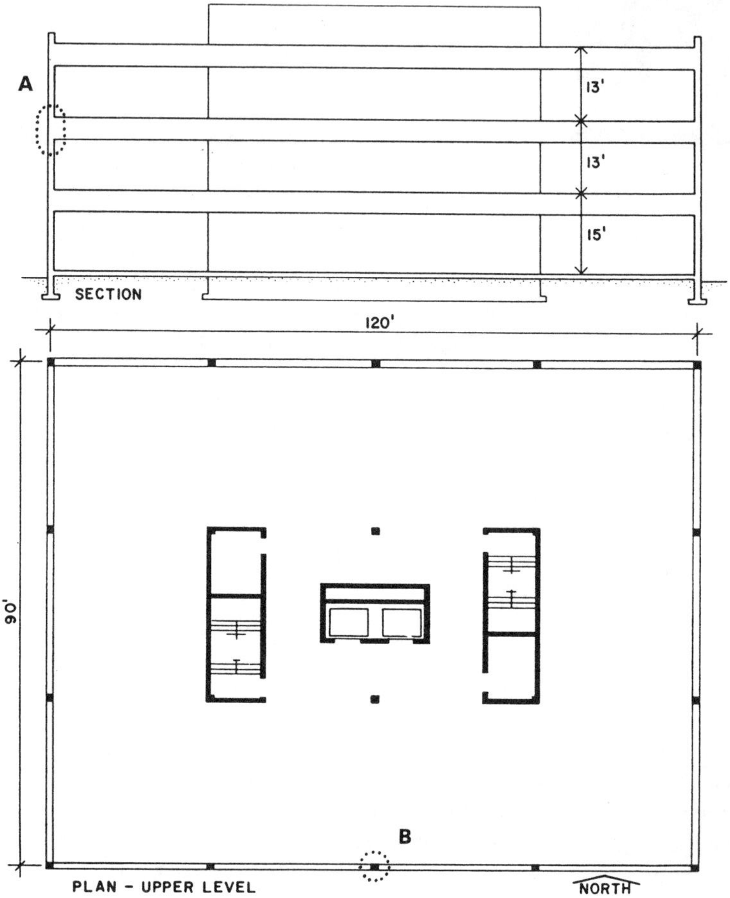

Figure 20.1 Building Three: general form.

ticular proprietary system for the curtain wall, interior modular partitioning, or an integrated ceiling system may establish a reference dimension.

For buildings built as investment properties, with speculative occupancies that may vary over the life of the building, it is usually desirable to accommodate future redevelopment of the building interior with some ease. For the basic construction, this means a design with as few permanent structural elements as possible. At a bare minimum, what is usually required is the construction of the major structure (columns, floors and roof), the exterior walls, and the interior walls that enclose stairs, elevators, rest rooms, and risers for building services. Everything else should be nonstructural or demountable in nature, if possible.

Spacing of columns on the building interior should be as wide as possible, basically to reduce the number of freestanding columns in the rented portion of the building plan. A column-free interior may be possible if the distance from a central core (grouped permanent elements) to the outside walls is not too far for a single span. Spacing of columns at the building perimeter does not affect this issue, so additional columns are sometimes used at this location to reduce their size for gravity loading or to develop a stiffer perimeter rigid frame system for lateral loads.

The space between the underside of suspended ceilings and the top of floor or roof structures above must typically contain many elements besides those of the basic construction. This usually represents a situation requiring major coordination for the integration of the space needs for the elements of the structural, HVAC, electrical, communication, lighting, and fire-fighting systems.

A major design decision that must often be made very early in the design process is that of the overall dimension of the space required for this collection of elements. Depth permitted for the spanning structure and the general level-to-level vertical building height will be established—and will not be easy to change later, if the detailed design of any of the enclosed systems indicates a need for more space.

Generous provision of the space for building elements makes the work of the designers of the various other building subsystems easier, but the overall effects on the building design must be considered. Extra height for the exterior walls, stairs, elevators, and service risers all result in additional cost, making tight control of the level-to-level distance very important.

A major architectural design issue for this building is the choice of a basic form of the construction of the exterior walls. For the column-framed structure, two elements must be integrated: the columns and the nonstructural infill wall. The basic form of the construction as shown in Figure. 20.2 involves the incorporation of the columns into the wall, with windows developed in horizontal strips between the columns. With the exterior column and spandrel covers developing a general continuous surface, the window units are thus developed as "punched" holes in the wall.

The windows in this example do not exist as parts of a continuous curtain wall system. They are essentially single individual units, placed in and supported by the general wall system. The curtain wall is developed as a stud-and-surfacing system not unlike the typical light wood stud wall

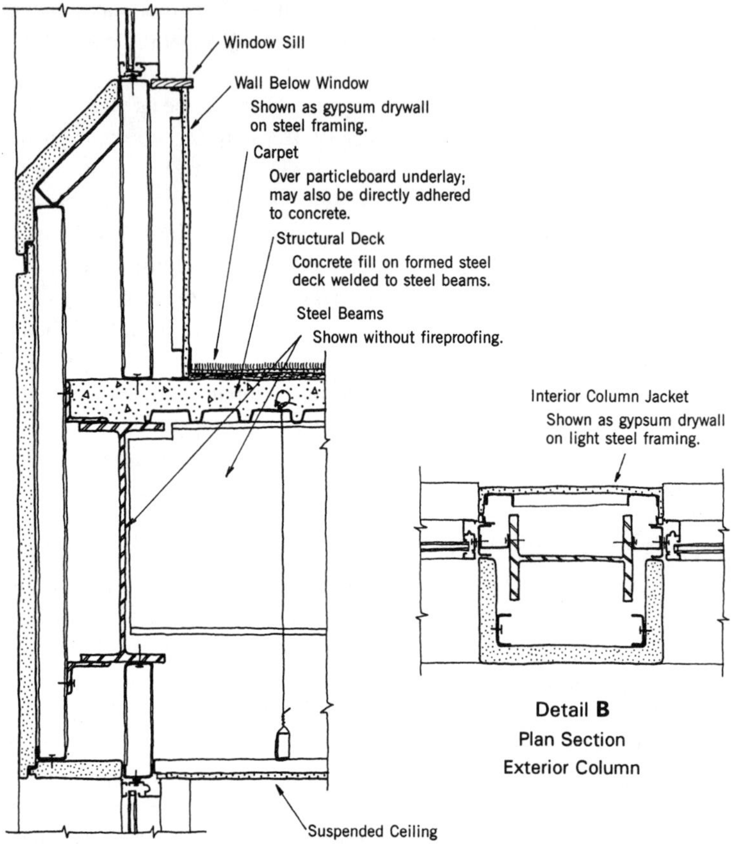

Detail **A**
Vertical Section
Upper Floor at Exterior Wall

Detail **B**
Plan Section
Exterior Column

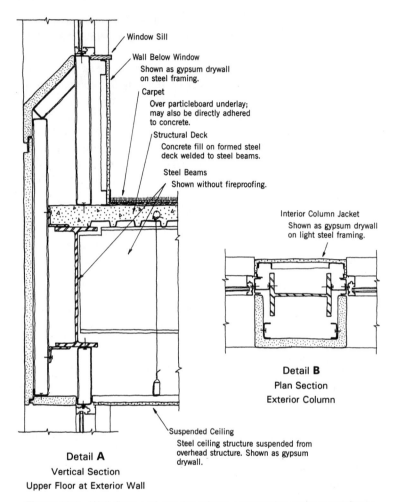

Figure 20.2 Wall, floor, and exterior column construction at the upper floors.

system in character. The studs in this case are light-gage steel; the exterior covering is a system of metal faced sandwich panel units; and the interior covering, where required, is gypsum drywall, attached to the metal studs with screws.

Detailing of the wall construction (as shown in detail A of Figure 20.2) results in a considerable interstitial void space. Although taken up partly with insulation materials, this space may easily contain elements for the electrical system or other services. In cold climates, a perimeter hot-water heating system would most likely be used, and it could be incorporated in the wall space shown here.

Design Criteria

The following are used for the design work:

Design codes: ASCE 2003 Standard (Ref.1) and 1997 UBC (Ref. 2)
Live loads:
  Roof: 20 psf, reducible as described in Section 4.1.
  Floor: From Table 4.2, 50 psf minimum for office areas, 100 psf for lobbies and corridors, 20 psf for movable partitions
Wind: Map speed of 90 mph, exposure B
Assumed construction loads:
  Floor finish: 5 psf
  Ceilings, lights, ducts: 15 psf
  Walls (average surface weight):
  Interior, permanent: 15 psf
  Exterior curtain wall: 25 psf
Steel for rolled shapes: ASTM A36, $F_y = 36$ ksi

20.2 STRUCTURAL ALTERNATIVES

Structural options for this example are considerable, including possibly the light wood frame if the total floor area and zoning requirements permit its use. Certainly, many steel frame, concrete frame, and masonry bearing wall systems are feasible. Choice of the structural elements will depend mostly on the desired plan form, the form of window arrangements, and the clear spans required for the building interior.

At this height and taller, the basic structure must usually be steel, reinforced concrete, or masonry. Options illustrated here include versions in steel and concrete.

Design of the structural system must take into account both gravity and lateral loads. Gravity requires developing horizontal spanning systems for the roof and upper floors and the stacking of vertical supporting elements. The most common choices for the general lateral bracing system are the following (see Figure 20.3):

Core shear wall system (Figure 20.3*a*). Use of solid walls around core elements (stairs, elevators, restrooms, duct shafts) produce a very rigid vertical structure; the rest of the construction may lean on this rigid core.

Truss-braced core. Similar to the shear wall core; trussed bents replace solid walls.

Perimeter shear walls (Figure 20.3*b*). Turns the building into a tube-like structure; walls may be structurally continuous and pierced by holes for windows and doors or may be built as individual, linked piers between vertical strips of openings.

Mixed exterior and interior shear walls or trussed bents. For some building plans, the perimeter or core systems may not be feasible, requiring use of some mixture of walls and/or trussed bents.

Full rigid frame bent system (Figure 20.3*c*). Uses all the available bents described by vertical planes of columns and beams.

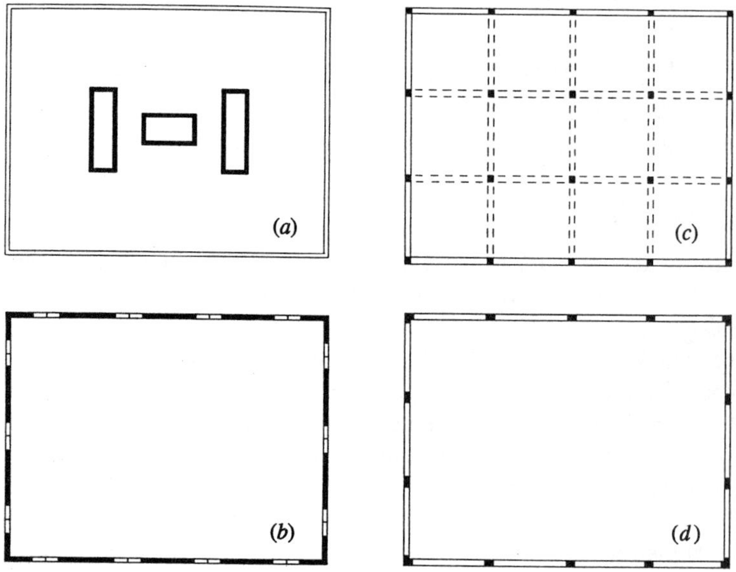

Figure 20.3 Options for the vertical elements of the lateral bracing system. (a) Braced core; shear walls or trussing. (b) Braced perimeter; shear walls or trussing. (c) Fully developed, three-dimensional, rigid frame. (d) Perimeter rigid frames.

Perimeter rigid frame bents (Figure 20.3*d*). Uses only the columns and spandrel beams in the exterior wall planes, resulting in only two bents in each direction for this building plan.

In the right circumstances, any of these systems may be acceptable for this size building. Each has some advantages and disadvantages from both structural and architectural design points of view.

Presented here are schemes for use of three lateral bracing systems: a truss-braced core, a rigid frame bent, and multistory shear walls. For the horizontal roof and floor structures, several schemes are also presented.

20.3 DESIGN OF THE STEEL STRUCTURE

Figure 20.4 shows a partial plan of a framing system for the typical upper floor that uses rolled steel beams spaced at a module related to the column spacing. As shown, the beams are 7.5 ft on center and the beams that are not on the column lines are supported by column line girders. Thus, three-fourths of the beams are supported by the girders and the remainder are supported directly by the columns. The beams in turn support a one-way spanning deck.

Within this basic system, there are a number of variables:

Beam spacing. Affects the deck span and the beam loading.

Deck. A variety available, as discussed later.

Beam/column relationship in plan. As shown, permits possible development of vertical bents in both directions.

Column orientation. The W shape has a strong axis and accommodates framing differently in different directions.

Fire protection. Various means, as related to codes and general building construction.

These issues and others are treated in the following discussions.

Inspection of the framing plan in Figure 20.4 reveals a few common elements of the system as well as several special beams required at the building core. The discussions that follow are limited to treatments of the common elements—that is, the members labeled "Beam" and "Girder" in Figure 20.4.

For the design of the speculative rental building, we must assume that different plan arrangements of the floors are possible. Thus, it is not completely possible to predict where there will be offices and where there

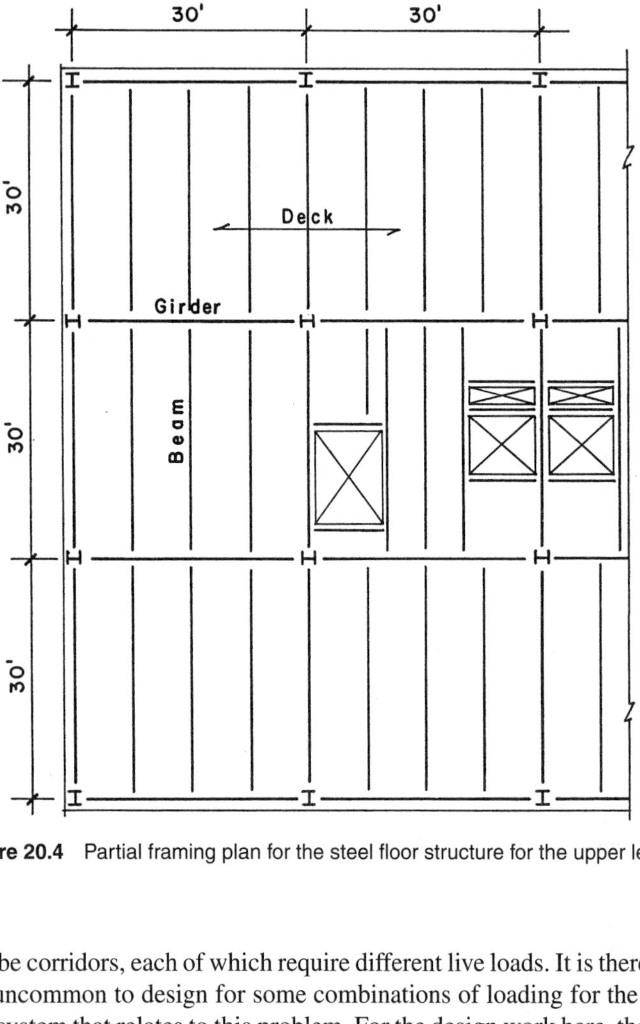

Figure 20.4 Partial framing plan for the steel floor structure for the upper levels.

will be corridors, each of which require different live loads. It is therefore not uncommon to design for some combinations of loading for the general system that relates to this problem. For the design work here, the following will be used:

For the deck, live load = 100 psf

For the beams, live load = 80 psf, with 20 psf added to dead load for movable partitions

For girders and columns, live load = 50 psf, with 20 psf added to dead load

The Structural Deck

Several options are possible for the floor deck. In addition to structural concerns, which include gravity loading and diaphragm action for lateral loads, consideration must be given to fire protection for the steel; to the accommodation of wiring, piping, and ducts; and to attachment of finish floor, roofing, and ceiling constructions. For office buildings, there are often networks for electrical power and communication that must be built into the ceiling, wall, and floor constructions.

If the structural floor deck is a concrete slab, either sitecast or precast, there is usually a nonstructural fill placed on top of the structural slab; power and communication networks may be buried in this fill. If a steel deck is used, closed cells of the formed sheet steel deck units may be used for some wiring, although this is no longer a common practice.

For this example, the selected deck is a steel deck with 1.5-in.-deep ribs, on top of which is cast a lightweight concrete fill with a minimum depth of 2.5 in. over the steel units. The unit average dead weight of this deck depends on the thickness of the sheet steel, the profile of the deck folds, and the unit density of the concrete fill. For this example, it is assumed that the average weight is 30 psf. Adding to this the assumed weight of the floor finish and suspended items, the total dead load for the deck design is thus 50 psf.

Although industry standards exist for these decks (see Ref. 7), data for deck design should be obtained from deck manufacturers.

The Common Beam

As shown in Figure 20.4, this beam spans 30 ft and carries a load strip that is 7.5 ft wide. The total peripheral load support area for the beam is thus $7.5 \times 30 = 225\ \text{ft}^2$. This allows for a reduced live load as follows (see Section 4.2):

$$L = L_o\left(0.25 + \frac{15}{\sqrt{K_{LL}A_T}}\right) = 80\left(0.25 + \frac{15}{\sqrt{2 \times 225}}\right) = 77\ \text{psf}$$

The beam loading is thus:

Live load = 7.5(77) = 578 lb/lineal ft (or plf)

Dead load = 7.5(50) = 525 plf + beam weight, say 560 plf

Total factored unit load = 1.2(560) + 1.6(578) = 672 +925 = 1597 plf

Total supported factored load = 1.597(30) = 48 kips

Assuming that the welding of the steel deck to the top flange of the beam provides almost continuous lateral bracing, the beam may be selected on the basis of flexural failure. For this load and span, Table 9.1 yields the following possible choices: W 16 × 45, W 18 × 40, or W 21 × 44. Actual choice may be affected by various considerations. For example, the table used does not incorporate concerns for deflection or lateral bracing. The deeper shape will obviously produce the least deflection, although in this case the live load deflection for the 16-in. shape is within the usual limit (see Figure 9.5). This beam becomes the typical member, with other beams being designed for special circumstances, including the column-line beams, the spandrels, and so on.

The Common Girder

Figure 20.5 shows the loading condition for the girder, as generated only by the supported beams. Although this ignores the effect of the weight of the girder as a uniformly distributed load, it is reasonable for use in an approximate design because the girder weight is a minor loading.

Note that the girder carries three beams and thus has a total load periphery of 3(225) = 675 ft^2. The reduced live load is thus (see Section 4.1):

$$L = (80 \times 675)\left(0.25 + \frac{15}{\sqrt{2 \times 675}}\right) = 35{,}100 \text{ lb, or } 35.1 \text{ kips}$$

The unit beam load for design of the girder is determined as follows:

Live load = 35.1/3 = 11.7 kips.

Dead load = 0.560(30) = 16.8 kips.

Factored unit load = 1.2(16.8) + 1.6(11.7) = 38.88 kips.

To account for beam weights, use 40 kips.

From Figure 20.5, maximum moment is 600 k-ft.

Selection of a member for this situation may be made using various data sources. Because this member is laterally braced at only 7.5-ft intervals, attention must be paid to this point. The maximum moment together with the laterally unbraced length can be used in Table 9.2 to determine acceptable choices. Possibilities are W 18 × 106, W 21 × 101, W 24 × 94, W 27 × 94 or W 30 × 90. The deeper members will have less deflection and will allow greater room for building service elements

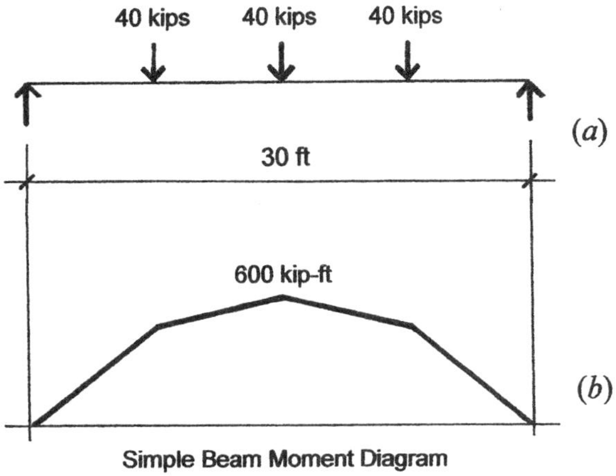

Figure 20.5 Loading condition for gravity load effects on the girder: (*a*) loads from the supported beams, (*b*) form of the simple beam moment diagram.

in the enclosed floor/ceiling space. However, shallower beams may reduce the required story height, resulting in cost savings.

Computation for deflections may be performed with formulas that recognize the true form of loading. However, approximate deflection values may be found using an equivalent load derived from the maximum moment, as discussed in Section 9.4. For this example, the equivalent uniform load (EUL) is obtained as follows:

$$M = \frac{WL}{8} = 600 \text{ kip-ft}$$

$$W = \frac{8M}{L} = \frac{8 \times 600}{30} = 160 \text{ kips}$$

This hypothetical uniformly distributed load may be used with the formula for deflection of a simple beam (see Figure 3.25) to find an approximate deflection. However, for a quick check, Figure 9.11 indicates that for this span all the previous choice options will have a total load deflection of less than 1/240 of the span.

Although deflection of individual elements should be investigated, there are wider issues regarding deflection, such as the following:

Bounciness of floors. This involves the stiffness and the fundamental period of spanning elements and may relate to the deck and/or the beams. In general, use of the static deflection limits usually ensures a reasonable lack of bounce, but just about anything that increases stiffness improves the situation.

Transfer of load to nonstructural walls. With the building construction completed, live load deflections of the structure may result in bearing of spanning members on nonstructural construction. Reducing deflections of the structure will help for this, but some special details may be required for attachment between the structure and the nonstructural construction.

Deflection during construction. The deflection of the girders plus the deflection of the beams adds up to a cumulative deflection at the center of a column bay. This may be critical for live load but can also create problems during construction. If the steel beams and steel deck are installed dead flat, then construction added later will cause deflection from the flat condition. In this example, that would include the concrete fill, which can cause a considerable deflection at the center of the column bay. One response is to camber (bow upward) the beams by bending them in the fabricating shop so that they deflect to a flat position under the dead load.

Column Design for Gravity Loads

Design of the steel columns must include considerations for both gravity and lateral loads. Gravity loads for individual columns are based on the column's *periphery*, which is usually defined as the area of supported surface on each level supported. Loads are actually delivered to the columns by the beams and girders, but the peripheral area is used for load tabulation and determination of live load reductions.

If beams are rigidly attached to columns with moment-resistive connections—as is done in development of rigid frame bents—then gravity loads will also cause bending moments and shears in the columns. Otherwise, the gravity loads are essentially considered only as axial compressive loads.

Involvement of the columns in development of resistance to lateral loads depends on the form of the lateral bracing system. If trussed bents are used, some columns will function as chords in the vertically cantilevered trussed bents, which will add some compressive forces and possibly cause some reversals with net tension in the columns. If columns are parts of rigid frame bents, the same chord actions will be involved, but the columns will also be subject to bending moments and shears from the rigid frame lateral actions.

Whatever the lateral force actions may do, the columns must also work for gravity load effects alone. In this part, this investigation is made and designs are completed without reference to lateral loads. This yields some reference selections, which can then be modified (but not reduced) when the lateral resistive system is designed. Later discussions in this chapter present designs for both a trussed bent system and a rigid frame system.

There are several different cases for the columns, due to the framing arrangements and column locations. For a complete design of all columns, it would be necessary to tabulate the loading for each different case. For illustration purpose here, tabulation is shown for a hypothetical interior column. The interior column illustrated assumes a general periphery of 900 sq ft of general roof or floor area. Actually, the floor plan in Figure 23.1 shows that all the interior columns are within the core area, so there is no such column. However, the tabulation yields a column that is general for the interior condition and can be used for approximate selection. As will be shown later, all the interior columns are involved in the lateral force systems, so this also yields a takeoff size selection for the design for lateral forces.

Table 20.1 is a common form of tabulation used to determine the column loads. For the interior columns, the table assumes the existence of a rooftop structure (penthouse) above the core, thus creating a fourth story for these columns.

Table 20.1 is organized to facilitate the following determinations:

1. Dead load on the periphery at each level, determined by multiplying the area by an assumed average dead load per sq ft. Loads determined in the process of design of the horizontal structure may be used for this estimate.
2. Live load on the periphery areas.
3. The reduced live load to be used at each story, based on the total supported periphery areas above that story.
4. Other dead loads directly supported, such as the column weight and any permanent walls within the load periphery.
5. The total load collected at each level.
6. A design load for each story, using the total accumulation from all levels supported.

TABLE 20.1 Service Load Tabulation for the Interior Column

		Load Tabulation (lb)	
Level Supported	Load Source and Computation	Dead Load	Live Load
Penthouse roof	Live load, not reduced = 20 psf × 225		4,500
225 ft²	Dead load = 40 psf × 225	9,000	
Building roof	Live load, not reduced = 20 psf × 675		13,500
675 ft²	Dead load = 40 psf × 675	27,000	
Penthouse floor	Live load = 100 psf × 225		22,500
225 ft²	Dead load = 50 psf × 225	11,250	
	Story loads + loads from above	47,250	40,500
	Reduced live load (50%)		20,250
Third floor	Live load = 50 psf × 900		45,000
900 ft²	Dead load = 70 psf × 900	63,000	
	Story loads + loads from above	110,250	85,000
	Reduced live load (50%)		42,500
Second floor	Story loads	63,000	45,000
(same as third)	Story loads + loads from above	173,250	130,500
	Reduced live load (50%)		65,250

For the entries in Table 20.1, the following assumptions were made:

Roof unit live load = 20 psf (reducible)

Roof dead load = 40 psf (estimated, based on the similar floor construction)

Penthouse floor live load = 100 psf (for equipment, average)

Penthouse floor dead load = 50 psf

Floor live load = 50 psf (reducible)

Floor dead load = 70 psf (including partitions)

Table 20.2 summarizes the design for the four-story column. For the pin-connected frame, a *K* factor of 1.0 is assumed and the full story heights are used as the unbraced column lengths.

Although column loads in the upper stories are quite low, and some small column sizes would be adequate for the loads, a minimum size of 10 in. is maintained for the W shapes for two reasons.

The first consideration involves the form of the horizontal framing members and the type of connections between the columns and the hor-

TABLE 20.2 Design of the Interior Column

Design Loads for Each Story (lb)	Possible Choices
Penthouse, unbraced height = 13 ft $P_u = \phi P_n = 1.2(9{,}000) + 1.6(4{,}500) = 10{,}800 + 7{,}200$ $= 18{,}000$ lb	W 8 × 24
Third Story, unbraced height = 13 ft $P_u = \phi P_n = 1.2(47{,}250) + 1.6(20{,}250) = 58{,}700 + 32{,}400$ $= 91{,}100$ lb	W 8 × 24, W 10 × 33
Second Story, unbraced height = 13 ft $P_u = \phi P_n = 1.2(110{,}250) + 1.6(42{,}500) = 132{,}300 + 68{,}000$ $= 200{,}300$ lb	W 8 ×31, W 10 × 45, W 12 × 45
First Story, unbraced height = 15 ft $P_u = \phi P_n = 1.2(173{,}250) + 1.6(65{,}250) = 207{,}900 + 104{,}400$ $= 312{,}300$ lb	W 8 ×58, W 10 × 54, W 12 × 53, W 14 × 53

izontal framing. All the H-shaped columns must usually facilitate framing in both directions, with beams connected both to column flanges and webs. With standard framing connections for field bolting to the columns, a minimum beam depth and flange width are required for practical installation of the connecting angles and bolts.

The second consideration involves the problem of achieving splices in the multistory column. If the building is too tall for a single-piece column, a splice must be used somewhere, and the stacking of one column piece on top of another to achieve a splice is made much easier if the two pieces are of the same nominal size group.

Add to this a possible additional concern relating to the problem of handling long pieces of steel during transportation to the site and erection of the frame. The smaller the members cross section, the shorter the piece that is feasible to handle.

For all of these reasons, a minimum column is often considered to be the W 10 × 33, which is the lightest shape in the group that has an 8-in.-wide flange. It is assumed that a splice occurs at 3 ft above the second-floor level (a convenient, waist-high distance for the erection crew), making two column pieces approximately 18 and 23 ft long. These lengths are readily available and quite easy to handle with the 10-in. nominal shape. On the basis of these assumptions, a possible choice would be for a W 10 × 33 for the penthouse and the third-story column and a W 10 × 54 for the lower two stories.

20.4 ALTERNATIVE FLOOR CONSTRUCTION WITH TRUSSES

A framing plan for the upper floor of Building Three is shown in Figure 20.6, indicating the use of open web steel joists and joist girders. Although this construction might be extended to the core and the exterior spandrels, it is also possible to retain the use of rolled shapes for these purposes. There are various possibilities for the development of lateral bracing with this scheme; one is the use of the same trussed core bents that were previously designed. Although somewhat more applicable to longer spans and lighter loads, this system is reasonably applicable to this situation as well.

One potential advantage of using the all truss framing for the horizontal structure is the higher degree of freedom of passage of building service elements within the enclosed space between ceilings and the sup-

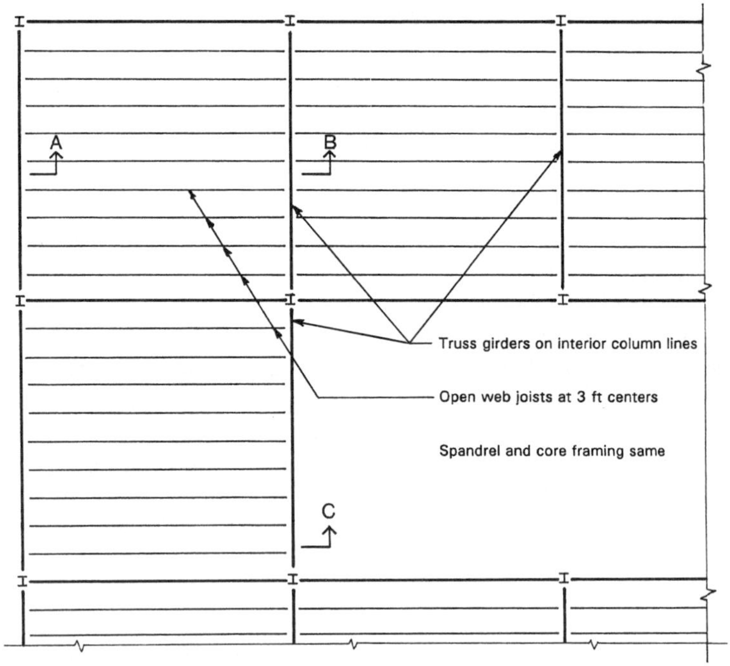

Figure 20.6 Building Three: partial framing plan at the upper floor using open-web joists and joist girders.

ported structure above. A disadvantage is the usual necessity for greater depth of the structure, adding to building height—a problem that increases with the number of stories.

Design of the Open Web Joists

General concerns and basic design for open web joists are presented in Section 9.10. Using the data for this example, a joist design is as follows:

Joists at 3 ft on center, span of 30 ft.

Dead load = 3(70) = 210 lb/ft not including joists.

Live load = 3(100) = 300 lb/ft not reduced. This is a high live load, but it permits location of a corridor anywhere on the plan and also reduces deflection and bounciness.

Total factored load = 1.2(210) + 1.6(300) = 252 + 480 = 732 lb/ft

Referring to Table 9.3, note that choices may be considered for any joist that will carry the total load of 732 lb/ft and a live load of 300 lb/ft on a span of 30 ft. The following choices are possible:

24K9 at 12 lb/ft, permitted load = 807 – 1.2(12) = 794 lb/ft

26K9, stronger than 24K9 and only 0.2 lb/ft heavier

28K8, stronger than 24K9 and only 0.7 lb/ft heavier

30K7, stronger than 24K9 and only 0.3 lb/ft heavier

All of the joists listed are economically equivalent. Choice would be made considering details of the general building construction. A shallower joist depth means a shorter story height and less overall building height. A deeper joist yields more open space in the floor construction for ducts, wiring, piping, and so on and also means less deflection and less floor bounce.

Design of the Joist Girders

Joist girders are also discussed in Section 9.10. Both the joists and the girders are likely to be supplied and erected by a single contractor. Although there are industry standards (see Ref. 6), consult the specific manufacturer for data regarding design and construction details for these products.

The pattern of the joist girder members is somewhat fixed and relates to the spacing of the supported joists. To achieve a reasonable proportion for the panel units of the truss, the dimension for the depth of the girder should be approximately the same as that for the joist spacing.

Considerations for design for the truss girder are as follows:

The assumed depth of the girder is 3 ft, which should be considered a *minimum* depth for this span (L/10). Any additional depth possible will reduce the amount of steel and also improve deflection responses. However, for floor construction in multistory buildings, this dimension is hard to bargain for.

Use a live load reduction of 40% (maximum with a live load of 50 psf). Thus, live load from one joist = (3 × 30)(0.6 × 50) = 2700 lb or 2.7 kips.

For dead load add a partition load of 20 psf to the construction load of 40 psf. Thus, dead load = (3 × 30)(60) = 5400 lb + joist weight of 12 lb/ft × 30 = 360 lb, total = 5400 + 360 = 5760 lb or 5.76 kips

Total factored load = 1.2(5.76) + 1.6(2.7) = 6.91 + 4.32 = 11.23 kips

Figure 20.7 shows a possible form for the joist girder. For this form and the computed data, the joist specification is as follows:

36G = a girder depth of 3 ft

10N = ten spaces between the joists

11.23K = the design factored joist load

Complete specification is thus 36G10N11.23K

Construction Details for the Truss Structure

Figure 20.8 shows some details for construction of the trussed system. The deck shown here is the same as that for the scheme with W-shape

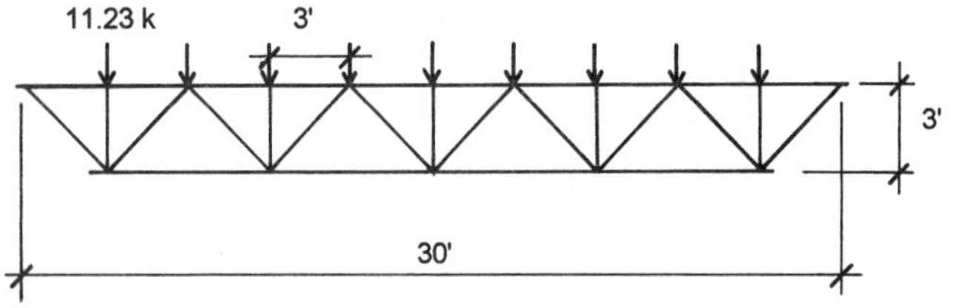

Figure 20.7 Layout form and data for the joist girder.

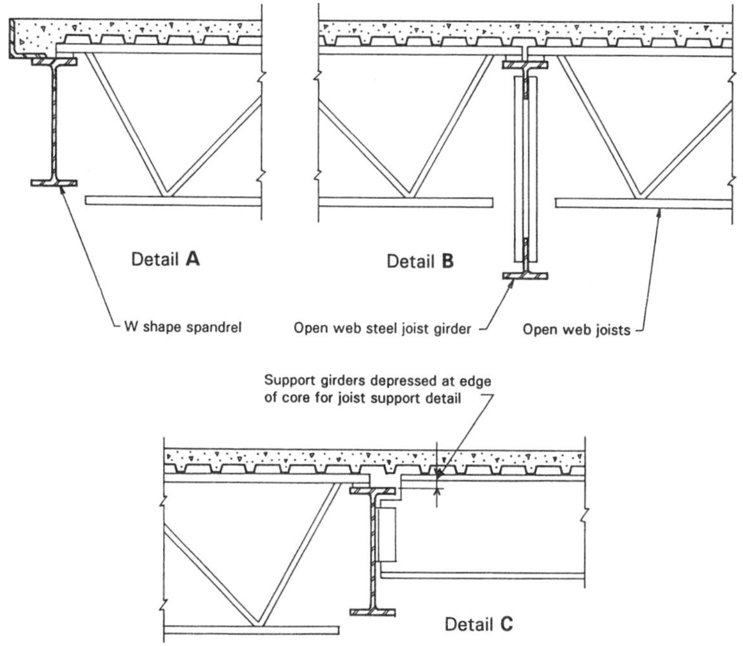

Figure 20.8 Details for the floor system with open-web joists and joist girders. For location of details see the framing plan in Fig. 20.6.

framing, although the shorter span may allow use of a lighter sheet steel deck. However, the deck must also be used for diaphragm action, which may limit the reduction.

Adding to the problem of overall height for this structure is the detail at the joist support, in which the joists must sit on top of the supporting members, whereas in the all W-shape system the beams and girders have their tops level.

With the closely spaced joists, ceiling construction may be directly supported by the bottom chords of the joists. This may be a reason for selection of the joist depth. However, it is also possible to suspend the ceiling from the deck, as is generally required for the all W-shape structure with widely spaced beams.

Another issue here is the usual necessity to use a fire-resistive ceiling construction, as it is not feasible to encase the joists or girders in fireproofing material.

20.5 DESIGN OF THE TRUSSED BENT FOR WIND

Figure 20.9 shows a partial framing plan for the core area, indicating the placement of some additional columns off the 30-ft grid. These columns are used together with the regular columns and some of the horizontal framing to define a series of vertical bents for the development of the trussed bracing system shown in Figure 20.10. With relatively slender diagonal members, it is assumed that the X-bracing behaves as if the tension diagonals function alone. There are thus considered to be four vertical, cantilevered, determinate trusses that brace the building in each direction.

With the symmetrical building exterior form and the symmetrically placed core bracing, this is a reasonable system for use in conjunction with the horizontal roof and upper floor structures to develop resistance to horizontal forces due to wind. The work that follows illustrates the design process, using criteria for wind loading from ASCE 2003 (Ref. 1).

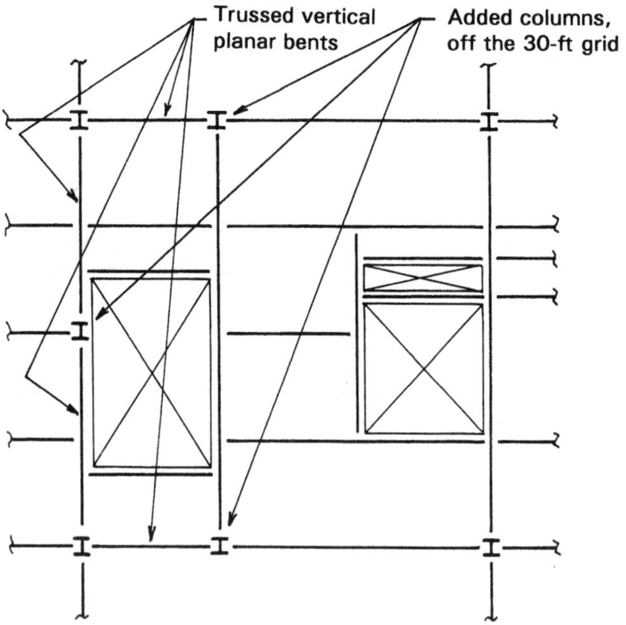

Figure 20.9 Modified framing plan for development of the trussed bents at the core.

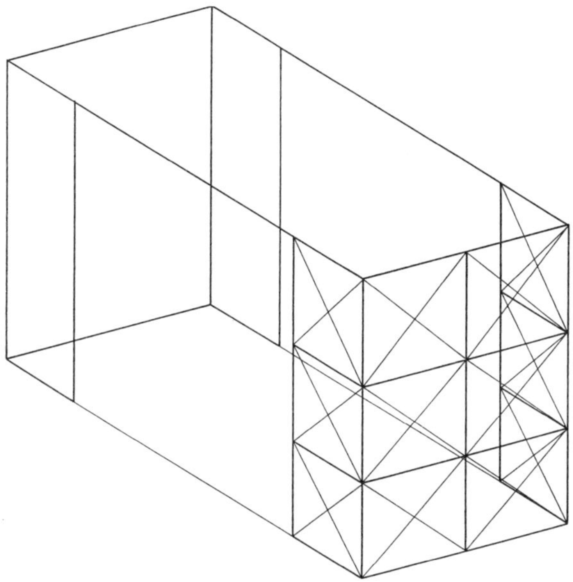

Figure 20.10 General form of the trussed bent bracing system at the core.

For the total wind force on the building, we will assume a base pressure of 15 psf, adjusted for height as described in ASCE 2003 (Ref. 1). The design pressures and their zones of application are shown in Figure 20.11.

For investigation of the lateral bracing system, the design wind pressures on the outside wall surface are distributed as edge loadings to the roof and floor diaphragms. These are shown as the forces H_1, H_2, and H_3 in Figure 20.11. The horizontal forces are next shown as loadings to one of the vertical truss bents in Figure 20.12*a*. For the bent loads, the total forces per bent is determined by multiplying the unit edge diaphragm load by the building width and dividing by the number of bracing bents for load in that direction. The bent loads are thus

$$H_1 = (165.5)(92)/4 = 3807 \text{ lb}$$

$$H_2 = (199.5)(92)/4 = 4589 \text{ lb}$$

$$H_3 = (210)(92)/4 = 4830 \text{ lb}$$

The truss loading, together with the reaction forces at the supports, are shown in Figure 20.12*b*. The internal forces in the truss members result-

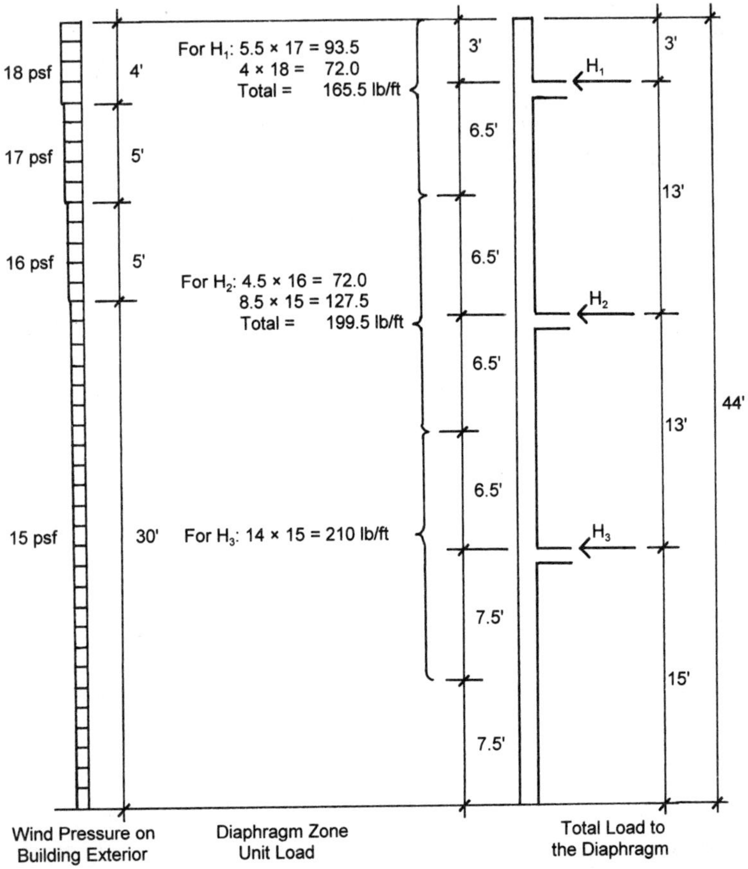

Figure 20.11 Building Three: development of the wind loads transferred to the upper level roof and floor deck diaphragms. Design wind pressures on the building exterior from ASCE 2002 (Ref. 1), assuming a base wind pressure of 15 psf. Diaphragm zones are defined by column mid-height points. Total loads on the diaphragm at each level are found by multiplying the diaphragm zone unit load per foot by the width of the building.

ing from this loading are shown in Figure 20.12*c*, with force values in pounds and sense indicated by *C* for compression and *T* for tension.

The forces in the diagonals may be used to design tension members, using the factored load combination that includes wind (see Section 4.2). The compression forces in the columns may be added to the gravity loads to see if this load combination is critical for the column design. The up-

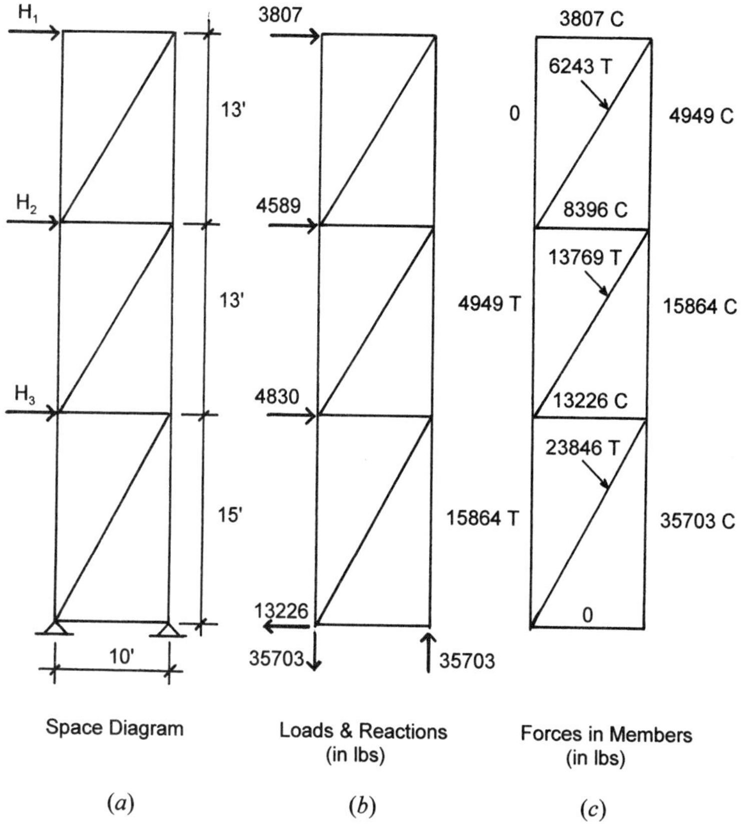

Figure 20.12 Building Three: investigation of one of the east-west core bents. (a) Layout and loading of the bent; loads shown are one fourth of the total diaphragm loads. (b) External forces (loads and reactions) on the cantilevered truss. (c) Internal forces in the truss members.

lift tension force at the column should be compared with the dead load to see if the column base needs to be designed for a tension anchorage force.

The horizontal forces should be added to the beams in the core framing, and an investigation should be done for the combined bending and compression. Because beams are often weak on their minor axis (y-axis), it may be practical to add some framing members at right angles to these beams to brace them against lateral buckling.

Design of the diagonals and their connections to the beam and column frame must be developed with consideration of the form of the elements

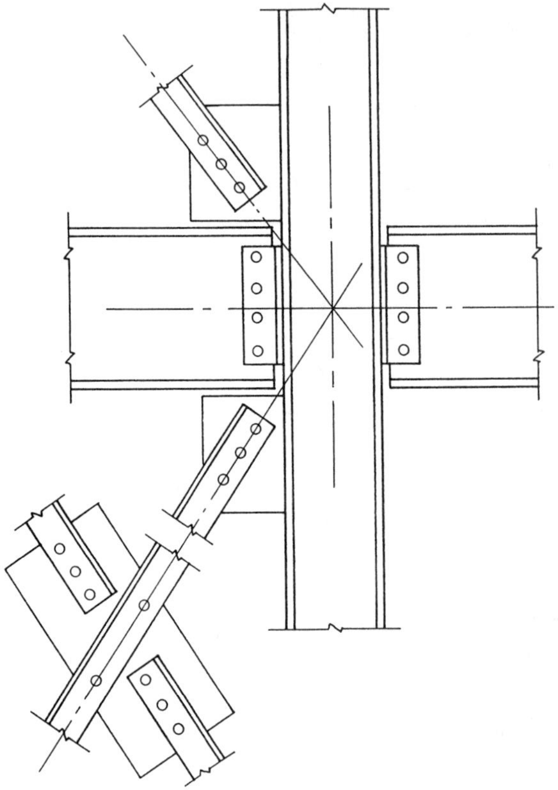

Figure 20.13 Details of the bent construction with bolted joints.

and some consideration for the wall construction in which they are imbedded. Figure 20.13 shows some possible details for the diagonals and the connections. A detail problem that must be solved is that of the crossing of the two diagonals at the middle of the bent. If we use double angles for the diagonals (a common truss form), the splice joint shown in Figure 20.13 is necessary. An option is to use either single angles or channel shapes for the diagonals, allowing the members to pass each other back-to-back at the center. The latter choice, however, involves some degree of eccentricity in the members and connections and a single shear load on the bolts, so it is not advisable if load magnitudes are high. For the tension member, a recommended minimum slenderness is represented by an *L*/*r* ratio of 300.

20.6 CONSIDERATIONS FOR A STEEL RIGID FRAME

The general nature of rigid frames is discussed in Section 3.12. A critical concern for multistory, multiple-bay frames is the lateral strength and stiffness of columns. Because the building must be developed to resist lateral forces in all directions, it becomes necessary in many cases to consider the shear and bending resistance of columns in two directions (north-south and east-west, for example). This presents a problem for W-shape columns because they have considerably greater resistance on their major (*x*-*x*) axis versus their minor (*y*-*y*) axis. Orientation of W-shape columns in plan thus sometimes becomes a major consideration in structural planning.

Figure 20.14*a* shows a possible plan arrangement for column orientation for Building Three, relating to the development of two major brac-

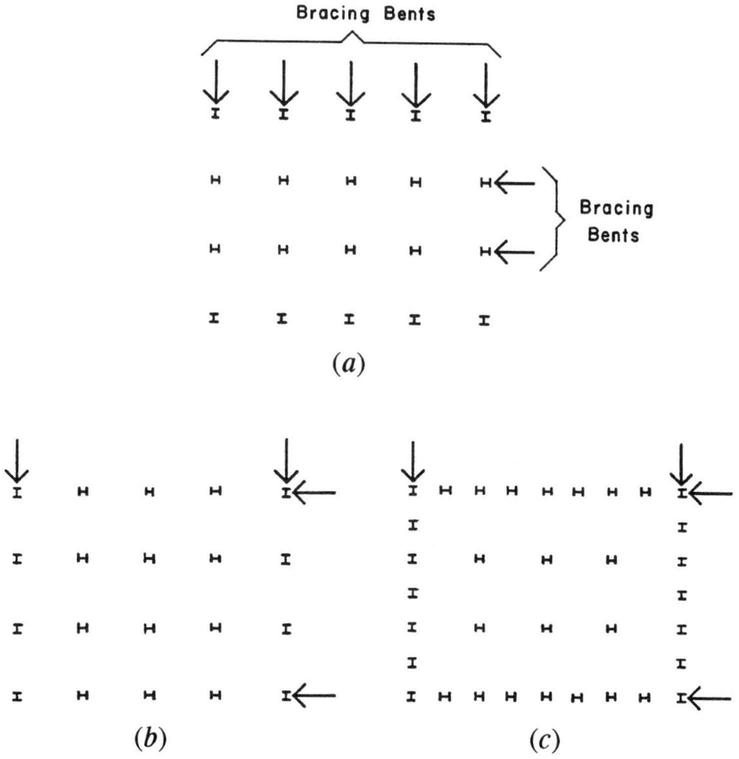

Figure 20.14 Building Three: optional arrangements for the steel W-shape columns for development of the rigid frame bents.

ing bents in the east-west direction and five shorter and less stiff bents in the north-south direction. The two stiff bents may well be approximately equal in resistance to the five shorter bents, giving the building a reasonably symmetrical response in the two directions.

Figure 20.14*b* shows a plan arrangement for columns designed to produce approximately symmetrical bents on the building perimeter. The form of such perimeter bracing is shown in Figure 20.15.

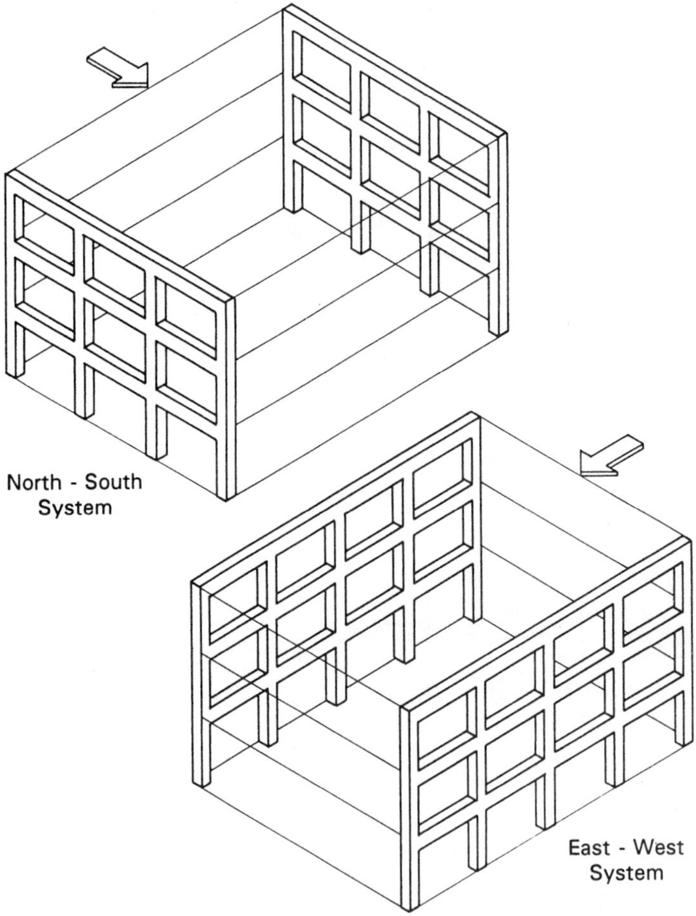

Figure 20.15 Building Three: form of the perimeter bent bracing system.

One advantage of perimeter bracing is the potential for using deeper (and thus stiffer) spandrel beams, because the restriction on depth that applies for interior beams does not exist at the exterior wall plane. Another possibility is to increase the number of columns at the exterior, as shown in Figure 20.14*c*—a possibility that does not compromise the building interior space. With deeper spandrels and closely spaced exterior columns, a very stiff perimeter bent is possible. In fact, such a bent may have very little flexing in the members, and its behavior approaches that of a pierced wall, rather than a flexible frame.

At the expense of requiring much stronger (and heavier and/or larger) columns and expensive moment-resistive connections, the rigid frame bracing offers architectural planning advantages with the elimination of solid shear walls or truss diagonals in the walls. However, the lateral deflection (drift) of the frames must be carefully controlled, especially with regard to damage to nonstructural parts of the construction.

20.7 CONSIDERATIONS FOR A MASONRY WALL STRUCTURE

An option for the construction of Building Three involves the use of structural masonry for development of the exterior walls. The walls are used for both vertical bearing loads and lateral shear wall functions. The choice of forms of masonry and details for the construction depend very much on regional considerations (climate, codes, local construction practices, etc.) and on the general architectural design. Major differences occur due to variations in the range in outdoor temperature extremes and the specific critical concerns for lateral forces.

General Considerations

Figure 20.16 shows a partial elevation of the masonry wall structure and a partial framing plan of the upper floors. The wood construction shown here is questionably acceptable for fire codes. The example is presented only to demonstrate the general form of the construction.

Plan dimensions for structures using CMUs (concrete blocks) must be developed so as to relate to the modular sizes of typical CMUs. Standard sizes are widely used, but individual manufacturers often have some special units or will accommodate requests for special shapes or sizes. However, whereas solid brick or stone units can be cut to produce precise, nonmodular dimensions, the hollow CMUs generally cannot. Thus, the dimensions for the CMU structure itself must be carefully developed

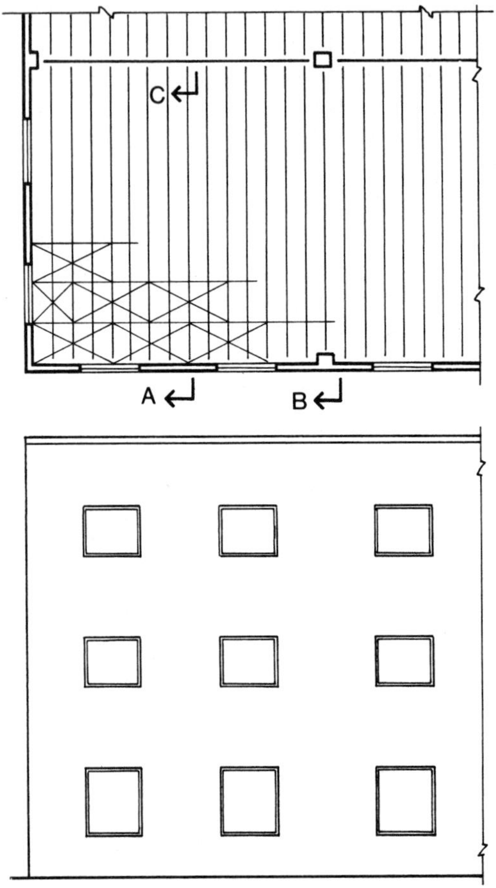

Figure 20.16 Building Three: partial framing plan for the upper floor and partial elevation for the masonry wall structure.

to have wall intersections, corners, ends, tops, and openings for windows and doors fall on the fixed modules. (See Figure18.7*a*.)

There are various forms of CMU construction. The one shown here is that widely used where either windstorm or earthquake risk is high. This is described as *reinforced masonry* and is produced to generally emulate reinforced concrete construction, with tensile forces resisted by steel reinforcement that is grouted into the hollow voids in the block construction. This construction takes the general form shown in Figure 20.17.

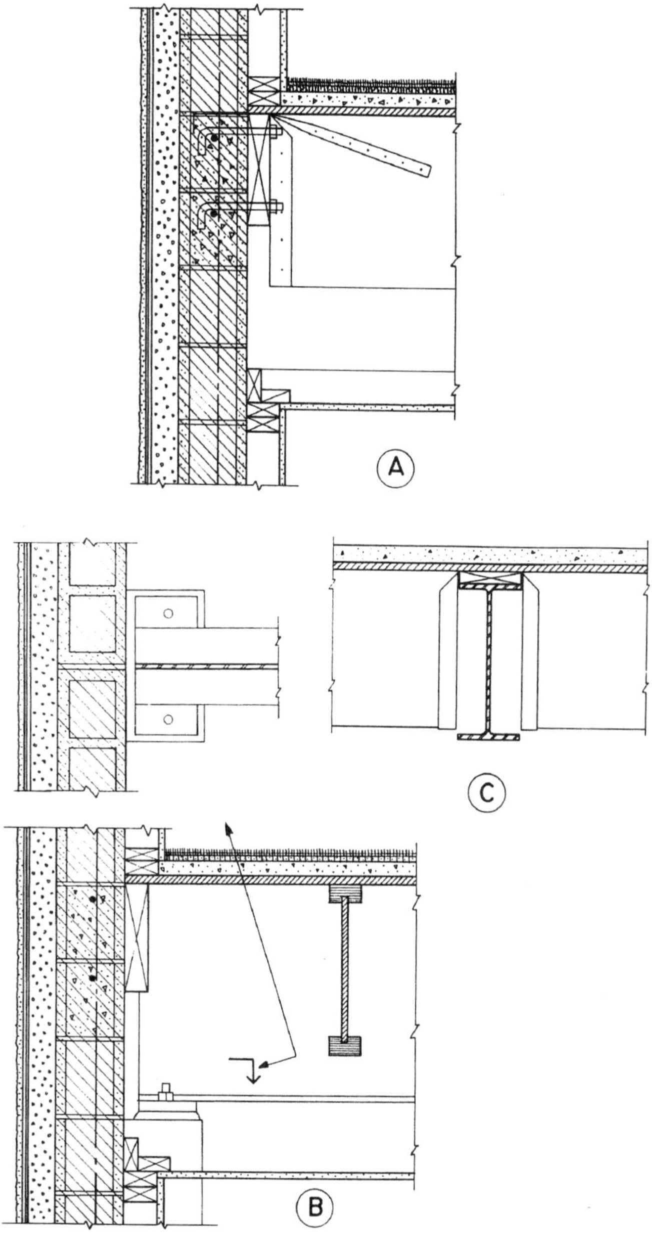

Figure 20.17 Details of the upper floor and exterior masonry wall construction.

Another consideration to be made for the general construction is that involving the relation of the structural masonry to the complete architectural development of the construction, regarding interior and exterior finishes, insulation, incorporation of wiring, and so on.

The Typical Floor

The floor framing system here uses column-line girders that support fabricated joists and a plywood deck. The girders could be glued laminated timber but are shown here as rolled steel shapes. Supports for the steel girders consist of steel columns on the interior and masonry pilaster columns at the exterior walls.

As shown in the details in Figure 20.17, the exterior masonry walls are used for direct support of the deck and the joists, through ledgers bolted and anchored to the interior wall face. With the plywood deck also serving as a horizontal diaphragm for lateral loads, the load transfers for both gravity and lateral forces must be carefully developed in the details for this construction.

The development of the girder support at the exterior wall is a bit tricky with the pilaster columns, which must not only provide support for the individual girders at each level but also maintain the vertical continuity of the column load from story to story. There are various options for the development of this detail. Detail C in Figure 20.17 indicates the use of a wide pilaster that virtually straddles the relatively narrow girder. The void created by the girder is thus essentially ignored, with the two outer halves of the pilaster bypassing it for a real vertical continuity.

Because the masonry structure in this scheme is used only for the exterior walls, the construction at the building core is free to be developed by any of the general methods shown for other schemes. If the steel girders and steel columns are used here, it is likely that a general steel framing system might be used for most of the core framing.

Attached only to its top, the supported construction does not provide very good lateral support for the steel girder in resistance to torsional buckling. It is advisable, therefore, to use a steel shape that is not too weak on its y-axis, generally indicating a critical concern for lateral unsupported length. Rotation of the girder at the supports might be resisted by the encasement in the pilaster column here.

The Masonry Walls

Buildings much taller than this have been achieved with structural masonry, so the feasibility of the system is well demonstrated. The vertical loads increase in lower stories, so it is expected that some increases in structural capacity will be achieved in lower portions of the walls. The two general means for increasing wall strength are to use thicker CMUs or to increase the amount of core grouting and reinforcement.

It is possible that the usual minimum structure—relating to code minimum requirements for the construction—may be sufficient for the top story walls, with increases made in steps for lower walls. Without increasing the CMU size, there is considerable range between the minimum and the feasible maximum potential for a wall.

It is common to use fully grouted walls (all cores filled) for CMU shear walls. In this scheme, that would technically involve using fully grouted construction for *all* of the exterior walls, which might likely rule against the economic feasibility of this scheme. Adding this to the concerns for thermal movements in the long walls might indicate the wisdom of using some control joints to define individual wall segments.

Design for Lateral Forces

A common solution for lateral bracing is the use of an entire masonry wall as a shear wall, with openings considered as producing the effect of a very stiff rigid frame. As for gravity loads, the total lateral shear force increases in lower stories. Thus, it is also possible to consider the use of the potential range for a wall from minimum construction (defining a minimum structural capacity) to the maximum possible strength with all voids grouted and some feasible upper limit for reinforcement.

The basic approach here is to design the required wall for each story, using the total shear at that story. In the end, however, the individual story designs must be coordinated for the continuity of the construction. However, it is also possible that the construction itself could be significantly altered in each story, if it fits with architectural design considerations.

Construction Details

There are many concerns for the proper detailing of the masonry construction to fulfill the shear wall functions. There are also many concerns for proper detailing to achieve the force transfers between the horizontal framing and the walls. The general framing plan for the upper floor is shown in Figure 20.16. The location of the details discussed here is indicated by the section marks on that plan.

Detail A. This shows the general exterior wall construction and the framing of the floor joists at the exterior wall. The wood ledger is used for vertical support of the joists, which are hung from steel framing de-

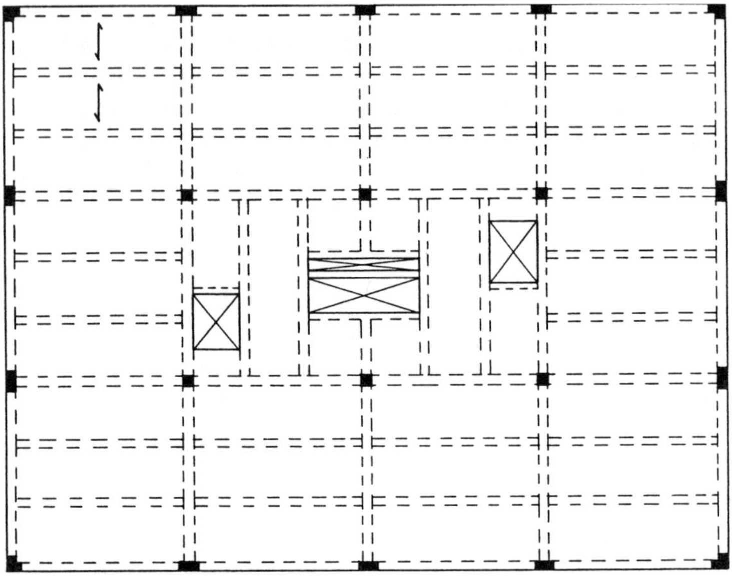

Figure 20.18 Building Three: framing plan for the concrete structure for the upper floor.

vices fastened to the ledger. The plywood deck is nailed directly to the ledger to transfer its horizontal diaphragm loads to the wall.

Outward forces on the wall must be resisted by anchorage directly between the wall and the joists. Ordinary hardware elements can be used for this, although the exact details depend on the type of forces (wind or seismic), their magnitude, the details of the joists, and the details of the wall construction. The anchor shown in the detail is really only symbolic.

General development of the construction here shows the use of a concrete fill on top of the floor deck, furred out wall surfacing with batt insulation on the interior wall side, and a ceiling suspended from the joists.

Detail B. This shows the section and plan details at the joint between the girder and the pilaster. The pilaster unavoidably creates a lump on the inside of the wall in this scheme. However, it is also possible to move the pilaster to the outside of the wall.

Detail C. This shows the use of the steel beam for support of the joists and the deck. After the wood lumber piece is bolted to the top of the steel beam, the attachment of the joists and the deck become essentially the same as it would be with a timber girder.

20.8 THE CONCRETE STRUCTURE

A structural framing plan for the upper floors in Building Three is presented in Figure 20.18, showing the use of a sitecast concrete slab and beam system. Support for the spanning structure is provided by concrete columns. The system for lateral bracing uses the exterior columns and spandrel beams as rigid frame bents at the building perimeter, as shown in Figure 20.15. This is a highly indeterminate structure for both gravity and lateral loads, and its precise engineering design would undoubtedly be done with a computer-aided design process. The presentation here treats the major issues and illustrates an approximate design using highly simplified methods.

Design of the Slab and Beam Floor Structure

For the floor structure, use $f'_c = 3$ ksi and $f_y = 40$ ksi. As shown in Figure 20.18, the floor framing system consists of a series of parallel beams at 10-ft centers that support a continuous, one-way spanning slab and are in turn supported by column-line girders or directly by columns. Although special beams are required for the core framing, the system is made up largely of repeated elements. The discussion here will focus on three of these elements: the continuous slab, the four-span interior beam, and the three-span spandrel girder.

Using the approximation method described in Section 14.1, the critical conditions for the slab, beam, and girder are shown in Figure 20.19. Use of these coefficients is reasonable for the slab and beam that support uniformly distributed loads. For the girder, however, the presence of major concentrated loads makes the use of the coefficients somewhat questionable. An adjusted method is thus described later for use with the girder. The coefficients shown in Figure 20.19 for the girder are for uniformly distributed loads only (the weight of the girder itself, for example).

Figure 20.20 shows a section of the exterior wall that illustrates the general form of the construction. The exterior columns and the spandrel beams are exposed to view. Use of the full available depth of the spandrel beams results in a much stiffened bent on the building exterior. As will be shown later, this is combined with the use of oblong-shaped columns at the exterior to create perimeter bents that will indeed absorb most of the lateral force on the structure.

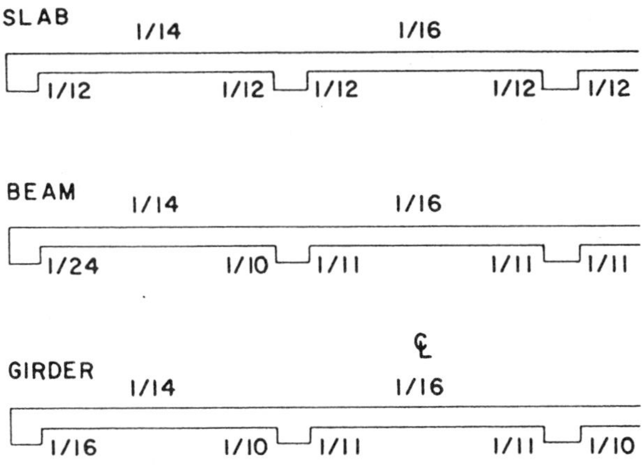

Figure 20.19 Approximate design factors for the slab and beam floor structure.

The design of the continuous slab is presented as the example in Section 14.1. The use of the 5-in. slab is based on assumed minimum requirements for fire protection. If a thinner slab is possible, the 9-ft clear span would not require this thickness based on limiting bending or shear conditions or recommendations for deflection control. If the 5-in. slab is used, however, the result will be a slab with a low percentage of steel bar weight per square foot—a situation usually resulting in lower cost for the structure.

The unit loads used for the slab design are determined as follows:

Floor live load: 100 psf (at the corridor) [4.79 kPa]

Floor dead load (see Table 4.1):

Carpet and pad at 5 psf

Ceiling, lights, and ducts at 15 psf

2-in. lightweight concrete fill at 18 psf

5-in.-thick slab at 62 psf

Total dead load: 100 psf [4.79 kPa]

With the slab determined, it is now possible to consider the design of one of the typical interior beams, loaded by a 10-ft-wide strip of slab, as shown in Figure 20.18. The supports for these beams are 30 ft on center. If the beams and columns are assumed to be a minimum of 12-in.-wide,

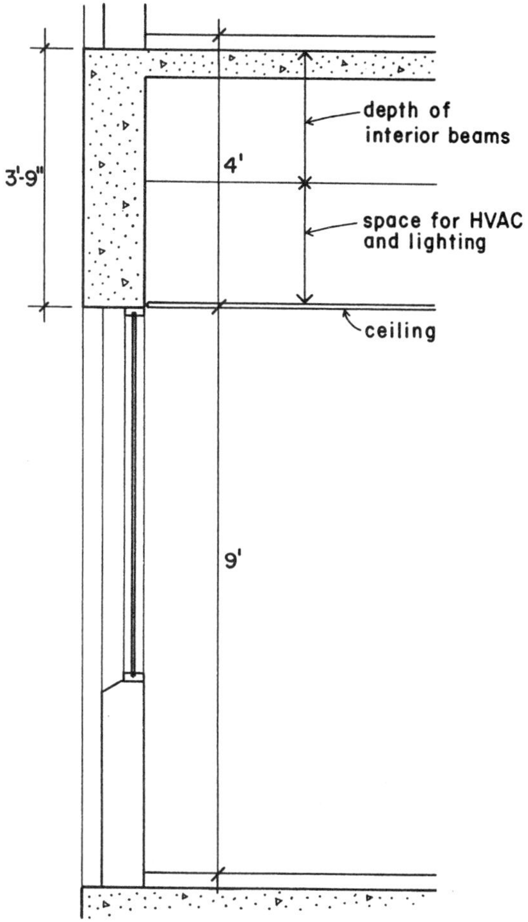

Figure 20.20 Building Three: section at the exterior wall with the concrete structure.

the clear span for the beam becomes 29 ft and its load periphery is 29 × 10 = 290 ft^2. Using the ASCE 2003 standard (Ref. 1) provisions for reduction of live load (see Section 14.1):

$$L = L_o\left(0.25 + \frac{15}{\sqrt{K_{LL}A_T}}\right) = 100\left(0.25 + \frac{15}{\sqrt{2 \times 290}}\right) = 87 \text{ psf}$$

The beam loading as a per ft unit load is determined as follows:

Live load = (87 psf)(10 ft) = 870 lb/ft [12.69 kN/m]

Dead load without the beam stem extending below the slab: (100 psf) × (10 ft) = 1000 lb/ft [14.6 kN/m]

Estimating a 12-in.-wide × 20-in.-deep beam stem extending below the bottom of the slab, the additional dead load becomes

$$\frac{12 \times 20}{144} \times 150 \text{ lb/ft}^3 = 250 \text{ lb/ft } [3.65 \text{ kN/m}]$$

The total dead load for the beam is thus 1000 + 250 = 1250 lb/ft, and the total uniformly distributed factored load for the beam is

$$w_u = 1.2(1250) + 1.6(870) = 1500 + 1392$$
$$= 2892 \text{ lb, or } 2.89 \text{ kips } [42.17 \text{ kN}]$$

Consider now the four-span continuous beam that is supported by the north-south column-line beams that are referred to as the girders. The approximation factors for design moments for this beam are given in Figure 20.19, and a summary of design data is given in Figure 20.21. Note that the design provides for tension reinforcement only, which is based on an assumption that the beam concrete section is adequate to prevent a critical bending compressive stress in the concrete. Using the strength method (see Section 13.3), the basis for this is as follows.

From Figure 20.19, the maximum bending moment in the beam is

$$M_u = \frac{wL^2}{10} = \frac{2.89 \times (29)^2}{10} = 243 \text{ kip-ft } [330 \text{ kN-m}]$$

$$M_r = \frac{M_u}{\phi} = \frac{243}{0.9} = 270 \text{ kip-ft } [366 \text{ kN-m}]$$

Then, using factors from Table 13.2 for a balanced section, the required value for bd^2 is determined as

$$bd^2 = \frac{M}{R} = \frac{270 \times 12}{1.149} = 2820$$

With the unit values as used for M and R, this quantity is in units of in.^{3}.

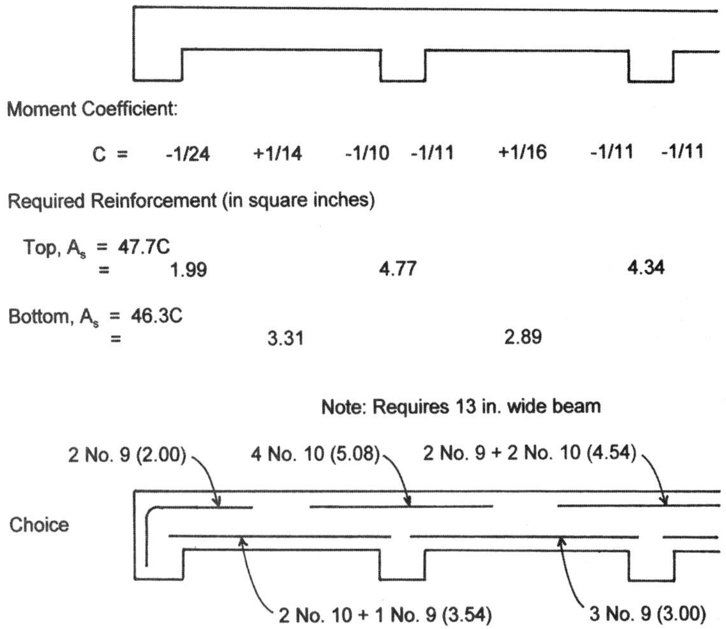

Figure 20.21 Building Three: summary of design for the four-span floor beam.

Various combinations of b and d may now be derived from this relationship of $bd^2 = 3351$, as demonstrated in Section 13.3. For this example, assuming a beam width of 12 in.:

$$d = \sqrt{\frac{2820}{12}} = 15.3 \text{ in.}$$

With minimum cover of 1.5 in., No. 3 stirrups, and moderate size bars for the tension reinforcement, an overall required beam dimension is obtained by adding approximately 2.5 in. to this derived value for the effective depth. Thus, any dimension selected that is at least 17.8 in. or more will ensure a lack of critical bending stress in the concrete. In most cases, the specified dimension is rounded off to the nearest full inch, in which case the overall beam height would be specified as 20 in. As discussed in Section 13.3, the balanced section is useful only for establishing a tension failure for the beam (yielding of the reinforcement).

Another consideration for choice of the beam depth is deflection control, as discussed in Section 13.3. From Table 13.10, a minimum overall height of $L/23$ is recommended for the end span of a continuous beam. This yields a minimum overall height of

$$h = \frac{29 \times 12}{23} = 15 \text{ in.}$$

Pushing these depth limits to their minimum is likely to result in high shear stress, a high percentage of reinforcement, and possible some excessive creep deflection. We will therefore consider the use of an overall height of 24 in., resulting in an approximate value of 24 – 2.5 = 21.5 inches for the effective depth d. Because this is quite close to the size assumed for dead load, no adjustment is made of the previously computed loading for the beam.

For the beams, the flexural reinforcement that is required in the top at the supports must pass either over or under the bars in the top of the girders. Figure 20.22 shows a section through the beam with an elevation of the girder in the background. It is assumed that the much heavier-loaded girder will be deeper than the beams, so the bar intersection problem does not exist in the bottoms of the intersecting members. At the top, however, the beam bars are run under the girder bars, favoring the heavier-loaded

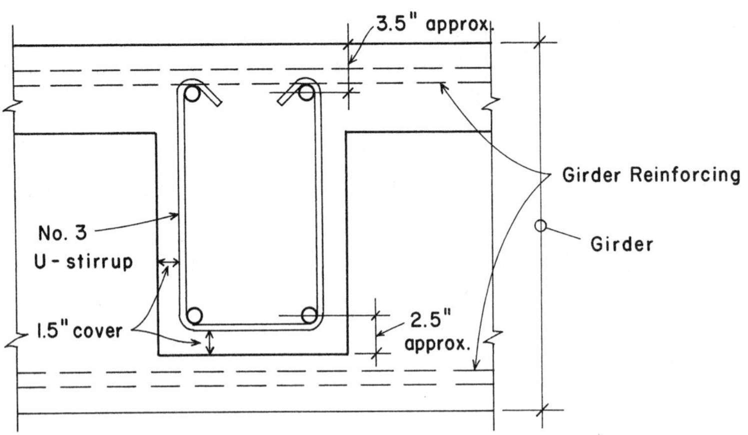

Figure 20.22 Considerations for layout of the reinforcement for the beams and girders.

girder. For an approximate consideration, an adjusted dimension of 3.5 to 4 in. should thus be subtracted from the overall beam height to obtain an effective depth for design of the beam. For the remainder of the computations, a value of 20 in. is used for the beam effective depth.

The beam cross section must also resist shear, and the beam dimensions should be verified to be adequate for this task before proceeding with design of the flexural reinforcement. Referring to Figure 14.2, the maximum shear force is approximated as 1.15 times the simple span shear of $wL/2$. For the beam, this produces a maximum shear of

$$V_u = 1.15\frac{wL}{2} = 1.15 \times \frac{2.89 \times 29}{2} = 48.2 \text{ kips [214 kN]}$$

As discussed in Section 13.5, this value may be reduced by the shear between the support and the distance of the beam effective depth from the support; thus:

$$\text{Design } V = 48.2 - \left(\frac{20}{12} \times 2.89\right) = 43.4 \text{ kips [193 kN]}$$

and the required maximum shear capacity is

$$V = \frac{43.4}{0.75} = 57.9 \text{ kips [258 kN]}$$

Using a d of 20 in. and b of 12 in., the critical shear capacity of the concrete alone is

$$V_c = 2\sqrt{f'_c}\,bd = 2\sqrt{3000}\,(12 \times 20) = 26{,}290 \text{ lb, or } 26.3 \text{ kips}$$

This leaves a shear force to be developed by the steel equal to

$$V'_s = 57.9 - 26.3 = 31.6 \text{ kips}$$

and the closest stirrup spacing at the beam end is

$$s = \frac{A_v f_y d}{V'_s} = \frac{0.22 \times 40 \times 20}{31.6} = 5.6 \text{ in. [142 mm]}$$

which is not an unreasonable spacing. For the approximate design shown in Figure. 20.21, the required area of steel at the points of support is determined as follows:

$$\text{Assume } a \text{ of 6 in., } jd = d - a/2 = 17 \text{ in.}$$

Then

$$A_s = \frac{M}{\phi f_y jd} = \frac{C \times 2.89 \times (29)^2 \times 12}{0.9\,(40 \times 17)} = 47.66C$$

At midspan points, the positive bending moments will be resisted by the slab and beam acting in T-beam action (see Section 13.3). For this condition, an approximate internal moment arm consists of $d - t/2$ and the required steel areas are approximated as

$$A_s = \frac{M}{\phi f_y \left(d - \dfrac{t}{2}\right)} = \frac{C \times 2.89 \times (29)^2 \times 12}{0.9\,(40 \times (20 - 2.5))} = 46.3C$$

Inspection of the framing plan in Figure 20.18 reveals that the girders on the north-south columns lines carry the ends of the beams as concentrated loads at their third points (10 ft from each support). The spandrel girders at the building ends carry the outer ends of the beams plus their own dead weight. In addition, all the spandrel beams support the weight of the exterior curtain walls. The form of the spandrels and the wall construction is shown in Figure 20.20.

The framing plan also indicates the use of widened columns at the exterior walls. Assuming a minimum width of 2 ft, the clear span of the spandrels thus becomes 28 ft. This much-stiffened bent, with very deep spandrel beams and widened columns, is used for lateral bracing, as discussed later in this section.

The spandrel beams carry a combination of uniformly distributed loads (spandrel weight plus wall) and concentrated loads (the beam ends). These loadings are determined as follows.

For reduction of the live load, the portion of floor loading carried is two times half the beam load, or approximately the same as one full beam: 290 sq ft. The design live load for the spandrel girders is thus reduced the same amount as it was for the beams. From the beam loading, therefore:

The total factored load from the beam is

$$P = 2.89 \text{ kip-ft} \times 30/2 \text{ ft} = 43.4 \text{ , say 44 kips}$$

The uniformly distributed load is basically all dead load, determined as

$$\text{Spandrel weight: } [(12)(45)/144)](150 \text{ pcf}) = 560 \text{ lb/ft}$$

$$\text{Wall weight: } (25 \text{ psf average})(9 \text{ ft high}) = 225 \text{ lb/ft}$$

$$\text{Total distributed load: } 560 + 225 = 785 \text{ lb, say } 0.8 \text{ kip/ft}$$

And the factored load is

$$w = 1.2(0.8) = 0.96 \text{ , say } 1.0 \text{ kip/ft}$$

For the distributed load, approximate design moments may be determined using the moment coefficients, as was done for the slab and beam. Values for this procedure are given in Figure 20.19. Thus:

$$M_u = C(w \times L^2) = C(1.0 \times 28^2) = 784C$$

The ACI Code does not permit use of coefficients for concentrated loads, but for an approximate design some adjusted coefficients may be derived from tabulated loadings for beams with third point load placement. Using these coefficients, the moments are

$$M_u = C(P \times L) = C(44 \times 28) = 1232C$$

Figure 20.23 presents a summary of the approximation of moments for the spandrel girder. This is, of course, only the gravity loading, which must be combined with effects of lateral loads for complete design of the bents. The design of the spandrel girder is therefore deferred until after the discussion of lateral loads later in this section.

Design of the Concrete Columns

The general cases for the concrete columns are as follows (see Figure 20.24):

1. The interior column, carrying primarily only gravity loads due to the stiffened perimeter bents

2. The corner columns, carrying the ends of the spandrel beams and functioning as the ends of the perimeter bents in both directions
3. The intermediate columns on the north and south sides, carrying the ends of the interior girders and functioning as members of the perimeter bents
4. The intermediate columns on the east and west sides, carrying the ends of the column-line beams and functioning as members of the perimeter bents

Summations of the design loads for the columns may be done from the data given previously. Because all columns will be subjected to combinations of axial load and bending moments, these gravity loads represent

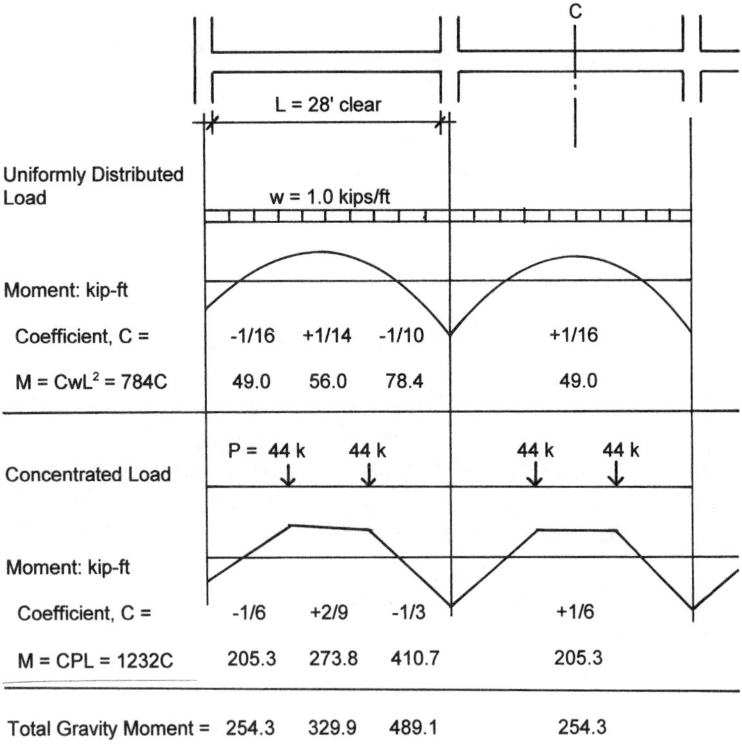

Figure 20.23 Factored gravity load effect on the spandrel girder.

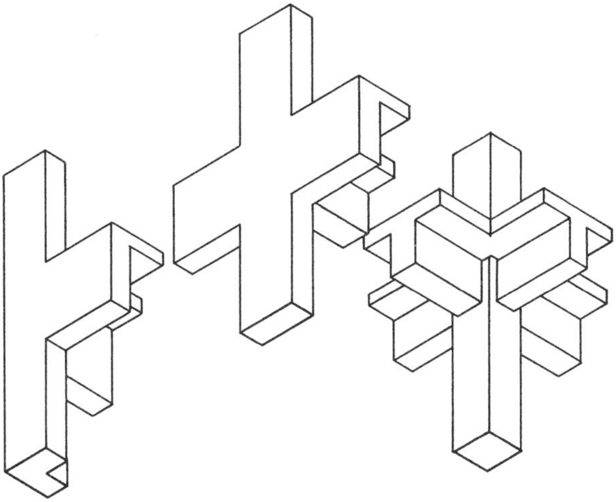

Figure 20.24 Relations between the columns and the floor framing.

only the axial compression action. Bending moments will be relatively low in magnitude on interior columns because they are framed into by beams on all sides. As discussed in Chapter 15, all columns are designed for a minimum amount of bending, so routine design, even when done for axial load alone, provides for some residual moment capacity. For an approximate design, therefore, it is reasonable to consider the interior columns for axial gravity loads only.

Figure 20.25 presents a summary of design for an interior column, using loads determined from a column load summation with the data given previously in this section. Note that a single size of 20 in. square is used for all three stories—a common practice permitting reuse of column forms for cost savings. Column load capacities indicated in Figure 20.25 were obtained from the graphs in Chapter 15.

A general cost-savings factor is the use of relatively low percentages of steel reinforcement. An economical column is therefore one with a minimum percentage (usually a threshold of 1% of the gross section) of reinforcement. However, other factors often affect design choices for columns. Some common ones as follows:

1. Architectural planning of building interiors. Large columns are often difficult to plan around in developing of interior rooms, corridors,

stair openings, and so on. Thus, the *smallest* feasible column sizes—obtained with maximum percentages of steel—are often desired.

2. Ultimate load response of lightly reinforced columns borders on brittle fracture failure, whereas heavily reinforced columns tend to have a yield form of ultimate failure. The yield character is especially desirable for rigid frame actions in general, and particularly for seismic loading conditions.

A general rule of practice in rigid frame design for lateral loadings (wind or earthquakes) is to prefer a form of ultimate response described as *strong column/weak beam failure*. In this example, this relates more to

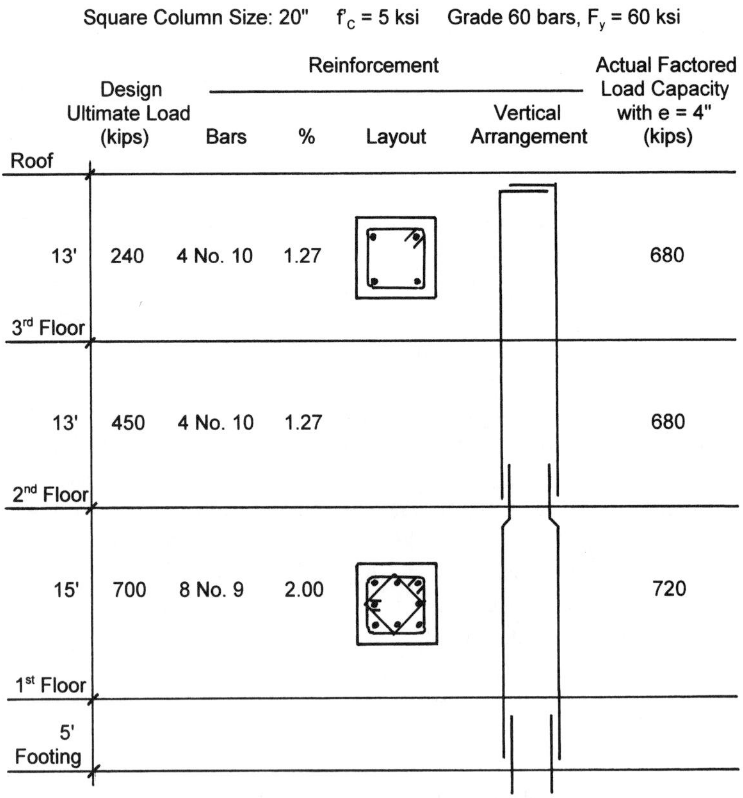

Figure 20.25 Design of the interior concrete column for gravity load only.

the columns in the perimeter bents but may also somewhat condition design choices for the interior columns because they will take *some* lateral loads when the building as a whole deflects sideways.

Column form may also be an issue that relates to architectural planning or to structural concerns. Round columns work well for some structural actions and may be quite economical for forming, but unless they are totally freestanding, they do not fit so well for planning of the rest of the building construction. Even square columns of large size may be difficult to plan around in some cases; an example is at the corners of stair wells and elevator shafts. T-shaped or L-shaped columns may be used in special situations.

Large bending moments in proportion to axial compression may also dictate some adjustment of column form or arrangement of reinforcement. When a column becomes essentially beamlike in its action, some of the practical considerations for beam design come into play. In this example, these concerns apply to the exterior columns to some degree.

For the intermediate exterior columns, there are four actions to consider:

1. The vertical compression due to gravity
2. Bending moment induced by the interior framing that intersects the wall. These columns are what provides the end resisting moments shown in Figure 20.21
3. Bending moments in the plane of the wall bent, induced by any unbalanced gravity load conditions (movable live loads) on the spandrels
4. Bending moments in the plane of the wall bents due to lateral loads

For the corner columns, the situation is similar to that for the intermediate exterior columns; that is, there is bending on both axes. Gravity loads will produce simultaneous bending on both axes, resulting in a net moment that is diagonal to the column. Lateral loads can cause the same effect, because neither wind nor earthquakes will work neatly on the building's major axes, even though this is how design investigation is performed.

Further discussion of the exterior columns is presented in the following considerations for lateral load effects.

Design for Lateral Forces

The major lateral force resisting systems for this structure are as shown in Figure 20.15. In truth, other elements of the construction will also resist lateral distortion of the structure, but by widening the exterior columns in the wall plane and using the very deep spandrel girders, the stiffness of these bents becomes considerable.

Whenever lateral deformation occurs, the stiffer elements will attract the force first. Of course, the stiffest elements may not have the necessary strength and will thus fail structurally, passing the resistance off to other resisting elements. Glass tightly held in flexible window frames, stucco on light wood structural frames, lightweight concrete block walls, or plastered partitions on light metal partition frames may thus be fractured first in lateral movements (as they often are). For the successful design of this building, the detailing of the construction should be carefully done to ensure that these events do not occur, in spite of the relative stiffness of the perimeter bents. In any event, the bents shown in Figure 20.15 will be designed for the entire lateral load. They thus represent the safety assurance for the structure, if not a guarantee against loss of construction.

With the same building profile, the wind loads on this structure will be the same as those determined for the steel structure in Section 20.4. As in the example in that section, the data given in Figure. 20.11 is used to determine the horizontal forces on the bracing bent as follows:

$$H_1 = 165.5(122)/2 = 11{,}096 \text{ lb, say } 10.1 \text{ kips/bent}$$

$$H_2 = 199.5(122)/2 = 12{,}170 \text{ lb, say } 12.2 \text{ kips/bent}$$

$$H_3 = 210(122)/2 = 12{,}810 \text{ lb, say } 12.8 \text{ kips/bent}$$

Figure 20.26*a* shows a profile of the north-south bent with these loads applied.

For an approximate analysis, consider the individual stories of the bent to behave as shown in Figure. 20.26*b*, with the columns developing an inflection point at their mid-height points. Because the columns are all deflected the same sideways distance, the shear force in a single column may be assumed to be proportionate to the relative stiffness of the column. If the columns all have the same stiffness, the total load at each story for this bent would simply be divided by 4 to obtain the column shear forces.

Even if the columns are all the same size, however, they may not all have the same resistance to lateral deflection. The end columns in the bent are slightly less restrained at their ends (top and bottom) because they are framed on only one side by a beam. For this approximation, therefore, it is assumed that the relative stiffness of the end columns is

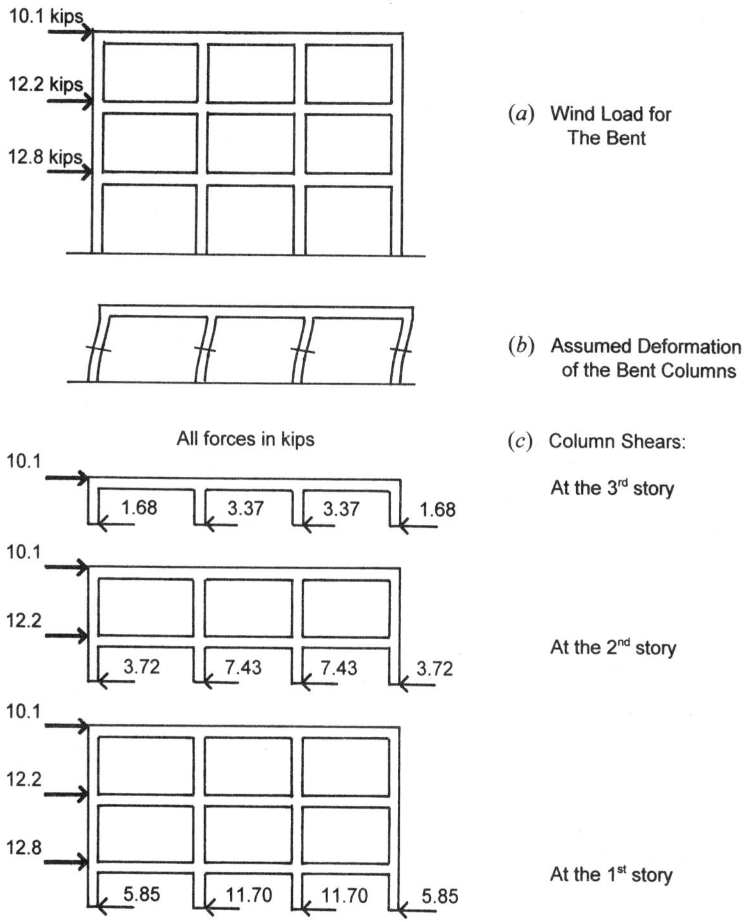

Figure 20.26 Aspects of lateral load response of the north-south perimeter bents.

one half that of the intermediate columns. Thus, the shear force in the end columns is one sixth of the total bent shear force and that in the intermediate column is one third of the total force. The column shears for each of the three stories is thus as shown in Figure 20.26*c*.

The column shear forces produce bending moments in the columns. With the column inflection points (points of zero moment) assumed to be at mid-height, the moment produced by a single shear force is simply the product of the force and half the column height. These column moments must be resisted by the end moments in the rigidly attached beams, and the actions are as shown in Figure 20.27. At each column/beam intersection, the sum of the column and beam moments must be balanced. Thus, the total of the beam moments may be equated to the total of the column moments, and the beam moments may be determined once the column moments are known.

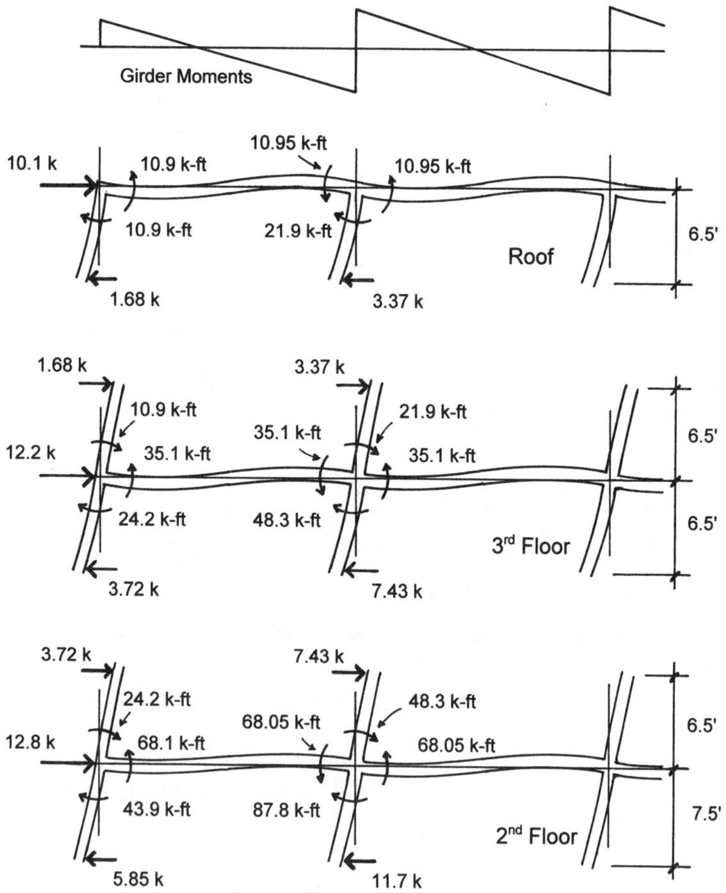

Figure 20.27 Investigation for column and girder bending moments in the north-south bents.

For example, at the second floor level of the intermediate column, the sum of the column moments from Figure 20.27 is

$$M = 48.3 + 87.8 = 136.1 \text{ kip-ft}$$

Assuming the two beams framing the column to have equal stiffness at their ends, the beams will share this moment equally, and the end moment in each beam is thus

$$M = \frac{136.1}{2} = 68.05 \text{ kip-ft}$$

as shown in the figure.

The data displayed in Figure 20.27 may now be combined with that obtained from gravity load analyses for a combined load investigation and the final design of the bent members.

Design of the Bent Columns

For the bent columns, the axial compression caused by gravity must first be combined with any moments induced by gravity for a gravity-only analysis. Then the gravity load actions are combined with the results from the lateral force analysis, using the usual adjustments for this combined loading. In the strength method, this occurs automatically through the use of the various load factors for the different loadings. In the stress method, the combined gravity and lateral load condition uses an adjusted stress—in this case, one increased by one third. Because our illustration uses strength methods, a comparison can be made between the effects produced by gravity only and those produced by the gravity plus lateral effects. This amounts to using the load adjustment factors for the two load combinations.

Gravity-induced moments for the girders are taken from the girder analysis in Figure 20.24 and are assumed to produce column moments as shown in Figure 20.28. The summary of design conditions for the corner and intermediate columns is given in Figure 20.29. The dual requirements for the columns are given in the bottom two lines of the table in Figure 20.29.

Note that a single column choice from Figure 15.8 is able to fulfill the requirements for all the columns. This is not unusual because the relationship between load magnitude and moment magnitude changes. A broader range of data for column choices is given in the extensive column design tables in various references.

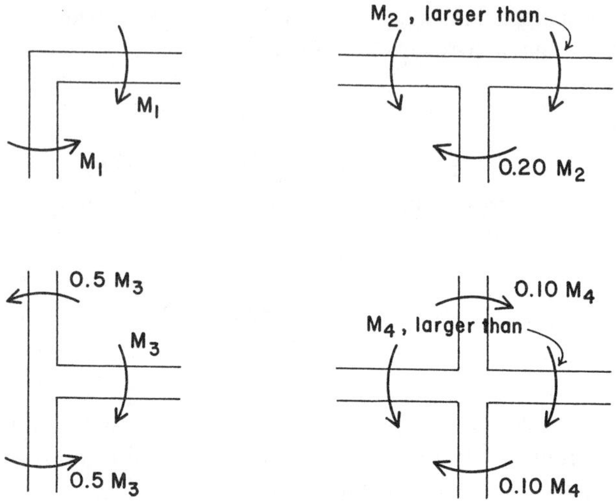

Figure 20.28 Assumptions for approximations of the distribution of bending moments in the bent columns due to gravity loading.

When bending moment is very high in comparison to the axial load (very large eccentricity), an effective approximate column design can be determined by designing a section simply as a beam with tension reinforcement; the reinforcement is then merely duplicated on both sides of the column.

Design of the Bent Girders

The spandrel girders must be designed for the same two basic load conditions as discussed for the columns. The summary of bending moments for the third-floor spandrel girder is shown in Figure 20.30. Values for the gravity moments are taken from Figure 20.23. Moment induced by wind is that shown in Figure 20.27. It may be noted from the data in Figure 20.30 that the effects of gravity loading prevail, and that wind loading is not a critical concern for the girder. This would most likely not be the case in lower stories of a much taller building, or possibly with a combined loading including major seismic effects.

Figure 20.31 presents a summary of design considerations for the third-floor spandrel girder. The construction assumed here is that shown

Note: Axial loads in kips, moments in kip-ft, dimensions in inches.

Story	Third	Second	First
Gravity Load Only:			
Axial Live Load	35	67	103
Axial Dead Load	83	154	240
Live Load Moment	32	38	38
Dead Load Moment	128	90	90
Ultimate Gravity Load: (1.2D + 1.6L)			
Axial Load	100+56 = 156	185+107 = 292	288+165 = 453
Moment	154+51 = 205	108+61 = 169	108+61 = 169
e	15.8 in.	7.0 in.	4.5 in.
Combined with Wind:			
Wind Load Moment	21.9	48.3	87.8
Ultimate Moment with Wind (1.2D+1.6W+L)	154+35+32 = 221	108+77+38 = 223	108+141+38 = 287
Ultimate Axial Load with Wind (1.2D + L)	100+35 = 135	185+67 = 252	288+103 = 391
e	19.6 in.	10.6 in.	8.8 in.
Choice of Column from Figure 15.8	?? Design as a beam for M = 221	14 × 20 6 NO. 9	14 × 24 6 No. 10

3rd story: Use 14×24, d = 21 in., $A_s = M_u/\phi f_y jd$ = (221×12)/0.9[60(0.9×21)] = 2.6 in.^{2}
Use same as 2nd story, 6 No. 9 bars (3 at each end). Use 14×24 for all stories.

Figure 20.29 Design of the north-south bent columns for combined gravity and lateral loading.

in Figure 20.21, with the very deep, exposed girder. Some attention should be given to the relative stiffness of the columns and girders, as discussed in Section 3.12. Keep in mind, however, that the girder is almost three times as long as the column and thus may have a considerably stiffer section without causing a disproportionate relationship to occur.

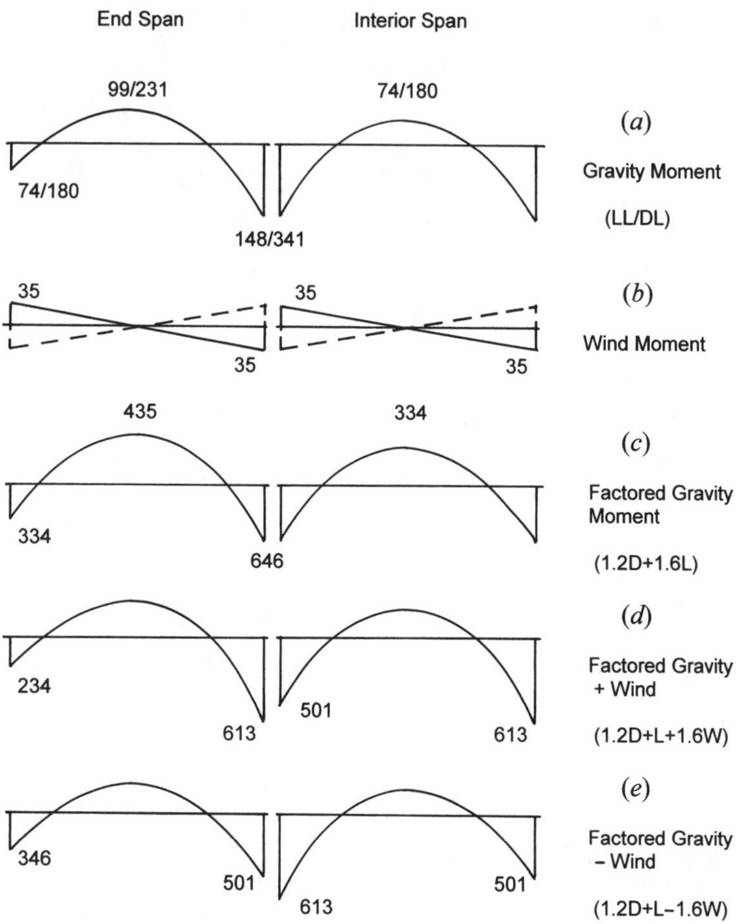

Figure 20.30 Combined gravity and lateral bending moments for the spandrel girders.

For computation of the required flexural reinforcement, the T-beam effect is ignored and an effective depth of 40 in. is assumed. Required areas of reinforcement may thus be derived as

$$A_s = \frac{M}{\phi f_y jd} = \frac{M \times 12}{0.9(40 \times 0.9 \times 40)} = 0.00926M$$

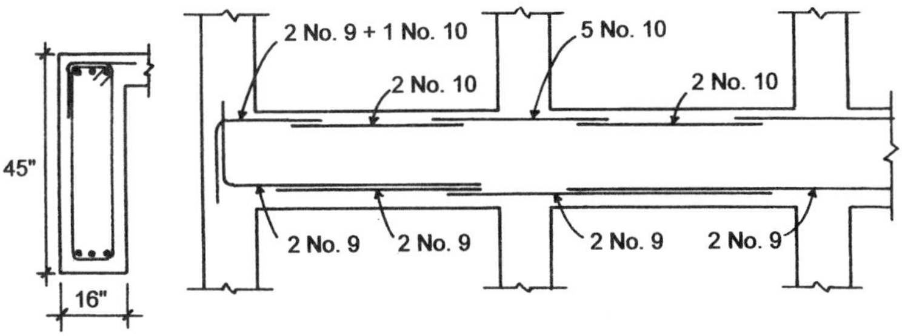

Design Ultimate Moment (See Figure 20.30)	+		435		334
	–	346		646	
Required Reinforcement $A_s = 0.00926$ M	top	3.20		5.98	
	bottom		4.03		3.09
Actual A_s	top	3.27	2.54	6.35	2.54
	bottom	2.00	4.00	2.00	4.00

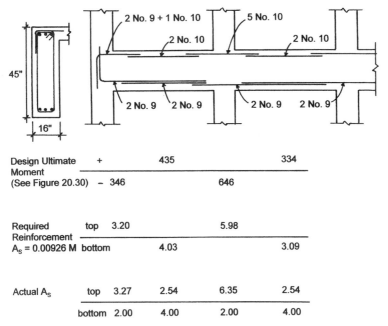

Figure 20.31 Design of the spandrel girder for the combined gravity and lateral loading effects.

Values determined for the various critical locations are shown in Figure 20.31. It is reasonable to consider the stacking of bars in two layers in such a deep section, but it is not necessary for the selection of reinforcement shown in the figure.

The very deep and relatively thin spandrel should be treated somewhat as a wall/slab, and thus the section in Figure 20.31 shows some additional horizontal bars at mid-height points. In addition, the stirrups shown should be of a closed form (see Figure13.14) to serve also as ties, vertical reinforcement for the wall/slab, and (with the extended top) as negative moment flexural reinforcement for the adjoining slab. In this situation, it would be advisable to use continuous stirrups at a maximum spacing of 18 in. or so for the entire girder span. Closer spacing may be necessary near the supports, if the end shear forces require it.

It is also advisable to use some continuous top and bottom reinforcement in spandrels. This relates to some of the following possible considerations:

1. Miscalculation of lateral effects, giving some reserved reversal bending capacity to the girders.
2. A general capability for torsional resistance throughout the beam length (intersecting beams produce this effect).
3. Something there to hold up the continuous stirrups.
4. Some reduction of long-term creep deflection with all sections doubly reinforced. Helps keep load off the window mullions and glazing.

20.9 DESIGN OF THE FOUNDATIONS

Unless site conditions require the use of a more complex foundation system, it is reasonable to consider the use of simple shallow bearing foundations (footings) for Building Three. Column loads will vary depending on which of the preceding structural schemes is selected. The heaviest loads are likely to come with the all-concrete structure in Section 20.8.

The most direct solution for concentrated column loads is a square footing, as described in Section 16.3. A range of sizes of these footings is given in Table 16.4. For a freestanding column, the choice is relatively simple, once an acceptable design pressure for the supporting soil is established.

Problems arise when the conditions at the base of a column involve other than a freestanding case for the column. In fact, this is the case for most of the columns in Building Three. Consider the structural plan as indicated in Figure. 20.20. All but two of the interior columns are adjacent to construction for the stair towers or the elevator shaft.

For the three-story building, the stair tower may not be a problem, although in some buildings these are built as heavy masonry or concrete towers and are used for part of the lateral bracing system. This might possibly be the case for the structure in Section 20.7.

Assuming that the elevator serves the lowest occupied level in the building, there will be a considerably deep construction below this level to house the elevator pit. If the interior footings are quite large, they may come very close to the elevator pit construction. In this case, the bottoms of these footings would need to be dropped to a level close to that of the bottom of the elevator pit. If the plan layout results in a column right at the edge of the elevator shaft, this is a more complicated problem because the elevator pit would need to be on top of the column footing.

For the exterior columns at the building edge, there are two special concerns. The first has to do with the necessary support for the exterior

(*a*)

(*b*)

(*c*)

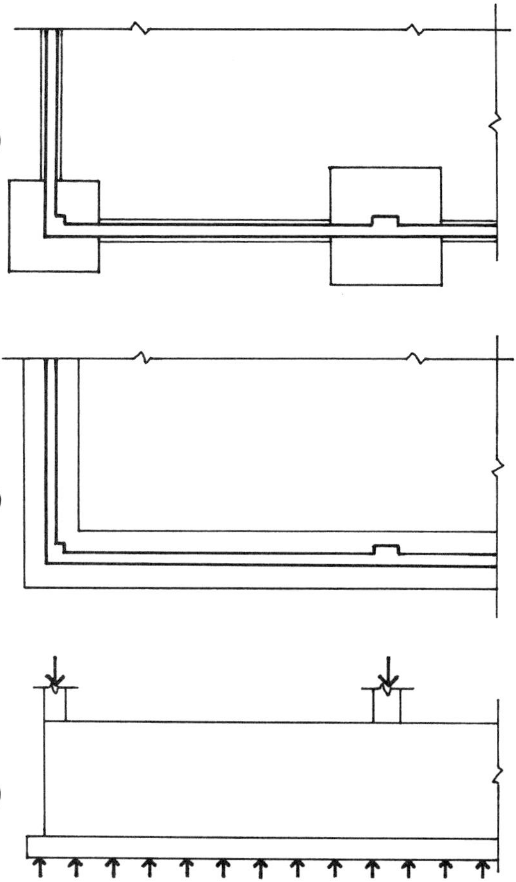

Figure 20.32 Building Three, considerations for the foundations. (*a*) partial plan with individual column footings; (*b*) partial plan with continuous wall footing; (*c*) use of a tall basement wall as a distribution girder.

building wall, which coincides with the column locations. If there is no basement, and the exterior wall is quite light in weight—possibly a metal curtain wall that is supported at each upper level—the exterior columns will likely get their own individual square footings and the wall will get a minor strip footing between the column footings. This scheme is shown in the partial foundation plan in Figure. 20.32*a*. Near the columns, the light wall will simply be supported by the column footings.

If the wall is very heavy, the solution in Figure. 20.32*a* may be less feasible, and it may be reasonable to consider the use of a wide strip footing that supports both the wall and the columns, as shown in Figure. 20.32*b*. The ability of the wall to serve as a distributing member for the uniform pressure on the strip footing must be considered, which brings up the other concern as well.

The other concern has to do with the presence or absence of a basement. If there is no basement, it may be theoretically possible to place footings quite close to the ground surface, with a minimal penetration of the general foundation construction below grade. If there is a basement, there will likely be a reasonably tall concrete wall at the building edge. Regardless of the wall construction above grade, it is reasonable to consider the use of the basement wall as a distributing member for the column loads, as shown in Figure. 20.32*c*. Again, this is only feasible for a low-rise building with relatively modest loads on the exterior columns.

Appendix A

PROPERTIES OF SECTIONS

This appendix deals with various geometric properties of planar (two-dimensional) areas. The areas referred to are the cross-sectional areas of structural members. These geometric properties are used in the analysis of stresses and deformations in the design of the structural members.

A.1 CENTROIDS

The *center of gravity* of a solid is the point at which all of its weight can be considered to be concentrated. Because a planar area has no weight, it has no center of gravity. The point in a planar area that corresponds to the center of gravity of a very thin plate of the same area and shape is called the *centroid* of the area. The centroid is a useful reference for various geometric properties of planar areas.

For example, when a beam is subjected to a bending moment, the materials in the beam above a certain plane in the beam are in compression and the materials below the plane are in tension. This plane is the *neutral stress plane*, also called the neutral surface or the zero stress plane (see Section 3.7). For a cross section of the beam, the intersection of the neutral stress plane is a line that passes through the centroid of the section and is called the *neutral axis* of the section. The neutral axis is very important for investigation of bending stresses in beams.

The centroid for symmetrical shapes is located on the axis of symmetry for the shape. If the shape is bisymmetrical—that is, it has two axes of symmetry—the centroid is at the intersection of these axes. Consider the rectangular area shown in Figure A.1*a*; obviously its centroid is at its geometric center and is quite easily determined.

(*Note:* Tables A.3 through A.8 and Figure A.11, referred to in the discussion that follows, are located at the end of this chapter.)

For more complex forms, such as those of rolled steel members, the centroid will also be on any axis of symmetry. And, as for the simple rectangle, if there are two axes of symmetry, the centroid is readily located.

For simple geometric shapes, such as those shown in Figure A.1, the location of the centroid is easily established. However, for more complex shapes, the centroid and other properties may have to be determined by computations. One method for achieving this is by use of the *statical moment*, defined as the product of an area times its distance from some reference axis. Use of this method is demonstrated in the following examples.

Example 1. Figure A.2 is a beam cross section that is unsymmetrical with respect to a horizontal axis (such as X-X in the figure). The area is symmetrical about its vertical centroidal axis, but the true location of the centroid requires the locating of the horizontal centroidal axis. Find the location of the centroid.

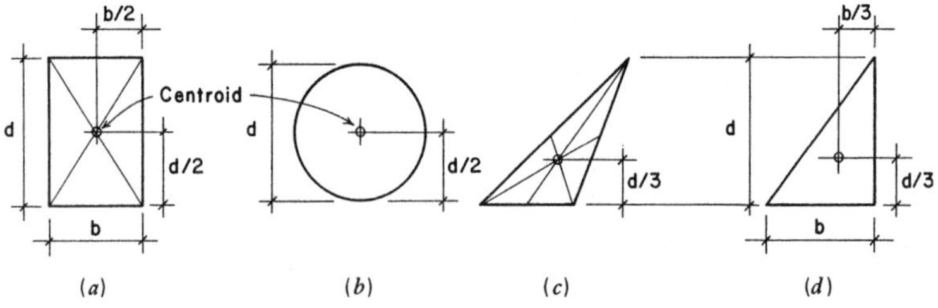

Figure A.1 Centroids of various planar shapes.

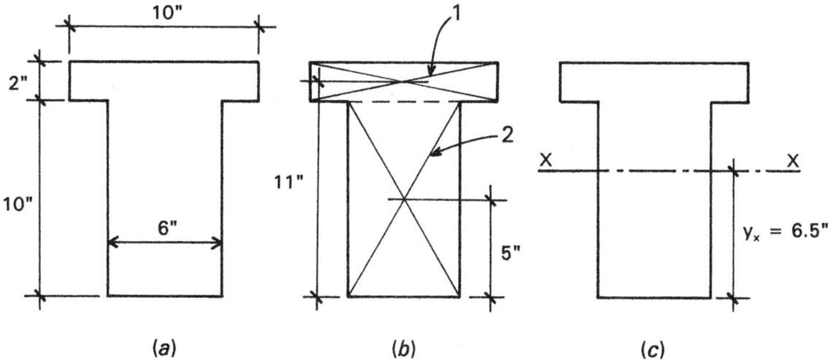

Figure A.2 Reference for Example 1.

Solution: Using the statical moment method, first divide the area into units for which the area and location of the centroid are readily determined. The division chosen here is shown in Figure A.2*b* with the two parts labeled 1 and 2.

The second step is to choose a reference axis about which to sum statical moments and from which the location of the centroid is readily measured. A convenient reference axis for this shape is one at either the top or bottom of the shape. With the bottom chosen, the distances from the centroids of the parts to this reference axis are shown in Figure A.2*b*.

We then determine the unit areas and their statical moments. This work is summarized in Table A.1, which shows the total area to be 80 in.^{2} and the total statical moment to be 520 in.^{3}. Dividing this moment by the total area produces the value of 6.5 in., which is the distance from the reference axis to the centroid of the whole shape, as shown in Figure A.2*c*.

Problems A.1.A–F. Find the location of the centroid for the cross-sectional areas shown in Figure A.3. Use the reference axes indicated, and compute the distances from the axes to the centroid, designated as c_x and c_y, as shown in Figure A.3*b*.

Table A.1 Summary of Computations for Centroid: Example 1

Part	Area (in.^{2})	y (in.)	$A \times y$ (in.^{3})
1	$2 \times 10 = 20$	11	220
2	$6 \times 10 = 60$	5	300
Σ	80		520

$$y_x = \frac{520}{80} = 6.5 \text{ in.}$$

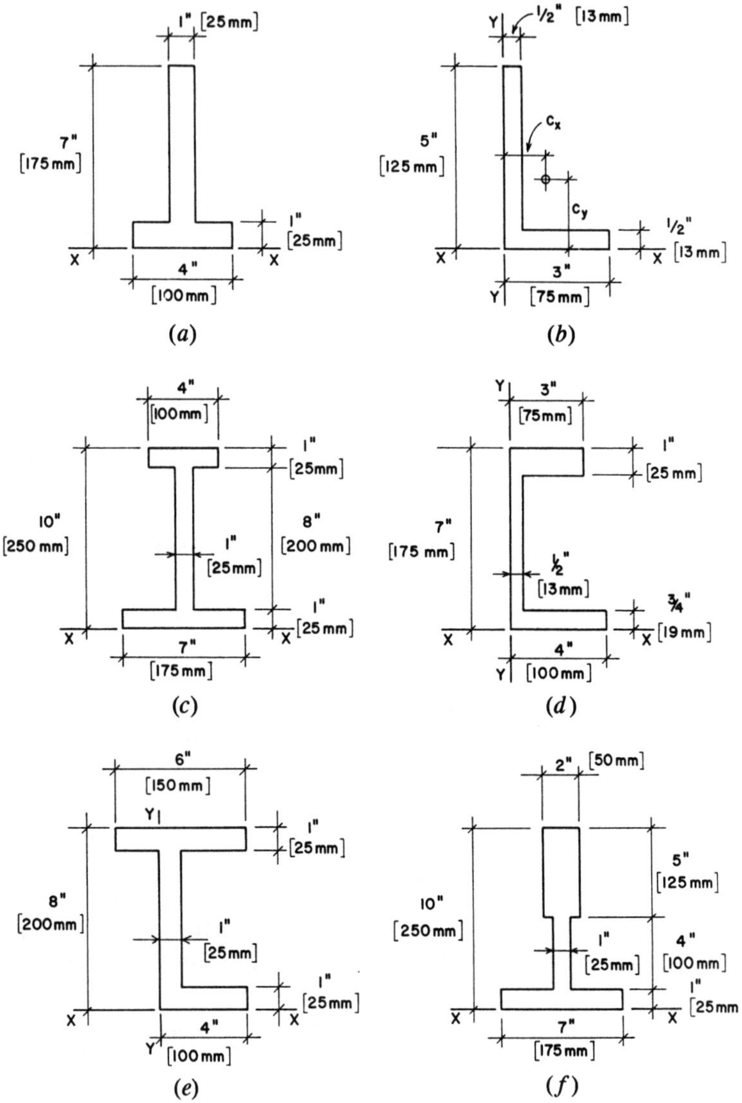

Figure A.3 Reference for Problem A.1.

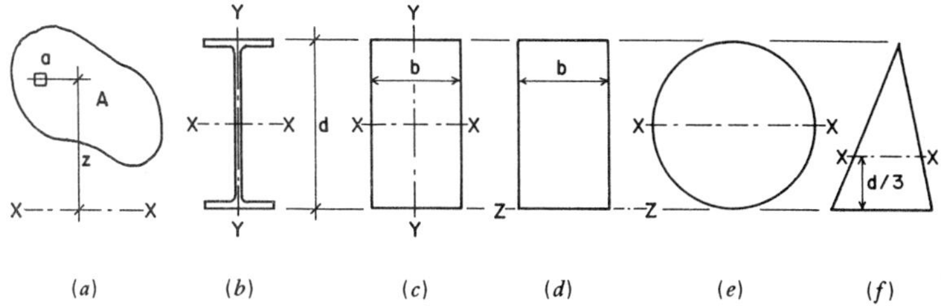

Figure A.4 Consideration of reference axis for the moment of inertia of various shapes of cross sections.

A.2 MOMENT OF INERTIA

Consider the area enclosed by the irregular line in Figure A.4*a*. In this area, designated *A*, a small unit area *a* is indicated at *z* distance from the axis marked X-X. If this unit area is multiplied by the square of its distance from the reference axis, the result is the quantity az^2. If all of the units of the area are thus identified and the sum of these products is made, the result is defined as the *second moment* or the *moment of inertia* of the area, designated as *I*. Thus:

$$\Sigma az^2 = I, \text{ or specifically } I_{X\text{-}X}$$

which is the moment of inertia of the area about the X-X axis.

The moment of inertia is a somewhat abstract item, less able to be visualized than area, weight, or center of gravity. It is nevertheless a real geometric property that becomes an essential factor for investigating stresses and deformations due to bending. Of particular interest is the moment of inertia about a centroidal axis and, most significantly, about a principal axis for the shape. Figures A.4*b, c, e,* and *f* indicate such axes for various shapes. An inspection of Tables A.3 through A.8 will reveal the properties of moment of inertia about the principal axes of the shapes in the tables.

Moment of Inertia of Geometric Figures

Values for moment of inertia can often be obtained from tabulations of structural properties. Occasionally, it is necessary to compute values for a given shape. This may be a simple shape, such as a square, rectangle, circular, or triangular area. For such shapes, simple formulas are derived to express the value for the moment of inertia.

Rectangles. Consider the rectangle shown in Figure A.4*c*. Its width is b and its depth is d. The two principal axes are X-X and Y-Y, both passing through the centroid of the area. For this case, the moment of inertia with respect to the centroidal axis X-X is

$$I_{X\text{-}X} = \frac{bd^3}{12}$$

and the moment of inertia with respect to the Y-Y axis is

$$I_{Y\text{-}Y} = \frac{db^3}{12}$$

Example 2. Find the value of the moment of inertia for a 6 in. × 12 in. wood beam about an axis through its centroid and parallel to the narrow dimension.

Solution: Referring to Table A.8, the actual dimensions of the section are 5.5 × 11.5 in. Then

$$I = \frac{bd^3}{12} = \frac{5.5 \times (11.5)^3}{12} = 697.1 \text{ in.}^4$$

which is in agreement with the value for $I_{X\text{-}X}$ in the table.

Circles. Figure A.4*e* shows a circular area with diameter d and axis X-X passing through its center. For the circular area the moment of inertia is

$$I = \frac{\pi d^4}{64}$$

Example 3. Compute the moment of inertia of a circular cross section, 10 in. in diameter, about its centroidal axis.

Solution: The moment of inertia is

$$I = \frac{\pi d^4}{64} = \frac{3.1416(10)^4}{64} = 490.9 \text{ in.}^4$$

Triangles. The triangle in Figure A.4*f* has a height h and a base width b. The moment of inertia about a centroidal axis parallel to the base is

$$I = \frac{bh^3}{36}$$

Example 4. If the base of the triangle in Figure A.4*f* is 12 in, wide and the height from the base is 10 in., find the value for the centroidal moment of inertia parallel to the base.

Solution: Using the given values in the formula

$$I = \frac{bh^3}{36} = \frac{12(10)^3}{36} = 333.3 \text{ in.}^4$$

Open and Hollow Shapes. Values of moment of inertia for shapes that are open or hollow may sometimes be computed by a method of subtraction. The following examples demonstrate this process. Note that this is possible only for shapes that are symmetrical.

Example 5. Compute the moment of inertia for the hollow box section shown in Figure. A.5*a* about a centroidal axis parallel to the narrow side.

Solution: Find first the moment of inertia of the shape defined by the outer limits of the box.

$$I = \frac{bd^3}{12} = \frac{6(10)^3}{12} = 500 \text{ in.}^4$$

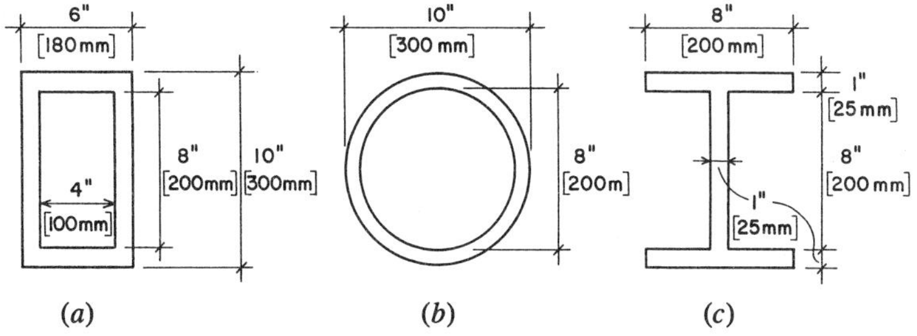

Figure A.5 Reference for Examples 5, 6, and 7.

Then find the moment of inertia for the shape defined by the void area.

$$I = \frac{4(8)^3}{12} = 170.7 \text{ in.}^4$$

The value for the hollow section is the difference, thus:

$$I = 500 - 170.7 = 329.3 \text{ in.}^4$$

Example 6. Compute the moment of inertia about the centroidal axis for the pipe section shown in Figure A.5*b*. The thickness of the shell is 1 in.

Solution: As in the preceding example, the two values may be found and subtracted. Or a single computation may be made as follows

$$I = \frac{\pi}{64}(d_o^4 - d_i^4) = \frac{3.1416}{64}(10^4 - 8^4) = 491 - 201 = 290 \text{ in.}^4$$

Example 7. Referring to Figure A.5*c*, compute the moment of inertia of the I-shape section about the centroidal axis parallel to the flanges.

Solution: This is essentially similar to the computation for Example 5. The two voids may be combined into a single one that is 7 in. wide. Thus:

$$I = \frac{8(10)^3}{12} - \frac{7(8)^3}{12} = 667 - 299 = 368 \text{ in.}^4$$

Note that this method can only be used when the centroids of the outer shape and the void coincide. For example, it cannot be used to find the moment of inertia for the I-shape about its vertical centroidal axis. For this computation the method discussed in the next section must be used.

A.3 TRANSFERRING MOMENTS OF INERTIA

Determination of the moment of inertia of unsymmetrical and complex shapes cannot be done by the simple processes illustrated in the preceding examples. An additional step that must be used is that involving the transfer of moment of inertia about a remote axis. The formula for achieving this transfer is as follows:

$$I = I_o + Az^2$$

where I = moment of inertia of the cross section about the required reference axis

I_o = moment of inertia of the cross section about its own centroidal axis, parallel to the reference axis

A = area of the cross section

z = distance between the two parallel axes

These relationships are illustrated in Figure A.6, where X-X is the centroidal axis of the area and Y-Y is the reference axis for the transferred moment of inertia.

Application of this principle is illustrated in the following examples.

Example 8. Find the moment of inertia of the T-shaped area in Figure A.7 about its horizontal (X-X) centroidal axis. (*Note:* The location of the centroid for this section was solved as Example 1 in Section A.1.)

Solution: A necessary first step in these problems is to locate the position of the centroidal axis if the shape is not symmetrical. In this case, the T-shape is symmetrical about its vertical axis, but not about the horizontal axis. Locating the position of the horizontal axis was the problem solved in Example 1 in Section A.1.

The next step is to break the complex shape down into parts for which centroids, areas, and centroidal moments of inertia are readily found. As was done in Example 1, the shape here is divided between the rectangular flange part and the rectangular web part.

The reference axis to be used here is the horizontal centroidal axis. Table A.2 summarizes the process of determining the factors for the par-

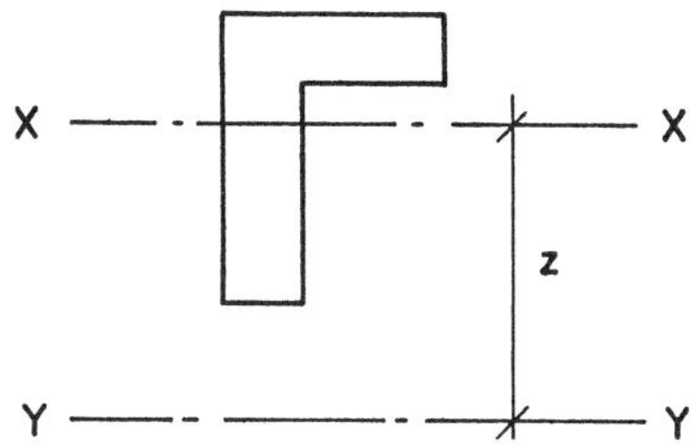

Figure A.6 Transfer of moment of inertia to a parallel axis.

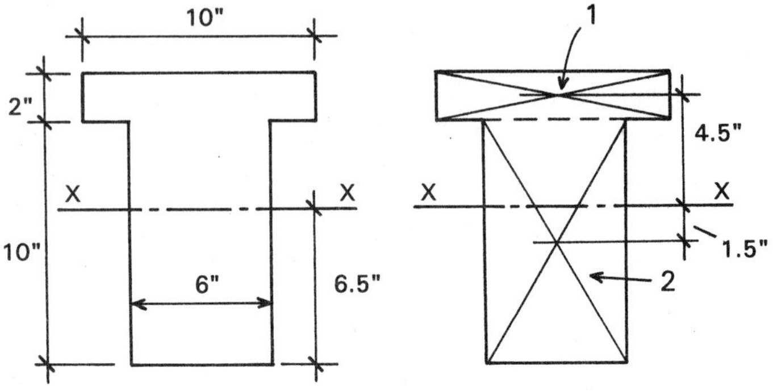

(a) (b)

Figure A.7 Reference for Example 8.

allel axis transfer process. The required value for I about the horizontal centroidal axis is determined to be 1046.7 in.^{4}.

A common situation in which this problem must be solved is in the case of structural members that are built up from distinct parts. One such section is that shown in Figure A.8, where a box-shaped cross section is composed by attaching two plates and two rolled channel sections. Even though this composite section is actually symmetrical about both its principal axes, and the locations of these axes are apparent, the values for moment of inertia about both axes must be determined by the parallel axis transfer process. The following example demonstrates the process.

Example 9. Compute the moment of inertia about the centroidal X-X axis of the built-up section shown in Figure A.8.

Solution: For this situation, the two channels are positioned so that their centroids coincide with the reference axis. Thus, the value of I_o for the channels is also their actual moment of inertia about the required refer-

Table A.2 Summary of Computations for Moment of Inertia: Example 9

Part	Area (in.^{2})	y (in.)	I_o (in.^{4})	$A \times y^2$ (in.^{4})	I_x (in.^{4})
1	20	4.5	$10(2)^3/12 = 6.7$	$20(4.5)^2 = 405$	411.7
2	60	1.5	$6(10)^3/12 = 500$	$60(1.5)^2 = 135$	635
Σ					1046.7

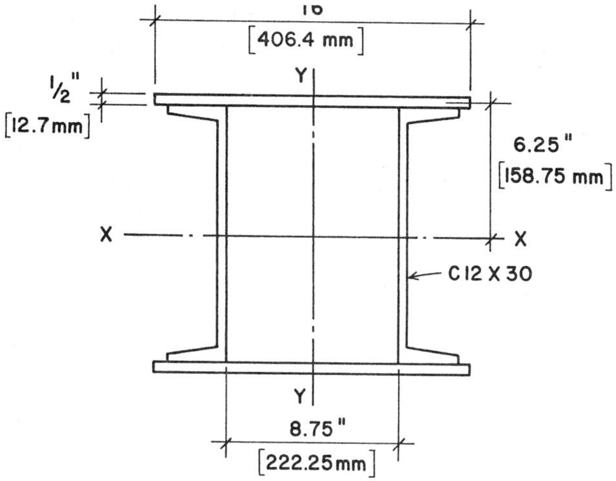

Figure A.8 Reference for Example 9.

ence axis, and their contributions to the required value here is simply two times their listed value for moment of inertia about their X-X axis, as given in Table A.4: 2(162) = 324 in.^{4}.

The plates have simple rectangular cross sections, and the centroidal moment of inertia of one plate is thus determined as

$$I_o = \frac{bd^3}{12} = \frac{16 \times (0.5)^3}{12} = 0.1667 \text{ in.}^4$$

The distance between the centroid of the plate and the reference X-X axis is 6.25 in. And the area of one plate is 8 in.^{2}. The moment of inertia for one plate about the reference axis is thus:

$$I_o + Az^2 = 0.1667 + (8)(6.25)^2 = 312.7 \text{ in.}^4$$

and the value for the two plates is twice this, or 625.4 in.^{4}.

Adding the contributions of the parts, the answer is 324 + 625.4 = 949.4 in.^{4}.

Problems A.3.A–F. Compute the moments of inertia about the indicated centroidal axes for the cross-sectional shapes in Figure A.9.

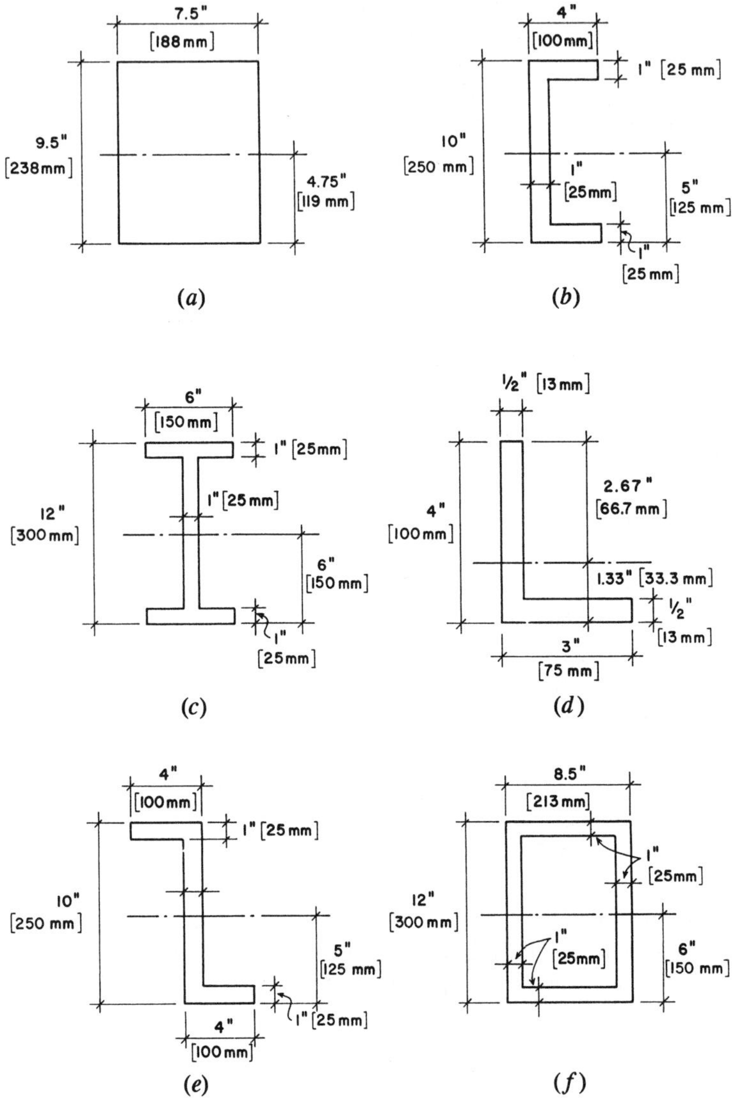

Figure A.9 Reference for Problem 4.3, part 1.

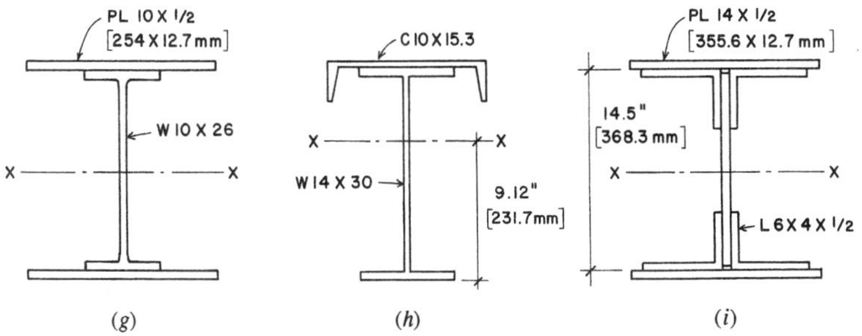

Figure A.10 Reference for Problem 4.3, part 2.

Problems A.3.G–I. Compute the moments of inertia with respect to the centroidal X-X axes for the built-up sections in Figure A.10. Make use of any appropriate data from the tables of properties for steel shapes.

A.4 MISCELLANEOUS PROPERTIES

Section Modulus

As noted in Section 3.7, the term I/c in the formula for flexural stress is called the *section modulus.* Using the section modulus permits a minor shortcut in the computations for flexural stress or the determination of the bending moment capacity of members. However, the real value of this property is in its measure of relative bending strength of members. As a geometric property, it is a direct index of bending strength for a given member cross section. Members of various cross section may thus be rank-ordered in terms of their bending strength strictly on the basis of their S values. Because of its usefulness, the value of S is listed together with other significant properties in the tabulations for steel and wood members.

For members of standard form (structural lumber and rolled steel shapes), the value of S may be obtained from tables similar to those presented at the end of this chapter. For complex forms not of standard form, the value of S must be computed, which is readily done once the centroidal axes are located and moments of inertia about the centroidal axes are determined.

Example 10. Verify the tabulated value for the section modulus of a 6×12 wood beam about the centroidal axis parallel to its narrow side.

Solution: From Table A.8, the actual dimensions of this member are 5.5 × 11.5 in. And the value for the moment of inertia is 697.068 in.^{4}. Then

$$S = \frac{I}{c} = \frac{697.068}{5.75} = 121.229 \text{ in.}^3$$

which agrees with the value in Table A.8.

Radius of Gyration

For design of slender compression members an important geometric property is the *radius of gyration*, defined as

$$r = \sqrt{\frac{I}{A}}$$

Just as with moment of inertia and section modulus values, the radius of gyration has an orientation to a specific axis in the planar cross section of a member. Thus, if the I used in the formula for r is that with respect to the X-X centroidal axis, then that is the reference for the specific value of r.

A value of r with particular significance is that designated as the *least radius of gyration*. Because this value will be related to the least value of I for the cross section, and because I is an index of the bending stiffness of the member, then the least value for r will indicate the weakest response of the member to bending. This relates specifically to the resistance of slender compression members to buckling. Buckling is essentially a sideways bending response, and its most likely occurrence will be on the axis identified by the least value of I or r. Use of these relationships is discussed for columns in Parts II and III.

Plastic Section Modulus

The plastic section modulus, designated Z, is used in a similar manner to the elastic stress section modulus S. The plastic modulus is used to determine the fully plastic stress moment capacity of a steel beam. Thus:

$$M_p = F_y \times Z$$

The use of the plastic section modulus is discussed in Section 9.2

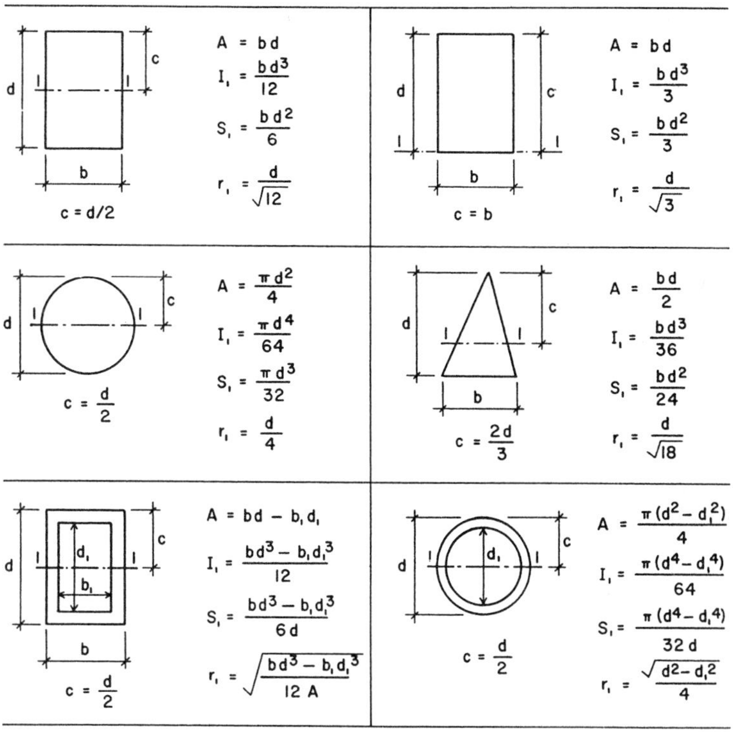

A = Area I = Moment of Inertia S = Section Modulus r = Radius of Gyration

Figure A.11 Properties of various geometric shapes of cross sections.

A.5 TABLES OF PROPERTIES OF SECTIONS

Figure A.11 presents formulas for obtaining geometric properties of various simple plane sections. Some of these may be used for single-piece structural members or for the building up of complex members.

Tables A.3 through A.8 present the properties of various plane sections. These are sections identified as those of standard industry-produced sections of wood and steel. Standardization means that the shapes and dimensions of the sections are fixed and each specific section is identified in some way.

Structural members may be employed for various purposes, and thus they may be oriented differently for some structural uses. Of note for any plane section are the *principal axes* of the section. These are the two, mutually perpendicular, centroidal axes for which the values will be greatest and least, respectively, for the section; thus, the axes are identified as the major and minor axes. If sections have an axis of symmetry, it will always be a principal axis—either major or minor.

For sections with two perpendicular axes of symmetry (rectangle, H, I, etc.), one axis will be the major axis and the other the minor axis. In the tables of properties, the listed values for I, S, and r are all identified as to a specific axis, and the reference axes are identified in a figure for the table.

Other values given in the tables are for significant dimensions, total cross-sectional area, and the weight of a 1-ft-long piece of the member. The weight of wood members is given in the table, assuming an average density for structural softwood of 35 lb/ft^3. The weight of steel members is given for W and channel shapes as part of their designation; thus, a W8 $\times$ 67 member weighs 67 lb/ft. For steel angles and pipes, the weight is given in the table, as determined from the density of steel at 490 lb/ft^3.

The designation of some members indicates their true dimensions. Thus, a 10-in. channel and a 6-in. angle have true dimensions of 10 and 6 in. For W shapes, pipe, and structural lumber, the designated dimensions are *nominal*, and the true dimensions must be obtained from the tables.

TABLE A.3 Properties of W Shapes

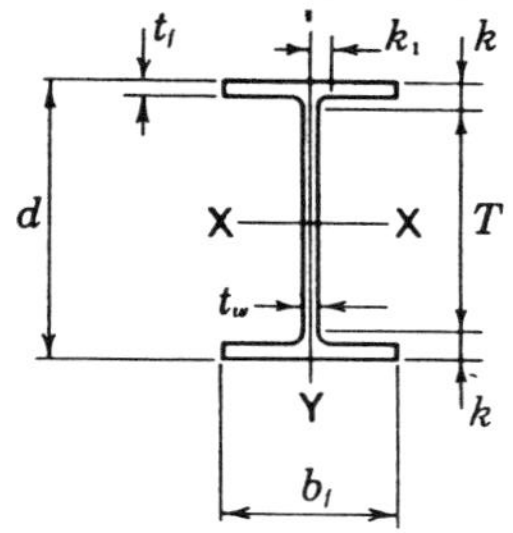

	Area	Depth	Web Thickness	Flange Width	Flange Thickness		Elastic Properties Axis X-X			Axis Y-Y			Plastic Modulus
Shape	A in.^{2}	d in.	t_w in.	b_f in.	t_f in.	k in.	I in.^{4}	S in.^{3}	r in.	I in.^{4}	S in.^{3}	r in.	Z_x in.^{3}
W30 × 116	34.2	30.01	0.565	10.495	0.850	1.625	4930	329	12.0	164	31.3	2.19	378
× 108	31.7	29.83	0.545	10.475	0.760	1.562	4470	299	11.9	146	27.9	2.15	346
× 99	29.1	29.65	0.520	10.450	0.670	1.437	3990	269	11.7	128	24.5	2.10	312
W27 × 94	27.7	26.92	0.490	9.990	0.745	1.437	3270	243	10.9	124	24.8	2.12	278
× 84	24.8	26.71	0.460	9.960	0.640	1.375	2850	213	10.7	106	21.2	2.07	244
W24 × 84	24.7	24.10	0.470	9.020	0.770	1.562	2370	196	9.79	94.4	20.9	1.95	224
× 76	22.4	23.92	0.440	8.990	0.680	1.437	2100	176	9.69	82.5	18.4	1.92	200
× 68	20.1	23.73	0.415	8.965	0.585	1.375	1830	154	9.55	70.4	15.7	1.87	177

(continued)

TABLE A.3 ***(Continued)***

Shape	A in.^{2}	d in.	t_w in.	b_f in.	t_f in.	k in.	I in.^{4}	S in.^{3}	r in.	I in.^{4}	S in.^{3}	r in.	Z_x in.^{3}
W21 × 83	24.3	21.43	0.515	8.355	0.835	1.562	1830	171	8.67	81.4	19.5	1.83	196
× 73	21.5	21.24	0.455	8.295	0.740	1.500	1600	151	8.64	70.6	17.0	1.81	172
× 57	16.7	21.06	0.405	6.555	0.650	1.375	1170	111	8.36	30.6	9.35	1.35	129
× 50	14.7	20.83	0.380	6.530	0.535	1.312	984	94.5	8.18	24.9	7.64	1.30	110
W18 × 86	25.3	18.39	0.480	11.090	0.770	1.437	1530	166	7.77	175	31.6	2.63	186
× 76	22.3	18.21	0.425	11.035	0.680	1.375	1330	146	7.73	152	27.6	2.61	163
× 60	17.6	18.24	0.415	7.555	0.695	1.375	984	108	7.47	50.1	13.3	1.69	123
× 55	16.2	18.11	0.390	7.530	0.630	1.312	890	98.3	7.41	44.9	11.9	1.67	112
× 50	14.7	17.99	0.355	7.495	0.570	1.250	800	88.9	7.38	40.1	10.7	1.65	101
× 46	13.5	18.06	0.360	6.060	0.605	1.250	712	78.8	7.25	22.5	7.43	1.29	90.7
× 40	11.8	17.90	0.315	6.015	0.525	1.187	612	68.4	7.21	19.1	6.35	1.27	78.4
W16 × 50	14.7	16.26	0.380	7.070	0.630	1.312	659	81.0	6.68	37.2	10.5	1.59	92.0
× 45	13.3	16.13	0.345	7.035	0.565	1.250	586	72.7	6.65	32.8	9.34	1.57	82.3
× 40	11.8	16.01	0.305	6.995	0.505	1.187	518	64.7	6.63	28.9	8.25	1.57	72.9
× 36	10.6	15.86	0.295	6.985	0.430	1.125	448	56.5	6.51	24.5	7.00	1.52	64.0
W14 × 216	62.0	15.72	0.980	15.800	1.560	2.250	2660	338	6.55	1030	130	4.07	390
× 176	51.8	15.22	0.830	15.650	1.310	2.000	2140	281	6.43	838	107	4.02	320
× 132	38.8	14.66	0.645	14.725	1.030	1.687	1530	209	6.28	548	74.5	3.76	234
× 120	35.3	14.48	0.590	14.670	0.940	1.625	1380	190	6.24	495	67.5	3.74	212
× 74	21.8	14.17	0.450	10.070	0.785	1.562	796	112	6.04	134	26.6	2.48	126

× 68	20.0	14.04	0.415	10.035	0.720	1.500	723	103	6.01	121	24.2	2.46	115
× 48	14.1	13.79	0.340	8.030	0.595	1.375	485	70.3	5.85	51.4	12.8	1.91	78.4
× 43	12.6	13.66	0.305	7.995	0.530	1.312	428	62.7	5.82	45.2	11.3	1.89	69.6
× 34	10.0	13.98	0.285	6.745	0.455	1.000	340	48.6	5.83	23.3	6.91	1.53	54.6
× 30	8.85	13.84	0.270	6.730	0.385	0.937	291	42.0	5.73	19.6	5.82	1.49	47.3
W12 × 136	39.9	13.41	0.790	12.400	1.250	1.937	1240	186	5.58	398	64.2	3.16	214
× 120	35.3	13.12	0.710	12.320	1.105	1.812	1070	163	5.51	345	56.0	3.13	186
× 72	21.1	12.25	0.430	12.040	0.670	1.375	597	97.4	5.31	195	32.4	3.04	108
× 65	19.1	12.12	0.390	12.000	0.605	1.312	533	87.9	5.28	174	29.1	3.02	96.8
× 53	15.6	12.06	0.345	9.995	0.575	1.250	425	70.6	5.23	95.8	19.2	2.48	77.9
× 45	13.2	12.06	0.335	8.045	0.575	1.250	350	58.1	5.15	50.0	12.4	1.94	64.7
× 40	11.8	11.94	0.295	8.005	0.515	1.250	310	51.9	5.13	44.1	11.0	1.93	57.5
× 30	8.79	12.34	0.260	6.520	0.440	0.937	238	38.6	5.21	20.3	6.24	1.52	43.1
× 26	7.65	12.22	0.230	6.490	0.380	0.875	204	33.4	5.17	17.3	5.34	1.51	37.2
W10 × 88	25.9	10.84	0.605	10.265	0.990	1.625	534	98.5	4.54	179	34.8	2.63	113
× 77	22.6	10.60	0.530	10.190	0.870	1.500	455	85.9	4.49	154	30.1	2.60	97.6
× 49	14.4	9.98	0.340	10.000	0.560	1.312	272	54.6	4.35	93.4	18.7	2.54	60.4
× 39	11.5	9.92	0.315	7.985	0.530	1.125	209	42.1	4.27	45.0	11.3	1.98	46.8
× 33	9.71	9.73	0.290	7.960	0.435	1.062	170	35.0	4.19	36.6	9.20	1.94	38.8
× 19	5.62	10.24	0.250	4.020	0.395	0.812	96.3	18.8	4.14	4.29	2.14	0.874	21.6
× 17	4.99	10.11	0.240	4.010	0.330	0.750	81.9	16.2	4.05	3.56	1.78	0.844	18.7

Source: Adapted from data in the *Manual of Steel Construction,* with permission of the publishers, American Institute of Steel Construction. This table is a sample from an extensive set of tables in the reference document.

TABLE A.4 Properties of American Standard Channels

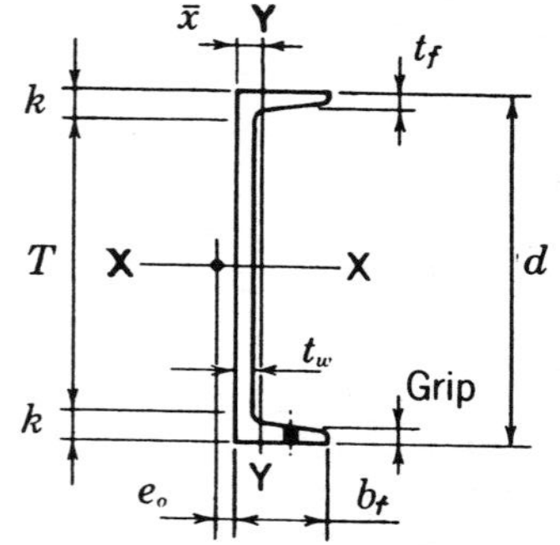

Shape	Area A (in.^{2})	Depth d (in.)	Web Thickness t_w (in.)	Flange Width b_f (in.)	Flange Thickness t_f (in.)	k (in.)	Elastic Properties, Axis X-X I (in.^{4})	Axis X-X S (in.^{3})	Axis X-X r (in.)	Axis Y-Y I (in.^{4})	Axis Y-Y S (in.^{3})	Axis Y-Y r (in.)	x^a (in.)	e_o^b (in.)
C15 × 50	14.7	15.0	0.716	3.716	0.650	1.44	404	53.8	5.24	11.0	3.78	0.867	0.798	0.583
× 40	11.8	15.0	0.520	3.520	0.650	1.44	349	46.5	5.44	9.23	3.37	0.886	0.777	0.767
× 33.9	9.96	15.0	0.400	3.400	0.650	1.44	315	42.0	5.62	8.13	3.11	0.904	0.787	0.896
C12 × 30	8.82	12.0	0.510	3.170	0.501	1.13	162	27.0	4.29	5.14	2.06	0.763	0.674	0.618
× 25	7.35	12.0	0.387	3.047	0.501	1.13	144	24.1	4.43	4.47	1.88	0.780	0.674	0.746
× 20.7	6.09	12.0	0.282	2.942	0.501	1.13	129	21.5	4.61	3.88	1.73	0.799	0.698	0.870

C10 × 30	8.82	10.0	0.673	3.033	0.436	1.00	103	20.7	3.42	3.94	1.65	0.669	0.649	0.369
× 25	7.35	10.0	0.526	2.886	0.436	1.00	91.2	18.2	3.52	3.36	1.48	0.676	0.617	0.494
× 20	5.88	10.0	0.379	2.739	0.436	1.00	78.9	15.8	3.66	2.81	1.32	0.692	0.606	0.637
× 15.3	4.49	10.0	0.240	2.600	0.436	1.00	67.4	13.5	3.87	2.28	1.16	0.713	0.634	0.796
C9 × 20	5.88	9.0	0.448	2.648	0.413	0.94	60.9	13.5	3.22	2.42	1.17	0.642	0.583	0.515
× 15	4.41	9.0	0.285	2.485	0.413	0.94	51.0	11.3	3.40	1.93	1.01	0.661	0.586	0.682
× 13.4	3.94	9.0	0.233	2.433	0.413	0.94	47.9	10.6	3.48	1.76	0.962	0.669	0.601	0.743
C8 × 18.75	5.51	8.0	0.487	2.527	0.390	0.94	44.0	11.0	2.82	1.98	1.01	0.599	0.565	0.431
× 13.75	4.04	8.0	0.303	2.343	0.390	0.94	36.1	9.03	2.99	1.53	0.854	0.615	0.553	0.604
× 11.5	3.38	8.0	0.220	2.260	0.390	0.94	32.6	8.14	3.11	1.32	0.781	0.625	0.571	0.697
C7 × 14.75	4.33	7.0	0.419	2.299	0.366	0.88	27.2	7.78	2.51	1.38	0.779	0.564	0.532	0.441
× 12.25	3.60	7.0	0.314	2.194	0.366	0.88	24.2	6.93	2.60	1.17	0.703	0.571	0.525	0.538
× 9.8	2.87	7.0	0.210	2.090	0.366	0.88	21.3	6.08	2.72	0.968	0.625	0.581	0.540	0.647
C6 × 13	3.83	6.0	0.437	2.157	0.343	0.81	17.4	5.80	2.13	1.05	0.642	0.525	0.514	0.380
× 10.5	3.09	6.0	0.314	2.034	0.343	0.81	15.2	5.06	2.22	0.866	0.564	0.529	0.499	0.486
× 8.2	2.40	6.0	0.200	1.920	0.343	0.81	13.1	4.38	2.34	0.693	0.492	0.537	0.511	0.599

[a]Distance to centroid of section.

[b]Distance to shear center of section.

Source: Adapted from data in the *Manual of Steel Construction*, with permission of the publishers, American Institute of Steel Construction. This table is a sample from an extensive set of tables in the reference document.

TABLE A.5 Properties of Single-Angle Shapes

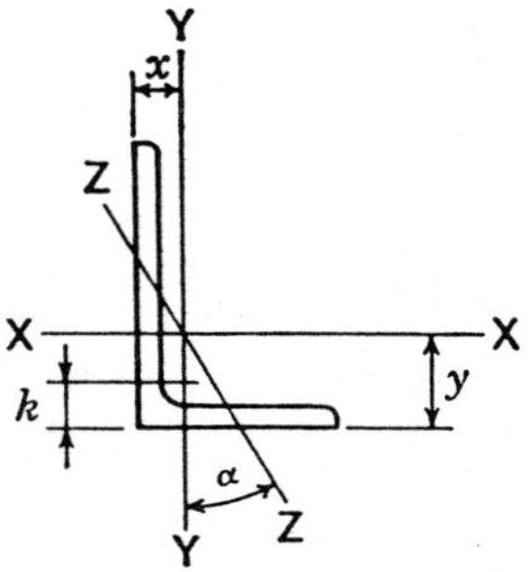

Size and Thickness (in.)	k (in.)	Weight per ft (lb.)	Area A in.^{2}	Axis X-X I in.^{4}	Axis X-X S in.^{3}	Axis X-X r in.	Axis X-X y in.	Axis Y-Y I in.^{4}	Axis Y-Y S in.^{3}	Axis Y-Y r in.	Axis Y-Y x in.	Axis Z-Z r in.	Axis Z-Z Tan α
8 × 8 × 1 – 1/8	1.75	56.9	16.7	98.0	17.5	2.42	2.41	98.0	17.5	2.42	2.41	1.56	1.000
× 1	1.62	51.0	15.0	89.0	15.8	2.44	2.37	89.0	15.8	2.44	2.37	1.56	1.000
8 × 6 × 3/4	1.25	33.8	9.94	63.4	11.7	2.53	2.56	30.7	6.92	1.76	1.56	1.29	0.551
× 1/2	1.00	23.0	6.75	44.3	8.02	2.56	2.47	21.7	4.79	1.79	1.47	1.30	0.558
6 × 6 × 5/8	1.12	24.2	7.11	24.2	5.66	1.84	1.73	24.2	5.66	1.84	1.73	1.18	1.000
× 1/2	1.00	19.6	5.75	19.9	4.61	1.86	1.68	19.9	4.61	1.86	1.68	1.18	1.000
6 × 4 × 5/8	1.12	20.0	5.86	21.1	5.31	1.90	2.03	7.52	2.54	1.13	1.03	0.864	0.435
× 1/2	1.00	16.2	4.75	17.4	4.33	1.91	1.99	6.27	2.08	1.15	0.987	0.870	0.440

× 3/8	0.87	12.3	3.61	13.5	3.32	1.93	1.94	4.90	1.60	1.17	0.941	0.877	0.446
5 × 3½ × ½	1.00	13.6	4.00	9.99	2.99	1.58	1.66	4.05	1.56	1.01	0.906	0.755	0.479
× 3/8	0.87	10.4	3.05	7.78	2.29	1.60	1.61	3.18	1.21	1.02	0.861	0.762	0.486
5 × 3 × ½	1.00	12.8	3.75	9.45	2.91	1.59	1.75	2.58	1.15	0.829	0.750	0.648	0.357
× 3/8	0.87	9.8	2.86	7.37	2.24	1.61	1.70	2.04	0.888	0.845	0.704	0.654	0.364
4 × 4 × ½	0.87	12.8	3.75	5.56	1.97	1.22	1.18	5.56	1.97	1.22	1.18	0.782	1.000
× 3/8	0.75	9.8	2.86	4.36	1.52	1.23	1.14	4.36	1.52	1.23	1.14	0.788	1.000
4 × 3 × ½	0.94	11.1	3.25	5.05	1.89	1.25	1.33	2.42	1.12	0.864	0.827	0.639	0.543
× 3/8	0.81	8.5	2.48	3.96	1.46	1.26	1.28	1.92	0.866	0.879	0.782	0.644	0.551
× 5/16	0.75	7.2	2.09	3.38	1.23	1.27	1.26	1.65	0.734	0.887	0.759	0.647	0.554
3½ × 3½ × 3/8	0.75	8.5	2.48	2.87	1.15	1.07	1.01	2.87	1.15	1.07	1.01	0.687	1.000
× 5/16	0.69	7.2	2.09	2.45	0.976	1.08	0.990	2.45	0.976	1.08	0.990	0.690	1.000
× 5/16	0.75	6.1	1.78	2.19	0.927	1.11	1.14	0.939	0.504	0.727	0.637	0.540	0.501
3 × 3 × 3/8	0.69	7.2	2.11	1.76	0.833	0.913	0.888	1.76	0.833	0.913	0.888	0.587	1.000
× 5/16	0.62	6.1	1.78	1.51	0.707	0.922	0.865	1.51	0.707	0.922	0.865	0.589	1.000
3 × 2½ × 3/8	0.75	6.6	1.92	1.66	0.810	0.928	0.956	1.04	0.581	0.736	0.706	0.522	0.676
× 5/16	0.69	5.6	1.62	1.42	0.688	0.937	0.933	0.898	0.494	0.744	0.683	0.525	0.680
3 × 2 × 3/8	0.69	5.9	1.73	1.53	0.781	0.940	1.04	0.543	0.371	0.559	0.539	0.430	0.428
× 5/16	0.62	5.0	1.46	1.32	0.664	0.948	1.02	0.470	0.317	0.567	0.516	0.432	0.435
2½ × 2½ × 3/8	0.69	5.9	1.73	0.984	0.566	0.753	0.762	0.984	0.566	0.753	0.762	0.487	1.000
× 5/16	0.62	5.0	1.46	0.849	0.482	0.761	0.740	0.849	0.482	0.761	0.740	0.489	1.000
2½ × 2 × 3/8	0.69	5.3	1.55	0.912	0.547	0.768	0.831	0.514	0.363	0.577	0.581	0.420	0.614
× 5/16	0.62	4.5	1.31	0.788	0.466	0.776	0.809	0.446	0.310	0.584	0.559	0.422	0.620

Source: Adapted from data in the *Manual of Steel Construction*, with permission of the publishers, American Institute of Steel Construction. This table is a sample from an extensive set of tables in the reference document.

TABLE A.6 Properties of Double-Angle Shapes with Long Legs Back to Back

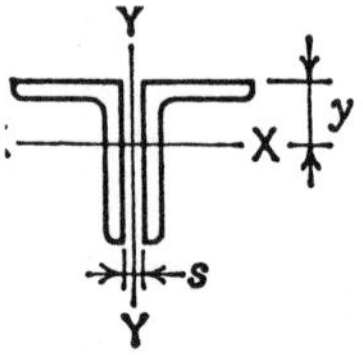

Size and Thickness	Weight per ft	Area A	Axis X-X				Axis Y-Y: Radii of Gyration Back to Back of Angles, in.		
			I	S	r	y			
(in.)	(lb)	(in.^{2})	(in.^{4})	(in.^{3})	(in.)	(in.)	0	$\frac{3}{8}$	$\frac{3}{4}$
$8 \times 6 \times 1$	88.4	26.0	161.0	30.2	2.49	2.65	2.39	2.52	2.66
$\times \frac{3}{4}$	67.6	19.9	126.0	23.3	2.53	2.56	2.35	2.48	2.62
$\times \frac{1}{2}$	46.0	13.5	88.6	16.0	2.56	2.47	2.32	2.44	2.57
$6 \times 4 \times \frac{3}{4}$	47.2	13.9	49.0	12.5	1.88	2.08	1.55	1.69	1.83
$\times \frac{1}{2}$	32.4	9.50	34.8	8.67	1.91	1.99	1.51	1.64	1.78
$\times \frac{3}{8}$	24.6	7.22	26.9	6.64	1.93	1.94	1.50	1.62	1.76
$5 \times 3\frac{1}{2} \times \frac{1}{2}$	27.2	8.00	20.0	5.97	1.58	1.66	1.35	1.49	1.63
$\times \frac{3}{8}$	20.8	6.09	15.6	4.59	1.60	1.61	1.34	1.46	1.60
$5 \times 3 \times \frac{1}{2}$	25.6	7.50	18.9	5.82	1.59	1.75	1.12	1.25	1.40
$\times \frac{3}{8}$	19.6	5.72	14.7	4.47	1.61	1.70	1.10	1.23	1.37
$\times \frac{5}{16}$	16.4	4.80	12.5	3.77	1.61	1.68	1.09	1.22	1.36
$4 \times 3 \times \frac{1}{2}$	22.2	6.50	10.1	3.78	1.25	1.33	1.20	1.33	1.48
$\times \frac{3}{8}$	17.0	4.97	7.93	2.92	1.26	1.28	1.18	1.31	1.45
$\times \frac{5}{16}$	14.4	4.18	6.76	2.47	1.27	1.26	1.17	1.30	1.44
$3\frac{1}{2} \times 2\frac{1}{2} \times \frac{3}{8}$	14.4	4.22	5.12	2.19	1.10	1.16	0.976	1.11	1.26
$\times \frac{5}{16}$	12.2	3.55	4.38	1.85	1.11	1.14	0.966	1.10	1.25
$\times \frac{1}{4}$	9.8	2.88	3.60	1.51	1.12	1.11	0.958	1.09	1.23
$3 \times 2 \times \frac{3}{8}$	11.8	3.47	3.06	1.56	0.940	1.04	0.777	0.917	1.07
$\times \frac{5}{16}$	10.0	2.93	2.63	1.33	0.948	1.02	0.767	0.903	1.06
$\times \frac{1}{4}$	8.2	2.38	2.17	1.08	0.957	0.993	0.757	0.891	1.04
$2\frac{1}{2} \times 2 \times \frac{3}{8}$	10.6	3.09	1.82	1.09	0.768	0.831	0.819	0.961	1.12
$\times \frac{5}{16}$	9.0	2.62	1.58	0.932	0.776	0.809	0.809	0.948	1.10
$\times \frac{1}{4}$	7.2	2.13	1.31	0.763	0.784	0.787	0.799	0.935	1.09

Source: Adapted from data in the *Manual of Steel Construction*, with permission of the publishers, American Institute of Steel Construction. This table is a sample from an extensive set of tables in the reference document.

TABLE A.7 Properties of Standard Weight Steel Pipe

Dimensions				Weight	Properties			
Nominal Diameter (in.)	Outside Diameter (in.)	Inside Diameter (in.)	Wall Thickness (in.)	per ft (lb)	*A* (in.^{2})	*I* (in.^{4})	*S* (in.^{3})	*r* (in.)
3	3.500	3.068	0.216	7.58	2.23	3.02	1.72	1.16
32	4.000	3.548	0.226	9.11	2.68	4.79	2.39	1.34
4	4.500	4.026	0.237	10.79	3.17	7.23	3.21	1.51
5	5.563	5.047	0.258	14.62	4.30	15.2	5.45	1.88
6	6.625	6.065	0.280	18.97	5.58	28.1	8.50	2.25
8	8.625	7.981	0.322	28.55	8.40	72.5	16.8	2.94
10	10.750	10.020	0.365	40.48	11.9	161	29.9	3.67
12	12.750	12.000	0.375	49.56	14.6	279	43.8	4.38

Source: Adapted from data in the *Manual of Steel Construction*, with permission of the publishers, American Institute of Steel Construction. This table is a sample from an extensive set of tables in the reference document.

TABLE A.8 Properties of Structural Lumber

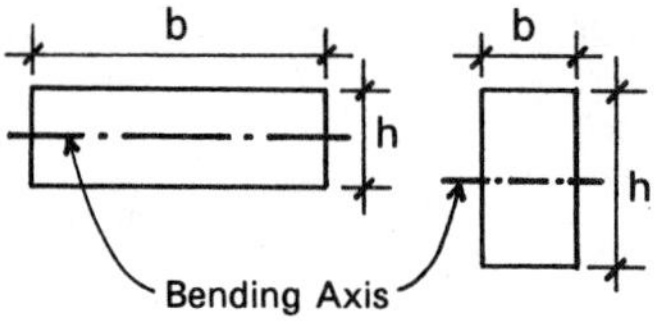

Dimensions (in.) Nominal b h	Actual b h	Area A (in.^{2})	Section Modulus S (in.^{3})	Moment of Inertia I (in.^{4})	Weight[a] (lb/ft)
2 × 3	1.5 × 2.5	3.75	1.563	1.953	0.9
2 × 4	1.5 × 3.5	5.25	3.063	5.359	1.3
2 × 6	1.5 × 5.5	8.25	7.563	20.797	2.0
2 × 8	1.5 × 7.25	10.875	13.141	47.635	2.6
2 × 10	1.5 × 9.25	13.875	21.391	98.932	3.4
2 × 12	1.5 × 11.25	16.875	31.641	177.979	4.1
2 × 14	1.5 × 13.25	19.875	43.891	290.775	4.8
3 × 2	2.5 × 1.5	3.75	0.938	0.703	0.9
3 × 4	2.5 × 3.5	8.75	5.104	8.932	2.1
3 × 6	2.5 × 5.5	13.75	12.604	34.661	3.3
3 × 8	2.5 × 7.25	18.125	21.901	79.391	4.4
3 × 10	2.5 × 9.25	23.125	35.651	164.886	5.6
3 × 12	2.5 × 11.25	28.125	52.734	296.631	6.8
3 × 14	2.5 × 13.25	33.125	73.151	484.625	8.1
3 × 16	2.5 × 15.25	38.125	96.901	738.870	9.3
4 × 2	3.5 × 1.5	5.25	1.313	0.984	1.3
4 × 3	3.5 × 2.5	8.75	3.646	4.557	2.1
4 × 4	3.5 × 3.5	12.25	7.146	12.505	3.0
4 × 6	3.5 × 5.5	19.25	17.646	48.526	4.7
4 × 8	3.5 × 7.25	25.375	30.661	111.148	6.2
4 × 10	3.5 × 9.25	32.375	49.911	230.840	7.9
4 × 12	3.5 × 11.25	39.375	73.828	415.283	9.6
4 × 14	3.5 × 13.25	46.375	102.411	678.475	11.3
4 × 16	3.5 × 15.25	53.375	135.661	1034.418	13.0
6 × 2	5.5 × 1.5	8.25	2.063	1.547	2.0
6 × 3	5.5 × 2.5	13.75	5.729	7.161	3.3
6 × 4	5.5 × 3.5	19.25	11.229	19.651	4.7
6 × 6	5.5 × 5.5	30.25	27.729	76.255	7.4
6 × 8	5.5 × 7.5	41.25	51.563	193.359	10.0
6 × 10	5.5 × 9.5	52.25	82.729	392.963	12.7
6 × 12	5.5 × 11.5	63.25	121.229	697.068	15.4
6 × 14	5.5 × 13.5	74.25	167.063	1127.672	18.0
6 × 16	5.5 × 15.5	85.25	220.229	1706.776	20.7
8 × 2	7.25 × 1.5	10.875	2.719	2.039	2.6

TABLE A.8 ***(Continued)***

Dimensions (in.)		Area	Section Modulus	Moment of Inertia	
Nominal *b h*	Actual *b h*	*A* (in.^{2})	*S* (in.^{3})	*I* (in.^{4})	Weight[a] (lb/ft)
8 × 3	7.25 × 2.5	18.125	7.552	9.440	4.4
8 × 4	7.25 × 3.5	25.375	14.802	25.904	6.2
8 × 6	7.5 × 5.5	41.25	37.813	103.984	10.0
8 × 8	7.5 × 7.5	56.25	70.313	263.672	13.7
8 × 10	7.5 × 9.5	71.25	112.813	535.859	17.3
8 × 12	7.5 × 11.5	86.25	165.313	950.547	21.0
8 × 14	7.5 × 13.5	101.25	227.813	1537.734	24.6
8 × 16	7.5 × 15.5	116.25	300.313	2327.422	28.3
8 × 18	7.5 × 17.5	131.25	382.813	3349.609	31.9
8 × 20	7.5 × 19.5	146.25	475.313	4634.297	35.5
10 × 10	9.5 × 9.5	90.25	142.896	678.755	21.9
10 × 12	9.5 × 11.5	109.25	209.396	1204.026	26.6
10 × 14	9.5 × 13.5	128.25	288.563	1947.797	31.2
10 × 16	9.5 × 15.5	147.25	380.396	2948.068	35.8
10 × 18	9.5 × 17.5	166.25	484.896	4242.836	40.4
10 × 20	9.5 × 19.5	185.25	602.063	5870.109	45.0
12 × 12	11.5 × 11.5	132.25	253.479	1457.505	32.1
12 × 14	11.5 × 13.5	155.25	349.313	2357.859	37.7
12 × 16	11.5 × 15.5	178.25	460.479	3568.713	43.3
12 × 18	11.5 × 17.5	201.25	586.979	5136.066	48.9
12 × 20	11.5 × 19.5	224.25	728.813	7105.922	54.5
12 × 22	11.5 × 21.5	247.25	885.979	9524.273	60.1
12 × 24	11.5 × 23.5	270.25	1058.479	12437.129	65.7
14 × 14	13.5 × 13.5	182.25	410.063	2767.922	44.3
16 × 16	15.5 × 15.5	240.25	620.646	4810.004	58.4

[a] Based on an assumed average density of 35 psf.

Source: Compiled from data in the *National Design Specification for Wood Construction* (Ref. 3), with permission of the publishers, National Forest Products Association.

REFERENCES

1. *Minimum Design Loads for Buildings and Other Structures*, SEI/ASCE 7–02, American Society of Civil Engineers, Reston, VA, 2002.
2. *Uniform Building Code, Volume 2: Structural Engineering Provisions*, International Conference of Building Officials, Whittier, CA, 1997.
3. *National Design Specification for Wood Construction*, American Forest and Paper Association, Washington, DC, 2001.
4. *Timber Construction Manual*, 5th ed., American Institute of Timber Construction, Wiley, Hoboken, NJ, 2005.
5. *Manual of Steel Construction, Load and Resistance Factor Design*, 3rd ed., American Institute of Steel Construction, Chicago, IL, 2001.
6. *Standard Specifications, Load Tables, and Weight Tables for Steel Joists and Joist Girders*, Steel Joist Institute, Myrtle Beach, SC, 2002.
7. *Steel Deck Institute Design Manual for Composite Decks, Form Decks, and Roof Decks*, Steel Deck Institute, St. Louis, MO, 2000.
8. *Building Code Requirements for Structural Concrete* (ACI 318–02) and Commentary (ACI 318R-02) American Concrete Institute, Detroit, MI, 2002.

9. James Ambrose and Dimitry Vergun, *Simplified Building Design for Wind and Earthquake Forces*, 3rd ed., Wiley, Hoboken, NJ, 1995.
10. James Ambrose, *Simplified Design of Building Foundations*, 2nd ed., Wiley, Hoboken, NJ, 1988.
11. Jack McCormac and James Nelson, *Structural Steel Design, LRFD Method*, 3rd ed., Pearson Education, Upper Saddle River, NJ, 2003.
12. Jack McCormac and James Nelson, *Design of Reinforced Concrete*, 6th ed., Wiley, Hoboken, NJ, 2005.
13. C. M. Uang, S. W. Wattar, and K. M. Leet, "Proposed Revision of the Equivalent Axial Load Method for LRFD Steel and Composite Beam—Column Design," *Engineering Journal*, AISC, Fall, 1990.

ANSWERS TO EXERCISE PROBLEMS

Answers are given here for some of the exercise problems for which computational work is required and a single correct answer exists. In most cases there are two exercise problems that are similar to each computational example problem in the text. In those cases the answer is given here for one of the related exercise problems.

Chapter 1

1.3.A. $R = 80.62$ lb, upward to the right, 29.74° from the horizontal
1.3.C. $R = 94.87$ lb, downward to the right, 18.43° from the horizontal
1.3.E. $R = 100$ lb, downward to the left, 53.13° from the horizontal
1.3.G. $R = 58.07$ lb, downward to the right, 7.49° from the horizontal
1.3.I. $R = 91.13$ lb, upward to the right, 9.495° from the horizontal
1.4.A. 141.4 lb T
1.4.C. 300 lb
1.5.A. Sample values: $CI = 2000C$, $IJ = 812.5T$, $JG = 1250T$
1.6.A. Same as 1.5.A.

Chapter 2

2.1.A. 3.33 in.^{2} [2150 mm^2]
2.1.C. 0.874 in., or 7/8 in. [22.2 mm]
2.1.E. 18.4 kips [81.7 kN]
2.1.G. 2.5 in.^{2} [1613 mm^2]
2.2.A. 19,333 lb [86 kN]
2.2.C. 29,550,000 psi [203 GPa]

Chapter 3

3.1.A. Sample: M about R_1 = + (500 × 4) + (400 × 6) + (600 × 10) − (650 × 16)
3.2.A. R_1 = 3593.75 lb [15.98 kN], R_2 = 4406.25 lb [19.60 kN]
3.2.C. R_1 = 7667 lb [34.11 kN], R_2 = 9333 lb [41.53 kN]
3.2.E. R_1 = 7143 lb [31.79 kN], R_2 = 11,857 lb [52.76 kN]
3.3.A. Maximum shear = 10 kips [44.5 kN]
3.3.C. Maximum shear = 1114 lb [4.956 kN]
3.3.E. Maximum shear = 9.375 kips [41.623 kN]
3.4.A. Maximum M = 60 kip-ft [80.1 kN-m]
3.4.C. Maximum M = 4286 ft-lb [5.716 kN-m]
3.4.E. Maximum M = 18.35 kip-ft [24.45 kN-m]
3.5.A. R_1 = 1860 lb [8.27 kN], maximum V = 1360 lb [6.05 kN], maximum, − M = 2000 ft-lb [2.66 kN-m], maximum + M = 3200 ft-lb [4.27 kN-m]
3.5.C. R_1 = 2760 lb [12.28 kN], maximum V = 2040 lb [9.07 kN], maximum − M = 2000 ft-lb [2.67 kN-m], maximum +M = 5520 ft-lb [7.37 kN-m]
3.5.E. Maximum V = 1500 lb [6.67 kN], maximum M = 12,800 ft-lb [17.1 kN-m]
3.5.G. Maximum V = 1200 lb [5.27 kN], maximum M = 8600 ft-lb [11.33 kN-m]
3.6.A. M = 32 kip-ft [43.4 kN-m]
3.6.C. M = 90 kip-ft [122 kN-m]
3.8.A. At neutral axis f_v = 811.4 psi; at junction of web and flange f_v = 175 psi and 700 psi
3.9.A R_1 = R_3 = 1200 lb [5.34 kN], R_2 = 4000 lb [17.79 kN], + M = 3600 ft-lb [4.99 kN-m], − M = 6400 ft-lb [8.68 kN-m]
3.9.C. R_1 = 7.67 kips, R_2 = 35.58 kips, R_3 = 12.75 kips, + M = 14.69 kip-ft and 40.64 kip-ft, − M = 52 kip-ft
3.9.E. R_1 = R_3 = 937.5 lb [4.17 kN], R_2 = 4125 lb [18.35 kN], + M = 7031 ft-lb [9.53 kN-m], − M = 13,500 ft-lb [18.31 kN-m]
3.9.G. R_1 = R_4 = 9600 lb [42.7 kN], R_2 = R_3 = 26,400 lb [117.4 kN], + M_1 = 46,080 ft-lb [62.48 kN-m], + M_2 = 14,400 ft-lb [19.53 kN-m], − M = 57,600 ft-lb [78.11 kN-m]
3.9.I. Maximum V = 8 kips, maximum + M = maximum − M = 44 kip-ft, inflection at 5.5 ft from end

3.10.A. $A_x = C_x = 2.143$ kips [9.643 kN], $A_y = 2.857$ kips [12.857 kN], $C_y = 7.143$ kips [32.143 kN]

3.10.C. $R_1 = 16$ kips [72 kN], $R_2 = 48$ kips [216 kN], maximum $+M = 64$ kip-ft [86.4 kN-m], maximum $-M = 80$ kip-ft [108 kN-m], inflection at pin location in both spans

3.10.E. $R_1 = 6.4$ kips [28.8 kN], $R_2 = 19.6$ kips [88.2 kN], $+M = 20.48$ kip-ft [27.7 kN-m] in end span and 24.4 kip-ft [33.1 kN-m] in center span, $-M = 25.6$ kip-ft [34.4 kN-m], inflection at 3.2 ft from R_2 in end span and 5 ft from support in center span

3.11.A (a) 3.04 ksf (b) 5.33 ksf

3.12.A. $R = 10$ kips up and 110 kip-ft counterclockwise

3.12.C. $R = 6$ kips to the left and 72 kip-ft counterclockwise

3.12.E. Left $R = 4.5$ kips down and 6 kips to left, right $R = 4.5$ kips up and 6 kips to left

Chapter 5

5.2.A. 3 × 16

5.2.C. (a) 2 × 12, (b) 2 × 12

5.3.A. $f_v = 83.1$ psi, less than allowable of 170 psi, beam is OK

5.3.C. $f_v = 68.65$ psi, less than allowable of 170 psi, beam is OK

5.4.A. Stress = 303 psi, allowable is 625 psi, beam is OK

5.5.A. $\Delta = 0.31$ in. [7.6 mm], allowable is 0.8 in., beam is OK

5.5.C. $\Delta = 0.23$ in. [6 mm], allowable is 0.75 in., beam is OK

5.5.E. Need $I = 911$ in.^{4}, lightest choice is 4 × 16

5.6.A. 2 × 10

5.6.C. 2 × 12

5.6.E. 2 × 8

5.6.G. 2 × 12

Chapter 6

6.1.A. 7.0 kips

6.1.C. 21.4 kips

6.2.A. 6 × 6

6.2.C. 10 × 10

6.4.A. 2 × 4s at 24 in. are OK

6.4.C. Very close to limit, but OK

Chapter 7

7.2.A. 1050 lb

7.2.C 1560 lb

Chapter 9:

9.2.A. 13.5%

9.3.A. 1) 849 kip-ft 2) 725 kip-ft 3) 520 kip-ft

9.4.A. W 14 × 26
9.4.C. W 10 × 26
9.4.E. W 12 × 22
9.4.G. W 10 × 19
9.4.I. W 16 × 36
9.5.A. (a) W 30 × 90 (b) W 30 × 108 (c) W 27 × 114
9.5.C. (a) W 24 × 62 (b) W 24 × 76 (c) W 24 × 76
9.6.A. 220 kips
9.6.C. 54.8 kips
9.7.A. (a) 0.794 in. (b) 0.9 in.
9.7.C. (a) 0.829 in. (b) 0.8 in.
9.8.A. (a) W 16 × 57 (b) W 10 × 88
9.8.C. (a) W 24 × 55 (b) W 18 × 86
9.8.E. (a) W 12 × 16 (b) W 10 × 19
9.8.G. (a) W 24 × 76 (b) W 21 × 83
9.10.A. 26K7
9.10.C. (a) 24K4 (b) 22K6

Chapter 10

10.3.A. 361 kips [1606 kN]
10.3.C. 440 kips [1957 kN]
10.4.A. W 8 × 31
10.4.C. W 10 × 68
10.4.E. 4 in. standard pipe
10.4.G. 6 in. standard pipe
10.4.I. 98 kips [436 kN]
10.4.K. HHS 6 × 6 × 3/16
10.4.M. 104 kips
10.4.O. 4 × 3 × 5/16
10.5.A. W 12 × 45
10.5.C. W 12 × 96
10.5.E. W 14 × 82

Chapter 11

11.3.A. 6 bolts, outer plates 1/2 in., middle plate 11/16 in.

Chapter 12

12.2.A WR20
12.2.C WR18
12.2.E IR22 or WR22

Chapter 13

13.3.A. Possible choice: b = 12 in., d = 18.2 in., requires 9.54 in.^{2}, use 7 No. 10 bars. However, width required to get bars into one layer is critical, least width is 20 in., must have wider beam or use two layers of bars, which would give you a beam with a height of 20 in.

13.3.C. From work for Problem 13.3.A, this section is underreinforced. Try a/d = 0.4, required A_s = 4.48 in.^{2}, actual a/d = 0.137, revised area = 3.85 in.^{2} and minimum required steel = 1.92 in.^{2} use 4 No. 9 bars, width required = 10 in.

13.4.A. For bending moment, approximate area of reinforcement = 3.840 in.^{2}. However, minimum reinforcement based on b_f is 5.52 in.^{2}.

13.4.C. Check balanced M_B = 718 kip-ft, use tension reinforcement of 5 No. 9, compressive reinforcement of 3 No. 7.

13.4.E. As with Problem 13.4.C, balanced moment capacity exceeds required moment. Can use 5 No. 11 bars for tension reinforcement, but width required is 17 in.; need wider beam or two layers of bars. Use compression reinforcement of 3 No. 9 bars.

13.5.A. For deflection use 8-in.-thick slab, referring to Figure 14.4, reinforce with No. 8 @ 16 in, No. 7 @ 12 in, No. 6 @ 9 in., or No. 5 @ 6 in., use temperature reinforcement of No. 4 at 12 in.

13.6.A. Possible choice for spacing: 1 at 6 in., 8 at 13 in.

13.6.C. Possible choice: 1 at 6, 5 at 13

13.7.A. Required development length in cantilever is 17 in., 34 in. provided, OK. Required development length in support is 13 in, 22 in. provided, anchorage as shown is adequate by code, but use a hook anyway.

13.7.C. Required development length is 4.9 in., but use the full length available (22 in.)

Chapter 14

14.1.A. For deflection need 6-in.-thick slab. Referring to Figure 14.4, left to right, with all No. 4 bars, use spacings of 16, 18, 13, 18, 16 in.

Chapter 15

15.3.A. From Figure 15.6, with factored load of 136 kips and e = 3 in.: 10-in. column, 4 No. 7 bars.

15.3.C. From Figure 15.7, with factored load of 480 kips and e = 6.2 in.: 20-in. column, 4 No. 10 bars

15.3.D. From Figure 15.8, with factored load of 136 kips and e = 3 in.: 12 × 16 column, 6 No. 7 bars, no savings (overstrong, but smallest choice from graph)

15.3.F. From Figure 15.8, with factored load of 480 kips and e = 6.2 in.: 14 × 24 column, 6 No. 10 bars, more steel but less concrete.

15.3.G. From Figure 15.9, with factored load of 136 kips and e = 3 in., 12-in.-diameter column, 4 No. 7 bars

15.3.I. From Figure 15.9, with factored load of 480 kips and e = 6.2 in., 24-in.-diameter column, 6 No. 10 bars

Chapter 16

16.2.A. Possible choice: w = 6 ft-9 in., h = 17 in., 7 No. 5 in long direction, No. 5 at 11 in. in short direction.

16.3.A. From Table 16.4, Possible choice: 9-ft 0-in. square, h = 21 in., 9 No. 8 each way. Design permits 8-ft 8-in., h = 20 in., 7 No. 8 in each direction.

Appendix

A.1.A. c_y = 2.6 in. [66 mm]

A.1.C. c_y = 4.2895 in. [107.24 mm]

A.1.E. c_y = 4.4375 in. [110.9 mm], c_x = 1.0625 in. [27.0 mm]

A.3.A. I = 535.86 in.^{4} [2.11×10^8 mm^4]

A.3.C. I = 447.33 in.^{4} [174.74×10^6 mm^4]

A.3.E. I = 205.33 in.^{4} [80.21×10^6 mm^4]

A.3.G. I = 438 in.^{4} [1.823×10^8 mm^4]

A.3.I. I = 1672.45 in.^{4} [6.96×10^8 mm^4]

INDEX